D0152566

FINITE ELEMENTS FOR STRUCTURAL ANALYSIS

PRENTICE-HALL CIVIL ENGINEERING AND ENGINEERING MECHANICS SERIES

W. J. Hall, editor

FINITE ELEMENTS FOR STRUCTURAL ANALYSIS

William Weaver, Jr.
Stanford University

Paul R. Johnston
Failure Analysis Associates, Palo Alto

PRENTICE-HALL, INC., ENGLEWOOD CLIFFS, NEW JERSEY 07632

Library of Congress Cataloging in Publication Data

Weaver, William, 1929–
Finite elements for structural analysis.

Includes bibliographies and index.
1. Finite element method. 2. Structures, Theory of.
I. Johnston, Paul R., 1953– . II. Title.
TA347.F5W4 1984 624.1'71 83-9705
ISBN 0-13-317099-3

To CONNIE and TERRY

Printed in the United States of America

10 9 8 7 6 5 4 3 2 1

Editorial/production supervision and interior design: Paul Spencer
Cover design: Edsal Enterprises
Manufacturing buyer: Anthony Caruso

ISBN 0-13-317099-3

Prentice-Hall International, Inc., *London*
Prentice-Hall of Australia Pty. Limited, *Sydney*
Editora Prentice-Hall do Brasil, Ltda., *Rio de Janeiro*
Prentice-Hall Canada Inc., *Toronto*
Prentice-Hall of India, Private Limited, *New Delhi*
Prentice-Hall of Japan, Inc., *Tokyo*
Prentice-Hall of Southeast Asia Pte. Ltd., *Singapore*
Whitehall Books Limited, *Wellington, New Zealand*

Contents

Preface

The finite-element method is the best approach available for the numerical analysis of continua. Although the method has been applied to other fields, this book is devoted to the analysis of solids. In general, these solids are structures that serve to resist applied loads and other influences. The theory requires discretization of a given structure into a network of finite elements and implementation of the analysis on a digital computer. Because this book is intended for the beginner, we have emphasized clarity in the presentation. The user should find the explanations easy to read, either as a student or as a structural engineer in industry. The techniques and computer programs described herein have been developed and used in a graduate-level course at Stanford for many years.

A person who wishes to learn the contents of this book should be interested in the analysis and design of structures in civil, mechanical, or aeronautical and astronautical engineering, and should have had the usual undergraduate mathematics and solid mechanics courses offered in engineering curricula. It is also necessary to know the fundamentals of matrix algebra and computer programming. Another subject that should be studied first is matrix analysis of framed structures, because this topic may be considered a subset of the finite-element method. It is not necessary to know beforehand the theories of elasticity, plates, shells, dynamics, or stability, but any previous exposure to those topics would be of value.

Chapter 1 introduces the basic concepts of finite elements. The principle of virtual work serves as the basis for deriving the equations of equilibrium for the nodes of a finite element as well as those of a discretized continuum. In this chapter we develop stiffnesses and equivalent nodal loads for several useful one-dimensional finite elements. Axis transformations, assemblage of elements, and methods of solving equations are covered for any discretized structure. Discussions of potential-energy theorems, convergence criteria, and computer programs conclude the chapter.

The topics in Chapter 2 consist of plane-stress and plane-strain analysis of two-dimensional continua. Stiffnesses and equivalent nodal loads are developed for triangular, rectangular, and quadrilateral finite elements. A computer program for plane-stress or plane-strain analysis with a triangular element is thoroughly documented, and a complete flow chart for the logic appears in Appendix D. Practical applications of this program conclude Chapter 2.

Isoparametric finite elements are introduced in Chapter 3 and applied in two-dimensional problems. Natural dimensionless coordinates and numerical integration characterize the derivations of isoparametric elements. Quadrilaterals constitute the primary group of isoparametric formulations, but triangular elements are also discussed. Computer programs using quadrilateral elements are described at the end of the chapter and are demonstrated with practical applications.

Finite elements for general and axisymmetric solids are developed in Chapters 4 and 5, respectively. Here the emphasis is upon isoparametric elements for three-dimensional problems. Discussions include hexahedra, tetrahedra, and ring elements (for axisymmetric solids) with various cross-sectional shapes. The computer programs for general solids use hexahedral elements, while those for axisymmetric solids have ring elements with quadrilateral cross sections.

Chapter 6 deals with finite elements for flexure in plates. The classical theory of plate bending is explained and is applied to rectangular, triangular, quadrilateral, and annular elements. In this case the computer program uses a quadrilateral element derived by specializing an isoparametric hexahedron.

Elements for general and axisymmetric shells are discussed in Chapters 7 and 8, respectively. Again, specialized hexahedra become the elements of primary interest for general and axisymmetric shells. The computer programs in these chapters utilize such elements for analyzing shell structures.

Vibrational analysis of discretized continua is the topic of Chapter 9. A general formula for the consistent mass matrix of a finite element is derived and applied to various elements discussed in previous chapters. Then the eigenvalue problem for vibrations is set up, and a technique for reducing the size of the problem is described. The program for vibrational analysis uses the plate-bending quadrilateral from Chapter 6.

Chapter 10 shows the application of finite elements to instability analysis. The initial stress stiffness matrix is derived for beams and plates in flexure. Then eigenvalue problems are established for linear instability and vibrations of beam and plate structures. In addition, an incremental procedure for nonlinear instability is described for other types of structures. The program at the end of this chapter applies the plate-bending element used in Chapter 9 to linear instability analysis.

Appendix A contains integration formulas for a line, a triangle, and a tetrahedron. Gaussian quadrature is explained in Appendix B, which gives a table of numerical integration constants. In Appendix C computer solution routines are developed and presented in the form of flow charts for factorization and solution of simultaneous equations in vector and skyline forms.

References and problems appear at the end of each chapter. A notation list, answers to problems, and a list of other texts are given at the end of the book.

The computer programs discussed in the last sections of Chapters 2 through 10 are all accompanied by various applications to finite-element networks. FORTRAN codes for these programs are available from the second author for a nominal fee.

We wish to express our appreciation to graduate students and teaching assistants at Stanford who have helped to make this book possible. Suzanne M. Bennett did all the typing on the manuscript, and her superior ability made that phase of the work most enjoyable. We also thank Professor Thomas R. Kane for his helpful suggestions. Computer funds for the course on finite elements were provided by Stanford University.

William Weaver, Jr.
Paul R. Johnston

CHAPTER

1

Introduction to Finite Elements

1.1 FINITE-ELEMENT CONCEPTS

A *finite element* is a subregion of a discretized continuum. It is of finite size (not infinitesimal) and usually has a simpler geometry than that of the continuum. The finite-element method enables us to convert a problem with an infinite number of degrees of freedom to one with a finite number in order to simplify the solution process. Although the original applications were in the area of solid mechanics, its usage has spread to many other fields having similar mathematical bases. In any case it is a computer-oriented method that must be implemented with appropriate digital computer programs.

This book is devoted solely to the topic of finite elements for the analysis of structures. Here the word "structures" implies any solids that are subjected to loads or other influences. Such influences cause deformations (or strains) throughout the continuum, accompanied by internal stresses and reactions at restrained points. The primary objectives of analysis by finite elements are to calculate approximately the stresses and deflections in a structure.

The *classical approach* for analyzing a solid requires finding a stress or displacement function that satisfies the differential equations of equilibrium, the stress-strain relationships, and the compatibility conditions at every point in the continuum, including the boundaries. Because these requirements are so restrictive, very few classical solutions have been found. Among those, the solutions are often infinite series that in practical calculations require truncation, leading to approximate results. Furthermore, discretization of the differential equations by the method of *finite differences* has the primary disadvantage that

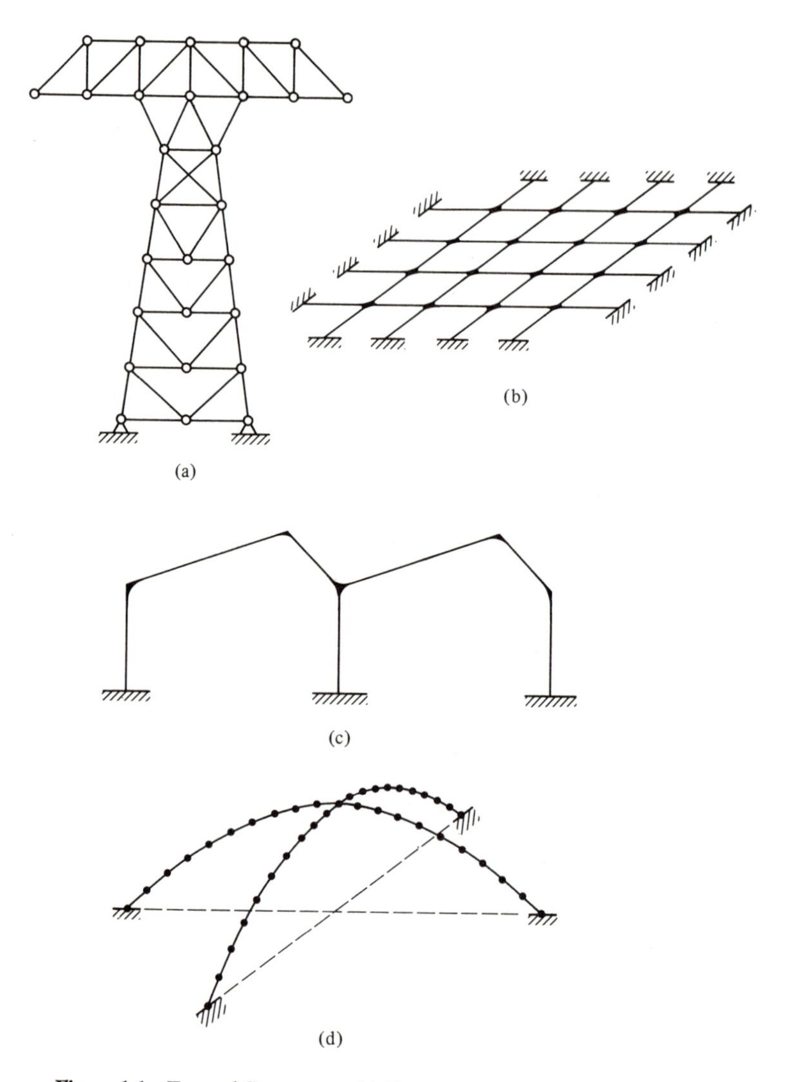

Figure 1.1 Framed Structures: (a) Truss (b) Grid (c) Frame (d) Arch

boundary conditions are difficult to satisfy. A secondary disadvantage is that accuracy of the results is usually poor.

On the other hand, the *finite-element approach* yields an approximate analysis based upon an assumed displacement field, a stress field, or a mixture of these within each element. Since the assumption of displacement functions is the technique most commonly used, the following steps suffice to describe this approach:

1. Divide the continuum into a finite number of subregions (or elements) of simple geometry (triangles, rectangles, and so on).

2. Select key points on the elements to serve as nodes, where conditions of equilibrium and compatibility are to be enforced.
3. Assume displacement functions within each element so that the displacements at each generic point are dependent upon nodal values.
4. Satisfy strain-displacement and stress-strain relationships within a typical element.
5. Determine stiffnesses and equivalent nodal loads for a typical element using work or energy principles.
6. Develop equilibrium equations for the nodes of the discretized continuum in terms of the element contributions.
7. Solve these equilibrium equations for the nodal displacements.
8. Calculate stresses at selected points within the elements.
9. Determine support reactions at restrained nodes if desired.

Many *structural applications* could be mentioned, and some of them are illustrated in Figs. 1.1 through 1.6. Figures 1.1(a) through (c) show examples of *framed structures*, for which the members framing between the joints are auto-

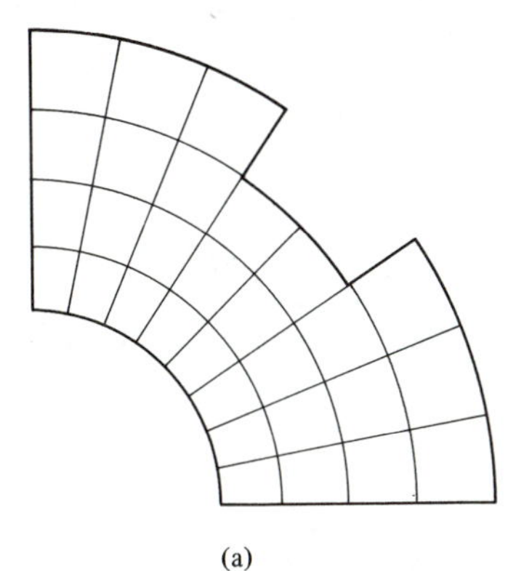

(a)

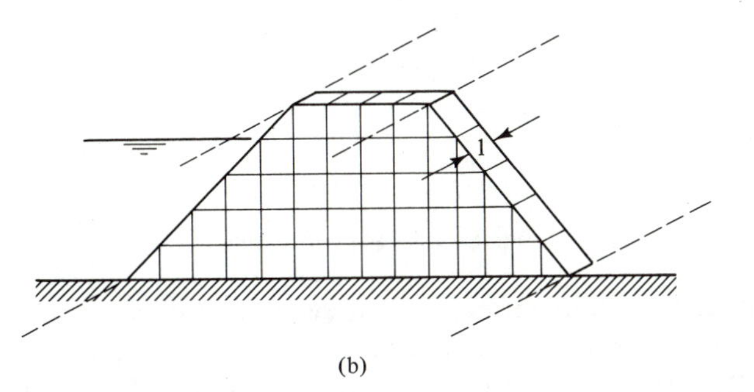

(b)

Figure 1.2 Two-Dimensional Continua: (a) Plane Stress (b) Plane Strain

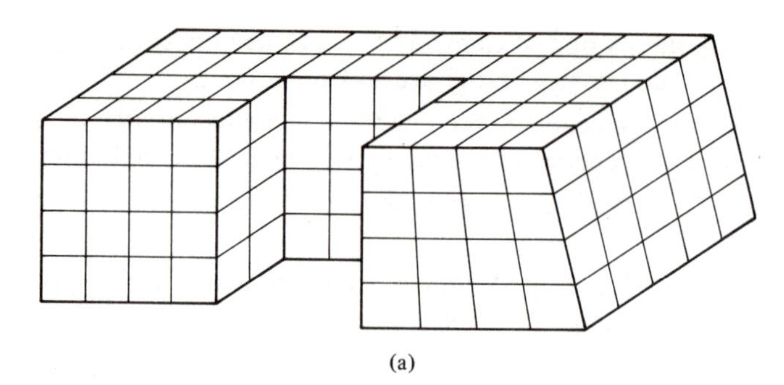

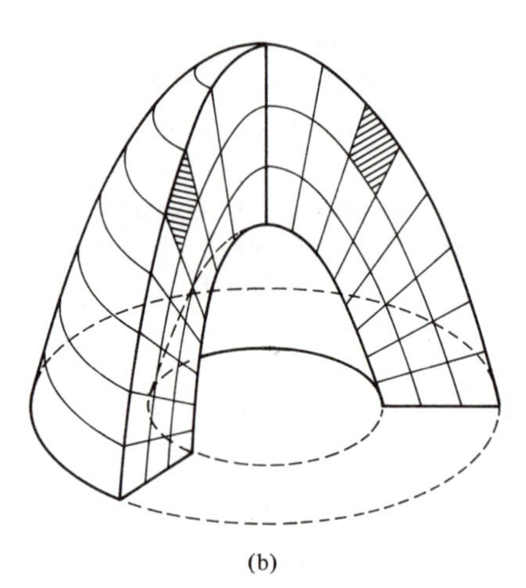

Figure 1.3 Three-Dimensional Continua: (a) General Solid (b) Axisymmetric Solid

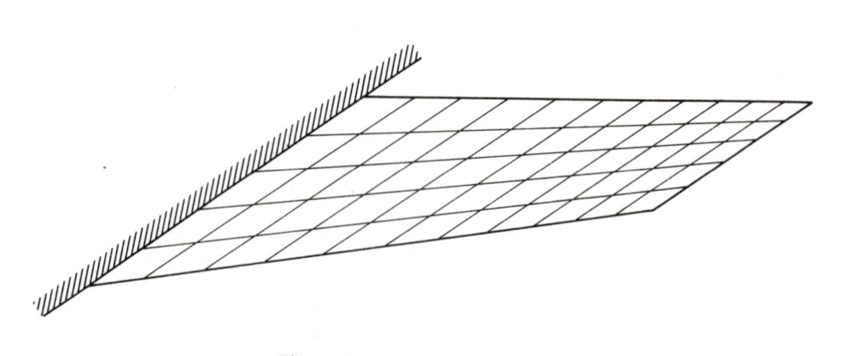

Figure 1.4 Plate in Bending

matically treated as finite elements. However, the arch structure in Fig. 1.1(d) would probably have to be discretized into short segments, as indicated. Figures 1.2(a) and (b) depict *two-dimensional continua* in states of *plane stress* and *plane strain*. Such two-dimensional problems are discussed thoroughly in Chapter 2. A discretized *general solid* (see Chapter 4) is illustrated in Fig. 1.3(a), and ring elements composing an *axisymmetric solid* (see Chapter 5) appear in Fig. 1.3(b). Next, in Fig. 1.4, we see a plate type of structure that may be subjected to bending. The theory of *flexure in plates* is covered in Chapter 6. Figures 1.5(a) and (b) depict a *general shell* and an *axisymmetric shell*, respectively. Chapters 7 and 8 describe elements for these types of structures. In addition, Chapters 9 and 10 deal with vibrations and instabilities of discretized continua.

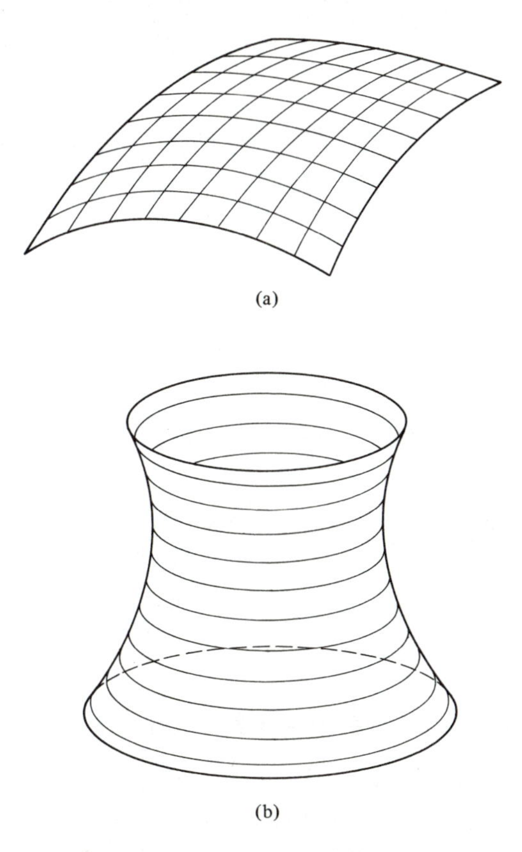

Figure 1.5 Shells: (a) General Shell (b) Axisymmetric Shell

It is also possible to discretize a structure into a mixture of different types of finite elements. For example, Fig. 1.6(a) shows a shear wall and frame composed of plane-stress rectangles for the wall and prismatic or tapered segments for the frame. Similarly, we see a stiffened shell in Fig. 1.6(b) represented by cylindrical shell elements that are reinforced with straight or curved flexural elements. Of course, a computer program for such problems must be able to use more than one type of finite element.

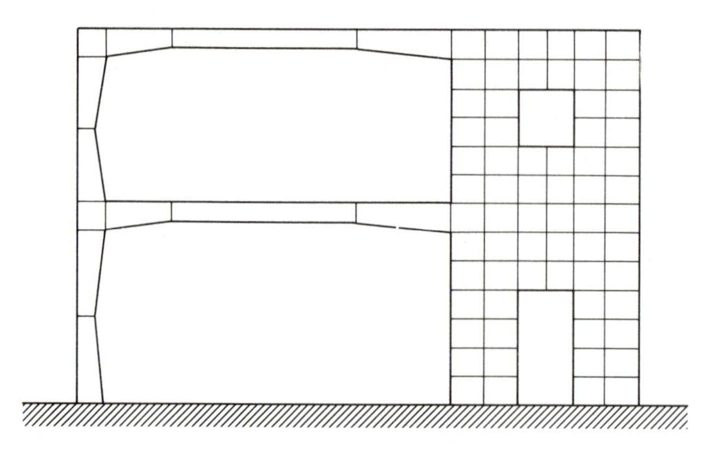

(a)

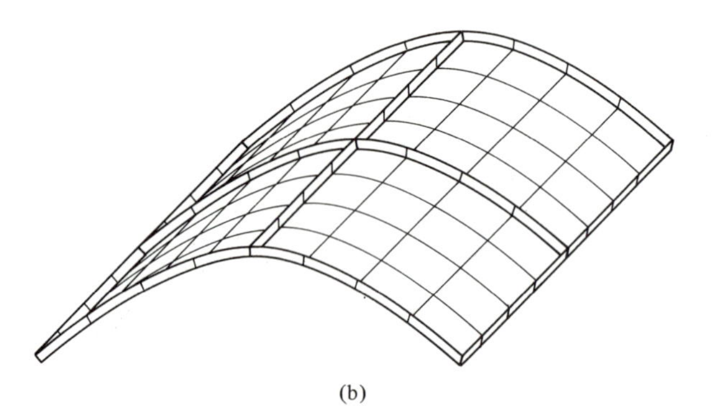

(b)

Figure 1.6 Combined Structures: (a) Frame and Shear Wall (b) Stiffened Shell

1.2 STRESSES AND STRAINS IN ELASTIC CONTINUA

In this book we assume that the continuum to be analyzed consists of an elastic material that undergoes small strains. Such strains and their corresponding stresses may be expressed with respect to some right-hand orthogonal coordinate system. For example, in a (rectangular) Cartesian set, the coordinates would be x, y, and z. And in a cylindrical coordinate system the coordinates would be r, θ, and z.

Figure 1.7 shows an infinitesimal element in Cartesian coordinates, where the edges are of length dx, dy, and dz. *Normal and shearing stresses* are indicated by arrows on the faces of the element. The normal stresses are labeled σ_x, σ_y, and σ_z, whereas the shearing stresses are named τ_{xy}, τ_{yz}, and so on. From equilibrium of the element, the following relationships are known:

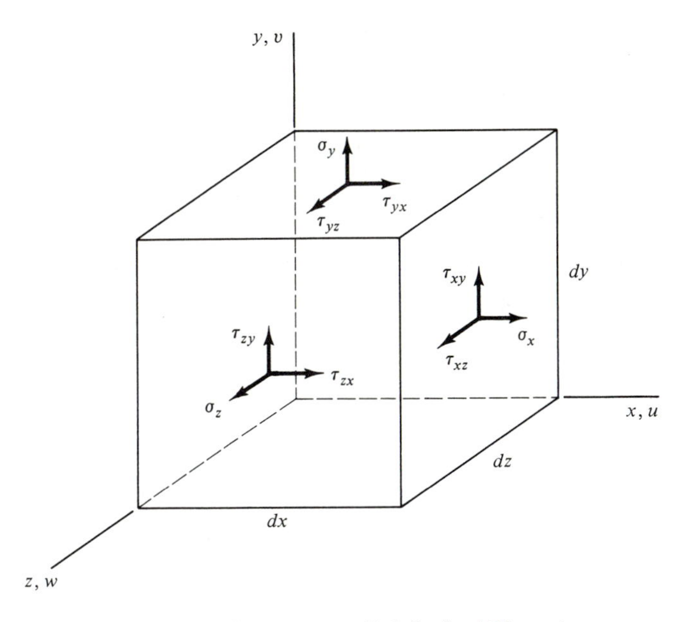

Figure 1.7 Stresses on an Infinitesimal Element

$$\tau_{xy} = \tau_{yx} \qquad \tau_{yz} = \tau_{zy} \qquad \tau_{zx} = \tau_{xz} \tag{a}$$

Thus, only three independent components of shearing stresses need be considered.

Corresponding to the stresses shown in Fig. 1.7 are *normal and shearing strains*. Normal strains ϵ_x, ϵ_y, and ϵ_z are defined as:

$$\epsilon_x = \frac{\partial u}{\partial x} \qquad \epsilon_y = \frac{\partial v}{\partial y} \qquad \epsilon_z = \frac{\partial w}{\partial z} \tag{b}$$

where u, v, and w are translations in the x, y, and z directions. Shearing strains γ_{xy}, γ_{yz}, and so on, are given by:

$$\begin{aligned} \gamma_{xy} &= \frac{\partial u}{\partial y} + \frac{\partial v}{\partial x} = \gamma_{yx} \\ \gamma_{yz} &= \frac{\partial v}{\partial z} + \frac{\partial w}{\partial y} = \gamma_{zy} \\ \gamma_{zx} &= \frac{\partial w}{\partial x} + \frac{\partial u}{\partial z} = \gamma_{xz} \end{aligned} \tag{c}$$

Hence, only three of the shearing strains are independent.

For convenience, the six independent stresses and the corresponding strains usually will be represented as column matrices (or vectors). Thus,

$$\boldsymbol{\sigma} = \begin{bmatrix} \sigma_1 \\ \sigma_2 \\ \sigma_3 \\ \sigma_4 \\ \sigma_5 \\ \sigma_6 \end{bmatrix} = \begin{bmatrix} \sigma_x \\ \sigma_y \\ \sigma_z \\ \tau_{xy} \\ \tau_{yz} \\ \tau_{zx} \end{bmatrix} \qquad \boldsymbol{\epsilon} = \begin{bmatrix} \epsilon_1 \\ \epsilon_2 \\ \epsilon_3 \\ \epsilon_4 \\ \epsilon_5 \\ \epsilon_6 \end{bmatrix} = \begin{bmatrix} \epsilon_x \\ \epsilon_y \\ \epsilon_z \\ \gamma_{xy} \\ \gamma_{yz} \\ \gamma_{zx} \end{bmatrix} \tag{d}$$

Strain-stress relationships (1)* for an isotropic material are drawn from the theory of elasticity, as follows:

$$\begin{aligned} \epsilon_x &= \frac{\sigma_x - \nu\sigma_y - \nu\sigma_z}{E} & \gamma_{xy} &= \frac{\tau_{xy}}{G} \\ \epsilon_y &= \frac{-\nu\sigma_x + \sigma_y - \nu\sigma_z}{E} & \gamma_{yz} &= \frac{\tau_{yz}}{G} \\ \epsilon_z &= \frac{-\nu\sigma_x - \nu\sigma_y + \sigma_z}{E} & \gamma_{zx} &= \frac{\tau_{zx}}{G} \end{aligned} \tag{e}$$

where

$$G = \frac{E}{2(1+\nu)}$$

In these expressions E = Young's modulus, G = shearing modulus, and ν = Poisson's ratio. In matrix format the relationships in Eqs. (e) may be written as:

$$\boldsymbol{\epsilon} = \mathbf{C}\ \boldsymbol{\sigma} \tag{1.2-1}$$

in which

$$\mathbf{C} = \frac{1}{E}\begin{bmatrix} 1 & -\nu & -\nu & 0 & 0 & 0 \\ -\nu & 1 & -\nu & 0 & 0 & 0 \\ -\nu & -\nu & 1 & 0 & 0 & 0 \\ 0 & 0 & 0 & 2(1+\nu) & 0 & 0 \\ 0 & 0 & 0 & 0 & 2(1+\nu) & 0 \\ 0 & 0 & 0 & 0 & 0 & 2(1+\nu) \end{bmatrix} \tag{1.2-2}$$

Matrix $\mathbf{C}$ is an operator that relates the strain vector $\boldsymbol{\epsilon}$ to the stress vector $\boldsymbol{\sigma}$. By the process of inversion, we can also obtain *stress-strain relationships* from Eq. (1.2-1), as follows:

$$\boldsymbol{\sigma} = \mathbf{E}\ \boldsymbol{\epsilon} \tag{1.2-3}$$

*Numbers in parentheses indicate references at the end of the chapter.

where

$$\mathbf{E} = \mathbf{C}^{-1} = \frac{E}{(1+\nu)(1-2\nu)} \begin{bmatrix} 1-\nu & \nu & \nu & 0 & 0 & 0 \\ \nu & 1-\nu & \nu & 0 & 0 & 0 \\ \nu & \nu & 1-\nu & 0 & 0 & 0 \\ 0 & 0 & 0 & \frac{1-2\nu}{2} & 0 & 0 \\ 0 & 0 & 0 & 0 & \frac{1-2\nu}{2} & 0 \\ 0 & 0 & 0 & 0 & 0 & \frac{1-2\nu}{2} \end{bmatrix} \tag{1.2-4}$$

Matrix $\mathbf{E}$ is an operator that relates the stress vector $\boldsymbol{\sigma}$ to the strain vector $\boldsymbol{\epsilon}$.

1.3 VIRTUAL WORK BASIS OF FINITE-ELEMENT METHOD

In this article the virtual work principle will be used to derive the finite-element method. Later, in Sec. 1.9, the method will also be developed on the basis of the potential-energy theorem. We begin with definitions, invoke the virtual work principle for a general element, and apply the method to an axial element.

Assume that a three-dimensional finite element exists in Cartesian coordinates x, y, and z. Let the *generic displacements* at any point within the element be expressed as the column vector $\mathbf{u}$:

$$\mathbf{u} = \{u, v, w\} \tag{1.3-1}$$

where u, v, and w are translations in the x, y, and z directions, respectively.*

If the element is subjected to *body forces*, such forces may be placed into a vector $\mathbf{b}$, as follows:

$$\mathbf{b} = \{b_x, b_y, b_z\} \tag{1.3-2}$$

Here the symbols b_x, b_y, and b_z represent components of forces (per unit of volume, area, or length) acting in the reference directions at a generic point.

Nodal displacements $\mathbf{q}$ will at first be considered as only translations in the x, y, and z directions. Thus, if n_{en} = number of element nodes,

$$\mathbf{q} = \{\mathbf{q}_i\} \qquad (i = 1, 2, \ldots, n_{en}) \tag{1.3-3}$$

where

$$\mathbf{q}_i = \{q_{xi}, q_{yi}, q_{zi}\} = \{u_i, v_i, w_i\} \tag{a}$$

However, other types of displacements, such as rotations (dv/dx, etc.) and curvatures (d^2v/dx^2, etc.), will also be utilized later.

*To save space, column vectors may be written in a row enclosed by braces { } and with commas separating the terms.

Similarly, *nodal actions* $\mathbf{p}$ will temporarily be taken as only forces in the x, y, and z directions at the nodes. That is,

$$\mathbf{p} = \{\mathbf{p}_i\} \qquad (i = 1, 2, \ldots, n_{en}) \tag{1.3-4}$$

in which

$$\mathbf{p}_i = \{p_{xi}, p_{yi}, p_{zi}\} \tag{b}$$

Other types of actions, such as moments, will be considered later.

For the type of finite-element method to which this book is devoted, certain assumed *displacement shape functions* relate generic displacements to nodal displacements, as follows:

$$\mathbf{u} = \mathbf{f}\ \mathbf{q} \tag{1.3-5}$$

In this expression the symbol $\mathbf{f}$ denotes a rectangular matrix containing the functions that make $\mathbf{u}$ completely dependent upon $\mathbf{q}$.

Strain-displacement relationships are obtained by differentiation of the generic displacements. This process may be expressed by forming a matrix $\mathbf{d}$, called a linear differential operator, and applying it with the rules of matrix multiplication. Thus,

$$\boldsymbol{\epsilon} = \mathbf{d}\ \mathbf{u} \tag{1.3-6}$$

In this equation the operator $\mathbf{d}$ expresses the strain vector $\boldsymbol{\epsilon}$ in terms of generic displacements in the vector $\mathbf{u}$ [see Eqs. (b) and (c) in Sec. 1.2]. Substitution of Eq. (1.3-5) into Eq. (1.3-6) yields:

$$\boldsymbol{\epsilon} = \mathbf{B}\ \mathbf{q} \tag{1.3-7}$$

where

$$\mathbf{B} = \mathbf{d}\ \mathbf{f} \tag{1.3-8}$$

Matrix $\mathbf{B}$ gives strains at any point within the element due to unit values of nodal displacements.

From Eq. (1.2-3) in Sec. 1.2, we have the matrix form of *stress-strain relationships*. That is,

$$\boldsymbol{\sigma} = \mathbf{E}\ \boldsymbol{\epsilon} \tag{1.3-9}$$

where $\mathbf{E}$ is a matrix relating stresses in $\boldsymbol{\sigma}$ to strains in $\boldsymbol{\epsilon}$. Substitution of Eq. (1.3-7) into Eq. (1.3-9) produces:

$$\boldsymbol{\sigma} = \mathbf{E}\ \mathbf{B}\ \mathbf{q} \tag{1.3-10}$$

in which the matrix product $\mathbf{E}\ \mathbf{B}$ gives stresses at a generic point due to unit values of nodal displacements.

Virtual work principle. *If a general structure in equilibrium is subjected to a system of small virtual displacements within a compatible state of deformation, the virtual work of external actions is equal to the virtual strain energy of internal stresses.*

If we apply this principle to a finite element, we have:

$$\delta U_e = \delta W_e \tag{1.3-11}$$

where δU_e is the virtual strain energy of internal stresses and δW_e is the virtual work of external actions on the element. To develop both of these quantities in detail, we assume a vector $\delta \mathbf{q}$ of small virtual displacements. Thus,

$$\delta \mathbf{q} = \{\delta \mathbf{q}_i\} \qquad (i = 1, 2, \ldots, n_{en}) \tag{c}$$

Then the resulting virtual generic displacements become [see Eq. (1.3-5)]:

$$\delta \mathbf{u} = \mathbf{f}\ \delta \mathbf{q} \tag{d}$$

Using the strain-displacement relationships in Eq. (1.3-7), we obtain:

$$\delta \boldsymbol{\epsilon} = \mathbf{B}\ \delta \mathbf{q} \tag{e}$$

Now the internal virtual strain energy δU_e can be written as:

$$\delta U_e = \int_V \delta \boldsymbol{\epsilon}^{\mathrm{T}} \boldsymbol{\sigma}\, dV \tag{f}$$

In addition, the external virtual work of nodal and body forces becomes:

$$\delta W_e = \delta \mathbf{q}^{\mathrm{T}} \mathbf{p} + \int_V \delta \mathbf{u}^{\mathrm{T}} \mathbf{b}\, dV \tag{g}$$

Substitution of Eqs. (f) and (g) into Eq. (1.3-11) produces:

$$\int_V \delta \boldsymbol{\epsilon}^{\mathrm{T}} \boldsymbol{\sigma}\, dV = \delta \mathbf{q}^{\mathrm{T}} \mathbf{p} + \int_V \delta \mathbf{u}^{\mathrm{T}} \mathbf{b}\, dV \tag{h}$$

Then we can substitute Eq. (1.3-9) for $\boldsymbol{\sigma}$ and use the transposes of Eqs. (d) and (e) to obtain:

$$\delta \mathbf{q}^{\mathrm{T}} \int_V \mathbf{B}^{\mathrm{T}} \mathbf{E}\ \boldsymbol{\epsilon}\, dV = \delta \mathbf{q}^{\mathrm{T}} \mathbf{p} + \delta \mathbf{q}^{\mathrm{T}} \int_V \mathbf{f}^{\mathrm{T}} \mathbf{b}\, dV \tag{i}$$

Next, substitute Eq. (1.3-7) for $\boldsymbol{\epsilon}$, and cancel $\delta \mathbf{q}^{\mathrm{T}}$ from both sides of Eq. (i).

$$\left(\int_V \mathbf{B}^{\mathrm{T}} \mathbf{E}\ \mathbf{B}\, dV\right) \mathbf{q} = \mathbf{p} + \int_V \mathbf{f}^{\mathrm{T}} \mathbf{b}\, dV \tag{j}$$

Equation (j) may be rewritten as:

$$\mathbf{K}\ \mathbf{q} = \mathbf{p} + \mathbf{p}_b \tag{1.3-12}$$

where

$$\mathbf{K} = \int_V \mathbf{B}^{\mathrm{T}} \mathbf{E}\ \mathbf{B}\, dV \tag{1.3-13}$$

and

$$\mathbf{p}_b = \int_V \mathbf{f}^{\mathrm{T}} \mathbf{b}\, dV \tag{1.3-14}$$

Matrix $\mathbf{K}$ in Eq. (1.3-13) is the *element stiffness matrix*, which contains fictitious actions at nodes due to unit values of nodal displacements. The vector $\mathbf{p}_b$ in Eq. (1.3-14) consists of *equivalent nodal loads* due to body forces in the vector $\mathbf{b}$.

The stresses and strains considered in the above derivation are due only to nodal displacements. If *initial strains* $\boldsymbol{\epsilon}_0$ exist, the total strains may be expressed as:

$$\boldsymbol{\epsilon} = \boldsymbol{\epsilon}_0 + \mathbf{C}\ \boldsymbol{\sigma} \tag{1.3-15}$$

in which $\mathbf{C}$ is a matrix of strain-stress relationships. That is, from Eq. (1.2-4) we have:

$$\mathbf{C} = \mathbf{E}^{-1} \tag{1.3-16}$$

Solving for the stress vector $\boldsymbol{\sigma}$ in Eq. (1.3-15) yields:

$$\boldsymbol{\sigma} = \mathbf{E}(\boldsymbol{\epsilon} - \boldsymbol{\epsilon}_0) \tag{1.3-17}$$

When this expression is used in place of $\boldsymbol{\sigma}$ in Eq. (h), the formulation leads to:

$$\mathbf{K}\ \mathbf{q} = \mathbf{p} + \mathbf{p}_b + \mathbf{p}_0 \tag{1.3-18}$$

where

$$\mathbf{p}_0 = \int_V \mathbf{B}^T\mathbf{E}\ \boldsymbol{\epsilon}_0\ dV \tag{1.3-19}$$

We may consider the vector $\mathbf{p}_0$ to be equivalent nodal loads due to initial strains, such as those caused by *temperature changes.*

Figure 1.8(a) shows an *axial element*, to which the finite-element method will now be applied. Indicated in the figure is a single generic translation u in the direction of x. Thus, from Eq. (1.3-1) we have:

$$\mathbf{u} = u$$

The corresponding body force is a single component b_x (force per unit length), acting in the x direction. Therefore, Eq. (1.3-2) gives:

$$\mathbf{b} = b_x$$

Nodal displacements q_1 and q_2 consist of translations in the x direction at nodes 1 and 2 [see Fig. 1.8(a)]. Thus, Eq. (1.3-3) becomes:

$$\mathbf{q} = \{q_1, q_2\} = \{u_1, u_2\}$$

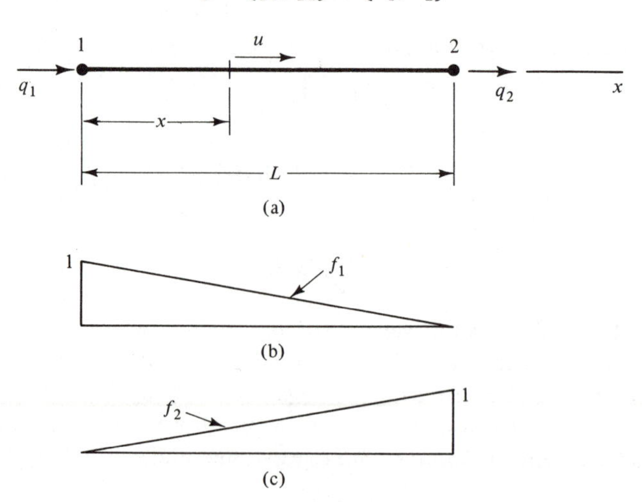

Figure 1.8 Axial Element

Corresponding nodal forces at points 1 and 2 are [see Eq. (1.3-4)]:

$$\mathbf{p} = \{p_1, p_2\} = \{p_{x1}, p_{x2}\}$$

We shall assume that the displacement u at any point within the element varies linearly with x, as follows:

$$u = c_1 + c_2 x \tag{k}$$

This expression is called a *displacement function.* It may be put into the form of a displacement shape function [see Eq. (1.3-5)] by evaluating the two undetermined constants c_1 and c_2. That is, for $x = 0$, we have $c_1 = q_1$; and for $x = L$, we find $q_2 = c_1 + c_2 L$. Thus, $c_2 = (q_2 - q_1)/L$. If we substitute these constants into Eq. (k), we obtain:

$$u = q_1 + \frac{q_2 - q_1}{L} x \tag{ℓ}$$

which is now in terms of the nodal displacements instead of the constants. Equation (ℓ) matches the form of Eq. (1.3-5) and can be rewritten as:

$$u = \begin{bmatrix} 1 - \dfrac{x}{L} & \dfrac{x}{L} \end{bmatrix} \begin{bmatrix} q_1 \\ q_2 \end{bmatrix} = \mathbf{f}\ \mathbf{q}$$

where

$$\mathbf{f} = [f_1 \quad f_2] = \begin{bmatrix} 1 - \dfrac{x}{L} & \dfrac{x}{L} \end{bmatrix} \tag{m}$$

These two displacement shape functions are illustrated in Figs. 1.8(b) and (c).

Strain-displacement relationships [see Eqs. (1.3-6), (1.3-7), and (1.3-8)] for the axial element consist of merely one derivative:

$$\boldsymbol{\epsilon} = \epsilon_x = \mathbf{d}\ \mathbf{u} = \frac{du}{dx} = \frac{d\mathbf{f}}{dx}\mathbf{q} = \mathbf{B}\ \mathbf{q}$$

Hence,

$$\mathbf{B} = \frac{d\mathbf{f}}{dx} = \frac{1}{L}[-1 \quad 1] \tag{n}$$

Similarly, stress-strain relationships [see Eqs. (1.3-9) and (1.3-10)] become merely:

$$\boldsymbol{\sigma} = \boldsymbol{\sigma}_x = \mathbf{E}\ \boldsymbol{\epsilon} = E\epsilon_x = E\mathbf{B}\ \mathbf{q}$$

Thus,

$$\mathbf{E} = E \quad \text{and} \quad E\mathbf{B} = \frac{E}{L}[-1 \quad 1] \tag{o}$$

Then element stiffnesses can be evaluated from Eq. (1.3-13), as follows:

$$\mathbf{K} = \int_V \mathbf{B}^{\mathrm{T}}\mathbf{E}\ \mathbf{B}\, dV = \frac{E}{L^2}\begin{bmatrix} -1 \\ 1 \end{bmatrix}[-1 \quad 1]\int_0^L \int_A dA\, dx$$

$$= \frac{EA}{L}\begin{bmatrix} 1 & -1 \\ -1 & 1 \end{bmatrix} \tag{p}$$

assuming that the cross-sectional area A is constant.

Figures 1.9(a) and (b) show a linearly varying distributed load b_x (force per unit length) that is defined to be:

$$b_x = b_1 + \frac{(b_2 - b_1)x}{L}$$

Due to this body force, Eq. (1.3-14) gives equivalent nodal loads as:

$$\mathbf{p}_b = \int_0^L \mathbf{f}^T b_x \, dx = \int_0^L \begin{bmatrix} 1 - \frac{x}{L} \\ \frac{x}{L} \end{bmatrix} \left[b_1 + \frac{(b_2 - b_1)x}{L} \right] dx$$

$$= \frac{L}{6} \begin{bmatrix} 2b_1 + b_2 \\ b_1 + 2b_2 \end{bmatrix} \tag{q}$$

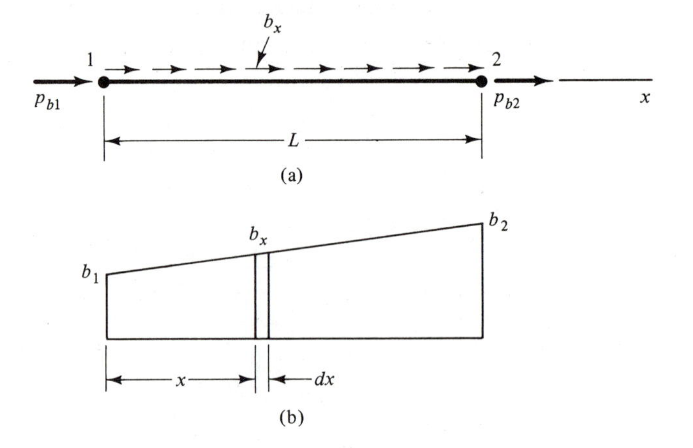

Figure 1.9 Distributed Axial Load

On the other hand, if the element were subjected to a uniform temperature change ΔT, the initial strain due to this influence would be:

$$\epsilon_0 = \epsilon_T = \alpha(\Delta T)$$

where α is the coefficient of thermal expansion. Then from Eq. (1.3-19) we have:

$$\mathbf{p}_0 = \mathbf{p}_T = \int_0^L \int_A \mathbf{B}^T \mathbf{E} \alpha(\Delta T) \, dA \, dx$$

$$= \frac{EA\alpha(\Delta T)}{L} \begin{bmatrix} -1 \\ 1 \end{bmatrix} \int_0^L dx$$

$$= EA\alpha(\Delta T) \begin{bmatrix} -1 \\ 1 \end{bmatrix} \tag{r}$$

If the temperature increases, these equivalent nodal loads act in the negative direction at node 1 and in the positive direction at node 2.

1.4 SHAPE FUNCTIONS AND GENERALIZED DISPLACEMENTS

In the preceding section we converted an assumed *displacement function* to a *displacement shape function* in a step-by-step manner. We will now formalize this process into a matrix operation. For this purpose, let the assumed displacement functions be expressed as a *geometric matrix* **g** times a vector of undetermined constants **c**, as follows:

$$\mathbf{u} = \mathbf{g}\ \mathbf{c} \tag{1.4-1}$$

Then evaluate the operator **g** at each of the nodes, so that:

$$\mathbf{q} = \mathbf{h}\ \mathbf{c} \tag{1.4-2}$$

where

$$\mathbf{h} = \{\mathbf{g}_i\} \qquad (i = 1, 2, \ldots, n_{en}) \tag{a}$$

and $\mathbf{g}_i$ represents matrix **g** evaluated at node *i*. Assuming that matrix **h** is square and nonsingular, solve for the constants **c** in Eq. (1.4-2):

$$\mathbf{c} = \mathbf{h}^{-1}\mathbf{q} \tag{1.4-3}$$

Substitute Eq. (1.4-3) into Eq. (1.4-1) to obtain:

$$\mathbf{u} = \mathbf{g}\ \mathbf{h}^{-1}\mathbf{q} \tag{b}$$

Comparing Eq. (b) with Eq. (1.3-5), we see that the desired shape functions are given by:

$$\mathbf{f} = \mathbf{g}\ \mathbf{h}^{-1} \tag{1.4-4}$$

For the axial element, we first cast the assumed displacement function [Eq. (k) from Sec. 1.3] into the form of Eq. (1.4-1). Thus,

$$u = [1 \quad x]\begin{bmatrix} c_1 \\ c_2 \end{bmatrix} \tag{c}$$

Therefore,

$$\mathbf{g} = [1 \quad x] \tag{d}$$

Evaluating matrix **g** at nodes 1 and 2 (see Eq. 1.4-2), we find:

$$\begin{bmatrix} q_1 \\ q_2 \end{bmatrix} = \begin{bmatrix} 1 & 0 \\ \hdashline 1 & L \end{bmatrix}\begin{bmatrix} c_1 \\ c_2 \end{bmatrix} \tag{e}$$

Notice that the matrix **h** is composed of two contributions, as indicated by the partition line. That is,

$$\mathbf{h} = \begin{bmatrix} 1 & 0 \\ \hdashline 1 & L \end{bmatrix} = \begin{bmatrix} \mathbf{g}_1 \\ \mathbf{g}_2 \end{bmatrix} \tag{f}$$

Inversion of **h** yields:

$$\mathbf{h}^{-1} = \begin{bmatrix} 1 & 0 \\ -\dfrac{1}{L} & \dfrac{1}{L} \end{bmatrix} \tag{g}$$

Then from Eq. (1.4-4) we have:

$$\mathbf{f} = \mathbf{g}\ \mathbf{h}^{-1} = \left[1 - \frac{x}{L} \quad \frac{x}{L}\right] \tag{h}$$

which is the same as Eq. (m) in Sec. 1.3.

The arbitrary coefficients $c_1, c_2, \ldots, c_n$ in assumed displacement functions may themselves be considered as *generalized displacements*. Relationships among nodal displacements, actions, stiffnesses, and their generalized counterparts will now be developed. Toward this end, we write the stiffness equation [see Eq. (1.3-18)] without body forces or initial strains:

$$\mathbf{K}\ \mathbf{q} = \mathbf{p} \tag{i}$$

Expanding matrix $\mathbf{K}$ in terms of matrices $\mathbf{g}$ and $\mathbf{h}$ yields:

$$\mathbf{h}^{-T} \int_V (\mathbf{d}\ \mathbf{g})^T \mathbf{E} (\mathbf{d}\ \mathbf{g})\, dV\, \mathbf{h}^{-1}\mathbf{q} = \mathbf{p} \tag{j}$$

Let

$$\mathbf{B}_c = \mathbf{d}\ \mathbf{g} \tag{1.4-5}$$

Premultiply both sides of Eq. (j) by $\mathbf{h}^T$, and substitute $\mathbf{B}_c$ for $\mathbf{d}\ \mathbf{g}$ in two places:

$$\int_V \mathbf{B}_c^T \mathbf{E}\ \mathbf{B}_c\, dV\, \mathbf{h}^{-1}\mathbf{q} = \mathbf{h}^T\mathbf{p} \tag{k}$$

This expression may be written more simply, as follows:

$$\mathbf{K}_c \mathbf{c} = \mathbf{p}_c \tag{1.4-6}$$

in which

$$\mathbf{c} = \mathbf{h}^{-1}\mathbf{q} \tag{1.4-3}$$

repeated

Also,

$$\mathbf{p}_c = \mathbf{h}^T\mathbf{p} \tag{1.4-7}$$

And

$$\mathbf{K}_c = \int_V \mathbf{B}_c^T \mathbf{E}\ \mathbf{B}_c\, dV \tag{1.4-8}$$

The vector $\mathbf{p}_c$ contains *generalized actions* corresponding to the generalized displacements $\mathbf{c}$, and the matrix $\mathbf{K}_c$ contains *generalized stiffnesses* relating $\mathbf{p}_c$ to $\mathbf{c}$. From Eq. (j) we see that:

$$\mathbf{K} = \mathbf{h}^{-T}\mathbf{K}_c\mathbf{h}^{-1} \tag{1.4-9}$$

And from Eq. (1.4-7) we find:

$$\mathbf{p} = \mathbf{h}^{-T}\mathbf{p}_c \tag{1.4-10}$$

If we reexamine the axial element with regard to using generalized displacements, we find that Eq. (1.4-3) gives:

$$\mathbf{c} = \mathbf{h}^{-1}\mathbf{q} = \begin{bmatrix} 1 & 0 \\ -\dfrac{1}{L} & \dfrac{1}{L} \end{bmatrix} \begin{bmatrix} q_1 \\ q_2 \end{bmatrix} = \begin{bmatrix} q_1 \\ \dfrac{q_2 - q_1}{L} \end{bmatrix} \tag{ℓ}$$

The first generalized displacement in Eq. (ℓ) implies a rigid-body mode, and the second denotes a uniform-strain mode. The corresponding generalized actions may be obtained from Eq. (1.4-7) as:

$$\mathbf{p}_c = \mathbf{h}^{\mathrm{T}}\mathbf{p} = \begin{bmatrix} 1 & 1 \\ 0 & L \end{bmatrix}\begin{bmatrix} p_1 \\ p_2 \end{bmatrix} = \begin{bmatrix} p_1 + p_2 \\ p_2 L \end{bmatrix} \tag{m}$$

Equation (1.4-8) provides the form of generalized stiffnesses, as follows:

$$\begin{aligned} \mathbf{K}_c &= \int_V \mathbf{B}_c^{\mathrm{T}}\mathbf{E}\ \mathbf{B}_c\, dV \\ &= \int_0^L \begin{bmatrix} 0 \\ 1 \end{bmatrix}[0 \quad 1]\, EA\, dx \\ &= EAL\begin{bmatrix} 0 & 0 \\ 0 & 1 \end{bmatrix} \end{aligned} \tag{n}$$

Thus, only one term in matrix $\mathbf{K}_c$ has a nonzero result. We can convert this array to matrix $\mathbf{K}$ using Eq. (1.4-9):

$$\begin{aligned} \mathbf{K} = \mathbf{h}^{-\mathrm{T}}\mathbf{K}_c\mathbf{h}^{-1} &= EAL\begin{bmatrix} 1 & -\frac{1}{L} \\ 0 & \frac{1}{L} \end{bmatrix}\begin{bmatrix} 0 & 0 \\ 0 & 1 \end{bmatrix}\begin{bmatrix} 1 & 0 \\ -\frac{1}{L} & \frac{1}{L} \end{bmatrix} \\ &= \frac{EA}{L}\begin{bmatrix} 1 & -1 \\ -1 & 1 \end{bmatrix} \end{aligned} \tag{o}$$

which is the same result found previously.

We conclude that integrations required for the element stiffness matrix are simplified if we keep the operator $\mathbf{h}^{-1}$ and its transpose outside the integral sign by using the generalized displacements $\mathbf{c}$. Similar statements pertain to the generation of equivalent nodal loads due to body forces and initial strains. Thus, from Eqs. (1.3-14) and (1.3-19) we obtain:

$$\mathbf{p}_b = \mathbf{h}^{-\mathrm{T}}\int_V \mathbf{g}^{\mathrm{T}}\mathbf{b}\, dV = \mathbf{h}^{-\mathrm{T}}\mathbf{p}_{bc} \tag{1.4-11}$$

and

$$\mathbf{p}_0 = \mathbf{h}^{-\mathrm{T}}\int_V \mathbf{B}_c^{\mathrm{T}}\mathbf{E}\boldsymbol{\epsilon}_0\, dV = \mathbf{h}^{-\mathrm{T}}\mathbf{p}_{0c} \tag{1.4-12}$$

which are in the same form as Eq. (1.4-10).

1.5 ONE-DIMENSIONAL ELEMENTS

In this section the characteristics of one-dimensional elements will be examined. Axial, flexural, and torsional elements with straight axes commonly appear in framed structures as primary framing members. An individual member may be subjected to one, two, or all three of these types of deformations, depending upon the type of frame (2).

Since the *axial element* (see Fig. 1.8) has already been discussed in preceding sections, the results found will be merely repeated here for completeness. The element stiffness matrix is:

$$\mathbf{K} = \frac{EA}{L}\begin{bmatrix} 1 & -1 \\ -1 & 1 \end{bmatrix} \tag{1.5-1}$$

Equivalent nodal loads due to a linearly varying axial body force b_x (see Fig. 1.9) are:

$$\mathbf{p}_b = \frac{L}{6}\begin{bmatrix} 2b_1 + b_2 \\ b_1 + 2b_2 \end{bmatrix} \tag{1.5-2}$$

Equivalent nodal loads due to a uniform temperature change ΔT were found to be:

$$\mathbf{p}_0 = \mathbf{p}_T = EA\alpha(\Delta T)\begin{bmatrix} -1 \\ 1 \end{bmatrix} \tag{1.5-3}$$

Of course, the stiffness matrix $\mathbf{K}$ given by Eq. (1.5-1) is unique for a prismatic element. However, there is an infinity of equivalent nodal load vectors, depending upon how the imposed influences vary along the length of the element.

Figure 1.10(a) shows a straight *flexural element*, for which the x-y plane is a principal plane of bending. Indicated in the figure is a single generic displacement v, which is a translation in the y direction. Thus,

$$\mathbf{u} = v$$

The corresponding body force is a single component b_y (force per unit length), acting in the y direction. Hence,

$$\mathbf{b} = b_y$$

At node 1 [see Fig. 1.10(a)] the two nodal displacements labeled q_1 and q_2 are a translation in the y direction and a small rotation in the z sense. The former is indicated by a single-headed arrow, and the latter by a double-headed arrow. Similarly, at node 2 the displacements numbered 3 and 4 are a translation and a small rotation, respectively. Therefore, the vector of nodal displacements becomes:

$$\mathbf{q} = \{q_1, q_2, q_3, q_4\} = \{v_1, \theta_{z1}, v_2, \theta_{z2}\}$$

in which

$$\theta_{z1} = \frac{dv_1}{dx} \qquad \theta_{z2} = \frac{dv_2}{dx}$$

These derivatives (or slopes) may be considered to be small rotations, even though they are actually rates of changes of translations at the nodes. Corresponding nodal actions at points 1 and 2 are:

$$\mathbf{p} = \{p_1, p_2, p_3, p_4\} = \{p_{y1}, M_{z1}, p_{y2}, M_{z2}\}$$

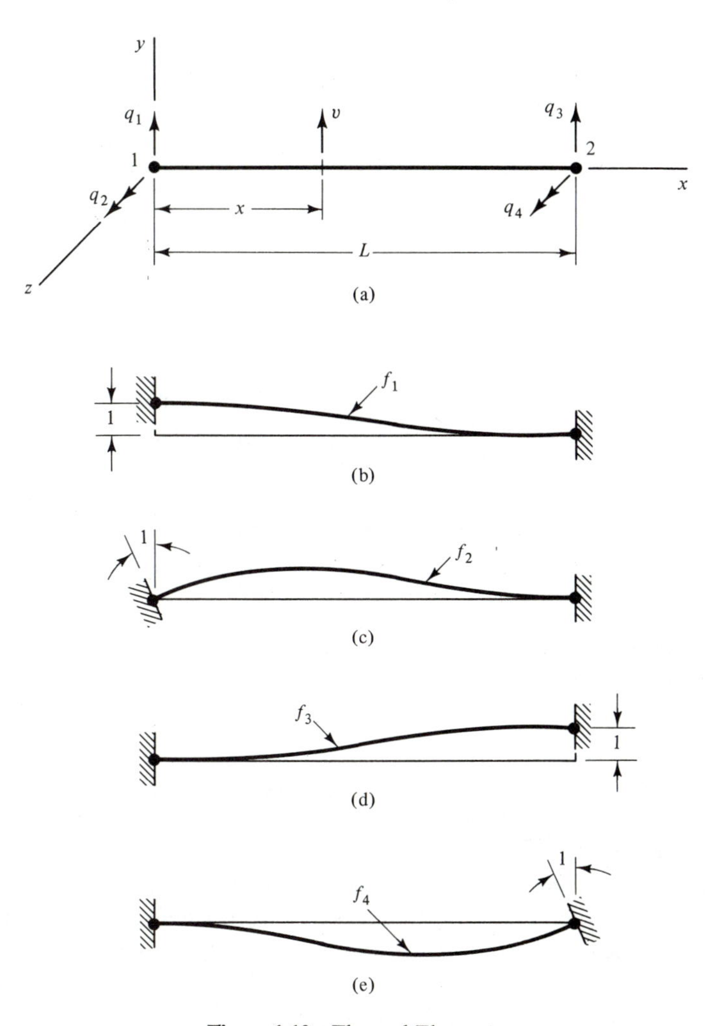

Figure 1.10 Flexural Element

The terms p_{y1} and p_{y2} denote forces in the y direction at nodes 1 and 2, and the symbols M_{z1} and M_{z2} represent moments in the z sense at those points.

Because there are four nodal displacements, a complete cubic displacement function may be assumed for the flexural element, as follows:

$$v = c_1 + c_2 x + c_3 x^2 + c_4 x^3 \tag{a}$$

Then the geometric matrix **g** becomes:

$$\mathbf{g} = [1 \quad x \quad x^2 \quad x^3] \tag{b}$$

In this case the second displacement at each node bears a differential relationship to the first. Therefore, we also need the derivative of **g** with respect to x.

$$\frac{d\mathbf{g}}{dx} = [0 \quad 1 \quad 2x \quad 3x^2] \tag{c}$$

Now matrix **h** can be formed for the two nodes as:

$$\mathbf{h} = \begin{bmatrix} \mathbf{g}_1 \\ \dfrac{d\mathbf{g}_1}{dx} \\ \hdashline \mathbf{g}_2 \\ \dfrac{d\mathbf{g}_2}{dx} \end{bmatrix} = \begin{bmatrix} 1 & 0 & 0 & 0 \\ 0 & 1 & 0 & 0 \\ \hdashline 1 & L & L^2 & L^3 \\ 0 & 1 & 2L & 3L^2 \end{bmatrix} \tag{d}$$

Inversion of matrix **h** yields:

$$\mathbf{h}^{-1} = \frac{1}{L^3}\begin{bmatrix} L^3 & 0 & 0 & 0 \\ 0 & L^3 & 0 & 0 \\ -3L & -2L^2 & 3L & -L^2 \\ 2 & L & -2 & L \end{bmatrix} \tag{e}$$

Premultiply $\mathbf{h}^{-1}$ by **g** to obtain displacement shape functions in matrix **f**, as follows:

$$\begin{aligned} \mathbf{f} &= \mathbf{g}\ \mathbf{h}^{-1} = [f_1 \quad f_2 \quad f_3 \quad f_4] \\ &= \frac{1}{L^3}[2x^3 - 3x^2L + L^3 \quad x^3L - 2x^2L^2 + xL^3 \quad -2x^3 + 3x^2L \quad x^3L - x^2L^2] \end{aligned} \tag{f}$$

These four shape functions are shown in Figs. 1.10(b), (c), (d), and (e). They represent the variations of v along the length due to unit values of the four nodal displacements q_1 through q_4.

Strain-displacement relationships can be developed for the flexural element if we assume that plane sections remain plane during deformation, as indicated in Fig. 1.11. The translation u in the x direction at any point on the cross section is:

$$u = -y\frac{dv}{dx} \tag{g}$$

Using this relationship, we obtain the following expression for flexural strain:

$$\epsilon_x = \frac{du}{dx} = -y\frac{d^2v}{dx^2} = -y\phi \tag{h}$$

in which ϕ represents the *curvature*.

$$\phi = \frac{d^2v}{dx^2} \tag{i}$$

From Eq. (h) we see that the linear differential operator **d** relating ϵ_x to v is:

$$\mathbf{d} = -y\frac{d^2}{dx^2} \tag{j}$$

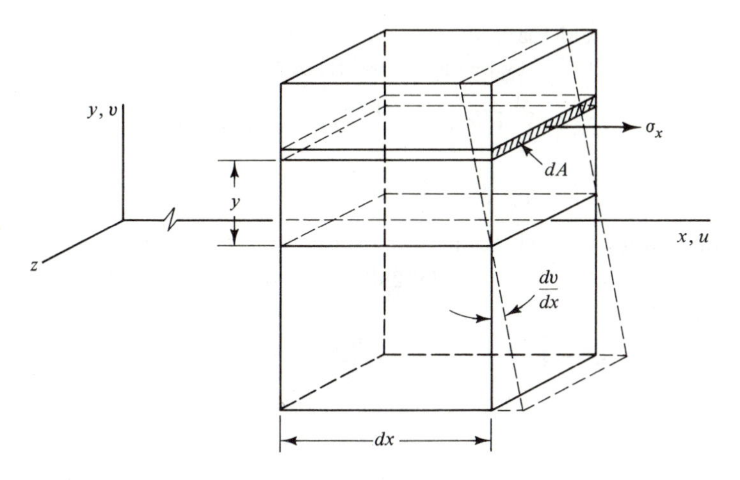

Figure 1.11 Flexural Deformations

Then Eq. (1.3-8) gives the strain-displacement matrix **B** as:

$$\mathbf{B} = \mathbf{d}\ \mathbf{f} = -\frac{y}{L^3}[12x - 6L \quad 6xL - 4L^2 \quad -12x + 6L \quad 6xL - 2L^2] \qquad \text{(k)}$$

Flexural stress σ_x (see Fig. 1.11) is related to flexural strain ϵ_x simply by:

$$\sigma_x = E\epsilon_x \qquad (\ell)$$

Hence,

$$\mathbf{E} = E \quad \text{and} \quad \mathbf{E}\ \mathbf{B} = E\mathbf{B} \qquad \text{(m)}$$

Element stiffnesses may now be obtained from Eq. (1.3-13), as follows:

$$\mathbf{K} = \int_V \mathbf{B}^{\mathrm{T}}\mathbf{E}\ \mathbf{B}\, dV$$

$$= \int_0^L \int_A \frac{Ey^2}{L^6} \begin{bmatrix} 12x - 6L \\ 6xL - 4L^2 \\ -12x + 6L \\ 6xL - 2L^2 \end{bmatrix} [12x - 6L \quad \cdots \quad \cdots \quad 6xL - 2L^2]\, dA\, dx$$

Multiplication and integration (with EI constant) yield:

$$\mathbf{K} = \frac{EI}{L^3} \begin{bmatrix} 12 & 6L & -12 & 6L \\ 6L & 4L^2 & -6L & 2L^2 \\ -12 & -6L & 12 & -6L \\ 6L & 2L^2 & -6L & 4L^2 \end{bmatrix} \qquad \text{(1.5-4)}$$

where

$$I = \int_A y^2\, dA \qquad \text{(n)}$$

represents the *moment of inertia of the cross section* with respect to the neutral axis.

Equivalent nodal loads due to a uniformly distributed force b_y per unit length [see Fig. 1.12(a)] may be calculated from Eq. (1.3-14) as:

$$\mathbf{p}_b = \int_0^L \mathbf{f}^{\mathrm{T}} b_y \, dx = \int_0^L \{f_1, f_2, f_3, f_4\} b_y \, dx$$

$$= \frac{b_y L}{12} \{6, L, 6, -L\} \tag{1.5-5}$$

For this integration the displacement shape functions f_1 through f_4 were obtained from Eq. (f). Similarly, due to a triangular loading [see Fig. 1.12(b)], the equivalent nodal loads are:

$$\mathbf{p}_b = \int_0^L \mathbf{f}^{\mathrm{T}} b_2 \frac{x}{L} \, dx = \frac{b_2 L}{60} \{9, 2L, 21, -3L\} \tag{1.5-6}$$

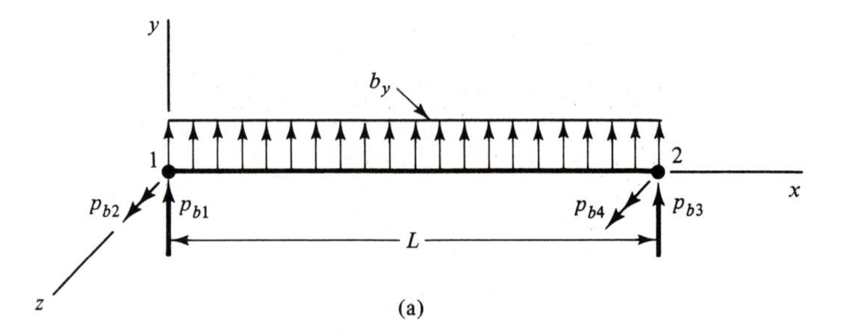

(a)

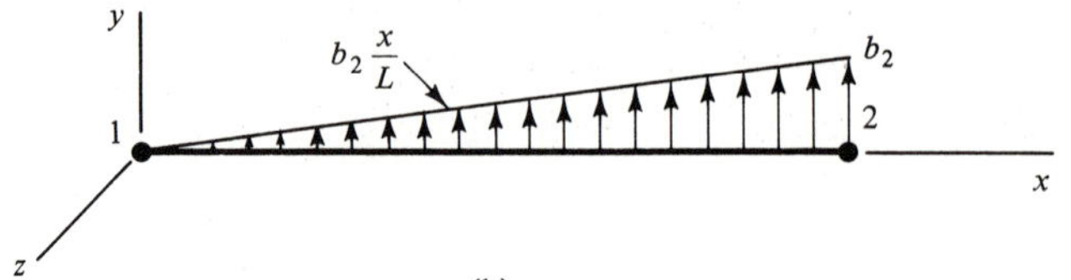

(b)

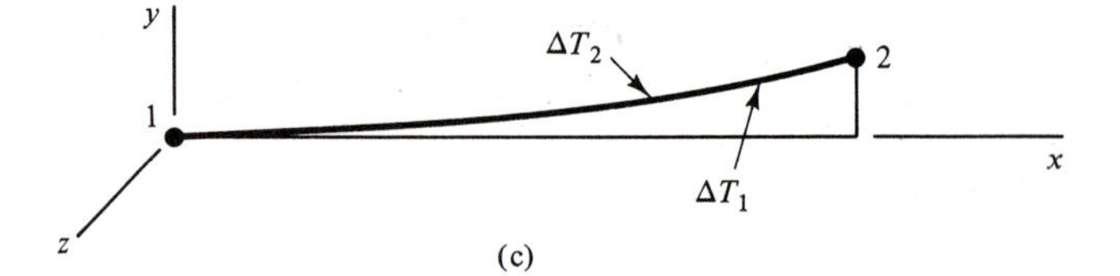

(c)

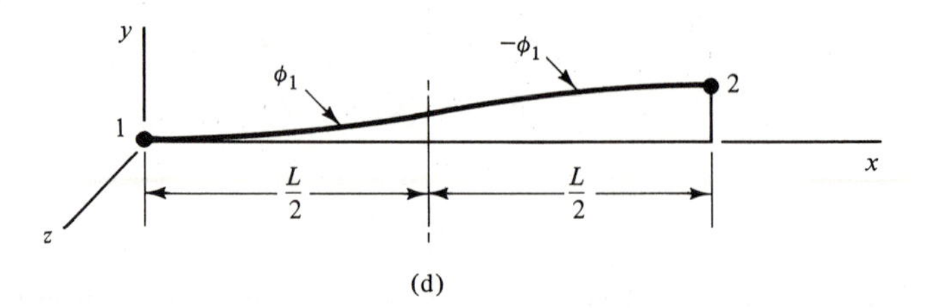

(d)

Figure 1.12 Flexural Element with Loads and Initial Strains

Assume that the flexural element is subjected to a linear variation of temperature from ΔT_1 at its lower surface to ΔT_2 at its upper surface, as indicated in Fig. 1.12(c). If $\Delta T_1 > \Delta T_2$ and the depth of the element is d, the temperature change at any point in the element is:

$$\Delta T = \frac{1}{2}(\Delta T_1 + \Delta T_2) - \frac{y}{d}(\Delta T_1 - \Delta T_2) \tag{o}$$

The first term in Eq. (o) causes axial strains that are not to be considered here. However, the second term produces flexural strains because it varies linearly with y. Thus,

$$\epsilon_{xT} = -\alpha \frac{y}{d}(\Delta T_1 - \Delta T_2) \tag{p}$$

Using this temperature strain in Eq. (1.3-19), we find:

$$\mathbf{p}_T = -\int_0^L \int_A \mathbf{B}^{\mathrm{T}} E\alpha \frac{y}{d}(\Delta T_1 - \Delta T_2)\, dA\, dx$$

After the matrix **B** from Eq. (k) is substituted and the integrations are performed (for constant EI), this expression yields:

$$\mathbf{p}_T = \frac{\alpha EI}{d}(\Delta T_1 - \Delta T_2)\{0, -1, 0, 1\} \tag{1.5-7}$$

The equivalent nodal loads in Eq. (1.5-7) consist of zero forces and equal and opposite moments at the two ends.

On the other hand, if the initial curvatures in Fig. 1.12(d) are known to exist, the initial strains in the element are:

$$\epsilon_{x0} = -y\phi_1 \qquad \left(0 \leq x \leq \frac{L}{2}\right)$$

and

$$\epsilon_{x0} = y\phi_1 \qquad \left(\frac{L}{2} \leq x \leq L\right)$$

In this instance, Eq. (1.3-19) produces:

$$\begin{aligned}\mathbf{p}_0 &= -\int_A \left(\int_0^{L/2} \mathbf{B}^{\mathrm{T}} Ey\phi_1\, dx - \int_{L/2}^L \mathbf{B}^{\mathrm{T}} Ey\phi_1\, dx\right) dA \\ &= \frac{3EI\phi_1}{2L}\{-2, -L, 2, -L\}\end{aligned} \tag{1.5-8}$$

Here we see that the forces are equal and opposite, while the moments are the same at the two ends.

Figure 1.13(a) depicts a *torsional element* that could be the shaft in a machine or a member in a grid. It has a single generic displacement θ_x, which is a small rotation about the x axis. Thus,

$$\mathbf{u} = \theta_x$$

Corresponding to this elastic displacement is a body action:

$$\mathbf{b} = m_x$$

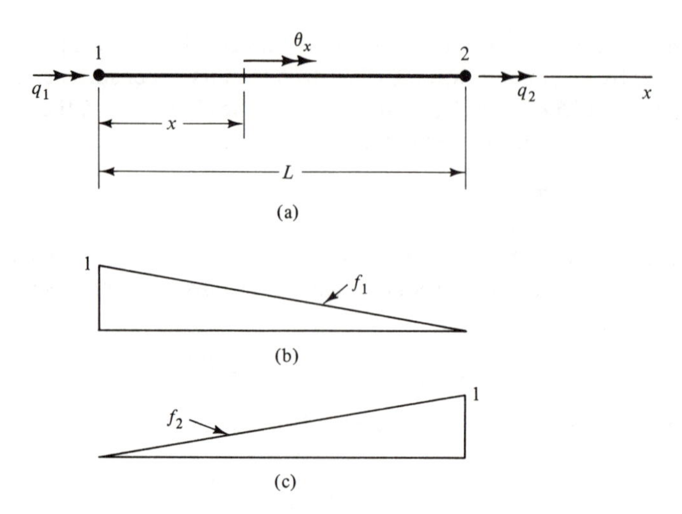

Figure 1.13 Torsional Element

which is a moment (per unit length) acting in the positive x sense. Nodal displacements [see Fig. 1.13(a)] consist of small axial rotations at nodes 1 and 2. Hence,

$$\mathbf{q} = \{q_1, q_2\} = \{\theta_{x1}, \theta_{x2}\}$$

Corresponding nodal actions at points 1 and 2 are:

$$\mathbf{p} = \{p_1, p_2\} = \{M_{x1}, M_{x2}\}$$

which are moments (or torques) acting in the x direction.

Because there are only two nodal displacements for the torsional element, a linear displacement function will be assumed. That is,

$$\theta_x = c_1 + c_2 x \tag{q}$$

As for the axial element, the displacement shape functions become:

$$\mathbf{f} = \mathbf{g}\ \mathbf{h}^{-1} = [f_1 \quad f_2] = \left[1 - \frac{x}{L} \quad \frac{x}{L}\right] \tag{r}$$

as shown in Figs. 1.13(b) and (c).

Strain-displacement relationships can be inferred for a torsional element with a circular cross section by examining Fig. 1.14. Assuming that radii remain straight during torsional deformation, we conclude that the shearing strain γ varies linearly with the radial distance r, as follows:

$$\gamma = r\frac{d\theta_x}{dx} = r\psi \tag{s}$$

where ψ is the *twist*, or rate of change of angular displacement. Thus,

$$\psi = \frac{d\theta_x}{dx} \tag{t}$$

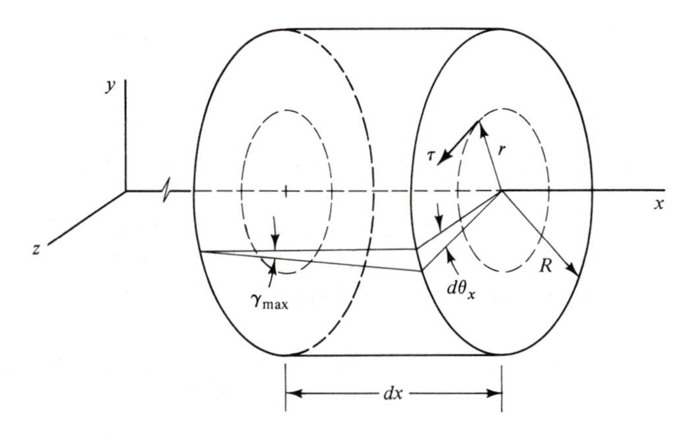

Figure 1.14 Torsional Deformations

It is also evident from Eq. (s) that the maximum value of the shearing strain occurs at the surface. Thus,

$$\gamma_{max} = R\psi$$

where R is the radius of the cross section (see Fig. 1.14). In addition, it is apparent from Eq. (s) that the linear differential operator **d** relating γ to θ_x is:

$$\mathbf{d} = r\frac{d}{dx} \tag{u}$$

Hence, the strain-displacement matrix **B** must be:

$$\mathbf{B} = \mathbf{d}\ \mathbf{f} = \frac{r}{L}[-1 \quad 1] \tag{v}$$

which is the same as for the axial element, except for the presence of r.

Shearing stress τ (see Fig. 1.14) is related to shearing strain γ in a torsional element by:

$$\tau = G\gamma \tag{w}$$

in which the symbol G denotes the shearing modulus of the material. Thus,

$$\mathbf{E} = G \quad \text{and} \quad \mathbf{E}\ \mathbf{B} = G\mathbf{B} \tag{x}$$

Torsional stiffnesses may now be found by applying Eq. (1.3-13), as follows:

$$\begin{aligned} \mathbf{K} &= \int_V \mathbf{B}^{\mathrm{T}}\mathbf{E}\ \mathbf{B}\, dV \\ &= \int_0^L \int_0^{2\pi} \int_0^R \frac{Gr^2}{L^2}\begin{bmatrix} -1 \\ 1 \end{bmatrix}[-1 \quad 1]\, r\, dr\, d\theta\, dx \\ &= \frac{GJ}{L}\begin{bmatrix} 1 & -1 \\ -1 & 1 \end{bmatrix} \end{aligned} \tag{1.5-9}$$

where GJ is constant. The *polar moment of inertia J* is defined as:

$$J = \int_0^{2\pi} \int_0^R r^3\, dr\, d\theta = \frac{\pi R^4}{2} \tag{y}$$

For noncircular cross sections the polar moment of inertia is replaced by a *torsion constant* (3).

Figure 1.15(a) shows a stepped torsional element of circular cross section with a radius of $2R$ over half the length and a radius of R in the other half. For this case the application of Eq. (1.3-13) to find stiffnesses yields:

$$\mathbf{K} = \frac{G}{L^2}\begin{bmatrix} 1 & -1 \\ -1 & 1 \end{bmatrix} \int_0^{2\pi} \left(\int_0^{L/2} \int_0^{2R} r^3\, dr\, dx + \int_{L/2}^{L} \int_0^{R} r^3\, dr\, dx \right) d\theta_x$$

$$= \frac{17GJ}{2L}\begin{bmatrix} 1 & -1 \\ -1 & 1 \end{bmatrix} \tag{1.5-10}$$

Comparing Eqs. (1.5-9) and (1.5-10), we see that the only difference is the constant multiplier.

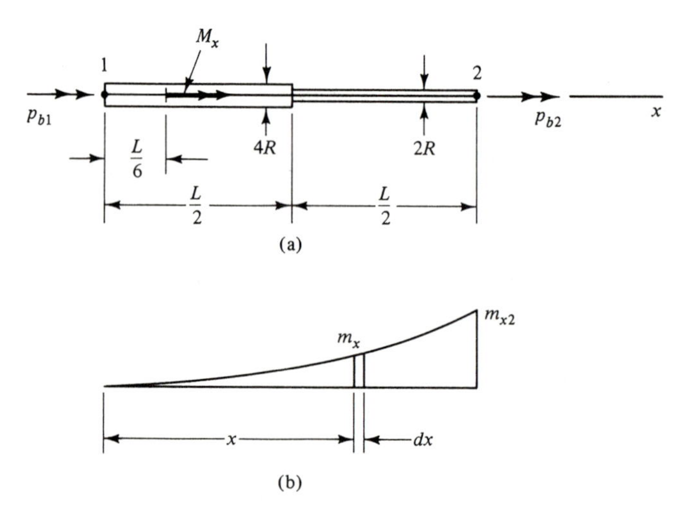

Figure 1.15 Stepped Torsional Element

Equivalent nodel loads due to a concentrated axial torque M_x at the location $x = L/6$ [see Fig. 1.15(a)] become:

$$\mathbf{p}_b = M_x(\mathbf{f}^{\mathrm{T}})_{x=L/6} = M_x\begin{bmatrix} 1 - \frac{1}{6} \\ \frac{1}{6} \end{bmatrix} = \frac{M_x}{6}\begin{bmatrix} 5 \\ 1 \end{bmatrix} \tag{1.5-11}$$

In this case the result is obtained by evaluating the displacement shape functions at the point $x = L/6$, and no integration [as in Eq. (1.3-14)] is required. Moreover, this result is the same regardless of whether the element is stepped, because the same shape functions are used in all instances.

Figure 1.15(b) shows a parabolically distributed axial torque (per unit length), defined as:

$$m_x = m_{x2}\frac{x^2}{L^2} \tag{z}$$

Due to this influence, the equivalent nodal loads calculated from Eq. (1.3-14) are:

$$\begin{aligned}
\mathbf{p}_b &= \int_0^L \mathbf{f}^{\mathrm{T}} m_{x2}\frac{x^2}{L^2}\,dx \\
&= \frac{m_{x2}}{L^2}\int_0^L \begin{bmatrix} 1 - \dfrac{x}{L} \\ \dfrac{x}{L} \end{bmatrix} x^2\,dx \\
&= \frac{m_{x2}L}{12}\begin{bmatrix} 1 \\ 3 \end{bmatrix}
\end{aligned} \tag{1.5-12}$$

which again are independent of the variation of the cross section.

1.6 GENERALIZED STRESSES AND STRAINS

Repetitious integrations over the cross sections of one-dimensional elements can be avoided by using generalized stress-strain operators. While this concept is trivial for an axial element, it can be useful for flexural and torsional elements.

Let us reconsider the flexural element (see Fig. 1.11) and integrate the moment of the σ_x stresses about the z axis to obtain M_z, as follows:

$$M_z = \int_A -\sigma_x y\,dA \tag{a}$$

Then substitute the stress-strain and the strain-displacement relationships from Eqs. (ℓ) and (h) in Sec. 1.5 to find:

$$M_z = E\phi\int_A y^2\,dA = EI\phi \tag{b}$$

If we take M_z as generalized (or integrated) stress and ϕ as generalized strain, the generalized stress-strain (or moment-curvature) operator $\bar{E}$ becomes:

$$\bar{E} = EI \tag{1.6-1}$$

which is the *flexural rigidity* of the cross section. Hence, from Eq. (b) we have:

$$M_z = \bar{E}\phi \tag{1.6-2}$$

In this approach the operator **d** [see Eq. (j) in Sec. 1.5] is devoid of the multiplier $-y$. In addition, the generalized matrix $\bar{\mathbf{B}}$ [Eq. (k) in Sec. 1.5 without the factor $-y$] may be used in place of matrix **B**. Thus,

$$\mathbf{B} = -y\bar{\mathbf{B}} \tag{c}$$

Then integration over the cross section for terms in matrix **K** becomes unnecessary. That is,

$$\mathbf{K} = \int_0^L \bar{\mathbf{B}}^{\mathrm{T}} \bar{E} \bar{\mathbf{B}}\, dx \tag{1.6-3}$$

This expression for **K** is equivalent to Eq. (1.3-13) used previously. Also, for any known state of initial curvature ϕ_0, Eq. (1.3-19) may be replaced with:

$$\mathbf{p}_0 = \int_0^L \bar{\mathbf{B}}^{\mathrm{T}} \bar{E} \phi_0\, dx \tag{1.6-4}$$

Turning now to the torsional element (see Fig. 1.14), we integrate the moment of the shearing stresses on the cross section about the x axis. Thus, we generate the torque M_x by the expression:

$$M_x = \int_0^{2\pi} \int_0^R \tau r^2\, dr\, d\theta \tag{d}$$

Substitution of the stress-strain and the strain-displacement relationships from Eqs. (w) and (s) in Sec. 1.5 produces:

$$M_x = G\psi \int_0^{2\pi} \int_0^R r^3\, dr\, d\theta = GJ\psi \tag{e}$$

For this element we can take M_x as generalized (or integrated) stress and ψ as generalized strain. Then the generalized stress-strain (or torque-twist) operator $\bar{G}$ becomes:

$$\bar{G} = GJ \tag{1.6-5}$$

which is the *torsional rigidity* of the cross section. Thus, from Eq. (e) we have:

$$M_x = \bar{G}\psi \tag{1.6-6}$$

By this method the operator **d** [see Eq. (u) in Sec. 1.5] does not include the multiplier r. Furthermore, the generalized matrix $\bar{\mathbf{B}}$ [Eq. (v) in Sec. 1.5, devoid of r] is used instead of matrix **B**. Hence,

$$\mathbf{B} = r\bar{\mathbf{B}} \tag{f}$$

From this point we can conclude that evaluations of the matrix **K** and the vector $\mathbf{p}_0$ do not require integrations over the cross section. Thus,

$$\mathbf{K} = \int_0^L \bar{\mathbf{B}}^{\mathrm{T}} \bar{G} \bar{\mathbf{B}}\, dx \tag{1.6-7}$$

and

$$\mathbf{p}_0 = \int_0^L \bar{\mathbf{B}}^{\mathrm{T}} \bar{G} \psi_0\, dx \tag{1.6-8}$$

where the symbol ψ_0 represents any state of initial twist.

Later in this chapter (Sec. 1.8) the process of assembly of elements and the calculation of nodal displacements will be explained. Subsequent to those steps we will want to find the state of stress within each element. This can be accomplished by using the following general formula from Sec. 1.3:

$$\boldsymbol{\sigma} = \mathbf{E}(\boldsymbol{\epsilon} - \boldsymbol{\epsilon}_0) \tag{1.3-17 repeated}$$

Substitution of Eq. (1.3-7) into this expression yields:

$$\boldsymbol{\sigma} = \mathbf{E}(\mathbf{B}\ \mathbf{q} - \boldsymbol{\epsilon}_0) \tag{1.6-9}$$

We can convert Eq. (1.6-9) into a special formula for generalized stress, depending upon the application. For example, a flexural element bears the following relationship between σ_x and M_z:

$$\sigma_x = -\frac{M_z y}{I} \tag{1.6-10}$$

Substituting this term as well as Eq. (c) and $\epsilon_{x0} = -y\phi_0$ into Eq. (1.6-9) produces:

$$-\frac{M_z y}{I} = -yE(\bar{\mathbf{B}}\ \mathbf{q} - \phi_0)$$

Thus,

$$M_z = \bar{E}(\bar{\mathbf{B}}\ \mathbf{q} - \phi_0) \tag{1.6-11}$$

This expression gives the moment at any point along the length of a flexural element in terms of the nodal displacements **q** and the initial curvature ϕ_0. Then the flexural stress at any point on the cross section can be obtained from Eq. (1.6-10).

For a torsional element the relationship between the shearing stress τ and the torque M_x is:

$$\tau = \frac{M_x r}{J} \tag{1.6-12}$$

This term as well as $E = G$, Eq. (f), and $\gamma_0 = r\psi_0$ can be substituted into Eq. (1.6-9), yielding:

$$\frac{M_x r}{J} = rG(\bar{\mathbf{B}}\ \mathbf{q} - \psi_0)$$

Hence,

$$M_x = \bar{G}(\bar{\mathbf{B}}\ \mathbf{q} - \psi_0) \tag{1.6-13}$$

Equation (1.6-13) is a formula for the torque at any point along the length of a torsional element in terms of the nodal displacements **q** and the initial twist ψ_0. Then the shearing stress at any point on the cross section can be found using Eq. (1.6-12).

1.7 AXIS TRANSFORMATIONS

Concepts of axis transformations prove to be very useful in mechanics. In this article the operations of translation and rotation of reference axes are described. These principles are applied to actions, displacements, and stiffnesses in a finite-element context.

We consider first the idea of *translation of axes*. By this technique it will be possible to transform a set of actions or displacements from one point to another for parallel sets of axes. Toward this end, Fig. 1.16(a) shows points j

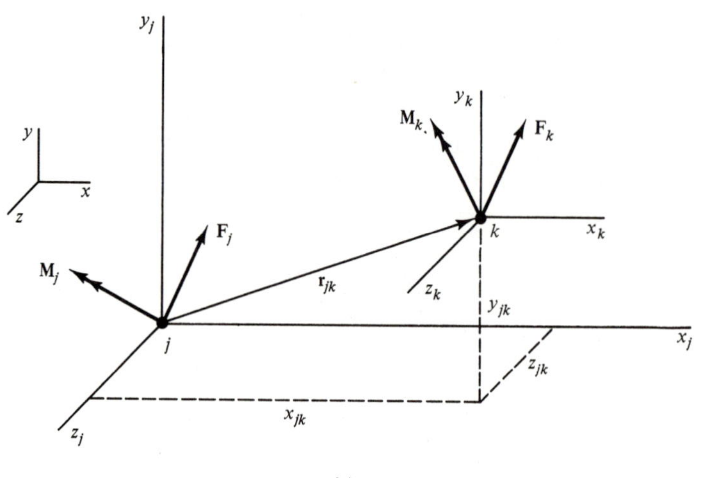

(a)

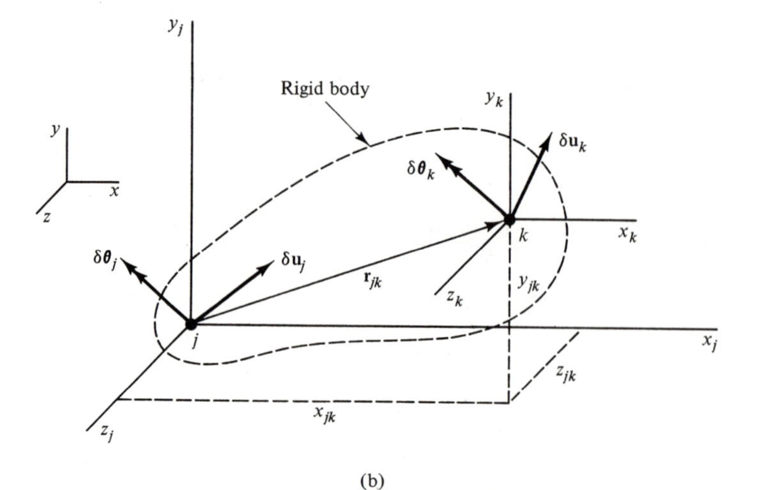

(b)

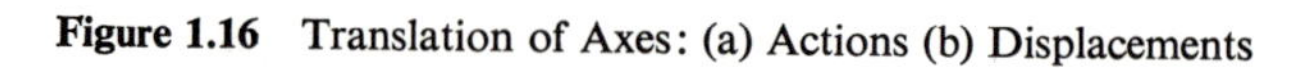
Figure 1.16 Translation of Axes: (a) Actions (b) Displacements

and k in a Cartesian coordinate system with reference axes x, y, and z. At point j is a force vector $\mathbf{F}_j$ and a moment vector $\mathbf{M}_j$ that are *statically equivalent* to $\mathbf{F}_k$ and $\mathbf{M}_k$ at point k. This state of equivalence may be expressed as:

$$\mathbf{F}_j = \mathbf{F}_k \tag{a}$$

and

$$\mathbf{M}_j = \mathbf{r}_{jk} \times \mathbf{F}_k + \mathbf{M}_k \tag{b}$$

As seen in the figure, $\mathbf{r}_{jk}$ is the location vector from j to k; and the scalar values of its components in the x, y, and z directions are x_{jk}, y_{jk}, and z_{jk}. The matrix operation that is equivalent to Eqs. (a) and (b) may be written as:

$$\mathbf{A}_j = \mathbf{T}_{jk}\mathbf{A}_k \tag{1.7-1}$$

where

$$\mathbf{A}_j = \{\mathbf{F}_j, \mathbf{M}_j\} = \{F_{xj}, F_{yj}, F_{zj}, M_{xj}, M_{yj}, M_{zj}\} \tag{c}$$

and

$$\mathbf{A}_k = \{\mathbf{F}_k, \mathbf{M}_k\} = \{F_{xk}, F_{yk}, F_{zk}, M_{xk}, M_{yk}, M_{zk}\} \tag{d}$$

Note that the components of $\mathbf{F}_j$, $\mathbf{M}_j$, $\mathbf{F}_k$, and $\mathbf{M}_k$ are all in directions parallel to the reference axes x, y, and z. In Eq. (1.7-1) the operator $\mathbf{T}_{jk}$ has the form:

$$\mathbf{T}_{jk} = \begin{bmatrix} \mathbf{I}_3 & \mathbf{O} \\ \mathbf{c}_{jk} & \mathbf{I}_3 \end{bmatrix} \tag{1.7-2}$$

It contains a 3×3 submatrix $\mathbf{c}_{jk}$ in the lower left-hand position, which consists of the scalar values of the components of $\mathbf{r}_{jk}$ arranged in a pattern for generating the vector product (or cross product) in Eq. (b). For this purpose, the following skew-symmetric matrix is required:

$$\mathbf{c}_{jk} = \begin{bmatrix} 0 & -z_{jk} & y_{jk} \\ z_{jk} & 0 & -x_{jk} \\ -y_{jk} & x_{jk} & 0 \end{bmatrix} \tag{1.7-3}$$

The other submatrices in $\mathbf{T}_{jk}$ are a 3×3 identity matrix $\mathbf{I}_3$ (appearing twice) and a 3×3 null matrix. Equation (1.7-1) may be used in analysis whenever static equivalents are required. The operator $\mathbf{T}_{jk}$ transforms a set of component actions $\mathbf{A}_k$ at point k into a parallel set $\mathbf{A}_j$ at point j, and it is called a *translation-of-axes transformation matrix*. The reverse transformation (from j to k) is:

$$\mathbf{A}_k = \mathbf{T}_{jk}^{-1}\mathbf{A}_j \tag{1.7-4}$$

in which

$$\mathbf{T}_{jk}^{-1} = \begin{bmatrix} \mathbf{I}_3 & \mathbf{O} \\ \mathbf{c}_{kj} & \mathbf{I}_3 \end{bmatrix} \tag{1.7-5}$$

and

$$\mathbf{c}_{kj} = -\mathbf{c}_{jk} \tag{1.7-6}$$

Kinematic relationships between the displacements of two points on a rigid body can also be expressed in operator form. Figure 1.16(b) shows points j and k on the same rigid body, which is subjected to a set of small displacements. These rigid-body displacements are represented at point j by the translation vector $\delta\mathbf{u}_j$ and the rotation vector $\delta\boldsymbol{\theta}_j$, while their *kinematic equivalents* at point k are labeled $\delta\mathbf{u}_k$ and $\delta\boldsymbol{\theta}_k$. The displacements at k may be computed from those at j by the relationships:

$$\delta\mathbf{u}_k = \delta\mathbf{u}_j + \delta\boldsymbol{\theta}_j \times \mathbf{r}_{jk} \tag{e}$$

and

$$\delta\boldsymbol{\theta}_k = \delta\boldsymbol{\theta}_j \tag{f}$$

In Eq. (e) the cross product can be rewritten as:

$$\delta\boldsymbol{\theta}_j \times \mathbf{r}_{jk} = -\mathbf{r}_{jk} \times \delta\boldsymbol{\theta}_j = \mathbf{r}_{kj} \times \delta\boldsymbol{\theta}_j \tag{g}$$

Then the matrix operation that is equivalent to Eqs. (e) and (f) becomes:

$$\mathbf{D}_k = \mathbf{T}_{jk}^{\mathrm{T}} \mathbf{D}_j \tag{1.7-7}$$

where

$$\mathbf{D}_j = \{\delta \mathbf{u}_j, \delta \boldsymbol{\theta}_j\} = \{\delta u_j, \delta v_j, \delta w_j, \delta \theta_{xj}, \delta \theta_{yj}, \delta \theta_{zj}\} \tag{h}$$

and

$$\mathbf{D}_k = \{\delta \mathbf{u}_k, \delta \boldsymbol{\theta}_k\} = \{\delta u_k, \delta v_k, \delta w_k, \delta \theta_{xk}, \delta \theta_{yk}, \delta \theta_{zk}\} \tag{i}$$

The operator $\mathbf{T}_{jk}^{\mathrm{T}}$ in Eq. (1.7-7) has the form:

$$\mathbf{T}_{jk}^{\mathrm{T}} = \begin{bmatrix} \mathbf{I}_3 & \mathbf{c}_{jk}^{\mathrm{T}} \\ \mathbf{O} & \mathbf{I}_3 \end{bmatrix} \tag{1.7-8}$$

in which

$$\mathbf{c}_{jk}^{\mathrm{T}} = -\mathbf{c}_{jk} = \mathbf{c}_{kj} = \begin{bmatrix} 0 & z_{jk} & -y_{jk} \\ -z_{jk} & 0 & x_{jk} \\ y_{jk} & -x_{jk} & 0 \end{bmatrix} \tag{1.7-9}$$

Here we see that transposition of $\mathbf{c}_{jk}$ merely changes its sign, which is an inherent property of a skew-symmetric matrix. In Eq. (1.7-7) the transposed operator $\mathbf{T}_{jk}^{\mathrm{T}}$ converts a set of component displacements $\mathbf{D}_j$ at point j into a parallel set $\mathbf{D}_k$ at point k, assuming that there are no relative displacements of the points. The reverse transformation is:

$$\mathbf{D}_j = \mathbf{T}_{jk}^{-\mathrm{T}} \mathbf{D}_k \tag{1.7-10}$$

Thus, the calculation of displacements at j in terms of those at k involves the transposed inverse of the operator used for obtaining actions at j from those at k [see Eq. (1.7-1)].

The concept of *rotation of axes* applies to a force, a moment, a translation, a small rotation, or a set of orthogonal coordinates. Figure 1.17 shows a force vector $\mathbf{F}$ and its components in the x, y, and z directions as well as its components in the directions of inclined axes x', y', and z'. The scalar values of the components in directions of the primed axes can be computed from those for the unprimed axes, as follows:

$$F_{x'} = (\mathbf{F}_x + \mathbf{F}_y + \mathbf{F}_z) \cdot \mathbf{i}' = \lambda_{11} F_x + \lambda_{12} F_y + \lambda_{13} F_z \tag{j}$$

$$F_{y'} = (\mathbf{F}_x + \mathbf{F}_y + \mathbf{F}_z) \cdot \mathbf{j}' = \lambda_{21} F_x + \lambda_{22} F_y + \lambda_{23} F_z \tag{k}$$

$$F_{z'} = (\mathbf{F}_x + \mathbf{F}_y + \mathbf{F}_z) \cdot \mathbf{k}' = \lambda_{31} F_x + \lambda_{32} F_y + \lambda_{33} F_z \tag{ℓ}$$

In these expressions the symbols $\mathbf{i}'$, $\mathbf{j}'$, and $\mathbf{k}'$ denote unit vectors in the directions of the primed axes, and $\lambda_{11}, \lambda_{12}, \ldots,$ are *direction cosines* of the primed axes with respect to the unprimed axes. (For example, λ_{12} is the direction cosine of axis x' with respect to axis y, and so on.) The matrix form of Eqs. (j), (k), and (ℓ) is:

$$\begin{bmatrix} F_{x'} \\ F_{y'} \\ F_{z'} \end{bmatrix} = \begin{bmatrix} \lambda_{11} & \lambda_{12} & \lambda_{13} \\ \lambda_{21} & \lambda_{22} & \lambda_{23} \\ \lambda_{31} & \lambda_{32} & \lambda_{33} \end{bmatrix} \begin{bmatrix} F_x \\ F_y \\ F_z \end{bmatrix} \tag{1.7-11}$$

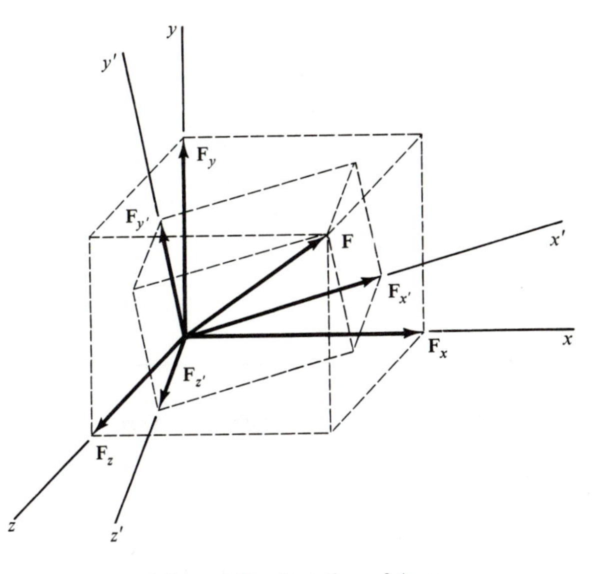

Figure 1.17 Rotation of Axes

Or,

$$\mathbf{F}' = \mathbf{R}\ \mathbf{F} \tag{1.7-12}$$

Matrix **R** is called the *rotation matrix*, consisting of direction cosines of the primed axes (with respect to the unprimed axes) listed rowwise. Hence, it is orthogonal, and the reverse transformation becomes:

$$\mathbf{F} = \mathbf{R}^{-1}\mathbf{F}' = \mathbf{R}^{\mathrm{T}}\mathbf{F}' \tag{1.7-13}$$

Simultaneous transformation of a force vector and a moment vector may be accomplished by:

$$\mathbf{A}' = \hat{\mathbf{R}}\ \mathbf{A} = \begin{bmatrix} \mathbf{R} & \mathbf{O} \\ \mathbf{O} & \mathbf{R} \end{bmatrix} \begin{bmatrix} \mathbf{F} \\ \mathbf{M} \end{bmatrix} \tag{1.7-14}$$

In this expression $\hat{\mathbf{R}}$ is a *rotation-of-axes transformation matrix* containing two identical rotation matrices in diagonal positions. The reverse transformation is:

$$\mathbf{A} = \hat{\mathbf{R}}^{-1}\mathbf{A}' = \hat{\mathbf{R}}^{\mathrm{T}}\mathbf{A}' = \begin{bmatrix} \mathbf{R}^{\mathrm{T}} & \mathbf{O} \\ \mathbf{O} & \mathbf{R}^{\mathrm{T}} \end{bmatrix} \begin{bmatrix} \mathbf{F}' \\ \mathbf{M}' \end{bmatrix} \tag{1.7-15}$$

If *local axes* for a finite element are not parallel to *global axes* for the whole structure, rotation-of-axes transformations must be applied to element stiffnesses, nodal displacements, and nodal loads. Thus, when the elements are assembled, the resulting nodal equilibrium equations will pertain to the global directions at each node. (It is also possible to apply translation-of-axes transformations in order to refer all nodal equilibrium equations to the same origin, but this type of operation is unnecessary.) Figure 1.18 shows a triangular element

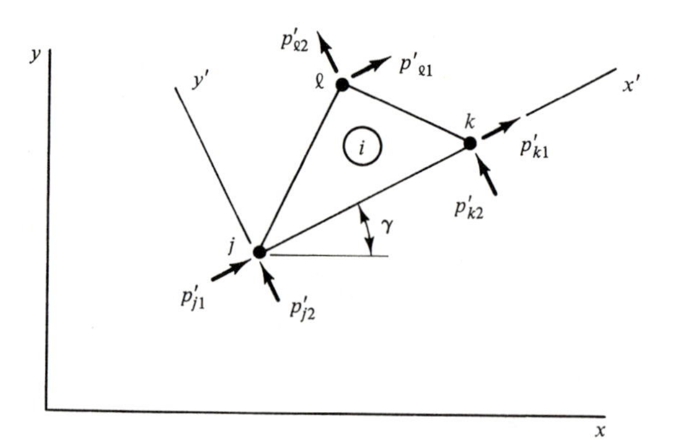

Figure 1.18 Triangle with Local Axes

i in the x-y plane. For local axes we can write:

$$\mathbf{K}'\mathbf{q}' = \mathbf{p}' \tag{1.7-16}$$

where the primes imply that nodal actions and displacements are in the directions of inclined axes x' and y'. Premultiply Eq. (1.7-16) with $\hat{\mathbf{R}}^{\mathrm{T}}$ and substitute $\mathbf{q}' = \hat{\mathbf{R}}\,\mathbf{q}$ to obtain:

$$\hat{\mathbf{R}}^{\mathrm{T}}\mathbf{K}'\hat{\mathbf{R}}\,\mathbf{q} = \hat{\mathbf{R}}^{\mathrm{T}}\mathbf{p}' \tag{m}$$

More briefly,

$$\mathbf{K}\,\mathbf{q} = \mathbf{p} \tag{1.7-17}$$

in which $\mathbf{K}$ is obtained from $\mathbf{K}'$ by the congruence transformation:

$$\mathbf{K} = \hat{\mathbf{R}}^{\mathrm{T}}\mathbf{K}'\hat{\mathbf{R}} \tag{1.7-18}$$

Also,

$$\mathbf{p} = \hat{\mathbf{R}}^{\mathrm{T}}\mathbf{p}' \tag{1.7-19}$$

In these equations the rotation transformation matrix $\hat{\mathbf{R}}$ is composed as follows:

$$\hat{\mathbf{R}} = \begin{bmatrix} \mathbf{R}_j & \mathbf{O} & \mathbf{O} \\ \mathbf{O} & \mathbf{R}_k & \mathbf{O} \\ \mathbf{O} & \mathbf{O} & \mathbf{R}_\ell \end{bmatrix} \tag{1.7-20}$$

where

$$\begin{aligned} \mathbf{R}_j = \mathbf{R}_k = \mathbf{R}_\ell &= \mathbf{R} \\ &= \begin{bmatrix} \lambda_{11} & \lambda_{12} \\ \lambda_{21} & \lambda_{22} \end{bmatrix} \\ &= \begin{bmatrix} \cos\gamma & \sin\gamma \\ -\sin\gamma & \cos\gamma \end{bmatrix} \end{aligned} \tag{1.7-21}$$

In this two-dimensional example the rotation matrix $\mathbf{R}$ is of size 2×2 instead of 3×3.

1.8 ASSEMBLAGE OF ELEMENTS AND SOLUTION

Assuming that stiffnesses and equivalent nodal loads have been obtained for individual elements, we can assemble them by applying the principle of virtual work to the whole structure. This principle (see Sec. 1.3) gives:

$$\delta U_s = \delta W_s \tag{1.8-1}$$

For a set of virtual nodal displacements $\delta\mathbf{D}_N$, the *internal virtual work* for the whole structure is:

$$\delta U_s = \delta\mathbf{D}_N^T\mathbf{S}_N\mathbf{D}_N \tag{a}$$

in which $\mathbf{D}_N$ represents a vector of actual *nodal displacements* and $\mathbf{S}_N$ is the *nodal stiffness matrix* for the whole structure. By summation,

$$\mathbf{S}_N = \sum_{i=1}^{n_e} \mathbf{K}_i \tag{b}$$

where n_e is the number of elements.

The *external virtual work* for the whole structure is:

$$\delta W_s = \delta\mathbf{D}_N^T(\mathbf{A}_N + \mathbf{A}_{Nb} + \mathbf{A}_{N0}) \tag{c}$$

In this expression $\mathbf{A}_N$ denotes a vector of actions applied at the nodes of the assemblage (or *applied nodal actions*). The other two action vectors, $\mathbf{A}_{Nb}$ and $\mathbf{A}_{N0}$, represent *equivalent nodal loads* due to body forces and initial strains, respectively. By summation, we have:

$$\mathbf{A}_N = \sum_{i=1}^{n_e} \mathbf{p}_i \qquad \mathbf{A}_{Nb} = \sum_{i=1}^{n_e} \mathbf{p}_{bi} \qquad \mathbf{A}_{N0} = \sum_{i=1}^{n_e} \mathbf{p}_{0i} \tag{d}$$

Substitution of Eqs. (a) and (c) into Eq. (1.8-1) and cancellation of $\delta\mathbf{D}_N^T$ yields:

$$\mathbf{S}_N\mathbf{D}_N = \mathbf{A}_N + \mathbf{A}_{Nb} + \mathbf{A}_{N0} \tag{1.8-2}$$

These are the *nodal equilibrium equations* for all the nodes of the structure, regardless of whether they are free or restrained.

In preparation for solving Eqs. (1.8-2), we can rearrange and partition them, as follows:

$$\begin{bmatrix} \mathbf{S}_{FF} & \mathbf{S}_{FR} \\ \mathbf{S}_{RF} & \mathbf{S}_{RR} \end{bmatrix} \begin{bmatrix} \mathbf{D}_F \\ \mathbf{D}_R \end{bmatrix} = \begin{bmatrix} \mathbf{A}_F \\ \mathbf{A}_R \end{bmatrix} \tag{1.8-3}$$

In this form the subscript F refers to *free nodal displacements*, and the subscript R denotes *restrained nodal displacements*. Expanding Eq. (1.8-3) gives:

$$\mathbf{S}_{FF}\mathbf{D}_F + \mathbf{S}_{FR}\mathbf{D}_R = \mathbf{A}_F \tag{1.8-4a}$$

and

$$\mathbf{S}_{RF}\mathbf{D}_F + \mathbf{S}_{RR}\mathbf{D}_R = \mathbf{A}_R \tag{1.8-4b}$$

Solving for $\mathbf{D}_F$ in Eq. (1.8-4a) yields:

$$\mathbf{D}_F = \mathbf{S}_{FF}^{-1}(\mathbf{A}_F - \mathbf{S}_{FR}\mathbf{D}_R) \tag{1.8-5}$$

Thus, *equivalent nodal loads due to restraint displacements* $\mathbf{D}_R$ are seen to be:

$$\mathbf{A}_{FR} = -\mathbf{S}_{FR}\mathbf{D}_R \tag{1.8-6}$$

After the free nodal displacements have been found, we can evaluate *support reactions* $\mathbf{A}_R$ from Eq. (1.8-4b) if desired. Then the stresses within each element may be found by:

$$\boldsymbol{\sigma}_i = \mathbf{E}_i(\mathbf{B}_i\mathbf{q}_i - \boldsymbol{\epsilon}_{0i}) \qquad (i = 1, 2, \ldots, n_e) \tag{1.8-7}$$

in which the vector $\mathbf{q}_i$ represents the nodal displacements for the ith element.

While the rearrangement of Eq. (1.8-2) as described above gives us a useful perspective, it is also possible to do the *solution without rearrangement*. This can be accomplished by modifying the stiffness and load matrices to convert the equations for support reactions into trivial displacement equations embedded within the complete set of equations. Then we can solve the whole set in place for the unknown nodal displacements as well as the known support displacements without having to rearrange and partition the matrices.

To show the technique, a small example will suffice. Suppose that a hypothetical structure has only four possible nodal displacements, as indicated by the following nodal equilibrium equations:

$$\begin{bmatrix} S_{N11} & S_{N12} & S_{N13} & S_{N14} \\ S_{N21} & S_{N22} & S_{N23} & S_{N24} \\ S_{N31} & S_{N32} & S_{N33} & S_{N34} \\ S_{N41} & S_{N42} & S_{N43} & S_{N44} \end{bmatrix} \begin{bmatrix} D_{N1} \\ D_{N2} \\ D_{N3} \\ D_{N4} \end{bmatrix} = \begin{bmatrix} A_{N1} \\ A_{N2} \\ A_{N3} \\ A_{N4} \end{bmatrix} \tag{1.8-8}$$

In addition, suppose that the third displacement is specified to be a nonzero support displacement $D_{N3} \neq 0$. Then the terms involving D_{N3} can be subtracted from both sides of Eq. (1.8-8), and the third equation can be replaced by the trivial expression $D_{N3} = D_{N3}$ to obtain:

$$\begin{bmatrix} S_{N11} & S_{N12} & 0 & S_{N14} \\ S_{N21} & S_{N22} & 0 & S_{N24} \\ 0 & 0 & 1 & 0 \\ S_{N41} & S_{N42} & 0 & S_{N44} \end{bmatrix} \begin{bmatrix} D_{N1} \\ D_{N2} \\ D_{N3} \\ D_{N4} \end{bmatrix} = \begin{bmatrix} A_{N1} - S_{N13}D_{N3} \\ A_{N2} - S_{N23}D_{N3} \\ D_{N3} \\ A_{N4} - S_{N43}D_{N3} \end{bmatrix} \tag{1.8-9}$$

These equations may now be solved for the four nodal displacements, including D_{N3}.

Negative terms on the right-hand side of Eq. (1.8-9) represent equivalent nodal loads due to the specified support displacement D_{N3}. Of course, if $D_{N3} = 0$ these equivalent nodal loads are also zero. Any number of specified support displacements can be handled in this manner. This technique precludes the calculation of support reactions by the matrix multiplication approach given by Eq. (1.8-4b). Instead, it is necessary to obtain the reactions from nodal actions for elements connected to the supports.

1.9 POTENTIAL-ENERGY BASES OF FINITE-ELEMENT METHOD

Most of the equations in this book are derived from the principle of virtual work. However, it is also recognized that the finite-element method may be derived from potential-energy theory. In this section two complementary approaches will be outlined. The discussion pertains only to continua for which strains and displacements are small and no energy is dissipated in the static loading process. That is, the external work of gradually applied loads is equal to the energy stored in the structure, and the system is said to be *conservative*.

The first approach to be discussed utilizes the *principle of stationary potential energy*. This principle says that among all the displacement states of a conservative system that satisfy compatibility and boundary restraints, those that also satisfy equilibrium make the potential energy stationary. The stationary value may be a maximum, a minimum, or a neutral point; and when it is a minimum, the equilibrium is stable (4). For a discretized continuum, the minimization of total potential energy may be characterized by the following expression:

$$\frac{\partial V_s}{\partial \mathbf{D}_N} = \frac{\partial}{\partial \mathbf{D}_N}(U_s - W_s) = \mathbf{O} \tag{1.9-1}$$

In this equation the symbol V_s represents the *total potential energy*. It is written as the difference between the *strain energy* U_s in the whole structure and the *potential energy* W_s *of the loads* at all nodes. Analysis by this method requires *assumed displacement functions* that satisfy compatibility at the nodes (*global compatibility*) and at element boundaries (*local compatibility*). Partial differentiation of the total potential energy with respect to nodal displacements $\mathbf{D}_N$ [see Eq. (1.9-1)] produces nodal equilibrium equations of the *stiffness method* (2). By virtue of these equations, *global equilibrium* is satisfied, but *local equilibrium* (between element boundaries) is not satisfied. This technique is similar to Ritz's method (5), but here the displacement functions apply to individual elements instead of the whole continuum. As in that method, this finite-element approach provides a lower bound on the strain energy.

The second approach involves the *principle of stationary complementary potential energy*. This theorem says that among all the stress states of a conservative system that satisfy equilibrium and boundary loads, those that also satisfy compatibility make the complementary potential energy stationary. When the stationary value is a minimum, the displacement state is stable (6). For a discretized continuum, the minimization of the total complementary potential energy can be stated as:

$$\frac{\partial V_s^*}{\partial \mathbf{A}_N} = \frac{\partial}{\partial \mathbf{A}_N}(U_s^* - W_s^*) = \mathbf{O} \tag{1.9-2}$$

The symbol V_s^* in this equation denotes the *total complementary potential energy*. It is the difference between the *complementary strain energy* U_s^* in the whole structure and the *complementary potential energy* W_s^* *of the displacements* at all nodes. Analysis in this manner requires assumed stress functions that satisfy equilibrium at the nodes (*global equilibrium*) and at element boundaries (*local equilibrium*). Partial differentiation of the total complementary potential energy with respect to nodal actions $\mathbf{A}_N$ [see Eq. (1.9-2)] yields nodal compatibility equations of the *flexibility method* (2). Through these equations, *global compatibility* is satisfied, but *local compatibility* (between element boundaries) is not satisfied. This method provides an upper bound on complementary strain energy. For a linearly elastic structure, we have $U_s = U_s^*$; and if both the potential-energy and the complementary potential-energy approaches are applied, the strain energy can be bracketed.

We shall now apply the principle of minimum potential energy to a continuum that is discretized by finite elements. Figure 1.19 shows a stress-strain diagram for a linearly elastic material with initial strain. For such a material the strain energy in element i is:

$$\begin{aligned} U_{ei} &= \int_{V_i} \int_0^{\boldsymbol{\epsilon}_i} \boldsymbol{\sigma}^{\mathrm{T}} d\boldsymbol{\epsilon}\, dV = \int_{V_i} \int_0^{\boldsymbol{\epsilon}_i} (\boldsymbol{\epsilon} - \boldsymbol{\epsilon}_{0i})^{\mathrm{T}} \mathbf{E}_i \, d\boldsymbol{\epsilon}\, dV \\ &= \tfrac{1}{2} \int_{V_i} \boldsymbol{\epsilon}_i^{\mathrm{T}} \mathbf{E}_i \boldsymbol{\epsilon}_i \, dV - \int_{V_i} \boldsymbol{\epsilon}_{0i}^{\mathrm{T}} \mathbf{E}_i \boldsymbol{\epsilon}_i \, dV \\ &= \tfrac{1}{2} \mathbf{q}_i^{\mathrm{T}} \mathbf{K}_i \mathbf{q}_i - \mathbf{q}_i^{\mathrm{T}} \mathbf{p}_{0i} \end{aligned} \qquad \text{(a)}$$

Summation of such terms for the whole structure produces:

$$\begin{aligned} U_s &= \sum_{i=1}^{n_e} \left(\tfrac{1}{2} \mathbf{q}_i^{\mathrm{T}} \mathbf{K}_i \mathbf{q}_i - \mathbf{q}_i^{\mathrm{T}} \mathbf{p}_{0i}\right) \\ &= \tfrac{1}{2} \mathbf{D}_N^{\mathrm{T}} \mathbf{S}_N \mathbf{D}_N - \mathbf{D}_N^{\mathrm{T}} \mathbf{A}_{N0} \end{aligned} \qquad \text{(b)}$$

The potential energy of the loads on an individual element is:

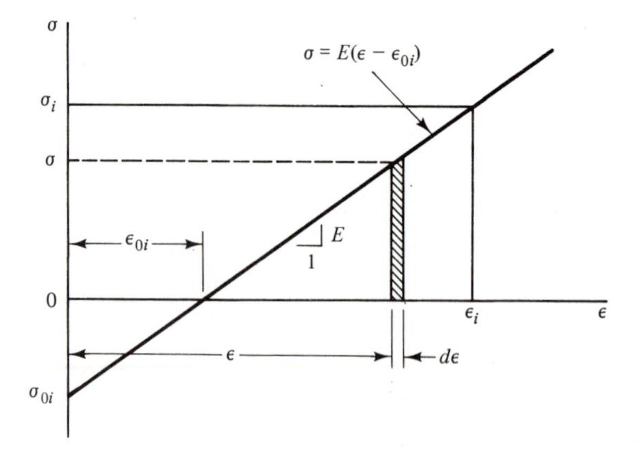

Figure 1.19 Linear Stress-Strain Diagram

$$W_{ei} = \mathbf{p}_i^{\mathrm{T}}\mathbf{q}_i + \mathbf{p}_{bi}^{\mathrm{T}}\mathbf{q}_i = \mathbf{q}_i^{\mathrm{T}}\mathbf{p}_i + \mathbf{q}_i^{\mathrm{T}}\mathbf{p}_{bi} \tag{c}$$

If we sum such terms for the whole structure, the result is:

$$W_s = \sum_{i=1}^{n_e} (\mathbf{q}_i^{\mathrm{T}}\mathbf{p}_i + \mathbf{q}_i^{\mathrm{T}}\mathbf{p}_{bi}) = \mathbf{D}_N^{\mathrm{T}}\mathbf{A}_N + \mathbf{D}_N^{\mathrm{T}}\mathbf{A}_{Nb} \tag{d}$$

Then the total potential energy for the whole structure becomes:

$$V_s = U_s - W_s = \tfrac{1}{2}\mathbf{D}_N^{\mathrm{T}}\mathbf{S}_N\mathbf{D}_N - \mathbf{D}_N^{\mathrm{T}}\mathbf{A}_{N0} - \mathbf{D}_N^{\mathrm{T}}\mathbf{A}_N - \mathbf{D}_N^{\mathrm{T}}\mathbf{A}_{Nb} \tag{1.9-3}$$

Differentiating V_s with respect to $\mathbf{D}_N$ and setting the results equal to zero yields:

$$\frac{\partial V_s}{\partial \mathbf{D}_N} = \mathbf{S}_N\mathbf{D}_N - \mathbf{A}_{N0} - \mathbf{A}_N - \mathbf{A}_{Nb} = \mathbf{O} \tag{1.9-4}$$

This expression is the same as Eq. (1.8-2), obtained by the principle of virtual work.

For a structure with only one free nodal displacement (one degree of freedom) and no initial strains or body forces, the expression for total potential energy [Eq. (1.9-3)] becomes simply:

$$V_s = U_s - W_s = \tfrac{1}{2}S_N D_N^2 - A_N D_N \tag{e}$$

For this case Fig. 1.20(a) shows plots of U_s and $-W_s$ versus D_N, and Fig. 1.20(b) shows their sum V_s. The value of D_N where V_s is a minimum represents the equilibrium position $(D_N)_{\mathrm{eq}}$. For systems with multiple nodal displacements (n degrees of freedom), diagrams similar to Fig. 1.20 cannot be drawn because the space is n-dimensional.

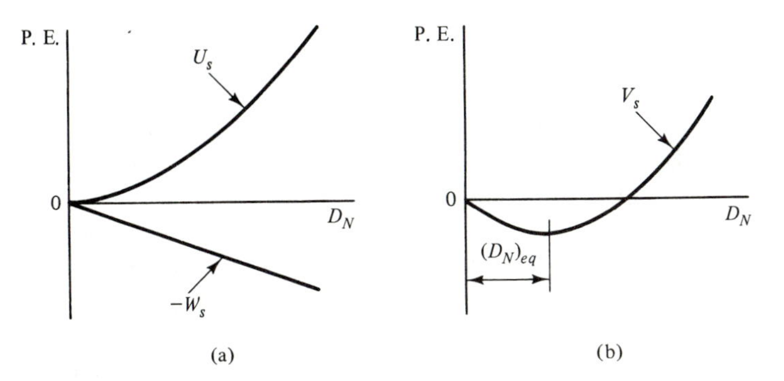

Figure 1.20 Potential Energy for System with 1 Degree of Freedom

Example

Figure 1.21 shows a beam composed of two flexural elements of equal lengths. The flexural rigidity of element 1 is $2EI$, and that of element 2 is EI. At node 2 two displacements, D_{N1} and D_{N2}, are free to occur. Let us apply Eq. (1.9-3) to this beam example, as follows:

$$V_s = \frac{1}{2}[D_{N1} \quad D_{N2}]\frac{EI}{L^3}\begin{bmatrix} 36 & -6L \\ -6L & 12L^2 \end{bmatrix}\begin{bmatrix} D_{N1} \\ D_{N2} \end{bmatrix} - [D_{N1} \quad D_{N2}]\begin{bmatrix} A_{N1} \\ A_{N2} \end{bmatrix}$$

$$= \frac{EI}{2L^3}(36D_{N1}^2 - 12LD_{N1}D_{N2} + 12L^2D_{N2}^2) - D_{N1}A_{N1} - D_{N2}A_{N2} \tag{f}$$

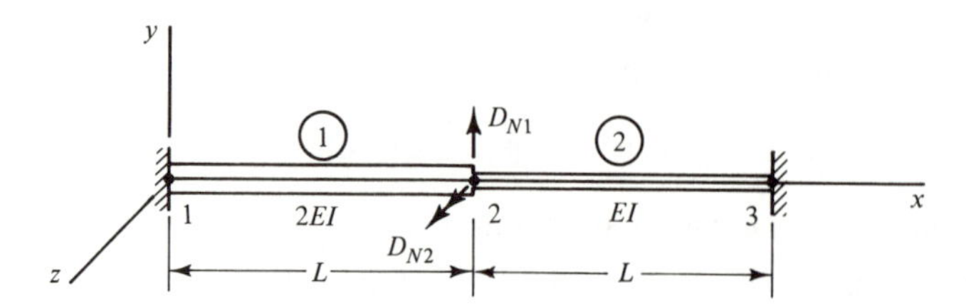

Figure 1.21 Beam Composed of Two Elements

assuming that there are no body forces or initial strains. Then application of Eq. (1.9-4) yields:

$$\frac{\partial V_s}{\partial D_{N1}} = \frac{EI}{L^3}(36D_{N1} - 6LD_{N2}) - A_{N1} = 0 \tag{g}$$

and

$$\frac{\partial V_s}{\partial D_{N2}} = \frac{EI}{L^3}(-6LD_{N1} + 12L^2D_{N2}) - A_{N2} = 0 \tag{h}$$

Casting these expressions into matrix form gives:

$$\frac{EI}{L^3}\begin{bmatrix} 36 & -6L \\ -6L & 12L^2 \end{bmatrix}\begin{bmatrix} D_{N1} \\ D_{N2} \end{bmatrix} = \begin{bmatrix} A_{N1} \\ A_{N2} \end{bmatrix} \tag{i}$$

Equation (i) constitutes the equilibrium equations for node 2 in standard form.

1.10 CONVERGENCE CRITERIA

The analysis of an elastic continuum by the method of finite elements must converge to the results implied by the exact theory as the network of elements is refined. If local and global compatibility are satisfied and nodal loads are consistent, a lower bound on strain energy is assured; and convergence is monotonic if the subdivision rules of Melosh (7) are followed.

Three convergence criteria are listed below. Only the first is essential (8) because the others are implied by the first. If interelement compatibility is not satisfied when the elements are of finite size, criterion A will guarantee such compatibility in the limit when elements are infinitesimal. In such cases convergence may not be monotonic, and lower bounds on strain energy are not obtained.

A. States of Constant Strain

For a given class of problem the element must be capable of modeling states of constant strain exactly. The types of strains involved are those given by the strain-displacement relationships. Thus, certain derivatives of assumed displacement functions must be nonzero. For example, in the axial element the derivative that must be nonzero is:

$$\epsilon_x = \frac{du}{dx} \neq 0 \tag{a}$$

Therefore, at least a linear displacement function for u must be assumed. For the flexural element the derivative that must be nonzero is:

$$\phi = \frac{d^2v}{dx^2} \neq 0 \tag{b}$$

which was characterized in Sec. 1.6 as generalized strain. Consequently, the displacement function assumed for v must be at least quadratic.

B. Rigid-Body Modes of Displacement

An element must be capable of displacing as a rigid body (for small displacements) without developing internal strains. This requirement may be considered as the extreme case of the constant-strain condition, with $\boldsymbol{\epsilon} = \mathbf{O}$.

To test an element for this capability, nodal displacements $\mathbf{q}_{RB}$ representing rigid body displacements may be premultiplied by matrix $\mathbf{B}$, and the results should be zeros. That is,

$$\mathbf{B}\ \mathbf{q}_{RB} = \mathbf{O} \tag{c}$$

which satisfies the criterion. Alternatively, premultiplication of vector $\mathbf{q}_{RB}$ with the element stiffness matrix $\mathbf{K}$ should also produce the null vector:

$$\mathbf{K}\ \mathbf{q}_{RB} = \mathbf{O} \tag{d}$$

C. Completeness and Balance of Assumed Functions

The assumed displacement functions must be complete, and it is also desirable that they be balanced. *Completeness* means that all terms of order less than that required by criterion A must be included in the assumed functions. For example, if a quadratic function is used for the flexural element, it must be:

$$v = c_1 + c_2x + c_3x^2 \tag{e}$$

with no terms omitted.

Balance in two- and three-dimensional elements is achieved by including terms of the same order for each generic displacement. For example, complete and balanced quadratic functions for a two-dimensional continuum are as follows:

$$u = c_1 + c_2x + c_3y + c_4x^2 + c_5xy + c_6y^2 \tag{f}$$

$$v = c_7 + c_8x + c_9y + c_{10}x^2 + c_{11}xy + c_{12}y^2 \tag{g}$$

The *Pascal triangle* shown in Fig. 1.22 serves as a guide to selection of terms for two-dimensional elements.

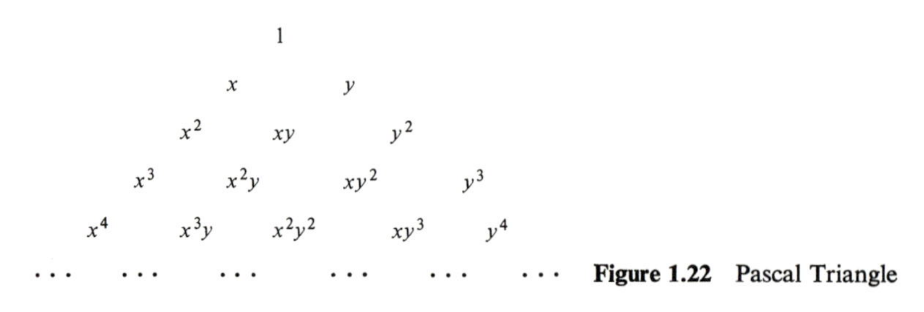

Figure 1.22 Pascal Triangle

1.11 COMPUTER PROGRAMS

In Chapters 2 through 8 computer programs for static analysis by finite elements will be discussed. These programs are all special-purpose algorithms that apply to only one class of problems, such as plane stress and plane strain. They are documented using FORTRAN-oriented flow charts, consisting of modular subprograms (or subroutines) that are called by a main program. The logic of these programs allows any number of problems to be analyzed in one run; and any number of loading systems can be processed for a given structure. When secondary influences (such as initial strains) are imposed, they are handled by determining equivalent nodal loads. The upper triangular "skyline" of the nodal stiffness matrix for the whole structure is stored as a singly subscripted variable (or vector). Equilibrium equations are solved (without rearrangement) for the nodal displacements using the subprograms SKYFAC and SKYSOL described in Appendix C. Subsequently, element stresses are computed from nodal displacements and initial values. Then the reactions at restrained points are obtained from the elements connected to such supports. All the computer programs for static analysis have the same outline, as follows:

1. Read and print structural data
   (a) Problem identification
   (b) Structural parameters
   (c) Nodal coordinates
   (d) Element information
   (e) Nodal restraints
   (f) Calculate displacement indexes
2. Generate structural stiffness matrix
   (a) Element stiffness matrix
   (b) Transfer to structural stiffness matrix
   (c) Modification of stiffness matrix for restraints
3. Factor structural stiffness matrix
4. Read and print load (and supplementary*) data
   (a) Load parameters

*Optional supplementary influences.

   (b) Nodal loads
   (c) Line loads*
   (d) Area loads*
   (e) Volume loads*
   (f) Temperature strains*
   (g) Prestrains*
   (h) Support displacements*
5. Solve nodal equilibrium equations
6. Calculate and print results
   (a) Nodal displacements
   (b) Element stresses
   (c) Support reactions

Programs for vibrational and instability analyses are described in Chapters 9 and 10. While such programs bear many similarities to those for static analysis, the main difference is that the algebraic eigenvalue problem must be solved instead of (or in addition to) simultaneous algebraic equations.

General-purpose programs that can handle various analyses for many types of structures are beyond the scope of this book. One of the most commonly used general-purpose programs in the public domain is NASTRAN (9), which was developed in the United States by the National Aeronautics and Space Administration (NASA).

REFERENCES

1. Timoshenko, S. P., and Goodier, J. N., *Theory of Elasticity*, 3d ed., McGraw-Hill, New York, 1970.
2. Weaver, W., Jr., and Gere, J. M., *Matrix Analysis of Framed Structures*, 2d ed., Van Nostrand-Reinhold, New York, 1980.
3. Timoshenko, S. P., *Strength of Materials*, Part II, Advanced Theory and Problems, 3d ed., Van Nostrand-Reinhold, New York, 1956.
4. Washizu, K., *Variational Methods in Elasticity and Plasticity*, 2d ed., Pergamon Press, New York, 1975.
5. Ritz, W., "Über eine neue Methode zur Lösung gewissen Variations–Probleme der mathematischen Physik," *J. Reine angew. Math.*, Vol. 135, 1909, pp. 1–61.
6. Przemieniecki, J. S., *Theory of Matrix Structural Analysis*, McGraw-Hill, New York, 1968.
7. Melosh, R. J., "Basis of Derivation of Matrices for the Direct Stiffness Method," *AIAA Jour.*, Vol. 1, No. 7, July 1963, pp. 1631–1637.
8. Bazeley, G. P., Cheung, Y. K., Irons, B. M., and Zienkiewicz, O. C., "Triangular Elements in Plate Bending: Conforming and Nonconforming Solutions," *Proc.*

*Optional supplementary influences.

Conf. Mat. Meth. Struc. Mech., AFIT, Wright-Patterson AF Base, Ohio, 1965, pp. 547–576.

9. MacNeal, R. H., ed., *The NASTRAN Theoretical Manual* (Level 16.0), NASA-SP-221(03), March 1976 (N79-27531, N.T.I.S.).

PROBLEMS

1.1-1. What is a finite element?

1.1-2. What are the primary objectives of analyzing structures by finite elements?

1.1-3. What is the basic assumption most commonly used to develop finite elements for structural analysis?

1.2-1. What are the definitions for terms in the following matrices? (1) $\boldsymbol{\sigma}$, (2) $\boldsymbol{\epsilon}$, (3) $\mathbf{C}$, and (4) $\mathbf{E}$.

1.3-1. What are the definitions for terms in the following matrices? (1) $\mathbf{u}$, (2) $\mathbf{b}$, (3) $\mathbf{q}$, (4) $\mathbf{p}$, (5) $\mathbf{f}$, (6) $\mathbf{d}$, (7) $\mathbf{B}$, (8) $\mathbf{K}$, (9) $\mathbf{p}_b$, (10) $\boldsymbol{\epsilon}_0$, (11) $\boldsymbol{\sigma}_0$, and (12) $\mathbf{p}_0$.

1.4-1. What are the definitions for terms in the following matrices? (1) $\mathbf{c}$, (2) $\mathbf{g}$, (3) $\mathbf{h}$, (4) $\mathbf{B}_c$, (5) $\mathbf{K}_c$, (6) $\mathbf{p}_c$, (7) $\mathbf{p}_{bc}$, and (8) $\mathbf{p}_{0c}$.

1.5-1. Find the 3×3 stiffness matrix $\mathbf{K}$ for an axial element having the quadratic displacement function $u = c_1 + c_2x + c_3x^2$ and three nodes, as shown in the figure. Assume that EA is constant.

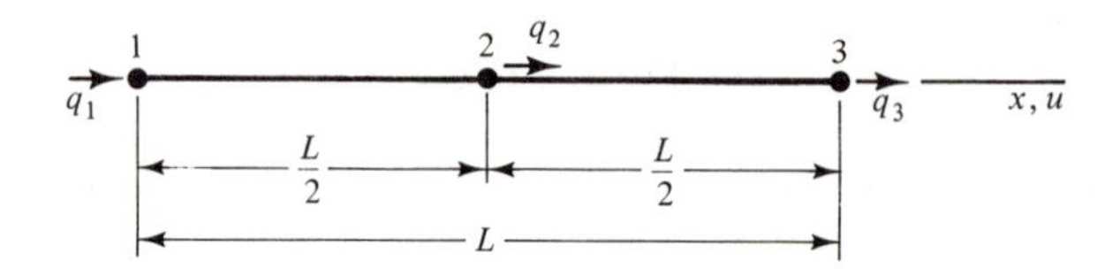

Problem 1.5-1

1.5-2. For the axial element in Prob. 1.5-1, determine the equivalent nodal loads p_{b1}, p_{b2}, and p_{b3} (see figure) due to a uniformly distributed axial force of magnitude b_x per unit length.

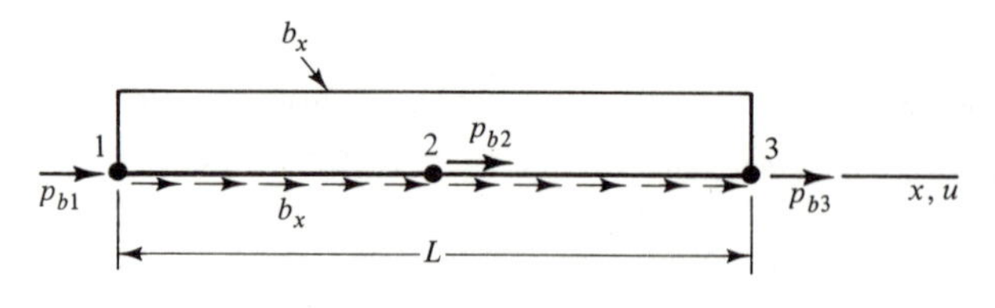

Problem 1.5-2

1.5-3. The figure shows a linearly varying axial force b_x applied to the element of Prob. 1.5-1. Find the equivalent nodal loads p_{b1}, p_{b2}, and p_{b3} due to this loading.

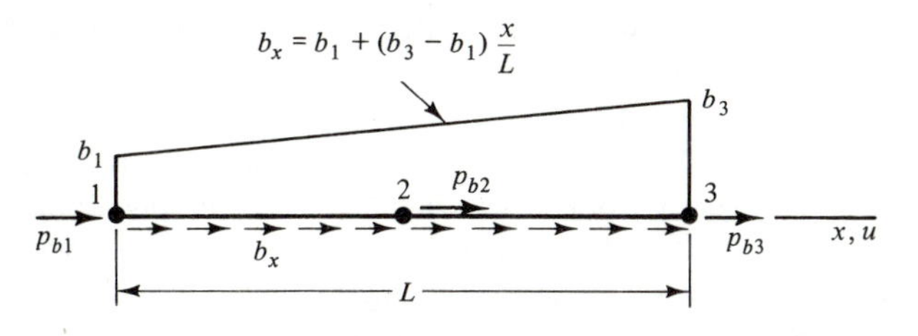

Problem 1.5-3

1.5-4. Suppose that the axial element in Prob. 1.5-1 is subjected to a parabolically varying temperature change ΔT, as indicated in the figure. Determine the equivalent nodal loads p_{T1}, p_{T2}, and p_{T3}.

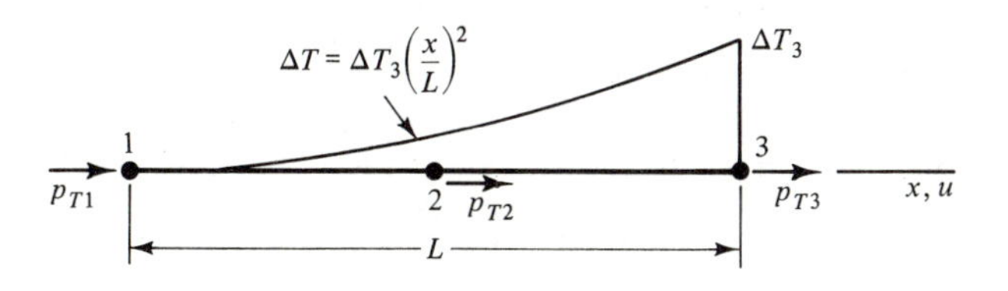

Problem 1.5-4

1.5-5. Derive the 4×4 stiffness matrix $\mathbf{K}$ for the flexural element using the generic displacements v and dv/dx (see figure), so that matrix $\hat{\mathbf{f}} = \{\mathbf{f}, d\mathbf{f}/dx\}$ is of size 2×4. Assume that EI is constant.

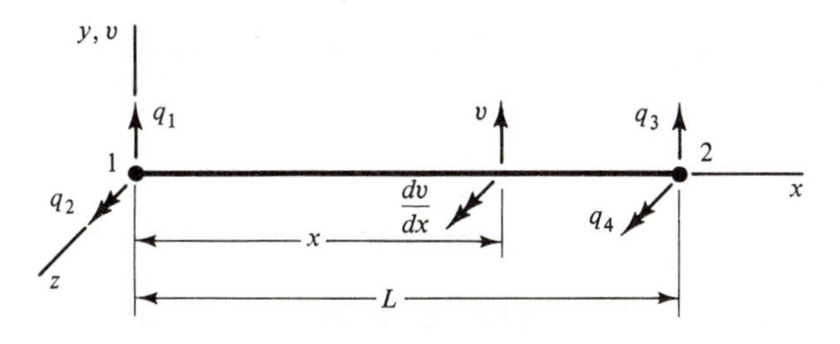

Problem 1.5-5

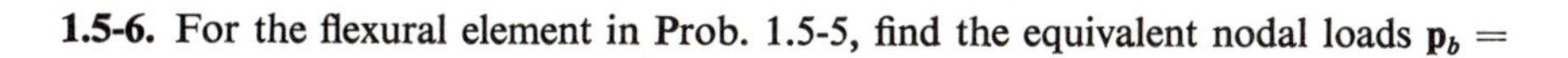
1.5-6. For the flexural element in Prob. 1.5-5, find the equivalent nodal loads $\mathbf{p}_b =$

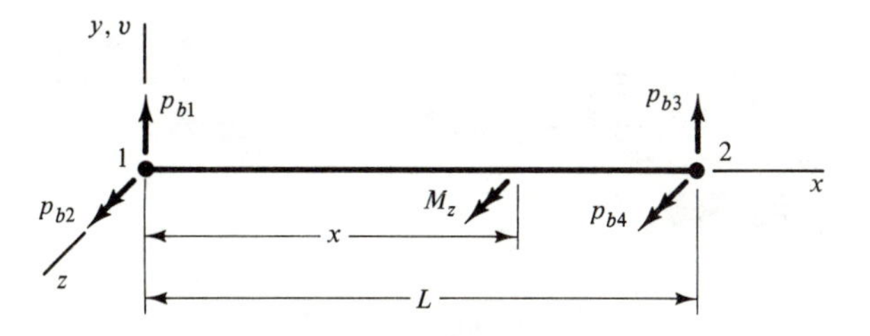

Problem 1.5-6

$\{p_{b1}, p_{b2}, p_{b3}, p_{b4}\}$ due to a concentrated moment M_z applied at the distance x from node 1, as shown in the figure.

1.5-7. The figure shows equal and opposite temperature changes at the upper and lower surfaces of the flexural element. At the upper surface the temperature change varies linearly from zero at node 1 to ΔT_2 at node 2; and at the lower surface the variation is from zero at node 1 to $-\Delta T_2$ at node 2. Determine the equivalent load vector $\mathbf{p}_T$ due to these temperature changes, assuming a linear temperature gradient through the depth d.

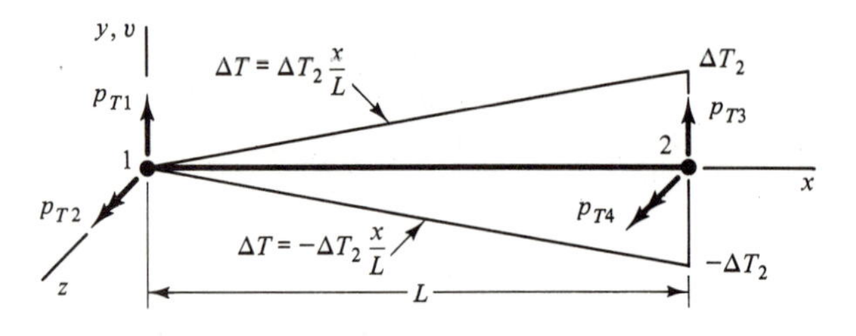

Problem 1.5-7

1.5-8. If the flexural element has initial curvature ϕ_0 that varies quadratically from zero at node 1 to ϕ_2 at node 2 (see figure), find the equivalent load vector $\mathbf{p}_0$ due to these deformations.

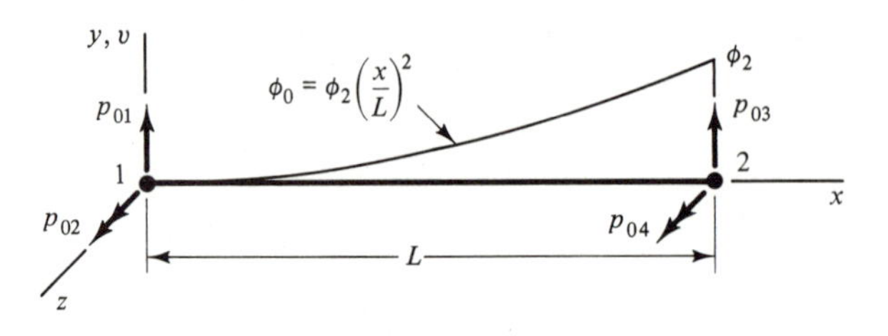

Problem 1.5-8

1.5-9. A prismatic flexural element on an elastic foundation appears in the figure. The modulus k of the elastic support has units of force per unit length per

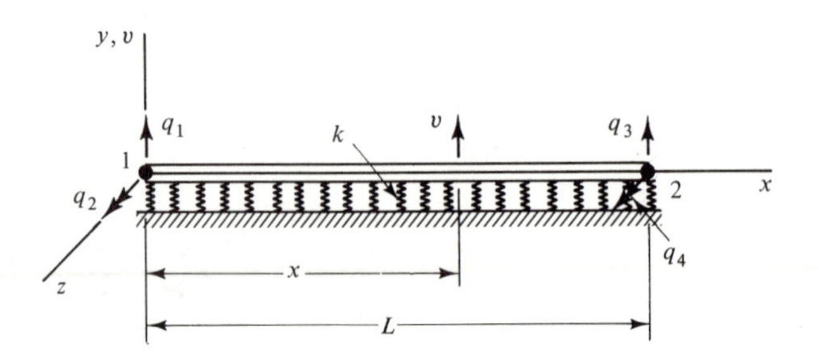

Problem 1.5-9

unit displacement. Determine a restraint stiffness matrix $\mathbf{K}_r$ for the effect of the foundation on this element, using the displacement shape functions in Eq. (f) of Sec. 1.5.

1.5-10. Using the concept described in Sec. 1.4 for generalized displacements $\mathbf{c}$, determine the stiffness matrix $\mathbf{K}_c$ for a prismatic flexural element. Then transform $\mathbf{K}_c$ to $\mathbf{K}$, as shown by Eq. (1.4-9).

1.5-11. In accordance with the concept of generalized displacements $\mathbf{c}$ discussed in Sec. 1.4, find the load vector $\mathbf{p}_c$ for the triangular loading on the flexural element in Fig. 1.12(b). Next, transform $\mathbf{p}_c$ to $\mathbf{p}$, as given by Eq. (1.4-10).

1.6-1. Derive again the 4×4 stiffness matrix $\mathbf{K}$ for a flexural element using moment M_z and curvature ϕ as generalized stress and strain. Assume that EI is constant.

1.6-2. Suppose that the flexural element has an initial curvature defined as $\phi_0 = \phi_2(x/L)^3$, where ϕ_2 is the curvature at node 2. Determine the equivalent load vector for this condition.

1.6-3. Rederive the 2×2 stiffness matrix $\mathbf{K}$ for a torsional element using moment M_x and twist ψ as generalized stress and strain. Assume that GJ is constant.

1.6-4. If the torsional element has initial twist ψ_0 that is constant along its length, find the equivalent load vector $\mathbf{p}_0$ due to this condition.

1.7-1. The circularly curved element in the figure lies in the x-y plane. Determine the actions p_{j1}, p_{j2}, and p_{j3} (two forces and a moment) at point j that are statically equivalent to the actions p_{k1}, p_{k2}, and p_{k3} at point k. For this purpose, apply the translation-and-rotation-of-axes concepts in Sec. 1.7.

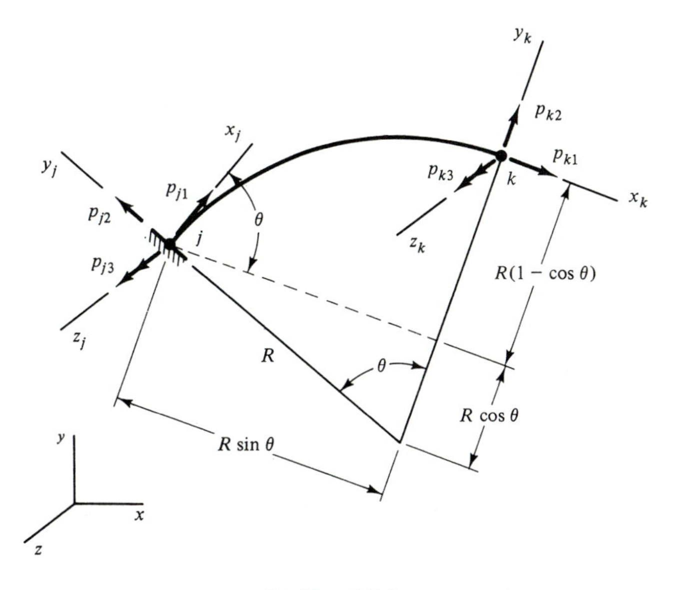

Problem 1.7-1

1.7-2. The bent rod (see figure) is subjected to small displacements $q_{A1}, q_{A2}, \ldots, q_{A6}$ (three translations and three rotations) at the support point A. Using a translation-of-axes transformation, find the displacements $q_{D1}, q_{D2}, \ldots, q_{D6}$ at point D.

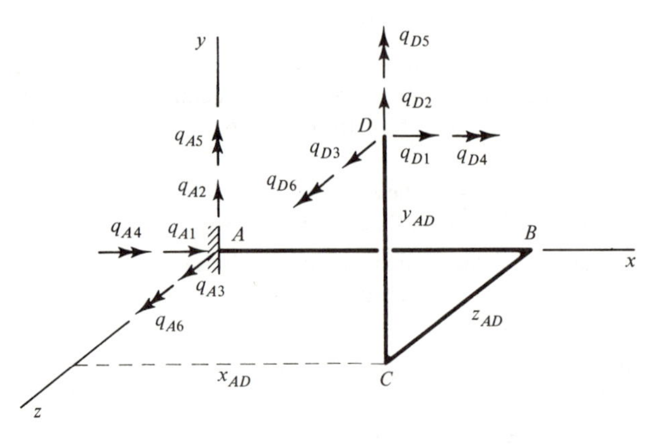

Problem 1.7-2

1.7-3. Part (a) of the figure shows an axial element in the x-y plane with its axis inclined at the angle γ from the x axis. Convert the 2×2 element stiffness matrix [see Eq. (1.5-1)] for the inclined axis x'_e to a 4×4 stiffness matrix for the axes x_e and y_e [see part (b) of the figure], which are parallel to axes x and y.

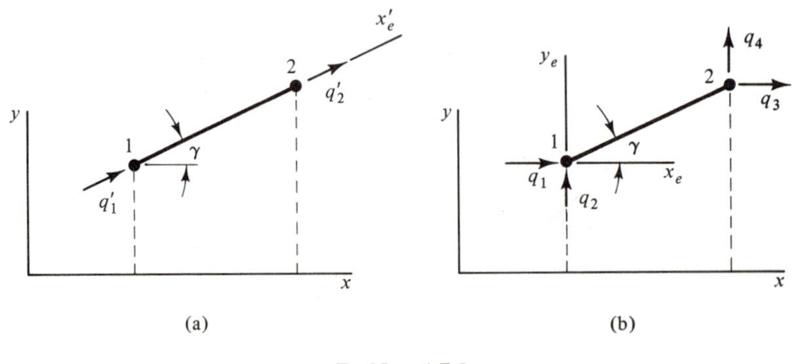

Problem 1.7-3

1.7-4. The plane frame element in the figure is a combination of an axial element and a flexural element. By rotation of axes, transform the 6×6 element stiffness matrix $\mathbf{K}'$ for the primed axes x'_e, y'_e, and z'_e to the matrix $\mathbf{K}$ for the unprimed axes x_e, y_e, and z_e, which are parallel to axes x, y, and z.

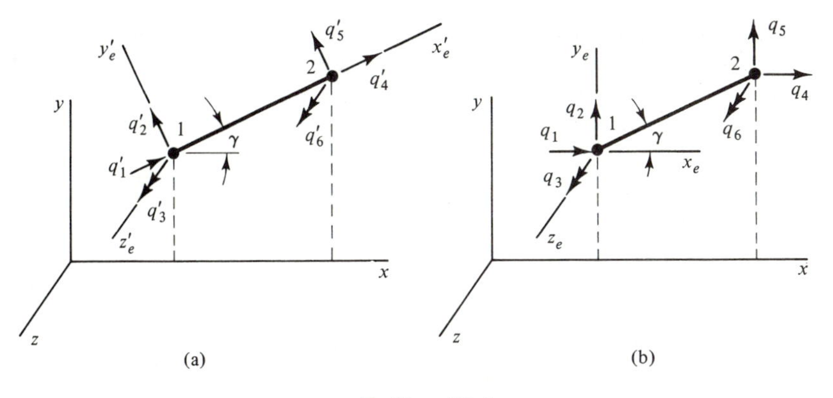

Problem 1.7-4

1.8-1. The figure shows a continuous beam composed of three prismatic beam elements having flexural rigidities $2EI$, EI, and $3EI$. Assemble the stiffness and load matrices in a form suitable for solution without rearrangement [see Eq. (1.8-9)]. Assume that all support displacements are zero.

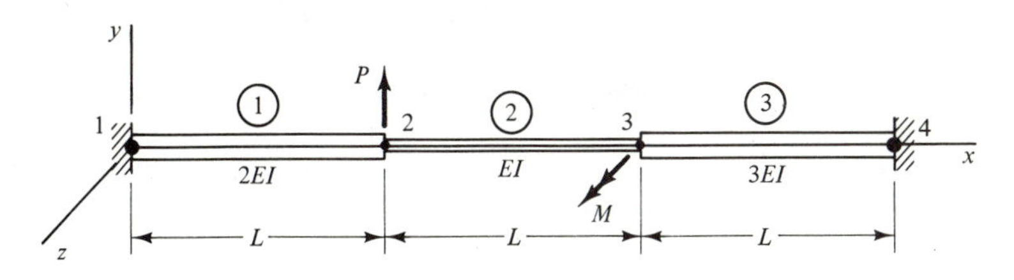

Problem 1.8-1

1.8-2. Assemble the stiffness and load matrices for the plane truss (see figure) in a form suitable for solution without rearrangement [see Eq. (1.8-9)]. The axial rigidity of each element is EA, and there are no support displacements.

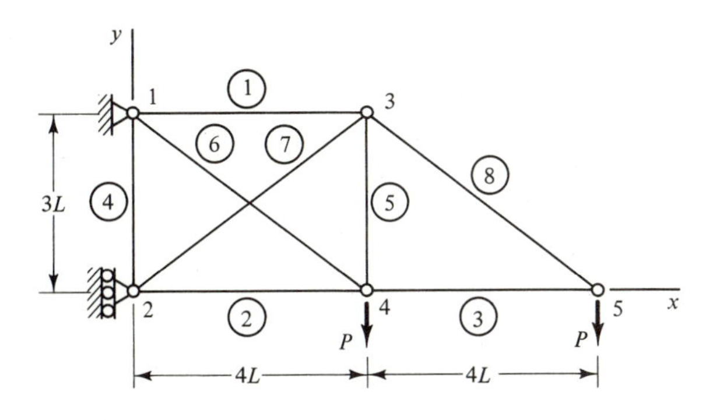

Problem 1.8-2

1.8-3. The plane frame in the figure has constant flexural rigidity EI and axial rigidity $EA = 100EI/L^2$ for each element. Assemble the stiffness and load matrices for this structure in a form suitable for solution without rearrangement [see Eq. (1.8-9)]. Assume that all support displacements are zero.

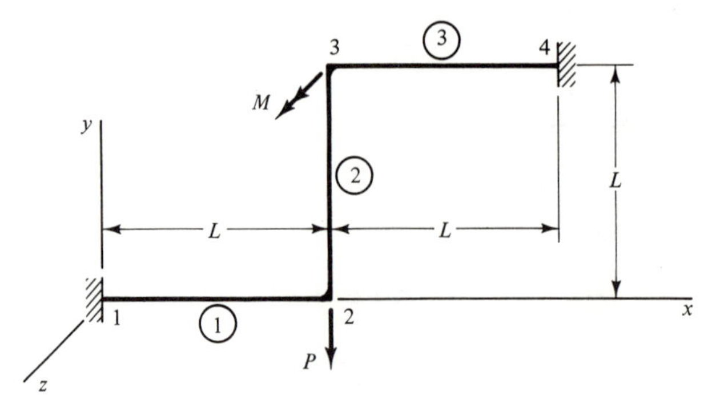

Problem 1.8-3

1.8-4. Repeat Prob. 1.8-1 with a support rotation of magnitude θ_z in the z sense at point 4.

1.8-5. Repeat Prob. 1.8-2 with a support translation of magnitude Δ_x in the x direction at point 2.

1.8-6. Repeat Prob. 1.8-3 with a support translation of magnitude Δ_y in the y direction at point 1.

CHAPTER

2

Plane Stress and Plane Strain

2.1 TWO-DIMENSIONAL STRESSES AND STRAINS

Many problems in *theory of elasticity* (1) are two-dimensional in nature. When forces are applied to a thin plate in its own plane, the state of stress and deformation within the plate is called *plane stress* [see Fig. 1.2(a)]. In this case only two dimensions (in the plane of the plate) are required for the analysis. On the other hand, a prismatic solid may be subjected to a constant condition of loading normal to its axis. If so, the solid can be analyzed as an infinity of two-dimensional slices of unit thickness [see Fig. 1.2(b)]. This type of problem is identified by the name *plane strain*. In the present section the topics of plane stress and plane strain are discussed in detail. Subsequently, the stress-strain relationships developed here are used to derive stiffnesses and equivalent loads for two-dimensional finite elements of various shapes.

To begin, we consider the two-dimensional state of stress indicated in Fig. 2.1. The infinitesimal element of size dx by dy has *normal stresses* σ_x and σ_y acting in the x and y directions, respectively. Also shown in the figure is the *shearing stress* τ_{xy}, which acts on the x edge in the y direction. It is accompanied by the *complementary shearing stress* τ_{yx}, acting on the y edge in the x direction. The three independent types of stresses may be assembled in a vector, as folows:

$$\boldsymbol{\sigma} = \{\sigma_x, \sigma_y, \tau_{xy}\} \tag{a}$$

Also of interest to us are *inclined stresses*. For this purpose Fig. 2.2 shows stresses in the directions of axes x' and y', which are inclined at the angle θ with axes x and y. The inclined normal stress $\sigma_{x'}$ may be obtained from equilib-

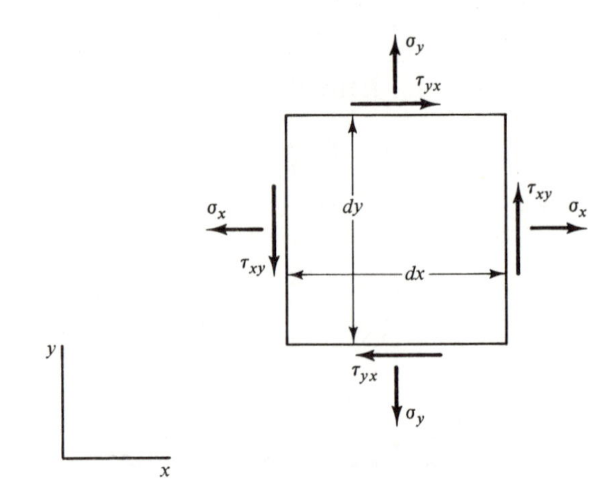

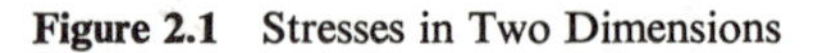

Figure 2.1 Stresses in Two Dimensions

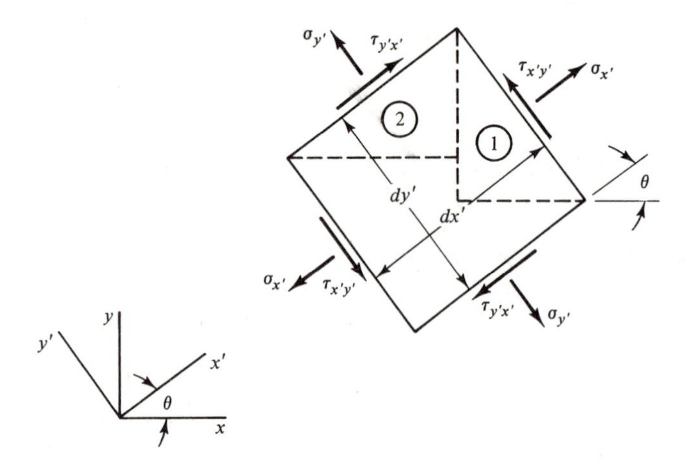

Figure 2.2 Stresses for Inclined Directions

rium of the triangular segment 1 of the inclined element in Fig. 2.2. For this purpose Fig. 2.3 gives an isolated view of segment 1, from which summation of the forces in the x' direction yields:

$$\sigma_{x'} = \sigma_x \cos^2 \theta + \sigma_y \sin^2 \theta + 2\tau_{xy} \sin \theta \cos \theta \tag{b}$$

Similarly, isolation of the triangular segment 2 in Fig. 2.2 and summation of forces in the y' direction leads to:

$$\sigma_{y'} = \sigma_x \sin^2 \theta + \sigma_y \cos^2 \theta - 2\tau_{xy} \sin \theta \cos \theta \tag{c}$$

Finally, summation of forces in the y' direction of Fig. 2.3 gives:

$$\tau_{x'y'} = -(\sigma_x - \sigma_y) \sin \theta \cos \theta + \tau_{xy}(\cos^2 \theta - \sin^2 \theta) \tag{d}$$

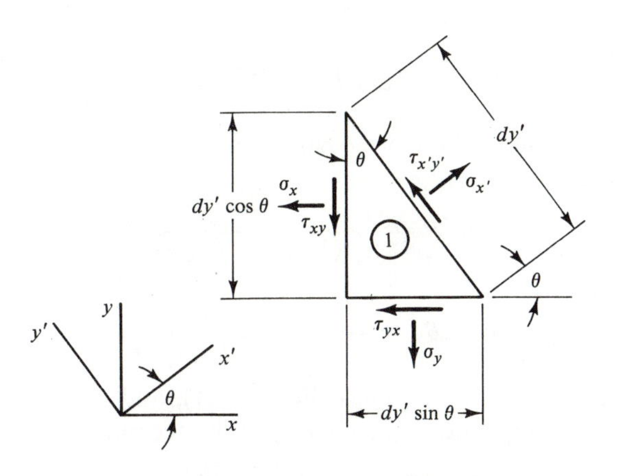

Figure 2.3 Equilibrium of Stresses

Altogether, the stresses for the primed directions may be obtained from those for the unprimed directions by the matrix transformation:

$$\boldsymbol{\sigma}' = \mathbf{T}_\theta \boldsymbol{\sigma} \tag{2.1-1}$$

in which the operator $\mathbf{T}_\theta$ has the form:

$$\mathbf{T}_\theta = \begin{bmatrix} \cos^2\theta & \sin^2\theta & 2\sin\theta\cos\theta \\ \sin^2\theta & \cos^2\theta & -2\sin\theta\cos\theta \\ -\sin\theta\cos\theta & \sin\theta\cos\theta & \cos^2\theta - \sin^2\theta \end{bmatrix} \tag{2.1-2}$$

This matrix serves the useful purpose of calculating inclined stresses for any angle of inclination θ with respect to the x and y directions.

Substitution of double-angle identities into Eqs. (b), (c), and (d) produces:

$$\sigma_{x'} = \frac{\sigma_x + \sigma_y}{2} + \frac{\sigma_x - \sigma_y}{2}\cos 2\theta + \tau_{xy}\sin 2\theta \tag{e}$$

$$\sigma_{y'} = \frac{\sigma_x + \sigma_y}{2} - \frac{\sigma_x - \sigma_y}{2}\cos 2\theta - \tau_{xy}\sin 2\theta \tag{f}$$

$$\tau_{x'y'} = -\frac{\sigma_x - \sigma_y}{2}\sin 2\theta + \tau_{xy}\cos 2\theta \tag{g}$$

Addition of Eqs. (e) and (f) results in:

$$\sigma_{x'} + \sigma_{y'} = \sigma_x + \sigma_y \tag{2.1-3}$$

This expression proves that the sum of normal stresses is invariant with respect to orientation.

It is possible to find the directions of *principal stresses* by differentiating Eq. (e) with respect to θ and setting the results equal to zero, as follows:

$$\frac{d\sigma_{x'}}{d\theta} = -(\sigma_x - \sigma_y)\sin 2\theta + 2\tau_{xy}\cos 2\theta = 0$$

Hence,

$$\tan 2\theta_P = \frac{2\tau_{xy}}{\sigma_x - \sigma_y} \tag{2.1-4}$$

where θ_P is the value of θ for principal normal stresses. We also observe from Eq. (2.1-4) that:

$$\sin 2\theta_P = \frac{2\tau_{xy}}{\sqrt{(2\tau_{xy})^2 + (\sigma_x - \sigma_y)^2}} \tag{h}$$

and

$$\cos 2\theta_P = \frac{\sigma_x - \sigma_y}{\sqrt{(2\tau_{xy})^2 + (\sigma_x - \sigma_y)^2}} \tag{i}$$

Substitution of Eqs. (h) and (i) into Eqs. (e), (f), and (g) yields the following expressions for principal normal stresses and the accompanying shearing stress:

$$\sigma_{P_1} = \frac{\sigma_x + \sigma_y}{2} + \sqrt{\left(\frac{\sigma_x - \sigma_y}{2}\right)^2 + \tau_{xy}^2} = \sigma'_{\max} \tag{2.1-5}$$

$$\sigma_{P_2} = \frac{\sigma_x + \sigma_y}{2} - \sqrt{\left(\frac{\sigma_x - \sigma_y}{2}\right)^2 + \tau_{xy}^2} = \sigma'_{\min} \tag{2.1-6}$$

$$\tau_{x'y'} = 0 \tag{2.1-7}$$

Thus, the shearing stress is zero when the normal stresses have principal values (maximum and minimum).

On the other hand, the *maximum shearing stress* $\tau'_{\max}$ occurs at an angle that can be found by differentiating Eq. (g) with respect to θ and setting the results equal to zero, as follows:

$$\frac{d\tau_{x'y'}}{d\theta} = -(\sigma_x - \sigma_y)\cos 2\theta - 2\tau_{xy}\sin 2\theta = 0$$

Hence,

$$\tan 2\theta_S = -\frac{\sigma_x - \sigma_y}{2\tau_{xy}} \tag{2.1-8}$$

where $\theta_S = \theta_P \pm \pi/4$ is the value of θ for which the shearing stress is maximum. Substitution of sine and cosine functions defined by Eq. (2.1-8) into Eq. (g) produces the following expression for maximum shearing stress:

$$\tau'_{\max} = \sqrt{\left(\frac{\sigma_x - \sigma_y}{2}\right)^2 + \tau_{xy}^2} \tag{2.1-9}$$

Similar substitutions into Eqs. (e) and (f) give the accompanying normal stresses as:

$$\sigma_{x'} = \sigma_{y'} = \frac{\sigma_x + \sigma_y}{2} \tag{2.1-10}$$

As mentioned above, these stresses always exist at the angle $\pm\pi/4$ from the directions of principal stresses.

Figure 2.4 shows an infinitesimal element that has been displaced by the amounts u and v in the x and y directions. Also shown on the figure are partial

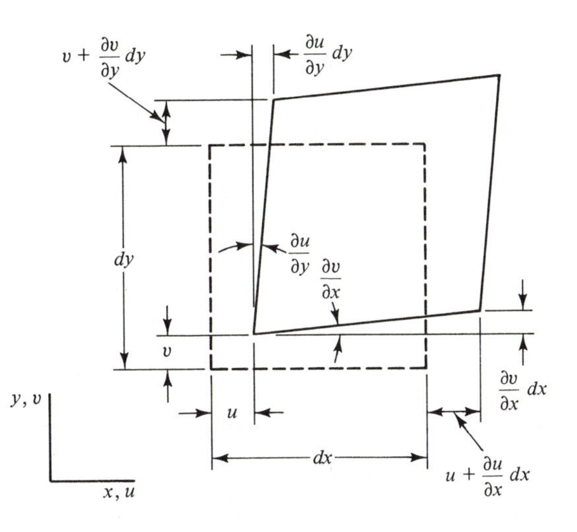

Figure 2.4 Strains in Two Dimensions

derivatives (or strains) that define incremental deformations of the element. The types of strains shown are:

$$\boldsymbol{\epsilon} = \{\epsilon_x, \epsilon_y, \gamma_{xy}\} \tag{j}$$

in which

$$\epsilon_x = \frac{\partial u}{\partial x} \qquad \epsilon_y = \frac{\partial v}{\partial y} \qquad \gamma_{xy} = \frac{\partial u}{\partial y} + \frac{\partial v}{\partial x} \tag{k}$$

The first two types, ϵ_x and ϵ_y, are *normal strains* in the x and y directions, and γ_{xy} is *shearing strain*. They correspond to the normal stresses and the shearing stress listed previously in Eq. (a). Altogether, the expressions in Eq. (k) are *strain-displacement relationships* expressing the strains in $\boldsymbol{\epsilon}$ in terms of the generic displacements u and v. As before, they may be written in matrix form as:

$$\boldsymbol{\epsilon} = \mathbf{d}\ \mathbf{u} \tag{1.3-6}$$

repeated

where

$$\mathbf{d} = \begin{bmatrix} \frac{\partial}{\partial x} & 0 \\ 0 & \frac{\partial}{\partial y} \\ \frac{\partial}{\partial y} & \frac{\partial}{\partial x} \end{bmatrix} \tag{2.1-11}$$

and

$$\mathbf{u} = \{u, v\} \tag{ℓ}$$

Of course, *strains for inclined directions* are also of interest. To investigate strains in the directions of x' and y' (see Fig. 2.2), we equate the *complementary virtual strain energy densities* for the primed and unprimed axes, as follows:

$$(\delta\boldsymbol{\sigma}')^{\mathrm{T}}\boldsymbol{\epsilon}' = (\delta\boldsymbol{\sigma})^{\mathrm{T}}\boldsymbol{\epsilon} \tag{m}$$

From Eq. (2.1-1) we have:

$$\delta\boldsymbol{\sigma}' = \mathbf{T}_\theta \delta\boldsymbol{\sigma} \tag{n}$$

Substitution of this expression into Eq. (m) gives:

$$(\delta\boldsymbol{\sigma})^{\mathrm{T}}\mathbf{T}_\theta^{\mathrm{T}}\boldsymbol{\epsilon}' = (\delta\boldsymbol{\sigma})^{\mathrm{T}}\boldsymbol{\epsilon}$$

Hence,

$$\boldsymbol{\epsilon} = \mathbf{T}_\theta^{\mathrm{T}}\boldsymbol{\epsilon}' \tag{2.1-12}$$

Multiplying Eq. (2.1-12) by $\mathbf{T}_\theta^{-\mathrm{T}}$ yields:

$$\boldsymbol{\epsilon}' = \mathbf{T}_\theta^{-T}\boldsymbol{\epsilon} \tag{2.1-13}$$

in which

$$\mathbf{T}_\theta^{-\mathrm{T}} = \begin{bmatrix} \cos^2\theta & \sin^2\theta & \sin\theta\cos\theta \\ \sin^2\theta & \cos^2\theta & -\sin\theta\cos\theta \\ -2\sin\theta\cos\theta & 2\sin\theta\cos\theta & \cos^2\theta - \sin^2\theta \end{bmatrix} \tag{2.1-14}$$

Thus, the operator $\mathbf{T}_\theta^{-\mathrm{T}}$ required for calculating $\boldsymbol{\epsilon}'$ from $\boldsymbol{\epsilon}$ is the transposed inverse of that for obtaining $\boldsymbol{\sigma}'$ from $\boldsymbol{\sigma}$ [see Eq. (2.1-1)].

There also exist *principal normal strains* ϵ_{P_1} and ϵ_{P_2}, corresponding to σ_{P_1} and σ_{P_2} [see Eqs. (2.1-4), (2.1-5), and (2.1-6)]. In addition, at the angle $\pm\pi/4$ from the directions of such principal strains, we have the *maximum shearing strain* $\gamma'_{\max}$, corresponding to $\tau'_{\max}$ [see Eqs. (2.1-8) and (2.1-9)].

Assuming an *isotropic material*, we shall develop relationships between stresses and strains for both plane stress and plane strain (1). The former is based on the assumptions that:

$$\sigma_z = \tau_{xz} = \tau_{yz} = 0 \qquad \gamma_{xz} = \gamma_{yz} = 0 \qquad \epsilon_z \neq 0 \tag{o}$$

Writing strains in terms of stresses, we have:

$$\begin{gathered} \epsilon_x = \frac{1}{E}(\sigma_x - \nu\sigma_y) \qquad \epsilon_y = \frac{1}{E}(-\nu\sigma_x + \sigma_y) \\ \gamma_{xy} = \frac{1}{G}\tau_{xy} = \frac{2(1+\nu)}{E}\tau_{xy} \end{gathered} \tag{p}$$

Also,

$$\epsilon_z = -\frac{\nu}{E}(\sigma_x + \sigma_y) \tag{q}$$

Solving for the stresses in Eqs. (p) in terms of the strains, we find:

$$\begin{gathered} \sigma_x = \frac{E}{1-\nu^2}(\epsilon_x + \nu\epsilon_y) \qquad \sigma_y = \frac{E}{1-\nu^2}(\nu\epsilon_x + \epsilon_y) \\ \tau_{xy} = \frac{E}{2(1+\nu)}\gamma_{xy} = \frac{E\lambda}{1-\nu^2}\gamma_{xy} \end{gathered} \tag{r}$$

in which

$$\lambda = \frac{1-\nu}{2}$$

As in Sec. 1.2, we can write the *strain-stress relationships* in Eqs. (p) as the matrix expression:

$$\boldsymbol{\epsilon} = \mathbf{C}\,\boldsymbol{\sigma} \tag{1.2-1 repeated}$$

where

$$\mathbf{C} = \frac{1}{E}\begin{bmatrix} 1 & -\nu & 0 \\ -\nu & 1 & 0 \\ 0 & 0 & 2(1+\nu) \end{bmatrix} \tag{2.1-15}$$

Similarly, the matrix form of Eqs. (r) is:

$$\boldsymbol{\sigma} = \mathbf{E}\,\boldsymbol{\epsilon} \tag{1.2-3 repeated}$$

where

$$\mathbf{E} = \mathbf{C}^{-1} = \frac{E}{1-\nu^2}\begin{bmatrix} 1 & \nu & 0 \\ \nu & 1 & 0 \\ 0 & 0 & \lambda \end{bmatrix} \tag{2.1-16}$$

represents the *stress-strain relationships* for the case of plane stress.

The case of plane strain is based on the assumptions that:

$$\epsilon_z = \gamma_{xz} = \gamma_{yz} = 0 \qquad \tau_{xz} = \tau_{yz} = 0 \qquad \sigma_z \neq 0 \tag{s}$$

When strains are written in terms of stresses, we get:

$$\begin{aligned} \epsilon_x &= \frac{1}{E}(\sigma_x - \nu\sigma_y - \nu\sigma_z) \\ \epsilon_y &= \frac{1}{E}(-\nu\sigma_x + \sigma_y - \nu\sigma_z) \\ \gamma_{xy} &= \frac{2(1+\nu)}{E}\tau_{xy} \end{aligned} \tag{t}$$

Also,

$$\epsilon_z = \frac{1}{E}(-\nu\sigma_x - \nu\sigma_y + \sigma_z) = 0$$

Hence,

$$\sigma_z = \nu(\sigma_x + \sigma_y) \tag{u}$$

which is linearly dependent upon σ_x and σ_y. Substitution of this expression for σ_z into the first two of Eqs. (t) yields:

$$\begin{aligned} \epsilon_x &= \frac{1+\nu}{E}[(1-\nu)\sigma_x - \nu\sigma_y] \\ \epsilon_y &= \frac{1+\nu}{E}[-\nu\sigma_x + (1-\nu)\sigma_y] \end{aligned} \tag{v}$$

As before, we can solve for the stresses σ_x, σ_y, and τ_{xy} in terms of the corresponding strains to obtain:

$$\sigma_x = \frac{E}{(1+\nu)(1-2\nu)}[(1-\nu)\epsilon_x + \nu\epsilon_y]$$
$$\sigma_y = \frac{E}{(1+\nu)(1-2\nu)}[\nu\epsilon_x + (1-\nu)\epsilon_y] \tag{w}$$
$$\tau_{xy} = \frac{E\gamma_{xy}}{2(1+\nu)}$$

The strain-stress operator **C** is found from Eqs. (v) and the third of Eqs. (t) as:

$$\mathbf{C} = \frac{1+\nu}{E}\begin{bmatrix} 1-\nu & -\nu & 0 \\ -\nu & 1-\nu & 0 \\ 0 & 0 & 2 \end{bmatrix} \tag{2.1-17}$$

In addition, the stress-strain operator **E** is seen from Eqs. (w) to be:

$$\mathbf{E} = \frac{E}{(1+\nu)(1-2\nu)}\begin{bmatrix} 1-\nu & \nu & 0 \\ \nu & 1-\nu & 0 \\ 0 & 0 & \frac{1-2\nu}{2} \end{bmatrix} \tag{2.1-18}$$

If we have *anisotropic materials*, the operators **E** in Eqs. (2.1-16) and (2.1-18) no longer apply. The case of *orthogonally anisotropic* (*orthotropic*) material will be discussed here, but generally anisotropic (skewed) materials will not be covered (2). For a material that is orthotropic in the x and y directions, the case of plane stress gives the following strain-stress relationships*:

$$\epsilon_x = \frac{1}{E_x}\sigma_x - \frac{\nu_{xy}}{E_y}\sigma_y \qquad \epsilon_y = -\frac{\nu_{yx}}{E_x}\sigma_x + \frac{1}{E_y}\sigma_y$$
$$\gamma_{xy} = \frac{1}{G_{xy}}\tau_{xy} \tag{x}$$

in which

$$\frac{1}{G_{xy}} \approx \frac{1+\nu_{yx}}{E_x} + \frac{1+\nu_{xy}}{E_y} \tag{y}$$

Hence,

$$\mathbf{C} = \begin{bmatrix} \frac{1}{E_x} & -\frac{\nu_{xy}}{E_y} & 0 \\ -\frac{\nu_{yx}}{E_x} & \frac{1}{E_y} & 0 \\ 0 & 0 & \frac{1}{G_{xy}} \end{bmatrix} \tag{2.1-19}$$

In general, there are four independent constants in matrix **C**, but use of an approximate value for $1/G_{xy}$ [see Eq. (y)] reduces the number of independent constants to three. Inversion of **C** yields:

*Here the symbol ν_{xy} denotes the strain in the x direction due to strain in the y direction, and ν_{yx} has the opposite meaning.

$$\mathbf{E} = \frac{1}{1 - \nu_{xy}\nu_{yx}} \begin{bmatrix} E_x & \nu_{xy}E_x & 0 \\ \nu_{yx}E_y & E_y & 0 \\ 0 & 0 & (1 - \nu_{xy}\nu_{yx})G_{xy} \end{bmatrix} \tag{2.1-20}$$

From the *reciprocal theorem*, we have:

$$\frac{E_y}{E_x} = \frac{\nu_{xy}}{\nu_{yx}} \tag{z}$$

Thus, the matrix **E** in Eq. (2.1-20) constitutes the stress-strain operator for plane stress in an orthotropic material.

On the other hand, the case of plane strain in an orthotropic material has the normal strain-stress relationships:

$$\begin{aligned} \epsilon_x &= \frac{1}{E_x}\sigma_x - \frac{\nu_{xy}}{E_y}\sigma_y - \frac{\nu_{xz}}{E_z}\sigma_z \\ \epsilon_y &= -\frac{\nu_{yx}}{E_x}\sigma_x + \frac{1}{E_y}\sigma_y - \frac{\nu_{yz}}{E_z}\sigma_z \end{aligned} \tag{a$'$}$$

Also

$$\epsilon_z = -\frac{\nu_{zx}}{E_x}\sigma_x - \frac{\nu_{zy}}{E_y}\sigma_y + \frac{1}{E_z}\sigma_z = 0$$

Thus,

$$\sigma_z = \frac{E_z}{E_x}\nu_{zx}\sigma_x + \frac{E_z}{E_y}\nu_{zy}\sigma_y \tag{b$'$}$$

Substitution of this expression for σ_z into Eqs. (a′) produces:

$$\begin{aligned} \epsilon_x &= \frac{1}{E_x}(1 - \nu_{xz}\nu_{zx})\sigma_x - \frac{1}{E_y}(\nu_{xy} + \nu_{xz}\nu_{zy})\sigma_y \\ \epsilon_y &= -\frac{1}{E_x}(\nu_{yx} + \nu_{yz}\nu_{zx})\sigma_x + \frac{1}{E_y}(1 - \nu_{yz}\nu_{zy})\sigma_y \end{aligned} \tag{c$'$}$$

Hence,

$$\mathbf{C} = \begin{bmatrix} \dfrac{d}{E_x} & -\dfrac{b}{E_y} & 0 \\ -\dfrac{c}{E_x} & \dfrac{a}{E_y} & 0 \\ 0 & 0 & \dfrac{1}{G_{xy}} \end{bmatrix} \tag{2.1-21}$$

in which

$$a = 1 - \nu_{yz}\nu_{zy} \qquad b = \nu_{xy} + \nu_{xz}\nu_{zy}$$
$$c = \nu_{yx} + \nu_{yz}\nu_{zx} \qquad d = 1 - \nu_{xz}\nu_{zx}$$

Inverting matrix **C** yields:

$$\mathbf{E} = \frac{1}{ad - bc} \begin{bmatrix} aE_x & bE_x & 0 \\ cE_y & dE_y & 0 \\ 0 & 0 & (ad - bc)G_{xy} \end{bmatrix} \tag{2.1-22}$$

From the reciprocal theorem, we find:

$$\frac{E_y}{E_x} = \frac{b}{c} \tag{d'}$$

Equation (2.1-22) gives the operator $\mathbf{E}$ for plane strain in an orthotropic material, where x and y are the *principal directions of orthotropy.*

Figure 2.5 indicates a two-dimensional orthotropic continuum with its principal material axes x_m and y_m inclined at the angle θ with respect to structural axes x_s and y_s. For this situation the *strain transformation* from structural directions to material directions can be written as:

$$\boldsymbol{\epsilon}_m = \mathbf{T}_{ms}\boldsymbol{\epsilon}_s \tag{2.1-23}$$

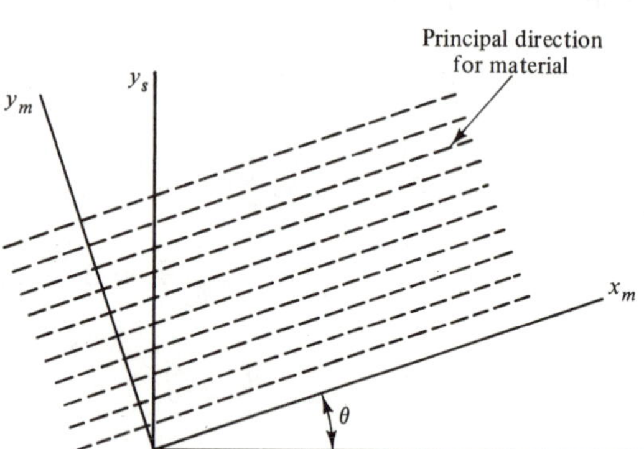

Figure 2.5 Orthotropic Material

In addition, the *stress transformation* from material directions to structural directions is:

$$\boldsymbol{\sigma}_s = \mathbf{T}_{ms}^{\mathrm{T}}\boldsymbol{\sigma}_m \tag{2.1-24}$$

The transformation operator $\mathbf{T}_{ms}$ used in these expressions is seen from Eqs. (2.1-13) and (2.1-23) to be:

$$\mathbf{T}_{ms} = \mathbf{T}_{\theta}^{-\mathrm{T}} \tag{2.1-25}$$

An appropriate *stress-strain transformation* also can be obtained by first writing stress-strain relationships for material directions, as follows:

$$\boldsymbol{\sigma}_m = \mathbf{E}_m\boldsymbol{\epsilon}_m \tag{2.1-26}$$

Now we substitute Eq. (2.1-23) into Eq. (2.1-26) and then substitute the latter into Eq. (2.1-24) to find:

$$\boldsymbol{\sigma}_s = \mathbf{T}_{ms}^{\mathrm{T}}\mathbf{E}_m\mathbf{T}_{ms}\boldsymbol{\epsilon}_s$$

We can rewrite this expression as:

$$\boldsymbol{\sigma}_s = \mathbf{E}_s\boldsymbol{\epsilon}_s \tag{2.1-27}$$

where

$$\mathbf{E}_s = \mathbf{T}_{ms}^{\mathrm{T}}\mathbf{E}_m\mathbf{T}_{ms} \tag{2.1-28}$$

Of course, this congruence transformation of $\mathbf{E}_m$ for material directions to $\mathbf{E}_s$ for structural directions preserves the symmetry of the stress-strain operator.

2.2 TRIANGULAR ELEMENTS

Finite elements with a triangular shape prove to be quite versatile for the purpose of discretizing any two-dimensional continuum. One of the earliest and best-known elements is the *constant strain triangle* described by Turner et al. (3). In the present section stiffnesses and equivalent nodal loads will be developed in detail for this triangle. In addition, a briefer description will be given for the *linear strain triangle* of Fraejis de Veubeke (4).

Figure 2.6 shows a constant strain triangle of thickness t, having the following generic displacements (translations) in the x-y plane:

$$\mathbf{u} = \{u, v\} \tag{a}$$

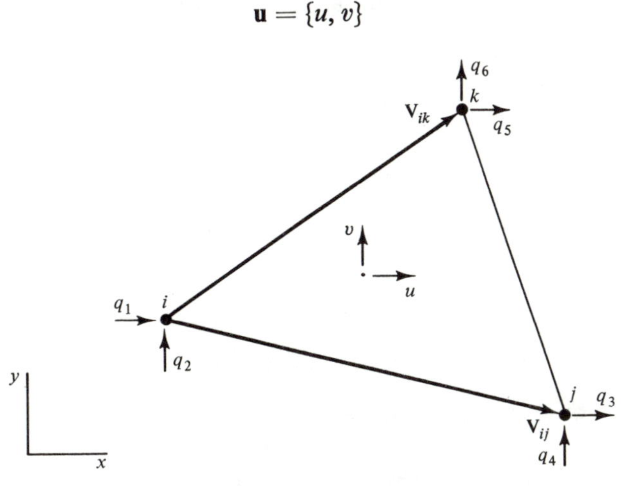

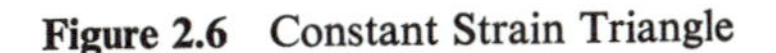
Figure 2.6 Constant Strain Triangle

Its three corners, points i, j, and k, serve as nodes that are numbered in a counterclockwise sequence. At each node there are two nodal translations in the directions of x and y, with the former preceding the latter. Thus,

$$\mathbf{q} = \{q_1, q_2, \ldots, q_6\} = \{u_1, v_1, \ldots, v_3\} \tag{b}$$

Displacement functions assumed for this element are:

$$u = c_1 + c_2 x + c_3 y \qquad v = c_4 + c_5 x + c_6 y \tag{c}$$

which are linear in both x and y. From these functions we can form the matrix $\mathbf{g}$ as:

$$\mathbf{g} = \begin{bmatrix} 1 & x & y & 0 & 0 & 0 \\ 0 & 0 & 0 & 1 & x & y \end{bmatrix} \tag{d}$$

Evaluating $\mathbf{g}$ at each node gives:

$$\mathbf{h} = \begin{bmatrix} \mathbf{g}_i \\ \mathbf{g}_j \\ \mathbf{g}_k \end{bmatrix} = \begin{bmatrix} 1 & x_i & y_i & 0 & 0 & 0 \\ 0 & 0 & 0 & 1 & x_i & y_i \\ \hdashline 1 & x_j & y_j & 0 & 0 & 0 \\ 0 & 0 & 0 & 1 & x_j & y_j \\ \hdashline 1 & x_k & y_k & 0 & 0 & 0 \\ 0 & 0 & 0 & 1 & x_k & y_k \end{bmatrix} \tag{e}$$

For the purpose of determining the inverse of $\mathbf{h}$, transform it to $\mathbf{h}_T$, as follows:

$$\mathbf{h}_T = \mathbf{T}\ \mathbf{h} = \begin{bmatrix} \mathbf{h}_1 & \mathbf{O} \\ \mathbf{O} & \mathbf{h}_1 \end{bmatrix} \tag{f}$$

In this expression the *rearrangement operator* $\mathbf{T}$ has the form:

$$\mathbf{T} = \begin{bmatrix} 1 & 0 & 0 & 0 & 0 & 0 \\ 0 & 0 & 1 & 0 & 0 & 0 \\ 0 & 0 & 0 & 0 & 1 & 0 \\ 0 & 1 & 0 & 0 & 0 & 0 \\ 0 & 0 & 0 & 1 & 0 & 0 \\ 0 & 0 & 0 & 0 & 0 & 1 \end{bmatrix} \tag{g}$$

And the submatrix $\mathbf{h}_1$, appearing twice within $\mathbf{h}_T$, becomes:

$$\mathbf{h}_1 = \begin{bmatrix} 1 & x_i & y_i \\ 1 & x_j & y_j \\ 1 & x_k & y_k \end{bmatrix} \tag{h}$$

The inverse matrix $\mathbf{h}^{-1}$ may then be obtained from:

$$\mathbf{h}^{-1} = \mathbf{h}_T^{-1}\mathbf{T} = \begin{bmatrix} \mathbf{h}_1^{-1} & \mathbf{O} \\ \mathbf{O} & \mathbf{h}_1^{-1} \end{bmatrix} \mathbf{T} \tag{i}$$

Within this expression, the submatrix inverse $\mathbf{h}_1^{-1}$ is:

$$\mathbf{h}_1^{-1} = \frac{\mathbf{h}_1^a}{|\mathbf{h}_1|} = \frac{(\mathbf{h}_1^c)^{\mathrm{T}}}{|\mathbf{h}_1|} = \frac{1}{|\mathbf{h}_1|} \begin{bmatrix} x_j y_k - x_k y_j & x_k y_i - x_i y_k & x_i y_j - x_j y_i \\ -y_{jk} & -y_{ki} & -y_{ij} \\ x_{jk} & x_{ki} & x_{ij} \end{bmatrix} \tag{j}$$

where $\mathbf{h}_1^a$ is the adjoint matrix of $\mathbf{h}_1$, $\mathbf{h}_1^c$ is its cofactor matrix, and $|\mathbf{h}_1|$ is its determinant. In Eq. (j) we also have the definitions $x_{ij} = x_j - x_i$, $y_{ij} = y_j - y_i$, and so on. Altogether, the inverse of $\mathbf{h}$ [as given by Eq. (i)] becomes:

$$\mathbf{h}^{-1} = \frac{1}{2A_{ijk}} \begin{bmatrix} x_j y_k - x_k y_j & 0 & x_k y_i - x_i y_k & 0 & x_i y_j - x_j y_i & 0 \\ -y_{jk} & 0 & -y_{ki} & 0 & -y_{ij} & 0 \\ x_{jk} & 0 & x_{ki} & 0 & x_{ij} & 0 \\ 0 & x_j y_k - x_k y_j & 0 & x_k y_i - x_i y_k & 0 & x_i y_j - x_j y_i \\ 0 & -y_{jk} & 0 & -y_{ki} & 0 & -y_{ij} \\ 0 & x_{jk} & 0 & x_{ki} & 0 & x_{ij} \end{bmatrix} \tag{k}$$

where A_{ijk} is the area of triangle ijk. Note that:

$$2A_{ijk} = |\mathbf{h}_1| = \begin{vmatrix} 1 & x_i & y_i \\ 1 & x_j & y_j \\ 1 & x_k & y_k \end{vmatrix}$$

$$= x_i y_j + x_j y_k + x_k y_i - x_i y_k - x_j y_i - x_k y_j$$

$$= x_{ij} y_k + x_{jk} y_i - x_{ik} y_j$$

$$= x_{ij} y_{ik} - x_{ik} y_{ij} \qquad (\ell)$$

This formula for twice the area can be verified by taking the absolute value of the vector (or cross) product between the vectors $\mathbf{V}_{ij}$ and $\mathbf{V}_{ik}$ in Fig. 2.6. That is,

$$2A_{ijk} = \text{Abs. } (\mathbf{V}_{ij} \times \mathbf{V}_{ik}) = x_{ij} y_{ik} - x_{ik} y_{ij}$$

Various other formulas can also be obtained by permuting the subscripts. With $\mathbf{h}^{-1}$ available, we can now determine the displacement shape functions $\mathbf{f}$ to be:

$$\mathbf{f} = \mathbf{g}\,\mathbf{h}^{-1} = \left[\begin{array}{cc:cc:cc} f_1 & 0 & f_2 & 0 & f_3 & 0 \\ 0 & f_1 & 0 & f_2 & 0 & f_3 \end{array}\right] \qquad (2.2\text{-}1a)$$

where

$$f_1 = \frac{1}{2A_{ijk}} (x_j y_k - x_k y_j - y_{jk} x + x_{jk} y)$$

$$f_2 = \frac{1}{2A_{ijk}} (x_k y_i - x_i y_k - y_{ki} x + x_{ki} y) \qquad (2.2\text{-}1b)$$

$$f_3 = \frac{1}{2A_{ijk}} (x_i y_j - x_j y_i - y_{ij} x + x_{ij} y)$$

Strain-displacement relationships for the triangle under consideration may be obtained using Eq. (2.1-11). Thus,

$$\mathbf{B} = \mathbf{d}\ \mathbf{f} = \begin{bmatrix} \dfrac{\partial}{\partial x} & 0 \\ 0 & \dfrac{\partial}{\partial y} \\ \dfrac{\partial}{\partial y} & \dfrac{\partial}{\partial x} \end{bmatrix} \mathbf{f} = \frac{1}{2A_{ijk}} \left[\begin{array}{cc:cc:cc} -y_{jk} & 0 & -y_{ki} & 0 & -y_{ij} & 0 \\ 0 & x_{jk} & 0 & x_{ki} & 0 & x_{ij} \\ x_{jk} & -y_{jk} & x_{ki} & -y_{ki} & x_{ij} & -y_{ij} \end{array}\right] \qquad (2.2\text{-}2)$$

We see that the terms in matrix $\mathbf{B}$ are all constants. For this reason the element is usually referred to as the *constant strain triangle*.

Assuming an isotropic material, we can write stress-strain relationships for either plane stress or plane strain by using a combination of Eqs. (2.1-16) and (2.1-18), as follows:

$$\mathbf{E} = \frac{E}{(1+\nu)e_2} \begin{bmatrix} e_1 & \nu & 0 \\ \nu & e_1 & 0 \\ 0 & 0 & e_3 \end{bmatrix} \qquad (2.2\text{-}3a)$$

Then the stress-displacement relationships are given by the product:

$$\mathbf{E}\ \mathbf{B} = \frac{E}{2A_{ijk}(1+\nu)e_2}\left[\begin{array}{cc|cc|cc} -e_1 y_{jk} & \nu x_{jk} & -e_1 y_{ki} & \nu x_{ki} & -e_1 y_{ij} & \nu x_{ij} \\ -\nu y_{jk} & e_1 x_{jk} & -\nu y_{ki} & e_1 x_{ki} & -\nu y_{ij} & e_1 x_{ij} \\ e_3 x_{jk} & -e_3 y_{jk} & e_3 x_{ki} & -e_3 y_{ki} & e_3 x_{ij} & -e_3 y_{ij} \end{array}\right] \tag{2.2-3b}$$

In this matrix expression the *elasticity constants* for plane stress are:

$$e_1 = 1 \qquad e_2 = 1 - \nu \qquad e_3 = \frac{e_2}{2}$$

whereas those for plane strain become:

$$e_1 = 1 - \nu \qquad e_2 = 1 - 2\nu \qquad e_3 = \frac{e_2}{2}$$

Having the necessary operators for stresses and strains, we can evaluate the *element stiffnesses* with Eq. (1.3-13).

$$\mathbf{K} = \int_V \mathbf{B}^{\mathrm{T}}\mathbf{E}\ \mathbf{B}\, dV = \mathbf{B}^{\mathrm{T}}\mathbf{E}\ \mathbf{B}A_{ijk}t = \mathbf{K}_1 + \mathbf{K}_2 \tag{m}$$

Here the matrix $\mathbf{K}$ is separated into the two parts, $\mathbf{K}_1$ and $\mathbf{K}_2$, which contain terms due to normal and shearing strains, respectively. These parts are listed in Table 2.1.

TABLE 2.1 Stiffness Matrix for Constant Strain Triangle

$$\mathbf{K}_1 = e_4 \begin{bmatrix} e_1 y_{jk}^2 & & & & & \\ -\nu x_{jk} y_{jk} & e_1 x_{jk}^2 & & & \text{Sym.} & \\ e_1 y_{ki} y_{jk} & -\nu y_{ki} x_{jk} & e_1 y_{ki}^2 & & & \\ -\nu x_{ki} y_{jk} & e_1 x_{ki} x_{jk} & -\nu x_{ki} y_{ki} & e_1 x_{ki}^2 & & \\ e_1 y_{ij} y_{jk} & -\nu y_{ij} x_{jk} & e_1 y_{ij} y_{ki} & -\nu y_{ij} x_{ki} & e_1 y_{ij}^2 & \\ -\nu x_{ij} y_{jk} & e_1 x_{ij} x_{jk} & -\nu x_{ij} y_{ki} & e_1 x_{ij} x_{ki} & -\nu x_{ij} y_{ij} & e_1 x_{ij}^2 \end{bmatrix}$$

$$\mathbf{K}_2 = e_5 \begin{bmatrix} x_{jk}^2 & & & & & \\ -x_{jk} y_{jk} & y_{jk}^2 & & & \text{Sym.} & \\ x_{jk} x_{ki} & -y_{jk} x_{ki} & x_{ki}^2 & & & \\ -x_{jk} y_{ki} & y_{jk} y_{ki} & -x_{ki} y_{ki} & y_{ki}^2 & & \\ x_{jk} x_{ij} & -y_{jk} x_{ij} & x_{ki} x_{ij} & -y_{ki} x_{ij} & x_{ij}^2 & \\ -x_{jk} y_{ij} & y_{jk} y_{ij} & -x_{ki} y_{ij} & y_{ki} y_{ij} & -x_{ij} y_{ij} & y_{ij}^2 \end{bmatrix}$$

$$e_4 = \frac{Et}{4A_{ijk}(1+\nu)e_2} \qquad e_5 = e_4 e_3$$

We now consider *equivalent nodal loads* for the constant strain triangle subjected to various influences. Figure 2.7(a) shows a uniformly distributed line loading b_x (force per unit length) acting in the x direction on edge ij. For the

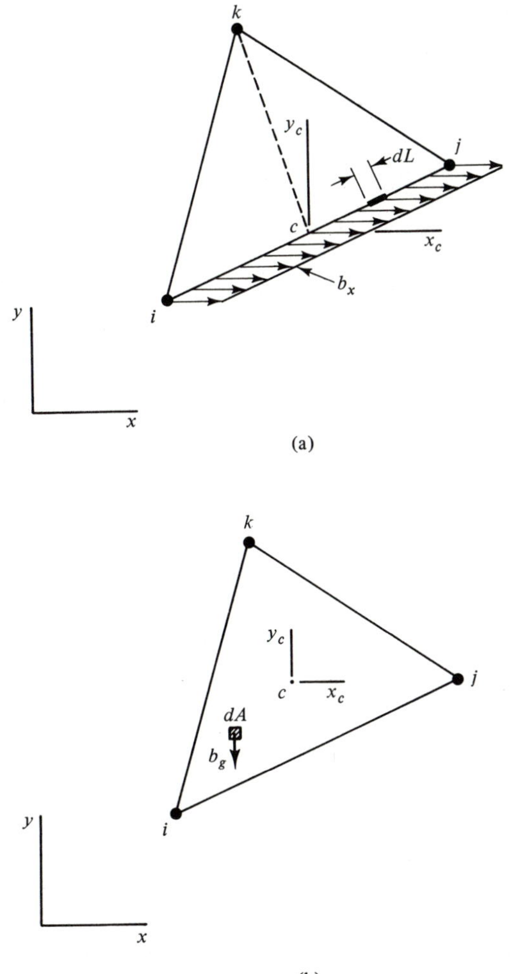

(b)

Figure 2.7 (a) Edge Load (b) Gravity Load

equivalent nodal loads due to this loading the integrations are convenient if the origin is taken at the midlength, point c [see Fig. 2.7(a) and Appendix A]. Thus,

$$\mathbf{p}_b = \mathbf{h}^{-\mathrm{T}} \int_L \mathbf{g}^{\mathrm{T}}\mathbf{b}\, dL = \mathbf{h}^{-\mathrm{T}} \int_L \begin{bmatrix} 1 & 0 \\ x & 0 \\ y & 0 \\ 0 & 1 \\ 0 & x \\ 0 & y \end{bmatrix} \begin{bmatrix} b_x \\ 0 \end{bmatrix} dL = \mathbf{h}^{-\mathrm{T}} b_x \begin{bmatrix} L_{ij} \\ 0 \\ 0 \\ 0 \\ 0 \\ 0 \end{bmatrix}$$

$$= \frac{b_x L_{ij}}{2A_{ijk}} \{x_j y_k - x_k y_j, 0, x_k y_i - x_i y_k, 0, x_i y_j - x_j y_i, 0\}$$

$$= \frac{b_x L_{ij}}{2A_{ijk}} \{A_{ijk}, 0, A_{ijk}, 0, 0, 0\}$$

$$= \frac{b_x L_{ij}}{2} \{1, 0, 1, 0, 0, 0\} \tag{n}$$

As a second example, let us consider the uniform gravity loading b_g (force per unit volume in the negative y direction) indicated in Fig. 2.7(b). In this case the integrations for the equivalent nodal loads are convenient if the origin is taken at the centroid, point c [see Fig. 2.7(b) and Appendix A]. Then,

$$\mathbf{p}_b = \mathbf{h}^{-T} \int_V \mathbf{g}^T \mathbf{b} \, dV = \mathbf{h}^{-T} \int_A \begin{bmatrix} 1 & 0 \\ x & 0 \\ y & 0 \\ 0 & 1 \\ 0 & x \\ 0 & y \end{bmatrix} \begin{bmatrix} 0 \\ -b_g \end{bmatrix} t \, dA = -\mathbf{h}^{-T} b_g t \begin{bmatrix} 0 \\ 0 \\ 0 \\ A_{ijk} \\ 0 \\ 0 \end{bmatrix}$$

$$= \frac{-b_g t}{2A_{ijk}} \left\{0, \frac{2A_{ijk}^2}{3}, 0, \frac{2A_{ijk}^2}{3}, 0, \frac{2A_{ijk}^2}{3}\right\}$$

$$= \frac{-b_g A_{ijk} t}{3} \{0, 1, 0, 1, 0, 1\} \tag{o}$$

Next, suppose that a constant strain triangle undergoes a uniform temperature increase ΔT. For this influence the equivalent nodal loads are:

$$\mathbf{p}_T = \int_V \mathbf{B}^T \mathbf{E} \, \boldsymbol{\epsilon}_T \, dV = \mathbf{B}^T \mathbf{E} \, \boldsymbol{\epsilon}_T A_{ijk} t$$

in which

$$\boldsymbol{\epsilon}_T = \alpha \, \Delta T\{1, 1, 0\}$$

Note that no integrations (beyond calculating the volume) are required for this state of constant strain. Furthermore, if the two-dimensional problem were plane strain instead of plane stress, the temperature strain vector $\boldsymbol{\epsilon}_T$ would have to be multiplied by the factor $(1 + \nu)$. For plane stress the equivalent nodal loads become:

$$\mathbf{p}_T = \frac{Et\alpha \, \Delta T}{2(1+\nu)e_2} \begin{bmatrix} -e_1 y_{jk} - \nu y_{jk} \\ \nu x_{jk} + e_1 x_{jk} \\ -e_1 y_{ki} - \nu y_{ki} \\ \nu x_{ki} + e_1 x_{ki} \\ -e_1 y_{ij} - \nu y_{ij} \\ \nu x_{ij} + e_1 x_{ij} \end{bmatrix} = \frac{Et\alpha \, \Delta T(e_1 + \nu)}{2(1+\nu)e_2} \begin{bmatrix} -y_{jk} \\ x_{jk} \\ -y_{ki} \\ x_{ki} \\ -y_{ij} \\ x_{ij} \end{bmatrix} \tag{p}$$

and for plane strain this result must be multiplied by $(1 + \nu)$.

Finally, if there exists a uniform initial shearing strain γ_0, the equivalent nodal loads are found from:

$$\mathbf{p}_0 = \int_V \mathbf{B}^{\mathrm{T}}\mathbf{E}\ \boldsymbol{\epsilon}_0\, dV = \mathbf{B}^{\mathrm{T}}\mathbf{E}\ \boldsymbol{\epsilon}_0 A_{ijk} t$$

where

$$\boldsymbol{\epsilon}_0 = \{0, 0, \gamma_0\}$$

Hence,

$$\mathbf{p}_0 = \frac{Et\gamma_0}{4(1+\nu)}\{x_{jk}, -y_{jk}, x_{ki}, -y_{ki}, x_{ij}, -y_{ij}\} \tag{q}$$

which is the same for both plane stress and plane strain.

Because the assumed displacement functions for the constant strain triangle are only linear, this element usually does not produce very accurate results for stresses. To achieve better accuracy with triangular elements, two choices are available. The first consists of refining the network of constant strain triangles, and the second is to change to a higher-order element. Ordinarily, the latter choice can lead to good improvement of accuracy without increasing the number of degrees of freedom in the analytical model. Therefore, we will examine the *linear strain triangle* of de Veubeke (4), as documented by McBean (5). Development of the properties of this element is similar to that for the constant strain triangle, but the steps are much more tedious. For this reason, only the most essential results will be presented.

The linear strain triangle in Fig. 2.8(a) has a constant thickness t and the following generic displacements:

$$\mathbf{u} = \{u, v\} \tag{r}$$

In addition to the three corner nodes (numbered 1, 2, and 3), there are also three midedge nodes (numbered 4, 5, and 6). In this case we number all the x-translations at the nodes before all the y-translations. Thus, the vector of nodal displacements becomes:

$$\mathbf{q} = \{u_1, u_2, \ldots, u_6, v_1, v_2, \ldots, v_6\} \tag{s}$$

For this element the assumed displacement functions are:

$$\begin{aligned} u &= c_1 + c_2 x + c_3 y + c_4 x^2 + c_5 xy + c_6 y^2 \\ v &= c_7 + c_8 x + c_9 y + c_{10} x^2 + c_{11} xy + c_{12} y^2 \end{aligned} \tag{t}$$

When these quadratic displacement functions (or the corresponding shape functions) are differentiated once (with respect to either x or y), they produce strain functions that vary linearly. For this reason the element is called the linear strain triangle.

If strain-displacement relationships for this element are evaluated at the nodes, we can find a *nodal strain matrix*. For example, at nodes 1, 2, and 3 we can write:

$$\boldsymbol{\epsilon} = \begin{bmatrix} \epsilon_x \\ \epsilon_y \\ \gamma_{xy} \end{bmatrix} = \begin{bmatrix} \mathbf{B}_x & \mathbf{O} \\ \mathbf{O} & \mathbf{B}_y \\ \mathbf{B}_y & \mathbf{B}_x \end{bmatrix} \begin{bmatrix} \mathbf{u} \\ \mathbf{v} \end{bmatrix} = \mathbf{B}\,\mathbf{q} \tag{2.2-4}$$

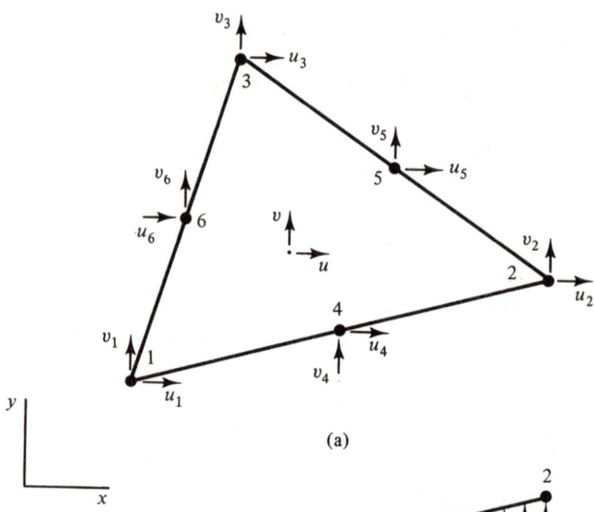

(a)

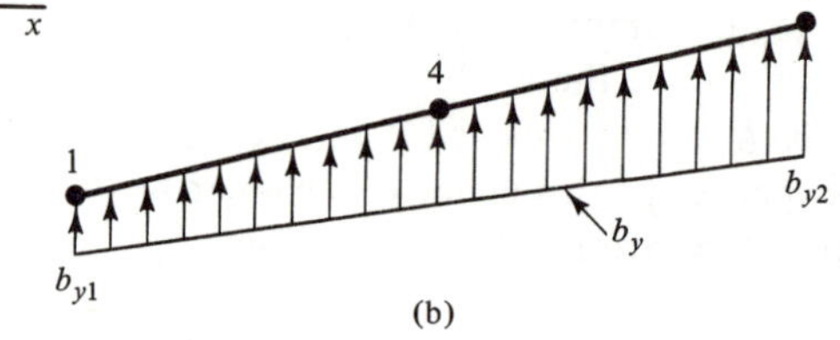

(b)

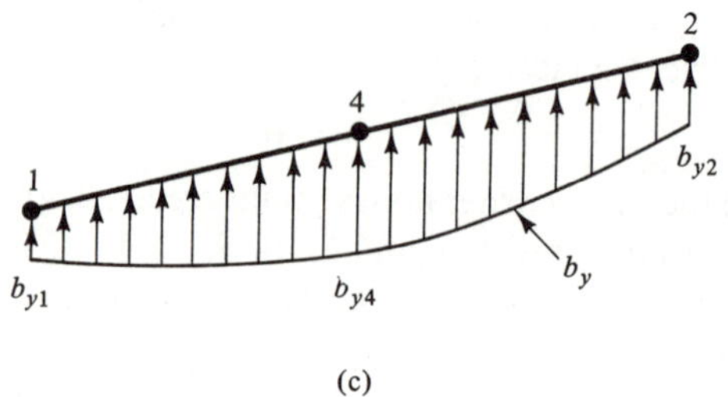

(c)

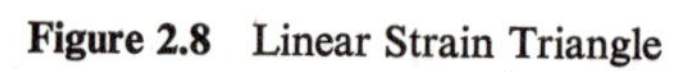

Figure 2.8 Linear Strain Triangle

where

$$\boldsymbol{\epsilon}_x = \begin{bmatrix} \epsilon_{x_1} \\ \epsilon_{x_2} \\ \epsilon_{x_3} \end{bmatrix} \qquad \boldsymbol{\epsilon}_y = \begin{bmatrix} \epsilon_{y_1} \\ \epsilon_{y_2} \\ \epsilon_{y_3} \end{bmatrix} \qquad \boldsymbol{\gamma}_{xy} = \begin{bmatrix} \gamma_{xy_1} \\ \gamma_{xy_2} \\ \gamma_{xy_3} \end{bmatrix} \tag{u}$$

And

$$\mathbf{B}_x = \frac{1}{2A}\begin{bmatrix} 3y_{32} & -y_{13} & -y_{21} & 4y_{13} & 0 & 4y_{21} \\ -y_{32} & 3y_{13} & -y_{21} & 4y_{32} & 4y_{21} & 0 \\ -y_{32} & -y_{13} & 3y_{21} & 0 & 4y_{13} & 4y_{32} \end{bmatrix} \tag{2.2-5a}$$

$$\mathbf{B}_y = \frac{1}{2A}\begin{bmatrix} 3x_{23} & -x_{31} & -x_{12} & 4x_{31} & 0 & 4x_{12} \\ -x_{23} & 3x_{31} & -x_{12} & 4x_{23} & 4x_{12} & 0 \\ -x_{23} & -x_{31} & 3x_{12} & 0 & 4x_{31} & 4x_{23} \end{bmatrix} \tag{2.2-5b}$$

in which A is the area of the triangle.

The stiffness matrix for the linear strain triangle (isotropic material in plane stress) appears in Table 2.2, where it has the following form:

$$\mathbf{K} = \frac{Et}{12A(1-\nu^2)} \begin{bmatrix} \mathbf{K}_A & & & \text{Sym.} \\ & \mathbf{K}_B & & \\ & & \mathbf{K}_C & \\ & & & \mathbf{K}_D \end{bmatrix} \quad \text{(v)}$$

Individual terms in this matrix were obtained using area coordinates for integration, as explained later in Sec. 3.2.

Figure 2.8(b) shows a linearly varying load (force per unit length) applied in the y direction to edge 1-2 of the linear strain triangle. Owing to this distributed body force, the equivalent nodal loads at points 1, 4, and 2 are:

$$\begin{bmatrix} p_{y1} \\ p_{y4} \\ p_{y2} \end{bmatrix} = \frac{L_{12}}{6} \begin{bmatrix} 1 & 0 \\ 2 & 2 \\ 0 & 1 \end{bmatrix} \begin{bmatrix} b_{y1} \\ b_{y2} \end{bmatrix} \quad \text{(w)}$$

As another example, a quadratically varying load (force per unit length, acting in the y direction) is applied to edge 1-2, as indicated in Fig. 2.8(c). The equivalent nodal loads at points 1, 4, and 2 due to this influence are:

TABLE 2.2 Stiffness Matrix for Linear Strain Triangle

$a_1 = x_3 - x_2$ $\quad$ $a_2 = x_1 - x_3$ $\quad$ $a_3 = x_2 - x_1$

$b_1 = y_2 - y_3$ $\quad$ $b_2 = y_3 - y_1$ $\quad$ $b_3 = y_1 - y_2$

$$c = \frac{1-\nu}{2} = \lambda$$

$K_A =$

	1	2	3
1	$3(b_1^2 + ca_1^2)$		
2	$-(b_1b_2 + ca_1a_2)$	$3(b_2^2 + ca_2^2)$	
3	$-(b_1b_3 + ca_1a_3)$	$-(b_2b_3 + ca_2a_3)$	$3(b_3^2 + ca_3^2)$
4	$4(b_1b_2 + ca_1a_2)$	$4(b_1b_2 + ca_1a_2)$	0
5	0	$4(b_2b_3 + ca_2a_3)$	$4(b_2b_3 + ca_2a_3)$
6	$4(b_1b_3 + ca_1a_3)$	0	$4(b_1b_3 + ca_1a_3)$
7	$3a_1b_1(\nu + c)$	$-(\nu a_1b_2 + ca_2b_1)$	$-(\nu a_1b_3 + ca_3b_1)$
8	$-(\nu a_2b_1 + ca_1b_2)$	$3a_2b_2(\nu + c)$	$-(\nu a_2b_3 + ca_3b_2)$
9	$-(\nu a_3b_1 + ca_1b_3)$	$-(\nu a_3b_2 + ca_2b_3)$	$3a_3b_3(\nu + c)$
10	$4(\nu a_2b_1 + ca_1b_2)$	$4(\nu a_1b_2 + ca_2b_1)$	0
11	0	$4(\nu a_3b_2 + ca_2b_3)$	$4(\nu a_2b_3 + ca_3b_2)$
12	$4(\nu a_3b_1 + ca_1b_3)$	0	$4(\nu a_1b_3 + ca_3b_1)$

TABLE 2.2 (cont.)

$\mathbf{K}_B =$	4	5	6
4	$8(b_1^2 + b_1b_2 + b_2^2) + 8c(a_1^2 + a_1a_2 + a_2^2)$		
5	$4(b_1b_2 + b_2^2 + b_2b_3 + 2b_1b_3) + 4c(a_1a_2 + a_2^2 + a_2a_3 + 2a_1a_3)$	$8(b_2^2 + b_2b_3 + b_3^2) + 8c(a_2^2 + a_2a_3 + a_3^2)$	
6	$4(b_1^2 + b_1b_2 + b_1b_3 + 2b_2b_3) + 4c(a_1^2 + a_1a_2 + a_1a_3 + 2a_2a_3)$	$4(b_1b_3 + b_2b_3 + b_3^2 + 2b_1b_2) + 4c(a_1a_3 + a_2a_3 + a_3^2 + 2a_1a_2)$	$8(b_1^2 + b_1b_3 + b_3^2) + 8c(a_1^2 + a_1a_3 + a_3^2)$
7	$4(va_1b_2 + ca_2b_1)$	0	$4(va_1b_3 + ca_3b_1)$
8	$4(va_2b_1 + ca_1b_2)$	$4(va_2b_3 + ca_3b_2)$	0
9	0	$4(va_3b_2 + ca_2b_3)$	$4(va_3b_1 + ca_1b_3)$
10	$4(2a_1b_1 + 2a_2b_2 + a_1b_2 + a_2b_1)(v + c)$	$4v(a_2b_3 + a_2b_2 + a_1b_2 + 2a_1b_3) + 4c(a_3b_2 + a_2b_2 + a_2b_1 + 2a_3b_1)$	$4v(a_1b_1 + a_2b_1 + a_1b_3 + 2a_2b_3) + 4c(a_1b_1 + a_1b_2 + a_3b_1 + 2a_3b_2)$
11	$4v(a_3b_2 + a_2b_2 + a_2b_1 + 2a_3b_1) + 4c(a_2b_3 + a_2b_2 + a_1b_2 + 2a_1b_3)$	$4(2a_2b_2 + 2a_3b_3 + a_2b_3 + a_3b_2)(v + c)$	$4v(a_3b_3 + a_2b_3 + a_3b_1 + 2a_2b_1) + 4c(a_3b_3 + a_3b_2 + a_1b_3 + 2a_1b_2)$
12	$4v(a_1b_2 + a_3b_1 + a_1b_1 + 2a_3b_2) + 4c(a_2b_1 + a_1b_3 + a_1b_1 + 2a_2b_3)$	$4v(a_1b_3 + a_3b_2 + a_3b_3 + 2a_1b_2) + 4c(a_3b_1 + a_2b_3 + a_3b_3 + 2a_2b_1)$	$4(2a_3b_3 + 2a_1b_1 + a_3b_1 + a_1b_3)(v + c)$

TABLE 2.2 (cont.)

$\mathbf{K}_C =$

	7	8	9
7	$3(a_1^2 + cb_1^2)$		
8	$-(a_1a_2 + cb_1b_2)$	$3(a_2^2 + cb_2^2)$	
9	$-(a_1a_3 + cb_1b_3)$	$-(a_2a_3 + cb_2b_3)$	$3(a_3^2 + cb_3^2)$
10	$4(a_1a_2 + cb_1b_2)$	$4(a_1a_2 + cb_1b_2)$	0
11	0	$4(a_2a_3 + cb_2b_3)$	$4(a_2a_3 + cb_2b_3)$
12	$4(a_1a_3 + cb_1b_3)$	0	$4(a_1a_3 + cb_1b_3)$

$\mathbf{K}_D =$

	10	11	12
10	$8(a_1^2 + a_1a_2 + a_2^2)$ $+ 8c(b_1^2 + b_1b_2 + b_2^2)$		
11	$4(a_1a_2 + a_2a_3 + a_2^2 + 2a_1a_3)$ $+ 4c(b_1b_2 + b_2b_3 + b_2^2 + 2b_1b_3)$	$8(a_2^2 + a_2a_3 + a_3^2)$ $+ 8c(b_2^2 + b_2b_3 + b_3^2)$	
12	$4(a_1a_2 + a_1a_3 + a_1^2 + 2a_2a_3)$ $+ 4c(b_1b_2 + b_1b_3 + b_1^2 + 2b_2b_3)$	$4(a_1a_3 + a_2a_3 + a_3^2 + 2a_1a_2)$ $+ 4c(b_1b_3 + b_2b_3 + b_3^2 + 2b_1b_2)$	$8(a_1^2 + a_1a_3 + a_3^2)$ $+ 8c(b_1^2 + b_1b_3 + b_3^2)$

$$\begin{bmatrix} p_{y1} \\ p_{y4} \\ p_{y2} \end{bmatrix} = \frac{L_{12}}{30} \begin{bmatrix} 4 & 2 & -1 \\ 2 & 16 & 2 \\ -1 & 2 & 4 \end{bmatrix} \begin{bmatrix} b_{y1} \\ b_{y4} \\ b_{y2} \end{bmatrix} \tag{x}$$

Equations (*w*) and (*x*) were developed by McBean (5).

2.3 RECTANGULAR ELEMENTS

In two-dimensional problems with rectangular boundaries (e.g., shear panels), rectangular elements serve a useful purpose. Among the formulations available, the *bilinear displacement rectangle*, developed by Melosh (6), is the most straightforward. Stiffnesses and equivalent nodal loads will be derived for this element, and a brief presentation of the *linear-cubic rectangle* of Oakberg and Weaver (7) will also be given.

A rectangular element of thickness t appears in Fig. 2.9(a). Also shown in the figure are *dimensionless centroidal coordinates*, defined as follows:

$$\xi = \frac{x}{a} \qquad \eta = \frac{y}{b} \tag{2.3-1}$$

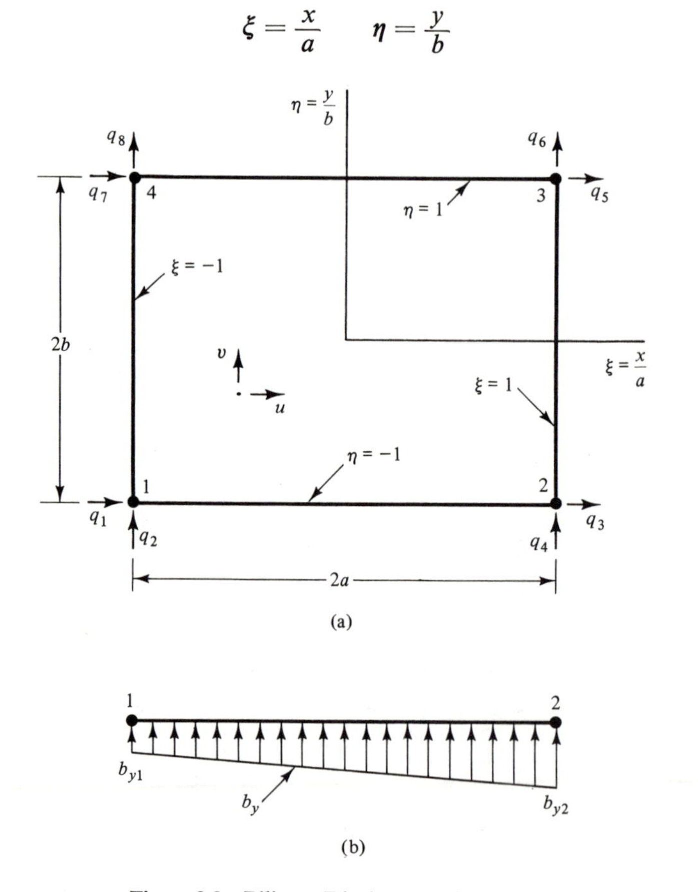

Figure 2.9 Bilinear Displacement Rectangle

in which a and b are half the width and height, respectively. Generic displacements for this element consist of translations in the x-y plane. Thus,

$$\mathbf{u} = \{u, v\} \tag{a}$$

Nodes 1, 2, 3, and 4 are designated at the corners, starting at the lower left and proceeding counterclockwise. Two nodal translations (in the x and y directions) appear at each node, and the vector of nodal displacements becomes:

$$\mathbf{q} = \{q_1, q_2, \ldots, q_8\} = \{u_1, v_1, \ldots, v_4\} \tag{b}$$

If the displacement functions assumed for this element are:

$$\begin{aligned} u &= c_1 + c_2\xi + c_3\eta + c_4\xi\eta \\ v &= c_5 + c_6\xi + c_7\eta + c_8\xi\eta \end{aligned} \tag{c}$$

then we see that they are bilinear in ξ and η. For this reason we refer to the element as the *bilinear displacement rectangle*. With these displacement functions, the matrix $\mathbf{g}$ becomes:

$$\mathbf{g} = \begin{bmatrix} 1 & \xi & \eta & \xi\eta & 0 & 0 & 0 & 0 \\ 0 & 0 & 0 & 0 & 1 & \xi & \eta & \xi\eta \end{bmatrix} \tag{d}$$

This matrix may be evaluated at each of the nodes to obtain:

$$\mathbf{h} = \begin{bmatrix} \mathbf{g}_1 \\ \mathbf{g}_2 \\ \mathbf{g}_3 \\ \mathbf{g}_4 \end{bmatrix} = \begin{bmatrix} 1 & -1 & -1 & 1 & 0 & 0 & 0 & 0 \\ 0 & 0 & 0 & 0 & 1 & -1 & -1 & 1 \\ \hdashline 1 & 1 & -1 & -1 & 0 & 0 & 0 & 0 \\ 0 & 0 & 0 & 0 & 1 & 1 & -1 & -1 \\ \hdashline 1 & 1 & 1 & 1 & 0 & 0 & 0 & 0 \\ 0 & 0 & 0 & 0 & 1 & 1 & 1 & 1 \\ \hdashline 1 & -1 & 1 & -1 & 0 & 0 & 0 & 0 \\ 0 & 0 & 0 & 0 & 1 & -1 & 1 & -1 \end{bmatrix} \tag{e}$$

Transforming the matrix $\mathbf{h}$ to $\mathbf{h}_T$, we have:

$$\mathbf{h}_T = \mathbf{T}\ \mathbf{h} = \begin{bmatrix} \mathbf{h}_1 & 0 \\ \mathbf{O} & \mathbf{h}_1 \end{bmatrix} \tag{f}$$

Here the rearrangement operator $\mathbf{T}$ must be:

$$\mathbf{T} = \begin{bmatrix} 1 & 0 & 0 & 0 & 0 & 0 & 0 & 0 \\ 0 & 0 & 1 & 0 & 0 & 0 & 0 & 0 \\ 0 & 0 & 0 & 0 & 1 & 0 & 0 & 0 \\ 0 & 0 & 0 & 0 & 0 & 0 & 1 & 0 \\ 0 & 1 & 0 & 0 & 0 & 0 & 0 & 0 \\ 0 & 0 & 0 & 1 & 0 & 0 & 0 & 0 \\ 0 & 0 & 0 & 0 & 0 & 1 & 0 & 0 \\ 0 & 0 & 0 & 0 & 0 & 0 & 0 & 1 \end{bmatrix} \tag{g}$$

And the submatrix $\mathbf{h}_1$ in Eq. (f) becomes:

$$\mathbf{h}_1 = \begin{bmatrix} 1 & -1 & -1 & 1 \\ 1 & 1 & -1 & -1 \\ 1 & 1 & 1 & 1 \\ 1 & -1 & 1 & -1 \end{bmatrix} \tag{h}$$

Inversion of $\mathbf{h}_1$ yields:

$$\mathbf{h}_1^{-1} = \frac{1}{4}\begin{bmatrix} 1 & 1 & 1 & 1 \\ -1 & 1 & 1 & -1 \\ -1 & -1 & 1 & 1 \\ 1 & -1 & 1 & -1 \end{bmatrix} = \frac{1}{4}\mathbf{h}_1^{\mathrm{T}} \tag{i}$$

Then the inverse of $\mathbf{h}$ can be evaluated as:

$$\mathbf{h}^{-1} = \mathbf{h}_T^{-1}\mathbf{T} = \frac{1}{4}\begin{bmatrix} 1 & 0 & 1 & 0 & 1 & 0 & 1 & 0 \\ -1 & 0 & 1 & 0 & 1 & 0 & -1 & 0 \\ -1 & 0 & -1 & 0 & 1 & 0 & 1 & 0 \\ 1 & 0 & -1 & 0 & 1 & 0 & -1 & 0 \\ 0 & 1 & 0 & 1 & 0 & 1 & 0 & 1 \\ 0 & -1 & 0 & 1 & 0 & 1 & 0 & -1 \\ 0 & -1 & 0 & -1 & 0 & 1 & 0 & 1 \\ 0 & 1 & 0 & -1 & 0 & 1 & 0 & -1 \end{bmatrix} = \frac{1}{4}\mathbf{h}^{\mathrm{T}} \tag{j}$$

Having $\mathbf{h}^{-1}$, we can now find the displacement shape functions in $\mathbf{f}$. Thus,

$$\mathbf{f} = \mathbf{g}\ \mathbf{h}^{-1} = \frac{1}{4}\left[\begin{array}{cc|cc|} (1-\xi)(1-\eta) & 0 & (1+\xi)(1-\eta) & 0 \\ 0 & (1-\xi)(1-\eta) & 0 & (1+\xi)(1-\eta) \end{array}\right.$$
$$\left.\begin{array}{cc|cc} (1+\xi)(1+\eta) & 0 & (1-\xi)(1+\eta) & 0 \\ 0 & (1+\xi)(1+\eta) & 0 & (1-\xi)(1+\eta) \end{array}\right] \tag{2.3-2a}$$

This expression may be stated more briefly as:

$$\mathbf{f}_i = \begin{bmatrix} f_i & 0 \\ 0 & f_i \end{bmatrix} \qquad (i = 1, 2, 3, 4) \tag{2.3-2b}$$

where

$$\begin{aligned} f_1 &= \tfrac{1}{4}(1-\xi)(1-\eta) \qquad & f_2 &= \tfrac{1}{4}(1+\xi)(1-\eta) \\ f_3 &= \tfrac{1}{4}(1+\xi)(1+\eta) \qquad & f_4 &= \tfrac{1}{4}(1-\xi)(1+\eta) \end{aligned} \tag{2.3-2c}$$

In order to calculate strains, we will need to differentiate the displacement shape functions with respect to x and y. However, those functions [see Eqs. (2.3-2c)] are expressed in terms of the dimensionless coordinates ξ and η. Therefore, we must use the chain rule for partial derivatives, as follows:

$$\frac{\partial}{\partial x} = \frac{\partial}{\partial \xi}\frac{\partial \xi}{\partial x} + \frac{\partial}{\partial \eta}\frac{\partial \eta}{\partial x} = \frac{1}{a}\frac{\partial}{\partial \xi} \qquad \frac{\partial}{\partial y} = \frac{\partial}{\partial \xi}\frac{\partial \xi}{\partial y} + \frac{\partial}{\partial \eta}\frac{\partial \eta}{\partial y} = \frac{1}{b}\frac{\partial}{\partial \eta} \tag{k}$$

The linear differential operator **d** becomes:

$$\mathbf{d} = \begin{bmatrix} \frac{\partial}{\partial x} & 0 \\ 0 & \frac{\partial}{\partial y} \\ \frac{\partial}{\partial y} & \frac{\partial}{\partial x} \end{bmatrix} = \begin{bmatrix} \frac{1}{a}\frac{\partial}{\partial \xi} & 0 \\ 0 & \frac{1}{b}\frac{\partial}{\partial \eta} \\ \frac{1}{b}\frac{\partial}{\partial \eta} & \frac{1}{a}\frac{\partial}{\partial \xi} \end{bmatrix} \tag{2.3-3}$$

Applying this operator to matrix **f**, we find that the matrix **B** is:

$$\mathbf{B} = \mathbf{d}\ \mathbf{f} = \frac{1}{4ab}\left[\begin{array}{cc|cc|cc|cc} -b(1-\eta) & 0 & b(1-\eta) & 0 & b(1+\eta) & 0 & -b(1+\eta) & 0 \\ 0 & -a(1-\xi) & 0 & -a(1+\xi) & 0 & a(1+\xi) & 0 & a(1-\xi) \\ -a(1-\xi) & -b(1-\eta) & -a(1+\xi) & b(1-\eta) & a(1+\xi) & b(1+\eta) & a(1-\xi) & -b(1+\eta) \end{array}\right] \tag{2.3-4a}$$

A more succinct version of this expression is:

$$\mathbf{B}_i = \begin{bmatrix} f_{i,x} & 0 \\ 0 & f_{i,y} \\ f_{i,y} & f_{i,x} \end{bmatrix} \qquad (i = 1, 2, 3, 4) \tag{2.3-4b}$$

The symbols $f_{i,x}$ and $f_{i,y}$ represent partial derivatives of f_i with respect to x and y, as follows:

$$\begin{aligned} f_{i,x} &= \frac{\partial f_i}{\partial x} = \frac{1}{a}\frac{\partial f_i}{\partial \xi} = \frac{1}{a} f_{i,\xi} \\ f_{i,y} &= \frac{\partial f_i}{\partial y} = \frac{1}{b}\frac{\partial f_i}{\partial \eta} = \frac{1}{b} f_{i,\eta} \end{aligned} \tag{2.3-4c}$$

If we assume that the material is orthotropic in the x and y directions, stress-strain relationships from Eqs. (2.1-20) and (2.1-22) can be written generally as:

$$\mathbf{E} = \begin{bmatrix} E_{11} & E_{12} & 0 \\ E_{12} & E_{22} & 0 \\ 0 & 0 & E_{33} \end{bmatrix} \tag{2.3-5a}$$

in which E_{11}, E_{12}, and so on, have definitions that were developed in Sec. 2.1. Then the stresses are obtained using:

$$\mathbf{E}\ \mathbf{B} = \frac{1}{4ab}\left[\begin{array}{cc|cc|cc|cc} -bE_{11}(1-\eta) & -aE_{12}(1-\xi) & bE_{11}(1-\eta) & -aE_{12}(1+\xi) & bE_{11}(1+\eta) & aE_{12}(1+\xi) & -bE_{11}(1+\eta) & aE_{12}(1-\xi) \\ -bE_{12}(1-\eta) & -aE_{22}(1-\xi) & bE_{12}(1-\eta) & -aE_{22}(1+\xi) & bE_{12}(1+\eta) & aE_{22}(1+\xi) & -bE_{12}(1+\eta) & aE_{22}(1-\xi) \\ -aE_{33}(1-\xi) & -bE_{33}(1-\eta) & -aE_{33}(1+\xi) & bE_{33}(1-\eta) & aE_{33}(1+\xi) & bE_{33}(1+\eta) & aE_{33}(1-\xi) & -bE_{33}(1+\eta) \end{array}\right] \tag{2.3-5b}$$

More briefly,

$$\mathbf{E}\ \mathbf{B}_i = \begin{bmatrix} E_{11}f_{i,x} & E_{12}f_{i,y} \\ E_{12}f_{i,x} & E_{22}f_{i,y} \\ E_{33}f_{i,y} & E_{33}f_{i,x} \end{bmatrix} \qquad (i = 1, 2, 3, 4) \tag{2.3-5c}$$

Element stiffnesses are conveniently evaluated using Eq. (1.4-8), as follows:

$$\mathbf{K}_c = \int_V \mathbf{B}_c^{\mathrm{T}}\mathbf{E}\ \mathbf{B}_c\, dV = abt \int_{-1}^{1}\int_{-1}^{1} \mathbf{B}_c^{\mathrm{T}}\mathbf{E}\ \mathbf{B}_c\, d\xi\, d\eta \tag{ℓ}$$

In this expression the matrix $\mathbf{B}_c$ is:

$$\mathbf{B}_c = \mathbf{d}\ \mathbf{g} = \frac{1}{ab}\begin{bmatrix} 0 & b & 0 & b\eta & 0 & 0 & 0 & 0 \\ 0 & 0 & 0 & 0 & 0 & 0 & a & a\xi \\ 0 & 0 & a & a\xi & 0 & b & 0 & b\eta \end{bmatrix} \tag{2.3-6}$$

Then we have:

$$\mathbf{B}_c^{\mathrm{T}}\mathbf{E}\ \mathbf{B}_c = \frac{1}{a^2b^2}\begin{bmatrix} 0 & & & & & & & \\ 0 & b^2E_{11} & & & & \text{Sym.} & & \\ 0 & 0 & a^2E_{33} & & & & & \\ 0 & b^2E_{11}\eta & a^2E_{33}\xi & \Phi_1 & & & & \\ 0 & 0 & 0 & 0 & 0 & & & \\ 0 & 0 & abE_{33} & abE_{33}\xi & 0 & b^2E_{33} & & \\ 0 & abE_{12} & 0 & abE_{12}\eta & 0 & 0 & a^2E_{22} & \\ 0 & abE_{12}\xi & abE_{33}\eta & \Phi_2 & 0 & b^2E_{33}\eta & a^2E_{22}\xi & \Phi_3 \end{bmatrix} \tag{m}$$

in which

$$\Phi_1 = b^2E_{11}\eta^2 + a^2E_{33}\xi^2$$
$$\Phi_2 = ab(E_{12} + E_{33})\xi\eta$$
$$\Phi_3 = a^2E_{22}\xi^2 + b^2E_{33}\eta^2$$

Integration of the product in Eq. (m) in accordance with Eq. (ℓ) yields:

$$\mathbf{K}_c = \frac{4t}{ab}\begin{bmatrix} 0 & & & & & & & \\ 0 & b^2E_{11} & & & & & & \\ 0 & 0 & a^2E_{33} & & \text{Sym.} & & & \\ 0 & 0 & 0 & C_1 & & & & \\ 0 & 0 & 0 & 0 & 0 & & & \\ 0 & 0 & abE_{33} & 0 & 0 & b^2E_{33} & & \\ 0 & abE_{12} & 0 & 0 & 0 & 0 & a^2E_{22} & \\ 0 & 0 & 0 & 0 & 0 & 0 & 0 & C_2 \end{bmatrix} \tag{n}$$

where

$$C_1 = \frac{b^2E_{11} + a^2E_{33}}{3} \qquad C_2 = \frac{a^2E_{22} + b^2E_{33}}{3}$$

Finally, from Eq. (1.4-9):

$$\mathbf{K} = \mathbf{h}^{-\mathrm{T}}\mathbf{K}_c\mathbf{h}^{-1} = \mathbf{K}_1 + \mathbf{K}_2 \tag{o}$$

Results of this calculation appear in Table 2.3.

TABLE 2.3 Stiffness Matrix for Bilinear Displacement Rectangle

$$\mathbf{K}_1 = \begin{bmatrix} 2s_1 & & & & & & \text{Sym.} & \\ s_3 & 2s_2 & & & & & & \\ -2s_1 & -s_3 & 2s_1 & & & & & \\ s_3 & s_2 & -s_3 & 2s_2 & & & & \\ -s_1 & -s_3 & s_1 & -s_3 & 2s_1 & & & \\ -s_3 & -s_2 & s_3 & -2s_2 & s_3 & 2s_2 & & \\ s_1 & s_3 & -s_1 & s_3 & -2s_1 & -s_3 & 2s_1 & \\ -s_3 & -2s_2 & s_3 & -s_2 & s_3 & s_2 & -s_3 & 2s_2 \end{bmatrix}$$

$$s_1 = \frac{tbE_{11}}{6a} \qquad s_2 = \frac{taE_{22}}{6b} \qquad s_3 = \frac{tE_{12}}{4}$$

$$\mathbf{K}_2 = \begin{bmatrix} 2s_4 & & & & & & \text{Sym.} & \\ s_6 & 2s_5 & & & & & & \\ s_4 & s_6 & 2s_4 & & & & & \\ -s_6 & -2s_5 & -s_6 & 2s_5 & & & & \\ -s_4 & -s_6 & -2s_4 & s_6 & 2s_4 & & & \\ -s_6 & -s_5 & -s_6 & s_5 & s_6 & 2s_5 & & \\ -2s_4 & -s_6 & -s_4 & s_6 & s_4 & s_6 & 2s_4 & \\ s_6 & s_5 & s_6 & -s_5 & -s_6 & -2s_5 & -s_6 & 2s_5 \end{bmatrix}$$

$$s_4 = \frac{taE_{33}}{6b} \qquad s_5 = \frac{tbE_{33}}{6a} \qquad s_6 = \frac{tE_{33}}{4}$$

A few examples will now be given for equivalent nodal loads on the bilinear displacement rectangle. Figure 2.9(b) shows a linearly varying load (force per unit length) applied to edge 1-2 in the y direction. For this case the vector of body forces is:

$$\mathbf{b} = \begin{bmatrix} 0 \\ \dfrac{b_{y1} + b_{y2}}{2} + \dfrac{(b_{y2} - b_{y1})\xi}{2} \end{bmatrix} \tag{p}$$

From Eq. (1.4-11), integration along edge $\eta = -1$ gives:

$$\begin{aligned} \mathbf{p}_{bc} &= \int_{-1}^{1} \mathbf{g}^{\mathrm{T}}\mathbf{b}a\,d\xi \\ &= a\left\{0, 0, 0, 0, b_{y1} + b_{y2}, \frac{b_{y2} - b_{y1}}{3}, -(b_{y1} + b_{y2}), -\frac{b_{y2} - b_{y1}}{3}\right\} \end{aligned} \tag{q}$$

Premultiplication of $\mathbf{p}_{bc}$ by $\mathbf{h}^{-\mathrm{T}}$ produces:

$$\mathbf{p}_b = \mathbf{h}^{-\mathrm{T}}\mathbf{p}_{bc} = \frac{a}{3}\{0, 2b_{y1} + b_{y2}, 0, b_{y1} + 2b_{y2}, 0, 0, 0, 0\} \tag{r}$$

Next, we consider a uniform gravity load b_g (force per unit volume), acting in the negative y direction. In this instance the body forces are:

$$\mathbf{b} = \{0, -b_g\} \tag{s}$$

Integration over the volume yields:

$$\mathbf{p}_{bc} = abt \int_{-1}^{1}\int_{-1}^{1} \mathbf{g}^{\mathrm{T}}\mathbf{b}\, d\xi\, d\eta = abt\{0, 0, 0, 0, -4b_g, 0, 0, 0\} \tag{t}$$

Then

$$\mathbf{p}_b = \mathbf{h}^{-\mathrm{T}}\mathbf{p}_{bc} = -abtb_g\{0, 1, 0, 1, 0, 1, 0, 1\} \tag{u}$$

For a uniform temperature increase ΔT, the vector of initial strains becomes:

$$\boldsymbol{\epsilon}_T = \{\alpha_1, \alpha_2, 0\}\,\Delta T \tag{v}$$

in which α_1 and α_2 are the temperature coefficients in the x and y directions. From Eq. (1.4-12), integration over the volume gives:

$$\mathbf{p}_{Tc} = abt \int_{-1}^{1}\int_{-1}^{1} \mathbf{B}_c^{\mathrm{T}}\mathbf{E}\; \boldsymbol{\epsilon}_T\, d\xi\, d\eta = 4t\,\Delta T\{0, d_1, 0, 0, 0, 0, d_2, 0\} \tag{w}$$

in which

$$d_1 = b(E_{11}\alpha_1 + E_{12}\alpha_2) \qquad d_2 = a(E_{12}\alpha_1 + E_{22}\alpha_2)$$

Premultiplying Eq. (w) by $\mathbf{h}^{-\mathrm{T}}$, we find:

$$\mathbf{p}_T = \mathbf{h}^{-\mathrm{T}}\mathbf{p}_{Tc} = t\,\Delta T\{-d_1, -d_2, d_1, -d_2, d_1, d_2, -d_1, d_2\} \tag{x}$$

This vector pertains only to the analysis of plane stress. For plane strain the results are more complicated.

As a last example, let the bilinear displacement rectangle have a uniform shearing prestrain γ_0. In this case the initial strain vector is:

$$\boldsymbol{\epsilon}_0 = \{0, 0, \gamma_0\} \tag{y}$$

Integration over the volume yields:

$$\mathbf{p}_{0c} = abt \int_{-1}^{1}\int_{-1}^{1} \mathbf{B}_c^{\mathrm{T}}\mathbf{E}\; \boldsymbol{\epsilon}_0\, d\xi\, d\eta = 4tE_{33}\gamma_0\{0, 0, a, 0, 0, b, 0, 0\} \tag{z}$$

Then

$$\mathbf{p}_0 = \mathbf{h}^{-\mathrm{T}}\mathbf{p}_{0c} = tE_{33}\gamma_0\{-a, -b, -a, b, a, b, a, -b\} \tag{a$'$}$$

To obtain better accuracy of the results, higher-order rectangular elements are also of interest. The *linear-cubic rectangle* has been formulated in detail by Barber (8), and only the highlights of that work are summarized here. For this purpose Fig. 2.10(a) shows a rectangular element of thickness t, having the generic displacements:

$$\mathbf{u} = \{u, v\} \tag{b$'$}$$

Because this element is intended primarily for analyzing shear walls in buildings, the nodal numbering sequence is left-to-right and top-to-bottom. At each node

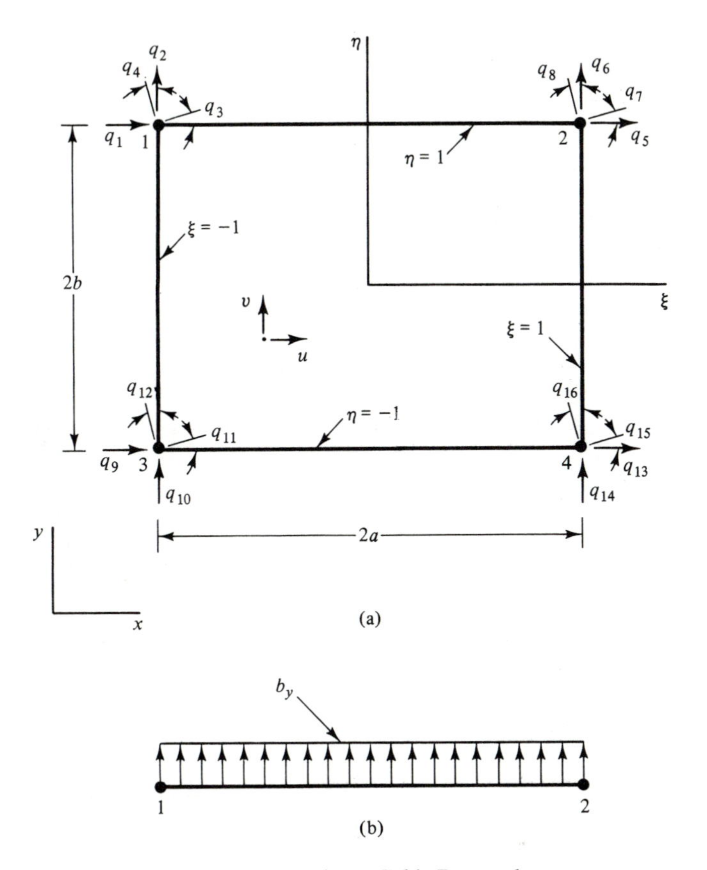

Figure 2.10 Linear-Cubic Rectangle

in the figure, four types of nodal displacements are indicated:

$$\mathbf{q}_i = \left\{u_i, v_i, \frac{\partial v_i}{\partial x}, -\frac{\partial u_i}{\partial y}\right\} \qquad (i = 1, 2, 3, 4) \tag{c'}$$

The first two terms in $\mathbf{q}_i$ are translations in the x and y directions, but the last two are partial derivatives representing slopes of horizontal and vertical lines. These translations and their rates of changes may be considered to be *nodal continuities* as well as displacements.

Displacement functions assumed for this element are:

$$\begin{aligned} u &= c_1 + c_2\xi + c_3\eta + c_4\xi\eta + c_5\eta^2 + c_6\xi\eta^2 + c_7\eta^3 + c_8\xi\eta^3 \\ v &= c_9 + c_{10}\xi + c_{11}\eta + c_{12}\xi\eta + c_{13}\xi^2 + c_{14}\xi^2\eta + c_{15}\xi^3 + c_{16}\xi^3\eta \end{aligned} \tag{d'}$$

The first of these functions is linear in ξ and cubic in η; the second is linear in η and cubic in ξ. For this reason the element is referred to as the linear-cubic rectangle. The cubic functions allow compatibility between any edge of the rectangle and a bordering flexural element. Displacement shape functions derived from Eq. (d′) may be expressed as:

$$f_i = \begin{bmatrix} f_{i1} & 0 & 0 & f_{i4} \\ 0 & f_{i2} & f_{i3} & 0 \end{bmatrix} \qquad (i = 1, 2, 3, 4) \tag{2.3-7}$$

where

$$i1 = 4i - 3 \qquad i2 = 4i - 2$$

$$i3 = 4i - 1 \qquad i4 = 4i$$

These functions are listed in Table 2.4, in which j corresponds to $i1$, $i2$, and so on at each node.

TABLE 2.4 Displacement Shape Functions and Derivatives for Linear-Cubic Rectangle

j	$8f_j$	$8f_{j,x}$	$8f_{j,y}$
1	$(1-\xi)(1+\eta)^2(2-\eta)$	$-(1+\eta)^2(2-\eta)/a$	$3(1-\xi)(1-\eta^2)/b$
2	$(1-\xi)^2(2+\xi)(1+\eta)$	$-3(1-\xi^2)(1+\eta)/a$	$(1-\xi)^2(2+\xi)/b$
3	$(1-\xi)^2(1+\xi)(1+\eta)a$	$-(1-\xi)(1+3\xi)(1+\eta)$	$(1-\xi)^2(1+\xi)a/b$
4	$(1-\xi)(1+\eta)^2(1-\eta)b$	$-(1+\eta)^2(1-\eta)b/a$	$(1-\xi)(1+\eta)(1-3\eta)$
5	$(1+\xi)(1+\eta)^2(2-\eta)$	$(1+\eta)^2(2-\eta)/a$	$3(1+\xi)(1-\eta^2)/b$
6	$(1+\xi)^2(2-\xi)(1+\eta)$	$3(1-\xi^2)(1+\eta)/a$	$(1+\xi)^2(2-\xi)/b$
7	$-(1+\xi)^2(1-\xi)(1+\eta)a$	$-(1+\xi)(1-3\xi)(1+\eta)$	$-(1+\xi)^2(1-\xi)a/b$
8	$(1+\xi)(1+\eta)^2(1-\eta)b$	$(1+\eta)^2(1-\eta)b/a$	$(1+\xi)(1+\eta)(1-3\eta)$
9	$(1-\xi)(1-\eta)^2(2+\eta)$	$-(1-\eta)^2(2+\eta)/a$	$-3(1-\xi)(1-\eta^2)/b$
10	$(1-\xi)^2(2+\xi)(1-\eta)$	$-3(1-\xi^2)(1-\eta)/a$	$-(1-\xi)^2(2+\xi)/b$
11	$(1-\xi)^2(1+\xi)(1-\eta)a$	$-(1-\xi)(1+3\xi)(1-\eta)$	$-(1-\xi)^2(1+\xi)a/b$
12	$-(1-\xi)(1-\eta)^2(1+\eta)b$	$(1-\eta)^2(1+\eta)b/a$	$(1-\xi)(1-\eta)(1+3\eta)$
13	$(1+\xi)(1-\eta)^2(2+\eta)$	$(1-\eta)^2(2+\eta)/a$	$-3(1+\xi)(1-\eta^2)/b$
14	$(1+\xi)^2(2-\xi)(1-\eta)$	$3(1-\xi^2)(1-\eta)/a$	$-(1+\xi)^2(2-\xi)/b$
15	$-(1+\xi)^2(1-\xi)(1-\eta)a$	$(1+\xi)(1-3\xi)(1-\eta)$	$(1+\xi)^2(1-\xi)a/b$
16	$-(1+\xi)(1-\eta)^2(1+\eta)b$	$-(1-\eta)^2(1+\eta)b/a$	$(1+\xi)(1-\eta)(1+3\eta)$

Also given in the last two columns of Table 2.4 are derivatives of the displacement shape functions with respect to x and y. These derivatives are the *strain functions* required to fill matrix **B**. The ith part of **B** takes the following form:

$$\mathbf{B}_i = \begin{bmatrix} f_{i1,x} & 0 & 0 & f_{i4,x} \\ 0 & f_{i2,y} & f_{i3,y} & 0 \\ f_{i1,y} & f_{i2,x} & f_{i3,x} & f_{i4,y} \end{bmatrix} \qquad (i = 1, 2, 3, 4) \tag{2.3-8}$$

in which

$$f_{i1,x} = \frac{1}{a} f_{i1,\xi} \qquad f_{i1,y} = \frac{1}{b} f_{i1,\eta} \qquad \text{(and so on)}$$

The 16×16 stiffness matrix for the linear-cubic rectangle is given in Table 2.5, assuming an isotropic material in plane stress. Terms in this table were derived as described before.

TABLE 2.5 Stiffness Matrix for Linear-Cubic Rectangle

$$
\mathbf{K} = \begin{bmatrix}
\alpha_1 & & & & & & & & & & & & & & & \\
-\alpha_2 & \beta_3 & & & & & & & & \text{Sym.} & & & & & & \\
\alpha_3 & \beta_4 & \theta_3 & & & & & & & & & & & & & \\
\alpha_4 & \beta_5 & -\theta_4 & \theta_8 & & & & & & & & & & & & \\
\alpha_5 & -\alpha_6 & -\alpha_3 & \alpha_7 & \alpha_1 & & & & & & & & & & & \\
\alpha_6 & \beta_6 & -\beta_7 & -\beta_8 & \alpha_2 & \beta_3 & & & & & & & & & & \\
-\alpha_3 & \beta_7 & -\theta_5 & \theta_4 & \alpha_3 & -\beta_4 & \theta_3 & & & & & & & & & \\
\alpha_7 & \beta_8 & \theta_4 & \theta_9 & \alpha_4 & -\beta_5 & -\theta_4 & \theta_8 & & & & & & & & \\
\alpha_8 & \alpha_6 & -\alpha_9 & -\alpha_0 & -\beta_1 & -\alpha_2 & \alpha_9 & -\beta_2 & \alpha_1 & & & & & & & \\
-\alpha_6 & \beta_9 & \beta_0 & -\beta_5 & -\alpha_2 & -\theta_1 & \theta_2 & -\beta_8 & \alpha_2 & \beta_3 & & & & & & \\
\alpha_9 & \beta_0 & \theta_6 & \theta_4 & -\alpha_9 & -\theta_2 & \theta_7 & -\theta_4 & -\alpha_3 & \beta_4 & \theta_3 & & & & & \\
\alpha_0 & -\beta_5 & \theta_4 & -\theta_0 & \beta_2 & \beta_8 & -\theta_4 & \phi_1 & -\alpha_4 & \beta_5 & -\theta_4 & \theta_8 & & & & \\
-\beta_1 & \alpha_2 & \alpha_9 & -\beta_2 & \alpha_8 & -\alpha_6 & -\alpha_9 & -\alpha_0 & \alpha_5 & \alpha_6 & \alpha_3 & -\alpha_7 & \alpha_1 & & & \\
\alpha_2 & -\theta_1 & -\theta_2 & \beta_8 & \alpha_6 & \beta_9 & -\beta_0 & \beta_5 & -\alpha_6 & \beta_6 & -\beta_7 & -\beta_8 & -\alpha_2 & \beta_3 & & \\
-\alpha_9 & \theta_2 & \theta_7 & -\theta_4 & \alpha_9 & -\beta_0 & \theta_6 & \theta_4 & \alpha_3 & \beta_7 & -\theta_5 & \theta_4 & -\alpha_3 & -\beta_4 & \theta_3 & \\
\beta_2 & -\beta_8 & -\theta_4 & \phi_1 & \alpha_0 & \beta_5 & \theta_4 & -\theta_0 & -\alpha_7 & \beta_8 & \theta_4 & \theta_9 & -\alpha_4 & -\beta_5 & -\theta_4 & \theta_8
\end{bmatrix}
$$

$\alpha_1 = \frac{13b}{35a} + \frac{a\mu}{5b}$ $\quad \alpha_2 = \frac{1+\nu}{8}$ $\quad \alpha_3 = \frac{a(1-3\nu)}{48}$ $\quad \alpha_4 = \frac{11b^2}{210a} + \frac{a\mu}{60}$ $\quad \alpha_5 = -\frac{13b}{35a} + \frac{a\mu}{10b}$

$\alpha_6 = \frac{1-3\nu}{8}$ $\quad \alpha_7 = -\frac{11b^2}{210a} + \frac{a\mu}{120}$ $\quad \alpha_8 = \frac{9b}{70a} - \frac{a\mu}{5b}$ $\quad \alpha_9 = \frac{a(1+\nu)}{48}$ $\quad \alpha_0 = -\frac{13b^2}{420a} + \frac{a\mu}{60}$

$\beta_1 = \frac{9b}{70a} + \frac{a\mu}{10b}$ $\quad \beta_2 = \frac{13b^2}{420a} + \frac{a\mu}{120}$ $\quad \beta_3 = \frac{13a}{35b} + \frac{b\mu}{5a}$ $\quad \beta_4 = \frac{11a^2}{210b} + \frac{b\mu}{60}$ $\quad \beta_5 = \frac{b(1-3\nu)}{48}$

$\beta_6 = \frac{9a}{70b} - \frac{b\mu}{5a}$ $\quad \beta_7 = -\frac{13a^2}{420b} + \frac{b\mu}{60}$ $\quad \beta_8 = \frac{b(1+\nu)}{48}$ $\quad \beta_9 = -\frac{13a}{35b} + \frac{b\mu}{10a}$ $\quad \beta_0 = -\frac{11a^2}{210b} + \frac{b\mu}{120}$

$\theta_1 = \frac{9a}{70b} + \frac{b\mu}{10a}$ $\quad \theta_2 = \frac{13a^2}{420b} + \frac{b\mu}{120}$ $\quad \theta_3 = \frac{a^3}{105b} + \frac{ab\mu}{45}$ $\quad \theta_4 = \frac{ab(1+\nu)}{288}$ $\quad \theta_5 = \frac{a^3}{140b} + \frac{ab\mu}{180}$

$\theta_6 = -\frac{a^3}{105b} + \frac{ab\mu}{90}$ $\quad \theta_7 = \frac{a^3}{140b} - \frac{ab\mu}{360}$ $\quad \theta_8 = \frac{b^3}{105a} + \frac{ab\mu}{45}$ $\quad \theta_9 = -\frac{b^3}{105a} + \frac{ab\mu}{90}$ $\quad \theta_0 = \frac{b^3}{140a} + \frac{ab\mu}{180}$

$\phi_1 = \frac{b^3}{140a} - \frac{ab\mu}{360}$ $\quad \mu = 1 - \nu$ All terms to be multiplied by $Et/(1-\nu^2)$.

As an example of equivalent nodal loads, consider those due to a uniformly distributed force (per unit length) applied in the y direction on edge 1-2 [see Fig. 2.10(b)]. In this case the vector $\mathbf{p}_b$ is found to be:

$$\mathbf{p}_b = \frac{b_y a}{3}\{0, 3, a, 0, 0, 3, -a, 0, \ldots, 0\} \qquad \text{(e')}$$

2.4 QUADRILATERAL ELEMENTS

The geometry of the quadrilateral is nearly as good as that of the triangle for modeling irregular boundary conditions. Moreover, if only the corners are taken to be nodes, the quadrilateral usually has more nodal displacements than the triangle. Hence, the assumed displacement functions can be of higher order. For these reasons quadrilaterals will tend to be favored over triangles in this book.

If desired, we can assemble quadrilaterals from constant strain triangles (3). For example, Fig. 2.11(a) shows a quadrilateral composed of two constant strain triangles T1 and T2. The stiffness matrix for the quadrilateral may be obtained by direct addition of those for the triangles, as follows:

$$\mathbf{K}_Q = \mathbf{K}_{T1} + \mathbf{K}_{T2} \qquad (2.4\text{-}1)$$

Of course, the 6×6 stiffness matrices for the triangles must be expanded to size 8×8 for the addition to be conformable. If two such quadrilaterals are formed from the same four nodes but different diagonals [see Fig. 2.11(b)], the

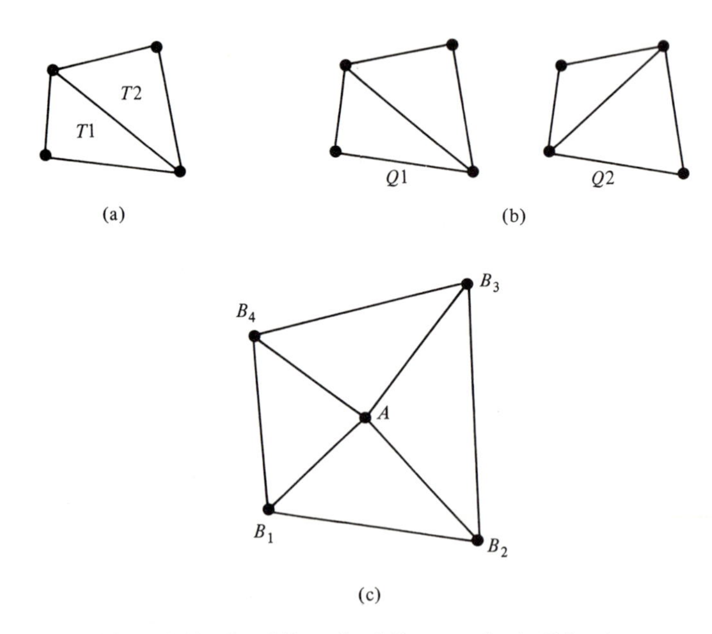

Figure 2.11 Quadrilaterals of Constant Strain Triangles

results can be averaged. Thus,

$$\mathbf{K}_Q = \tfrac{1}{2}(\mathbf{K}_{Q1} + \mathbf{K}_{Q2}) \tag{2.4-2}$$

Similar statements can also be made about equivalent nodal loads.

When four constant strain triangles are assembled, the resulting quadrilateral has an interior node, as illustrated in Fig. 2.11(c). If that node is labeled type A and the exterior nodes are labeled type B, the assembled equilibrium equations may be written as:

$$\begin{bmatrix} \mathbf{K}_{AA} & \mathbf{K}_{AB} \\ \mathbf{K}_{BA} & \mathbf{K}_{BB} \end{bmatrix} \begin{bmatrix} \mathbf{q}_A \\ \mathbf{q}_B \end{bmatrix} = \begin{bmatrix} \mathbf{p}_A \\ \mathbf{p}_B \end{bmatrix} \tag{a}$$

Rewriting this matrix equation in two parts, we have:

$$\mathbf{K}_{AA}\mathbf{q}_A + \mathbf{K}_{AB}\mathbf{q}_B = \mathbf{p}_A \tag{b}$$

$$\mathbf{K}_{BA}\mathbf{q}_A + \mathbf{K}_{BB}\mathbf{q}_B = \mathbf{p}_B \tag{c}$$

Now we solve for the vector $\mathbf{q}_A$ in Eq. (b) and substitute it into Eq. (c) to obtain:

$$\mathbf{K}^*_{BB}\mathbf{q}_B = \mathbf{p}^*_B \tag{2.4-3}$$

In this expression the collected coefficients of $\mathbf{q}_B$ are:

$$\mathbf{K}^*_{BB} = \mathbf{K}_{BB} - \mathbf{K}_{BA}\mathbf{K}^{-1}_{AA}\mathbf{K}_{AB} \tag{2.4-4}$$

and the modified nodal loads of type **B** are:

$$\mathbf{p}^*_B = \mathbf{p}_B - \mathbf{K}_{BA}\mathbf{K}^{-1}_{AA}\mathbf{p}_A \tag{2.4-5}$$

Because the matrices in Eq. (2.4-3) are smaller than those in Eq. (a), the reduction process described above is called *matrix condensation.* This technique serves many useful purposes in the analysis of solids and structures by finite elements.

It is also possible to assemble quadrilaterals from linear strain triangles (5). Figure 2.12(a) shows a quadrilateral element composed of four linear strain triangles. When displacements at interior nodes (of type A) are eliminated

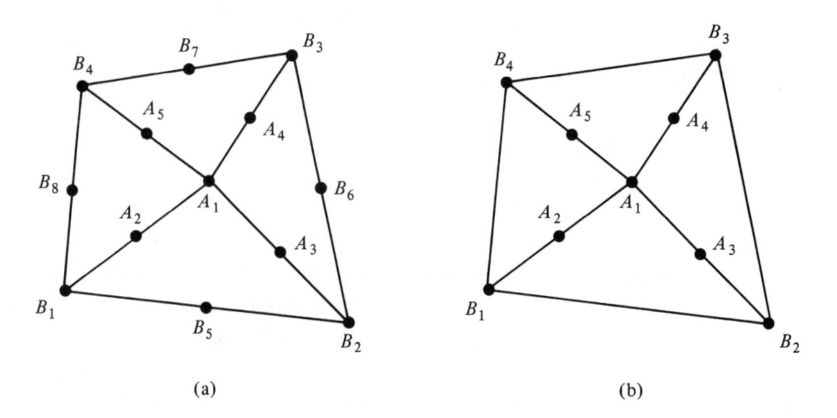

Figure 2.12 Quadrilaterals of Linear Strain Triangles

by matrix condensation, there remain eight exterior nodes (of type B). Consequently, the resulting element has a stiffness matrix of size 16×16. However, a modified version of size 8×8 can be created (9), as indicated in Fig. 2.12(b). The latter element is formed by introducing constraints to make displacements at midedge nodes equal to the averages of those at corner nodes. This restriction halves the independent type B displacements and produces constant strains at exterior edges. Then the matrix condensation may proceed as described before.

Although it can be advantageous to form quadrilaterals from triangles, an independent formulation is usually preferred. In the next chapter the bilinear displacement quadrilateral similar to the corresponding rectangle is developed and then extended to higher-order elements.

2.5 PROGRAM PSCST AND APPLICATIONS

In this section we discuss a computer program named PSCST, which analyzes two-dimensional continua in plane stress or plane strain. For this purpose it is assumed that a given continuum has been discretized using constant strain triangles (see Sec. 2.2). All such triangles in the analytical model have the same thickness t, and the material is taken to be linearly elastic, homogeneous, and isotropic.

A flow chart for the main program and the subprograms called by Program PSCST appears in Appendix D. Notes in the right-hand margin are intended to help the reader interpret the logic of the flow chart. Items enclosed in double boxes are subprograms called by the main program. For other static analysis programs, the logic in the main program need not change. Note that the steps in Subprograms SKYFAC and SKYSOL are shown in Appendix C. However, the remaining subprograms (SDATA, STIFF, LDATA, and RESUL) would be different for every program, as indicated by the asterisks.

Table 2.6 is a list of program notation, and Table 2.7 shows how data must be prepared for Program PSCST. The upper part of the latter table contains structural data that are read and printed by *Subprogram SDATA*. Then *Subprogram STIFF* generates the "skyline" in the upper triangle of the nodal stiffness matrix as a singly subscripted variable by assessing contributions from element stiffnesses. Within the same subprogram, this stiffness matrix is modified to avoid the necessity of rearrangement (due to restraints) before it is submitted to Subprogram SKYFAC for factorization. *Subprogram LDATA* reads, prints, and processes the load and supplementary data listed in the lower part of Table 2.7. Next, Subprogram SKYSOL is used to solve the nodal equilibrium equations for the unknown displacements. Finally, *Subprogram RESUL* calculates and prints the results, which include nodal displacements, element stresses, and restraint reactions.

Figure 2.13 shows the basic loading condition for a typical node k in the analytical model, consisting of forces $(A_N)_{2k-1}$ and $(A_N)_{2k}$ applied in the x and y

TABLE 2.6 Program Notation

Identifier	Definition	Identifier	Definition
AN( )	Actions at nodes	NED	Number of elements with displacements specified
AR( )	Support reactions	NEL	Number of elements with line loads
BL1,BL2, . . .	Intensities of line loads	NEP	Number of elements with prestrain
BS1,BS2, . . .	Intensities of surface loads	NES	Number of elements with surface loads
BV1,BV2, . . .	Intensities of volume loads	NET	Number of elements with temperature strain
DN( )	Displacements at nodes	NEV	Number of elements with volume loads
DR(1),DR(2), . . .	Displacements of restraints	NLN	Number of loaded nodes
E	Elasticity modulus	NLS	Number of loading systems
E1,E2, . . .	Elasticity constants	NN	Number of nodes
I,J,K	Indexes	NND	Number of nodal displacements
ID( )	Displacement indexes	NNR	Number of nodal restraints
IN( ), . . .	Indexes for nodes of element	NRL( )	Nodal restraint list
IR,IC	Row and column indexes	NRN	Number of restrained nodes
IPS	Indicator for plane stress or plane	NS	Number of terms in the skyline of matrix SN
	strain	PR	Poisson's ratio
ISN	Structure number	PS1,PS2, . . .	Prestrains
LN	Loading number	SE(,)	Element stiffness matrix
NC( )	Number of nonzero terms in the columns of the	SN( )	Nodal stiffness matrix
	skyline of matrix SN	SX,SY, . . .	Stresses
ND( )	Indexes for diagonals in matrix SN	T	Thickness
NDF	Number of degrees of freedom	TS	Temperature strain (TS = $\alpha \Delta T$)
NE	Number of elements	X( ),Y( ),Z( )	Nodal coordinates

TABLE 2.7 Preparation of Data for Program PSCST

Data	No. of Lines	Items on Data Lines
STRUCTURAL DATA		
(a) Problem identification	1	Descriptive title
(b) Structural parameters	1	NN,NE,NRN,NLS,IPS,E,PR,T
(c) Nodal coordinates	NN	J,X(J),Y(J)
(d) Element information*	NE	I,IN(I),JN(I),KN(I)
(e) Nodal restraint list	NRN	K,NRL(2K−1),NRL(2K)
LOAD DATA		
(a) Load parameters	1	NLN,NEL,NEV,NET,NEP,NED
(b) Nodal loads	NLN	K,AN(2K−1),AN(2K)
(c) Line loads†	NEL	I,J,BL1,BL2,BL3,BL4
(d) Volume loads†	NEV	I,BV1,BV2
(e) Temperature strain†	NET	I,TS
(f) Prestrains†	NEP	I,PS1,PS2,PS3
(g) Support displacements†	NED	I,DR(1),DR(2), . . . ,DR(6)

*Nodes must be in counterclockwise sequence.
†Optional supplementary influences.

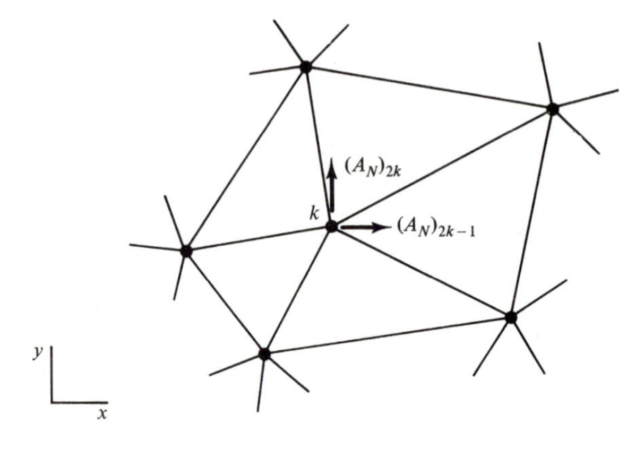

Figure 2.13 Nodal Loads for Element CST

directions, respectively. In addition to this direct loading, however, Table 2.7 contains various optional supplementary influences that can be accommodated by the program. For example, if linearly varying line loads are applied in the x and y directions along edge ij [see Fig. 2.14(a) and (b)], item (c) in the load data (see Table 2.7) gives the necessary parameters. In this case the components of force (per unit inclined length) vary from b_{L1} to b_{L2} in the x direction and from b_{L3} to b_{L4} in the y direction. Item (d) in the load data of Table 2.7 indicates also the possibility of volume loading (force per unit volume). Here the notation implies that the ith element may have a uniformly distributed force b_{V1} applied

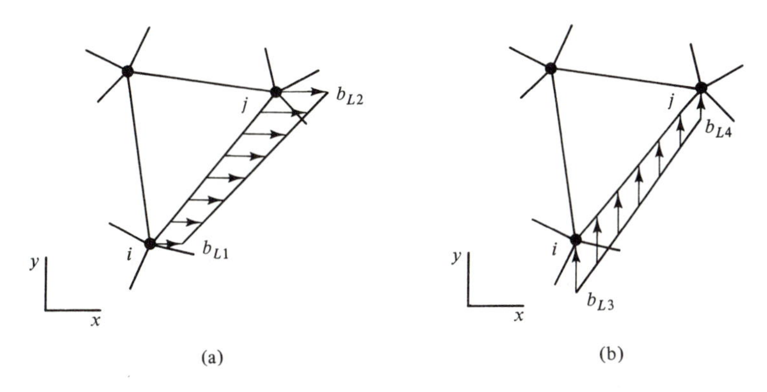

Figure 2.14 Line Loads for Element CST

in the x direction and a uniformly distributed force b_{Y2} applied in the y direction. Next, item (e) shows the possibility of handling a uniform temperature strain TS $= \alpha \Delta T$ in the ith element. In addition, three types of prestrains can be specified, as implied by item (f). That is, we may have: PS1 = uniform normal strain in the x direction, PS2 = uniform normal strain in the y direction, and PS3 = uniform shearing strain. Finally, item (g) covers the possibility of support displacements at the nodes of any element. The information needed is the element number, followed by DR(1) through DR(6), denoting the six possible restraint displacements, taken in x, y sequence at each of the nodes i, j, and k.

Inside the Subprogram LDATA the optional supplementary influences mentioned above are all converted to equivalent nodal loads. These equivalent loads are then added to the actual nodal loads to produce a vector of combined loads to be used in the solution. Of course, the initial strains due to temperature changes and specified prestrains must be included for the final calculation of element stresses in Subprogram RESUL.

Several applications of Program PSCST are described in the following paragraphs. These examples will be analyzed again at the end of the next chapter, using quadrilateral elements.

Example 1

To demonstrate various features of Program PSCST, we first consider a square plate composed of four constant strain triangles, as indicated in Fig. 2.15(a). With no particular system of units in mind, assume the following values for the structural parameters:

$$L = 1 \qquad E = 1 \qquad \nu = 0.3 \qquad t = 0.1$$

If forces are applied in the plane of the plate, we have a plane stress problem; so the parameter IPS is set equal to zero.

Figure 2.15(b) shows a variety of loading conditions and other influences. Loading 1 consists of two nodal forces, each of value 0.5, applied in the x direction at points 2 and 4. Loading 2 has a uniformly distributed force (per unit length) of intensity 1.0, applied in the x direction on edge 2-4 of triangle number 2. On the other hand, loading

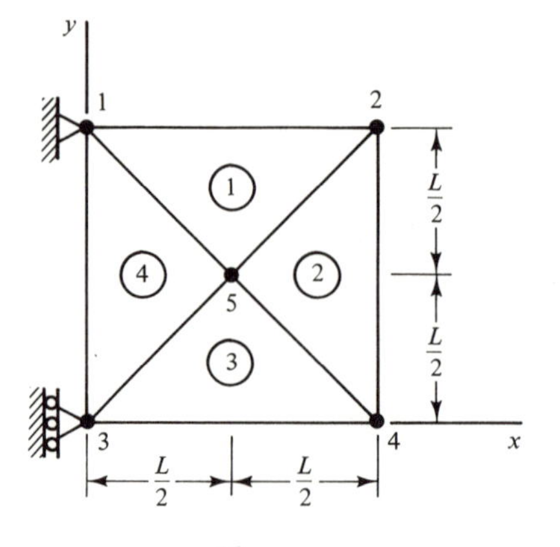

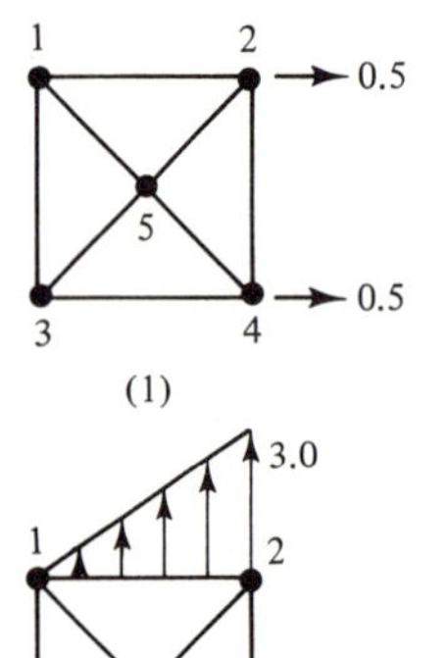

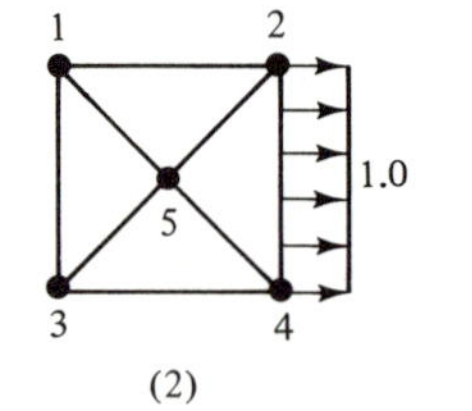

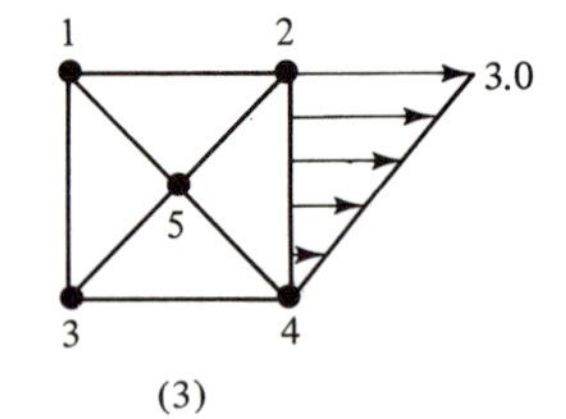

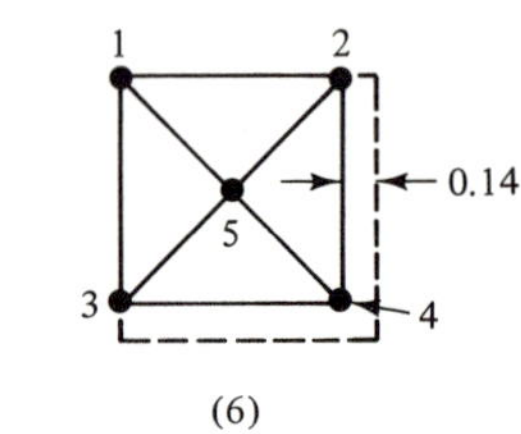

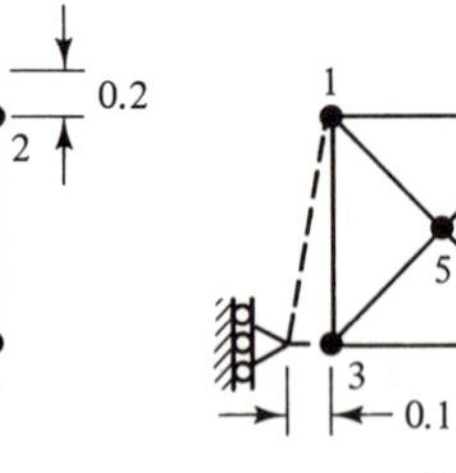

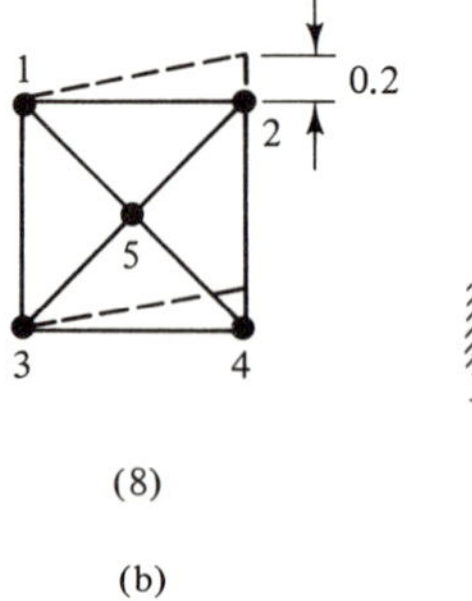

Figure 2.15 Example 1: (a) CST Network (b) Loadings

3 involves a triangular distribution of force (per unit length) of maximum intensity 3.0 on the same edge and in the same direction. Loading 4 shows the same triangular distribution of force (per unit length), but in the y direction on edge 1-2 of triangle 1. In loading 5 we have a uniformly distributed force (per unit volume) of intensity 3.0, applied in the x direction throughout the plate. Loading 6 involves a uniform increase of temperature ΔT, causing a uniform initial normal strain $\alpha \Delta T$ of 0.14 in each element. Similarly, loading 7 shows a uniform normal prestrain of 0.28 in all elements for both the x and y directions. In addition, loading 8 is due to a uniform shearing prestrain of 0.2 in all elements. Finally, loading 9 is caused by a support displacement of 0.1 in the negative x direction at node 3, which directly affects elements 3 and 4.

Table 2.8 shows a line print of the computer output for this example. Appearing on the first page of the table is an "echo" print of the structural data read by the computer. This is followed by the results of calculations for each of the nine loading cases. Inspection of these results shows that the program is operating correctly for a plate subjected to plane stress. Moreover, a similar run with the parameters $t = 1$ and IPS $= 1$ would show that plane strain analyses are performed equally well.

Example 2

Figure 2.16(a) shows a thin, square plate with a circular hole at its center. The plate is subjected to uniformly distributed edge loadings b_x (force per unit length) in the x direction. Thus, we have a condition of plane stress that can be analyzed using Program PSCST with the parameter IPS set equal to zero. Assume that the material is aluminum, and let the values of other parameters be as follows:

$$L = 8 \text{ in.} \qquad d = 1 \text{ in.} \qquad t = 0.1 \text{ in.}$$

$$b_x = 1 \text{ k/in.} \qquad E = 1 \times 10^4 \text{ k/in.}^2 \qquad \nu = 0.3$$

where the units are U.S. standard.

The primary objective of this example is to construct a graph of the variation of the normal stress σ_x along line AB in Fig. 2.16(a). We know from the theory of elasticity (1) that stress concentrations occur in the vicinity of a hole. Therefore, it is necessary to use a network of elements that is refined near the hole, as illustrated in Fig. 2.16(b). Because the problem is doubly symmetric, we need only analyze one quarter of the plate. For this purpose it is necessary to use roller restraints at nodes on axes x and y to prevent translations across the planes of symmetry.

The upper part of Fig. 2.17 shows a plot of the ratio $\sigma_x t/b_x$ along line AB, as obtained using Program PSCST. Note that the ratio approaches the value of 3.0 at the edge of the hole, which is the known theoretical solution (1). Also given in the lower part of Fig. 2.17 is a graph of the ratio $\sigma_y t/b_x$ along line CD. This ratio approaches the value of -1.0 at the edge of the hole, as it should. For the graphs in Fig. 2.17, the plotted points are located at the midedges of triangles bordering lines AB and CD.

Example 3

The concrete culvert in Fig. 2.18(a) represents a plane-strain problem. Its geometry consists of half a hexagon with a semicircular opening. A uniformly distributed loading b_y (force per unit length) is applied to the top edge in the negative y direction. Values

TABLE 2.8 Computer Output for Example 1

```
PROGRAM PSCST

***  EXAMPLE 1:  FOUR ELEMENT SQUARE PLATE IN PLANE STRESS  ***

STRUCTURAL PARAMETERS
   NN   NE  NRN  NLS  IPS           E            PR            T
    5    4    2    9    0  1.0000D 00  3.0000D-01  1.0000D-01

PLANE STRESS

COORDINATES OF NODES
 NODE           X           Y
    1  0.0000D-01  1.0000D 00
    2  1.0000D 00  1.0000D 00
    3  0.0000D-01  0.0000D-01
    4  1.0000D 00  0.0000D-01
    5  5.0000D-01  5.0000D-01

ELEMENT INFORMATION
ELEM.    I    J    K
    1    2    1    5
    2    4    2    5
    3    4    5    3
    4    5    1    3

NODAL RESTRAINTS
 NODE   R1   R2
    1    1    1
    3    1    0

NUMBER OF DEGREES OF FREEDOM =    7
NUMBER OF NODAL RESTRAINTS   =    3
NUMBER OF TERMS IN SN        =   33

**********  LOADING NUMBER    1  **********
  NLN  NEL  NEV  NET  NEP  NED
    2    0    0    0    0    0

ACTIONS AT NODES
 NODE         AN1         AN2
    2  5.0000D-01  0.0000D-01
    4  5.0000D-01  0.0000D-01

NODAL DISPLACEMENTS
 NODE         DN1         DN2
    1  0.0000D-01  0.0000D-01
    2  1.0000D 01 -3.5527D-15
    3 -0.0000D-01  3.0000D 00
    4  1.0000D 01  3.0000D 00
    5  5.0000D 00  1.5000D 00

ELEMENT STRESSES
ELEM.          SX          SY         SXY          SZ
    1  1.0000D 01  2.4401D-15 -1.1956D-15  0.0000D-01
    2  1.0000D 01  4.8801D-16 -1.0248D-15  0.0000D-01
    3  1.0000D 01  0.0000D-01 -8.5402D-16  0.0000D-01
    4  1.0000D 01  2.9281D-15 -1.1956D-15  0.0000D-01

SUPPORT REACTIONS
 NODE         AR1         AR2
```

TABLE 2.8 (cont.)

```
    1 -5.0000D-01  1.7521D-16
    3 -5.0000D-01  0.0000D-01

**********  LOADING NUMBER    2  **********
  NLN  NEL  NEV  NET  NEP  NED
    0    1    0    0    0    0

LINE LOADS
    I    J          BL1          BL2          BL3          BL4
    2    4  1.0000D 00  1.0000D 00  0.0000D-01  0.0000D-01

NODAL DISPLACEMENTS
 NODE          DN1          DN2
    1  0.0000D-01  0.0000D-01
    2  1.0000D 01 -3.1086D-15
    3 -0.0000D-01  3.0000D 00
    4  1.0000D 01  3.0000D 00
    5  5.0000D 00  1.5000D 00

ELEMENT STRESSES
ELEM.           SX           SY          SXY           SZ
    1  1.0000D 01  1.9520D-15 -1.1956D-15  0.0000D-01
    2  1.0000D 01  4.8801D-16 -1.0248D-15  0.0000D-01
    3  1.0000D 01  0.0000D-01 -6.8321D-16  0.0000D-01
    4  1.0000D 01  2.4401D-15 -1.0248D-15  0.0000D-01

SUPPORT REACTIONS
 NODE          AR1          AR2
    1 -5.0000D-01  1.5699D-16
    3 -5.0000D-01  0.0000D-01

**********  LOADING NUMBER    3  **********
  NLN  NEL  NEV  NET  NEP  NED
    0    1    0    0    0    0

LINE LOADS
    I    J          BL1          BL2          BL3          BL4
    2    4  3.0000D 00  0.0000D-01  0.0000D-01  0.0000D-01

NODAL DISPLACEMENTS
 NODE          DN1          DN2
    1  0.0000D-01  0.0000D-01
    2  2.3775D 01 -8.7750D 00
    3 -0.0000D-01  4.5000D 00
    4  6.2250D 00 -4.2750D 00
    5  7.5000D 00 -2.5000D-02

ELEMENT STRESSES
ELEM.           SX           SY          SXY           SZ
    1  2.3250D 01 -1.7500D 00 -1.5372D-15  0.0000D-01
    2  1.5000D 01 -4.8801D-16  1.7500D 00  0.0000D-01
    3  6.7500D 00  1.7500D 00  3.4161D-16  0.0000D-01
    4  1.5000D 01 -1.4640D-15 -1.7500D 00  0.0000D-01

SUPPORT REACTIONS
 NODE          AR1          AR2
    1 -1.0000D 00 -1.3878D-16
    3 -5.0000D-01  0.0000D-01

**********  LOADING NUMBER    4  **********
```

TABLE 2.8 (cont.)

```
  NLN  NEL  NEV  NET  NEP  NED
    0    1    0    0    0    0

LINE LOADS
    I    J          BL1           BL2           BL3           BL4
    1    2  0.00000D-01   0.00000D-01   0.00000D-01   3.00000D 00

NODAL DISPLACEMENTS
 NODE          DN1          DN2
    1  0.00000D-01  0.00000D-01
    2 -1.75500D 01  6.11000D 01
    3 -0.00000D-01  1.75500D 01
    4  1.75500D 01  4.35500D 01
    5  4.22500D 00  2.63250D 01

ELEMENT STRESSES
ELEM.          SX           SY          SXY           SZ
    1 -1.65000D 01  3.50000D 00  1.35000D 01  0.00000D-01
    2 -3.50000D 00  1.65000D 01  6.50000D 00  0.00000D-01
    3  1.65000D 01 -3.50000D 00  6.50000D 00  0.00000D-01
    4  3.50000D 00 -1.65000D 01  1.35000D 01  0.00000D-01

SUPPORT REACTIONS
 NODE          AR1          AR2
    1  1.00000D 00 -1.50000D 00
    3 -1.00000D 00  0.00000D-01

**********  LOADING NUMBER    5  **********
  NLN  NEL  NEV  NET  NEP  NED
    0    0    4    0    0    0

VOLUME LOADS
ELEM.          BV1          BV2
    1  3.00000D 00  0.00000D-01
    2  3.00000D 00  0.00000D-01
    3  3.00000D 00  0.00000D-01
    4  3.00000D 00  0.00000D-01

NODAL DISPLACEMENTS
 NODE          DN1          DN2
    1  0.00000D-01  0.00000D-01
    2  1.50000D 00  4.22500D-01
    3 -0.00000D-01  8.72500D-01
    4  1.50000D 00  4.50000D-01
    5  1.18870D 00  4.36250D-01

ELEMENT STRESSES
ELEM.          SX           SY          SXY           SZ
    1  1.50000D 00  2.74510D-16 -1.75000D-01  0.00000D-01
    2  6.75000D-01  1.75000D-01 -1.28100D-16  0.00000D-01
    3  1.50000D 00  0.00000D-01  1.75000D-01  0.00000D-01
    4  2.32500D 00 -1.75000D-01 -2.88230D-16  0.00000D-01

SUPPORT REACTIONS
 NODE          AR1          AR2
    1 -1.50000D-01  2.86230D-17
    3 -1.50000D-01  0.00000D-01

**********  LOADING NUMBER    6  **********
  NLN  NEL  NEV  NET  NEP  NED
```

TABLE 2.8 (cont.)

```
     0     0     0     4     0     0

 TEMPERATURE STRAIN
 ELEM.          TS
      1   1.4000D-01
      2   1.4000D-01
      3   1.4000D-01
      4   1.4000D-01

 NODAL DISPLACEMENTS
  NODE          DN1           DN2
      1   0.0000D-01   0.0000D-01
      2   1.4000D-01   2.7756D-16
      3  -0.0000D-01  -1.4000D-01
      4   1.4000D-01  -1.4000D-01
      5   7.0000D-02  -7.0000D-02

 ELEMENT STRESSES
 ELEM.           SX            SY           SXY            SZ
      1  -2.0817D-16  -1.1102D-16   4.2701D-17   0.0000D-01
      2  -1.1102D-16  -8.3267D-17   1.0675D-17   0.0000D-01
      3   5.5511D-17  -9.7145D-17   4.2701D-17   0.0000D-01
      4  -4.1633D-17  -1.2490D-16   7.4727D-17   0.0000D-01

 SUPPORT REACTIONS
  NODE          AR1           AR2
      1   6.1461D-18  -6.3426D-18
      3  -6.1257D-18   0.0000D-01

 **********  LOADING NUMBER     7  **********
   NLN   NEL   NEV   NET   NEP   NED
     0     0     0     0     4     0

 PRESTRAINS
 ELEM.          PS1           PS2           PS3
      1   2.8000D-01   2.8000D-01   0.0000D-01
      2   2.8000D-01   2.8000D-01   0.0000D-01
      3   2.8000D-01   2.8000D-01   0.0000D-01
      4   2.8000D-01   2.8000D-01   0.0000D-01

 NODAL DISPLACEMENTS
  NODE          DN1           DN2
      1   0.0000D-01   0.0000D-01
      2   2.8000D-01   3.1919D-16
      3  -0.0000D-01  -2.8000D-01
      4   2.8000D-01  -2.8000D-01
      5   1.4000D-01  -1.4000D-01

 ELEMENT STRESSES
 ELEM.           SX            SY           SXY            SZ
      1  -1.2490D-16  -6.9389D-17   5.3376D-17   0.0000D-01
      2  -2.7756D-17   2.7756D-17   1.0675D-17   0.0000D-01
      3   1.6653D-16   1.3878D-17   4.2701D-17   0.0000D-01
      4   5.5511D-17  -6.9389D-17   9.6077D-17   0.0000D-01

 SUPPORT REACTIONS
  NODE          AR1           AR2
      1   5.4210D-18  -6.0715D-18
      3  -7.8063D-18   0.0000D-01
```

TABLE 2.8 (cont.)

```
**********  LOADING NUMBER    8  **********
  NLN  NEL  NEV  NET  NEP  NED
    0    0    0    0    4    0

PRESTRAINS
ELEM.           PS1           PS2           PS3
    1  0.0000D-01  0.0000D-01  2.0000D-01
    2  0.0000D-01  0.0000D-01  2.0000D-01
    3  0.0000D-01  0.0000D-01  2.0000D-01
    4  0.0000D-01  0.0000D-01  2.0000D-01

NODAL DISPLACEMENTS
 NODE           DN1           DN2
    1  0.0000D-01  0.0000D-01
    2  9.3675D-17  2.0000D-01
    3 -0.0000D-01 -1.3878D-16
    4 -1.0842D-16  2.0000D-01
    5 -3.5562D-17  1.0000D-01

ELEMENT STRESSES
ELEM.           SX            SY           SXY            SZ
    1  1.2124D-16  9.1883D-17 -9.7145D-17  0.0000D-01
    2  4.3654D-17 -4.2415D-17 -4.1633D-17  0.0000D-01
    3 -1.0084D-16  2.5258D-17 -6.9389D-17  0.0000D-01
    4 -3.2407D-17  1.2906D-16 -1.1102D-16  0.0000D-01

SUPPORT REACTIONS
 NODE           AR1           AR2
    1 -7.3446D-18  1.0772D-17
    3  7.8217D-18  0.0000D-01

**********  LOADING NUMBER    9  **********
  NLN  NEL  NEV  NET  NEP  NED
    0    0    0    0    0    2

SUPPORT DISPLACEMENTS
ELEM.           DR1           DR2           DR3           DR4           DR5           DR6
    3  0.0000D-01  0.0000D-01  0.0000D-01  0.0000D-01 -1.0000D-01  0.0000D-01
    4  0.0000D-01  0.0000D-01  0.0000D-01  0.0000D-01 -1.0000D-01  0.0000D-01

NODAL DISPLACEMENTS
 NODE           DN1           DN2
    1  0.0000D-01  0.0000D-01
    2 -2.9490D-17 -1.0000D-01
    3 -1.0000D-01  7.0256D-17
    4 -1.0000D-01 -1.0000D-01
    5 -5.0000D-02 -5.0000D-02

ELEMENT STRESSES
ELEM.           SX            SY           SXY            SZ
    1 -4.4417D-17 -4.9754D-17  3.4027D-17  0.0000D-01
    2 -1.5632D-17  2.3066D-17  6.6720D-18  0.0000D-01
    3  7.4250D-17  1.6203D-17  2.9023D-17  0.0000D-01
    4  4.5465D-17 -5.6617D-17  5.6379D-17  0.0000D-01

SUPPORT REACTIONS
 NODE           AR1           AR2
    1  2.3429D-18 -4.8562D-18
    3 -5.1964D-18  0.0000D-01
```

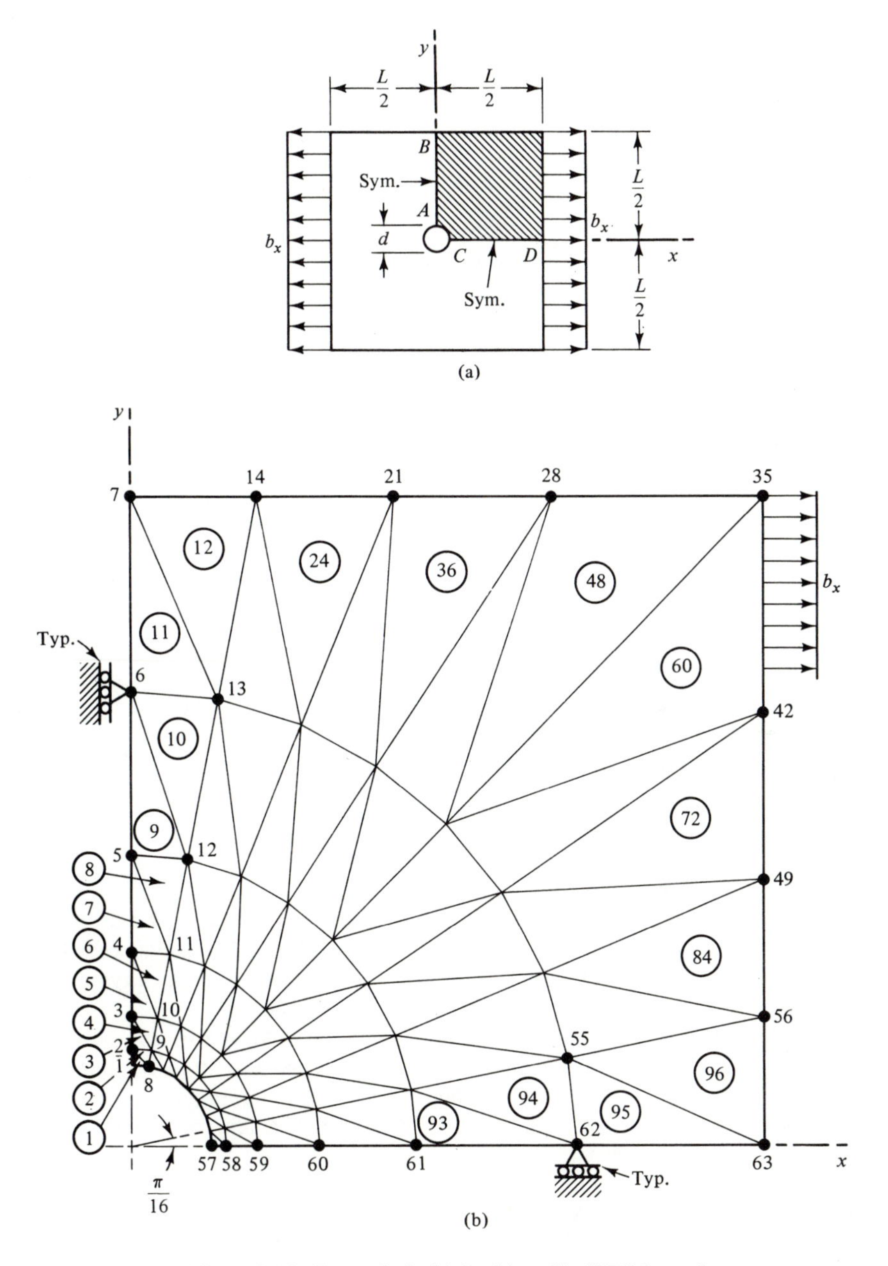

Figure 2.16 Example 2: (a) Problem (b) CST Network

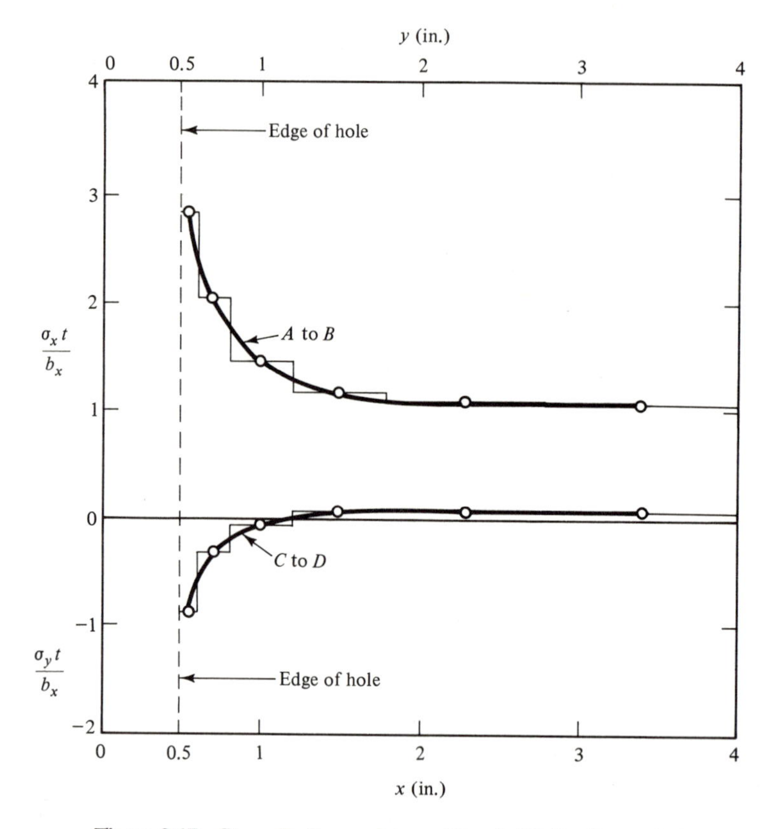

Figure 2.17 Stress Ratios on Lines *AB* and *CD* for Example 2

of the physical parameters are:

$$L = 1 \text{ m} \qquad t = 1 \text{ m} \qquad b_y = 5 \times 10^3 \text{ kN/m}$$
$$E = 2 \times 10^7 \text{ kN/m}^2 \qquad \nu = 0.3$$

for which S.I. units are used.

To analyze half the problem, we discretize the part on the right side of the centerline, as shown by the network of triangles in Fig. 2.18(b). Restraints needed for this analytical model consist of rollers at nodes on the y axis (in a plane of symmetry) and pinned supports at nodes on the x axis (to fix the base points).

For the purpose of design, we shall investigate the following stress variations: (1) normal stress σ_x along the line *AB* in Fig. 2.18(a), (2) normal stress $\sigma_{y'}$ along the line *CD*, and (3) normal stress σ_y along the line *EF*. To find approximate values of these stresses, it is necessary to run Program PSCST with the data for the network in Fig. 2.18(b) and the parameter IPS = 1.

Graphs of the stress ratios σ_x/b_y (on line *AB*), $\sigma_{y'}/b_y$ (on line *CD*), and σ_y/b_y (on line *EF*) appear in Fig. 2.19. For the first and third graphs, the plotted points are located at the midedges of triangles bordering on lines *AB* and *EF*. However, the points

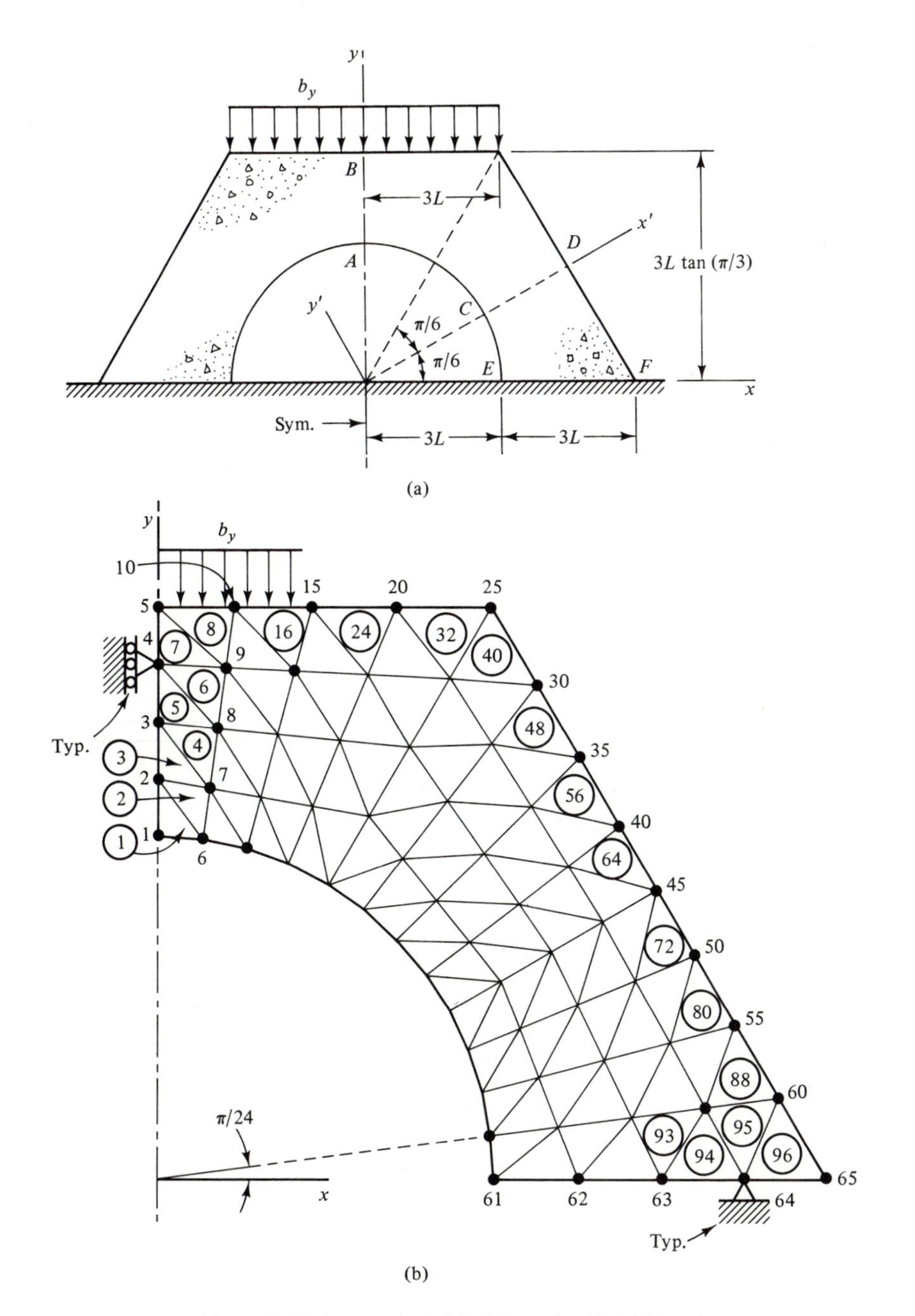

Figure 2.18 Example 3: (a) Culvert (b) CST Network

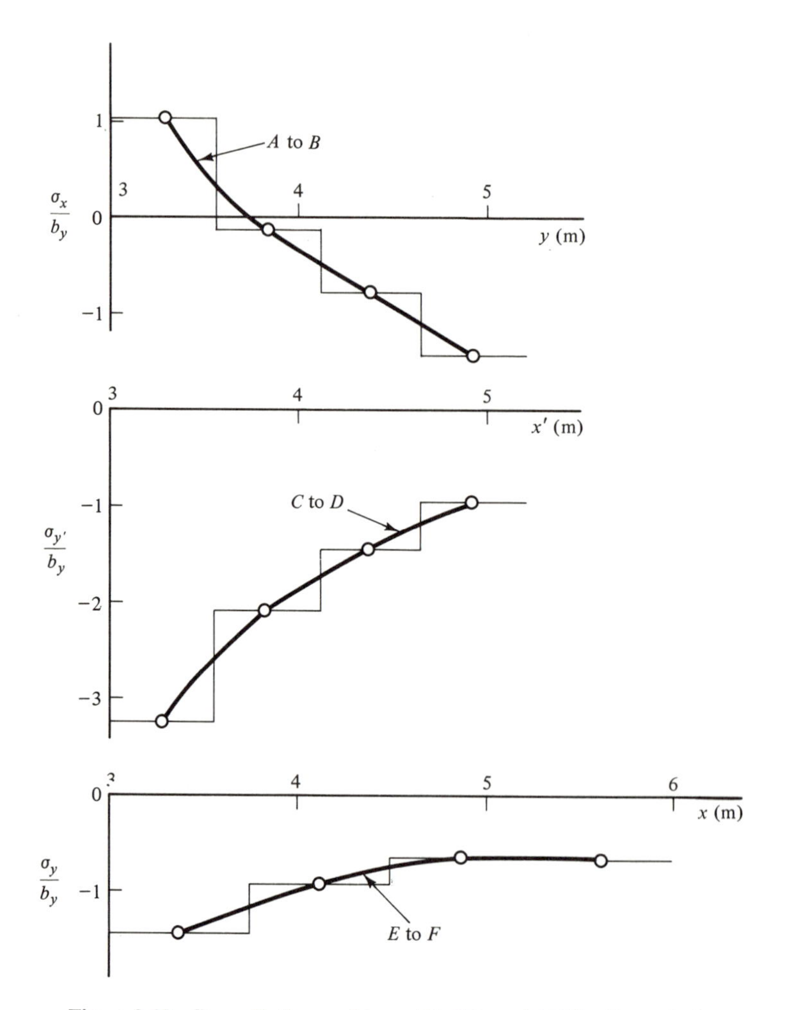

Figure 2.19 Stress Ratios on Lines *AB*, *CD*, and *EF* for Example 3

for the second graph are taken as average values from triangles bordering on both sides of line *CD*. We see from these plots that the normal stresses chosen tend to be high near the opening. It is also apparent that tensile reinforcement would be required in the vicinity of point *A*.

REFERENCES

1. Timoshenko, S. P., and Goodier, J. N., *Theory of Elasticity*, 3d ed., McGraw-Hill, New York, 1970.
2. Lekhnitskii, S. G., *Theory of Elasticity of an Anisotropic Body*, Translation from Russian by P. Fern, Holden-Day, San Francisco, 1963.

3. Turner, M. J., Clough, R. W., Martin, H. C., and Topp, L. J., "Stiffness and Deflection Analysis of Complex Structures," *Journal of the Aeronautical Sciences*, Vol. 23, No. 9, September 1956, pp. 805–823.
4. Fraejis de Veubeke, B., "Displacement and Equilibrium Models in the Finite Element Method," Chapter 9 of *Stress Analysis*, ed. O. C. Zienkiewicz and G. S. Hollister, Wiley, New York, 1965.
5. McBean, R. P., "Analysis of Stiffened Plates by the Finite Element Method," Ph.D. Thesis, Dept. of Civil Engineering, Stanford University, 1968.
6. Melosh, R. J., "Basis of Derivation of Matrices for the Direct Stiffness Method," *AIAA Journal*, Vol. 1, No. 7, July 1963, pp. 1631–1637.
7. Oakberg, R. G., and Weaver, W., Jr., "Analysis of Frames with Shear Walls by Finite Elements," Proceedings of ASCE Specialty Conference, *Applications of Finite Elements in Civil Engineering*, Vanderbilt University, November 13–14, 1969.
8. Barber, R. B., "Refined Rectangular Finite Element for Plane Stress Analysis," Engineer Thesis, Dept. of Civil Engineering, Stanford University, 1970.
9. Johnson, P. C., "The Analysis of Thin Shells by a Finite Element Procedure," Ph.D. Thesis, Dept. of Civil Engineering, University of California, Berkeley, 1968.

PROBLEMS

2.2-1. The figure shows a constant strain triangle with a small rigid-body rotation $\delta\theta$ about node i. Fill the nodal displacement vector $\mathbf{q}_{RB}$, and check to see whether all strains are zero (by premultiplication with the matrix $\mathbf{B}$).

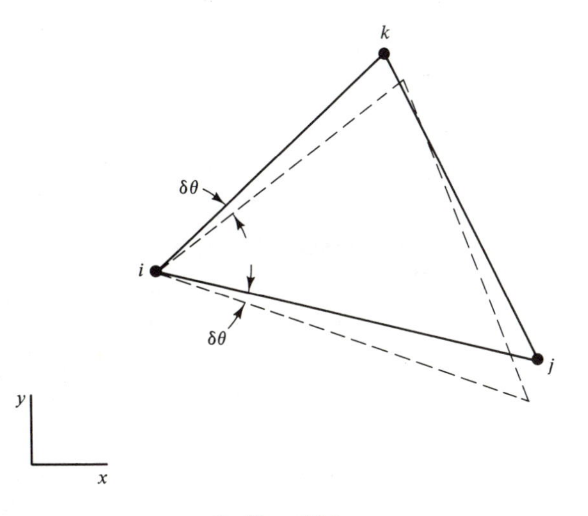

Problem 2.2-1

2.2-2. Determine equivalent nodal loads in the vector $\mathbf{p}_r$ due to the restraint translation u_{rj} in the x direction at node j of a constant strain triangle (see figure).

Problem 2.2-2

2.2-3. Suppose that a constant strain triangle has a rigid-body translation u_{RB} in the x direction. Check to see whether the product of **K** (from Table 2.1) and the nodal displacement vector $\mathbf{q}_{RB}$ yields a vector of zeros.

2.2-4. Assume that the coordinates of nodes i, j, and k for a constant strain triangle are as given in the figure. Find the equivalent nodal loads due to a concentrated force P applied at point m.

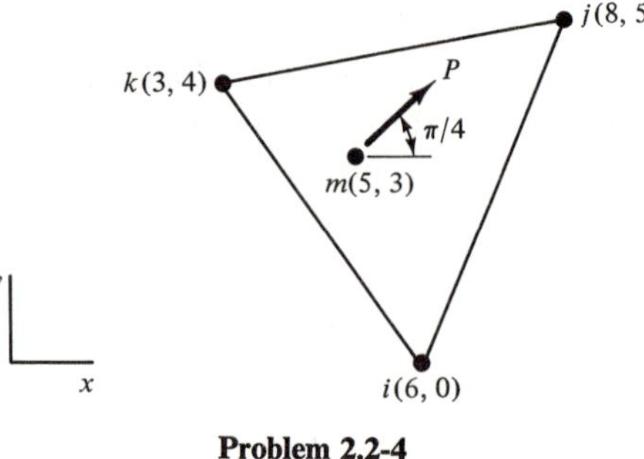

Problem 2.2-4

2.2-5. For the constant strain triangle, determine the equivalent nodal loads due to a linearly varying force (per unit length) applied in the x direction along edge jk, as shown in the figure.

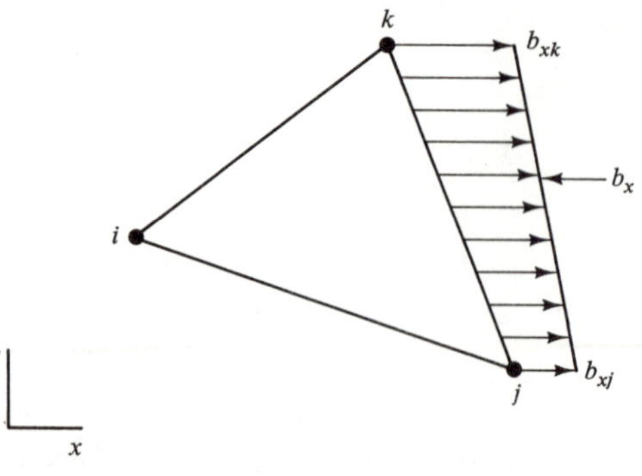

Problem 2.2-5

2.3-1. Verify the terms in the first column of the stiffness matrix for the bilinear displacement rectangle (see Table 2.3).

2.3-2. The figure shows a bilinear displacement rectangle with a small rigid-body rotation $\delta\theta$ about its centroid. Construct the nodal displacement vector $\mathbf{q}_{RB}$, and check to see whether premultiplication with the matrix $\mathbf{B}$ produces a null vector of strains.

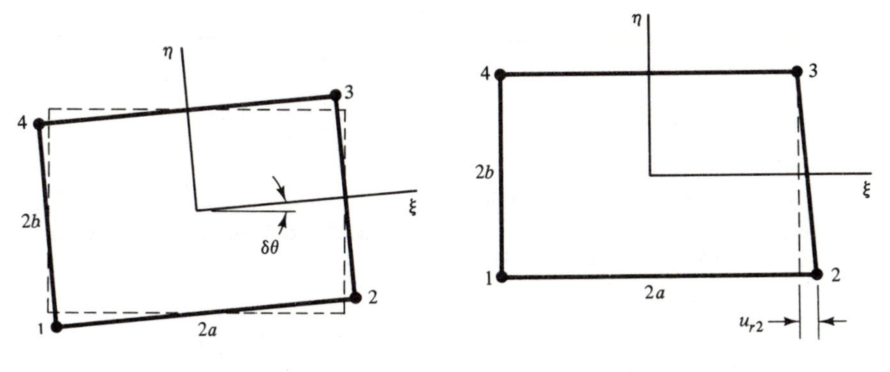

Problem 2.3-2 **Problem 2.3-3**

2.3-3. Assume a restraint translation u_{r2} in the x direction at node 2 of a bilinear displacement rectangle (see figure), and find the equivalent nodal loads in the vector $\mathbf{p}_r$.

2.3-4. Let a bilinear displacement rectangle have a rigid-body translation v_{RB} in the y direction. Verify the fact that the product of $\mathbf{K}$ (from Table 2.3) and the nodal displacement vector $\mathbf{q}_{RB}$ gives a null result.

2.3-5. For the bilinear displacement rectangle, derive the equivalent nodal loads due to the parabolically distributed force $b_y = b_{y0}(1 - \xi^2)$ per unit length, applied in the η direction along edge 1-2 in the figure.

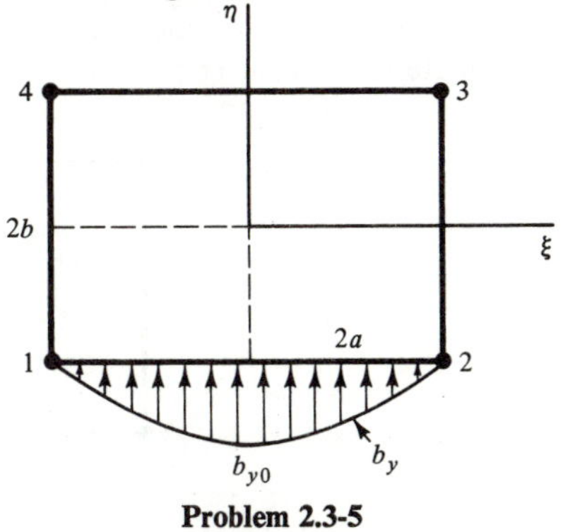

Problem 2.3-5

2.3-6. Suppose that a bilinear displacement rectangle is subjected to a temperature change that varies linearly from edge 1-4 to edge 2-3, as indicated in the figure. Develop equivalent nodal loads due to such a temperature variation, assuming isotropic material in plane stress.

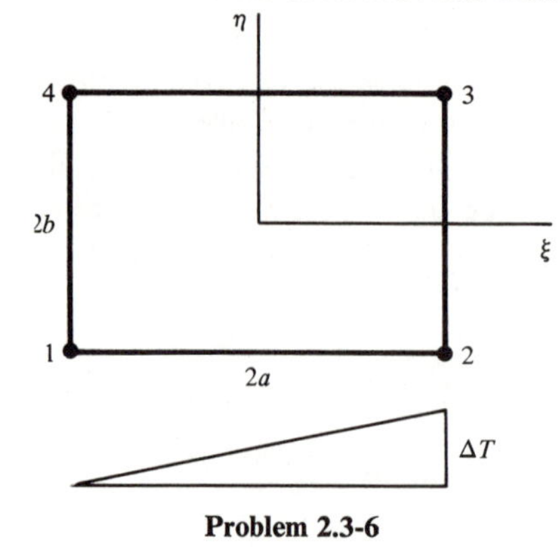

Problem 2.3-6

2.3-7. For a linear-cubic rectangle, find the equivalent nodal loads (four at each node) due to a concentrated force P applied in the ξ direction at point m, where ξ, $\eta = (-0.5, -0.5)$, as shown in the figure.

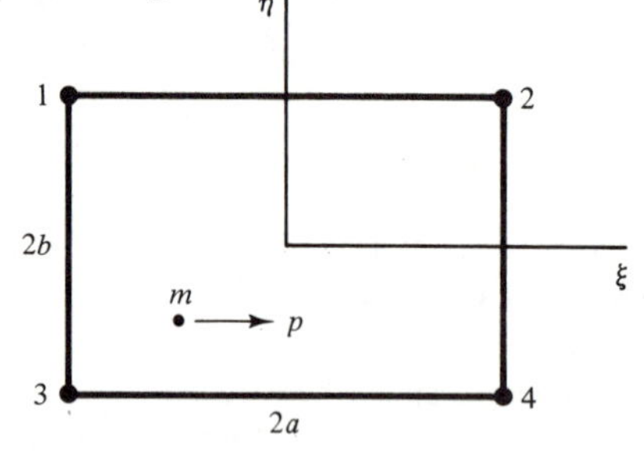

Problem 2.3-7

2.3-8. Assume that a linear-cubic rectangle has a temperature change that varies linearly from edge 1-3 to edge 2-4, as indicated in the figure. Derive equivalent nodal loads (four at each node) due to this temperature variation, assuming isotropic material in plane stress.

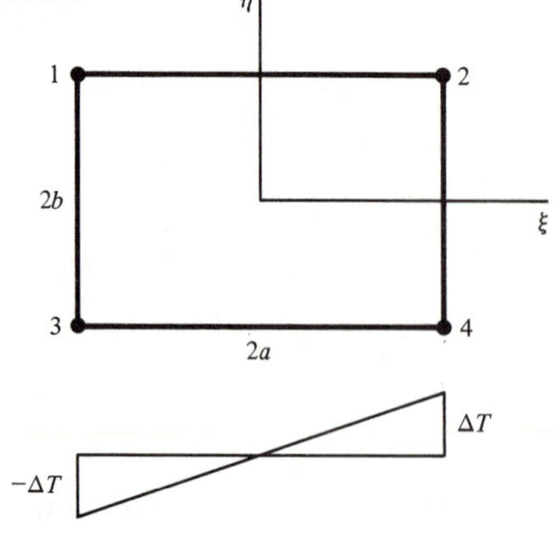

Problem 2.3-8

CHAPTER

3

Isoparametric Formulations

3.1 INTRODUCTION

A finite element is said to be *isoparametric* if the same interpolation formulas define both the geometric and the displacement shape functions. Such elements satisfy both geometric and displacement compatibility conditions. If the geometric interpolation functions are of lower order than the displacement shape functions, the element is called *subparametric*. If the reverse is true, the element is referred to as *superparametric* (1).

Because isoparametric elements are usually curved, they tend to be more suitable than subparametric elements for modeling geometric boundary conditions. However, strain-displacement relationships are complicated by the fact that generic displacements are expressed in terms of *local coordinates*, whereas differentiations with respect to the *global coordinates* are required. Also, it becomes necessary to employ numerical integration whenever explicit integrations are impossible.

In this chapter two-dimensional isoparametric elements are discussed. One-dimensional isoparametric elements were developed in Chapter 1 (axial and torsional elements). Three-dimensional isoparametric elements will be formulated in Chapter 4 for the analysis of general solids.

3.2 NATURAL COORDINATES

Geometric characteristics of certain finite elements lead to the use of dimensionless (or *natural*) coordinate systems in place of Cartesian coordinates. Formulations for triangles, quadrilaterals, and their three-dimensional counterparts are notable in this respect. Derivatives and integrals required in the development of stiffnesses and equivalent nodal loads are simplified by the use of coordinates that are consonant with the local geometry.

Figure 3.1 shows a *line element*, for which the location of any point 3 is given by dimensionless *length coordinates*, as follows:

$$\xi_1 = \frac{L_1}{L} \qquad \xi_2 = \frac{L_2}{L} \tag{a}$$

From the figure we see that:

$$L_1 + L_2 = L \tag{b}$$

Therefore,

$$\xi_1 + \xi_2 = 1 \tag{3.2-1}$$

which shows that ξ_1 and ξ_2 are dependent upon each other. It is possible to express the global coordinate x in terms of the local coordinates ξ_1 and ξ_2. Thus,

$$x = \xi_1 x_1 + \xi_2 x_2 \tag{3.2-2}$$

Conversely, we may write the local coordinates in terms of the global coordinate by solving for ξ_1 and ξ_2 in Eqs. (3.2-1) and (3.2-2), as follows:

$$\begin{bmatrix} \xi_1 \\ \xi_2 \end{bmatrix} = \frac{1}{L} \begin{bmatrix} x_2 & -1 \\ -x_1 & 1 \end{bmatrix} \begin{bmatrix} 1 \\ x \end{bmatrix} \tag{3.2-3}$$

The expressions given by Eq. (3.2-3) are plotted in Figs. 3.1(b) and (c). In that form we can see that the natural coordinates ξ_1 and ξ_2 are the same as the

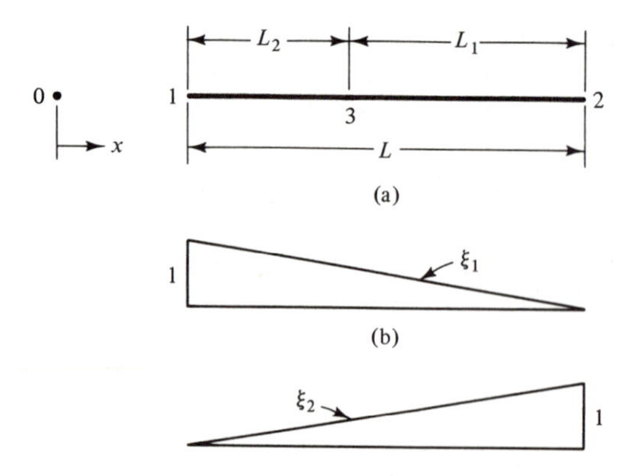

Figure 3.1 Natural Coordinates for a Line

displacement shape functions f_1 and f_2 for the axial element (see Fig. 1.8). Therefore, the axial element with two nodes and a linear displacement function is recognized to be isoparametric.

Differentiation of a function $f(\xi_1, \xi_2)$ with respect to x by the chain rule of differential calculus has the form:

$$\frac{\partial f}{\partial x} = \frac{\partial f}{\partial \xi_1}\frac{\partial \xi_1}{\partial x} + \frac{\partial f}{\partial \xi_2}\frac{\partial \xi_2}{\partial x} \tag{3.2-4}$$

From Eq. (3.2-3) the partial derivatives of ξ_1 and ξ_2 with respect to x are:

$$\frac{\partial \xi_1}{\partial x} = -\frac{1}{L} \qquad \frac{\partial \xi_2}{\partial x} = \frac{1}{L} \tag{c}$$

Substitution of these terms into Eq. (3.2-4) gives:

$$\frac{\partial f}{\partial x} = \frac{1}{L}\left(-\frac{\partial f}{\partial \xi_1} + \frac{\partial f}{\partial \xi_2}\right) \tag{3.2-5}$$

Integration of polynomial terms in the length coordinates ξ_1 and ξ_2 is done conveniently by the following expression (2):

$$\int_{x_1}^{x_2} \xi_1^a \xi_2^b \, dx = \int_0^1 \xi_1^a (1 - \xi_1)^b L \, d\xi_1 = \frac{a!\, b!}{(a + b + 1)!} L \tag{3.2-6}$$

In this formula $a!$ represents the factorial product $a(a - 1)(a - 2) \ldots (1)$ and so on. When it occurs, the factorial product $0!$ is defined to be unity.

Example 1

$$\int_L \xi_1^2 \xi_2^3 \, dL = \frac{2!\, 3!}{6!} L = \frac{L}{60}$$

Example 2

$$\begin{aligned}
\int_L x^2 \, dL &= \int_L (\xi_1 x_1 + \xi_2 x_2)^2 \, dL \\
&= \int_L (\xi_1^2 x_1^2 + 2\xi_1 x_1 \xi_2 x_2 + \xi_2^2 x_2^2) \, dL \\
&= \frac{2!}{3!} x_1^2 L + 2\frac{1}{3!} x_1 x_2 L + \frac{2!}{3!} x_2^2 L \\
&= (x_1^2 + x_1 x_2 + x_2^2)\frac{L}{3}
\end{aligned}$$

If the origin of x is located at the center of the line element, then $x_1 = -L/2, x_2 = L/2$, and $x_1 + x_2 = 0$. Thus, $(x_1 + x_2)^2 = 0$ leads to:

$$x_1 x_2 = -\frac{x_1^2 + x_2^2}{2}$$

Substitution of this expression into the integral above yields:

$$\int_L x^2 \, dL = (x_1^2 + x_2^2)\frac{L}{6}$$

Or,

$$\int_L x^2 \, dL = \frac{L^3}{12}$$

These results are the same as those given in Appendix A.

A *triangular element* of area A appears in Fig. 3.2(a). Its corners are numbered 1, 2, and 3; and any point 4 on the triangle may be located by dividing it into subtriangles having areas A_1, A_2, and A_3. Dimensionless *area coordinates* for the triangle are defined as:

$$\xi_1 = \frac{A_1}{A} \qquad \xi_2 = \frac{A_2}{A} \qquad \xi_3 = \frac{A_3}{A} \tag{d}$$

By inspection, we see that:

$$A_1 + A_2 + A_3 = A \tag{e}$$

Thus,

$$\xi_1 + \xi_2 + \xi_3 = 1 \tag{3.2-7}$$

which shows that ξ_1, ξ_2, and ξ_3 are interdependent. Figure 3.2(b) indicates that $\xi_1 = 1$ at point 1 and that $\xi_1 = 0$ along edge 2-3. Also indicated is a linear variation of ξ_1 from point 1 to the opposite edge, and similarly for ξ_2 and ξ_3. When the global coordinates x and y are written in terms of the local coordinates, we have:

$$\begin{aligned} x &= \xi_1 x_1 + \xi_2 x_2 + \xi_3 x_3 \\ y &= \xi_1 y_1 + \xi_2 y_2 + \xi_3 y_3 \end{aligned} \tag{3.2-8}$$

On the other hand, we can express the local coordinates in terms of x and y by solving for ξ_1, ξ_2, and ξ_3 in Eqs. (3.2-7) and (3.2-8), as follows:

$$\begin{bmatrix} \xi_1 \\ \xi_2 \\ \xi_3 \end{bmatrix} = \frac{1}{2A} \begin{bmatrix} x_2 y_3 - x_3 y_2 & -y_{23} & x_{23} \\ x_3 y_1 - x_1 y_3 & -y_{31} & x_{31} \\ x_1 y_2 - x_2 y_1 & -y_{12} & x_{12} \end{bmatrix} \begin{bmatrix} 1 \\ x \\ y \end{bmatrix} \tag{3.2-9}$$

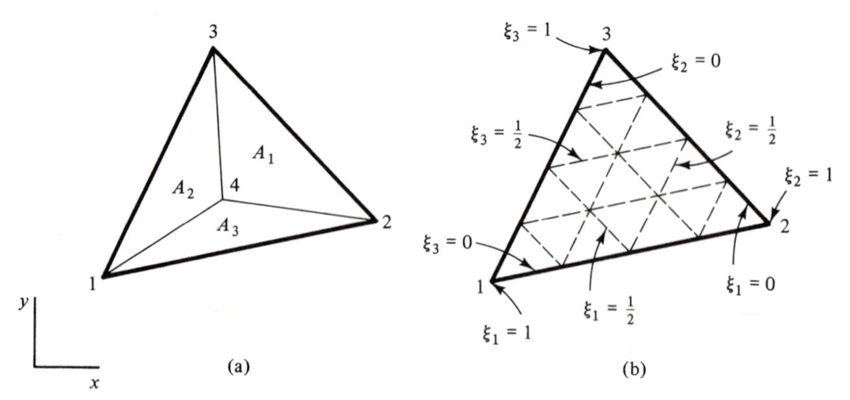

Figure 3.2 Natural Coordinates for a Triangle

Comparison of Eq. (3.2-9) with Eqs. (2.2-1b) shows that the natural coordinates ξ_1, ξ_2, and ξ_3 are the same as the displacement shape functions f_1, f_2, and f_3 for the constant strain triangle. Therefore, we can conclude that the constant strain triangular element (see Sec. 2.2) is isoparametric. To simplify Eq. (3.2-9), we define A_{ij} to be the area of the triangle having the nodes i and j and the origin of the global coordinates as vertices. In addition, we let:

$$\begin{array}{lll} a_1 = x_{23} & a_2 = x_{31} & a_3 = x_{12} \\ b_1 = -y_{23} & b_2 = -y_{31} & b_3 = -y_{12} \end{array} \tag{f}$$

Using these definitions in Eq. (3.2-9), we rewrite it as:

$$\begin{bmatrix} \xi_1 \\ \xi_2 \\ \xi_3 \end{bmatrix} = \frac{1}{2A} \begin{bmatrix} 2A_{23} & b_1 & a_1 \\ 2A_{31} & b_2 & a_2 \\ 2A_{12} & b_3 & a_3 \end{bmatrix} \begin{bmatrix} 1 \\ x \\ y \end{bmatrix} \tag{3.2-10}$$

If we have a function $f(\xi_1, \xi_2, \xi_3)$ to be differentiated with respect to x and y, the chain rule gives:

$$\begin{aligned} \frac{\partial f}{\partial x} &= \frac{\partial f}{\partial \xi_1}\frac{\partial \xi_1}{dx} + \frac{\partial f}{\partial \xi_2}\frac{\partial \xi_2}{\partial x} + \frac{\partial f}{\partial \xi_3}\frac{\partial \xi_3}{\partial x} = \sum_{i=1}^{3}\frac{\partial f}{\partial \xi_i}\frac{\partial \xi_i}{\partial x} \\ \frac{\partial f}{\partial y} &= \frac{\partial f}{\partial \xi_1}\frac{\partial \xi_1}{\partial y} + \frac{\partial f}{\partial \xi_2}\frac{\partial \xi_2}{\partial y} + \frac{\partial f}{\partial \xi_3}\frac{\partial \xi_3}{\partial y} = \sum_{i=1}^{3}\frac{\partial f}{\partial \xi_i}\frac{\partial \xi_i}{\partial y} \end{aligned} \tag{3.2-11}$$

From Eq. (3.2-10) we see that:

$$\frac{\partial \xi_i}{\partial x} = \frac{b_i}{2A} \qquad \frac{\partial \xi_i}{\partial y} = \frac{a_i}{2A} \tag{g}$$

Therefore, Eqs. (3.2-11) become:

$$\frac{\partial f}{\partial x} = \frac{1}{2A}\sum_{i=1}^{3} b_i \frac{\partial f}{\partial \xi_i} \qquad \frac{\partial f}{\partial y} = \frac{1}{2A}\sum_{i=1}^{3} a_i \frac{\partial f}{\partial \xi_i} \tag{3.2-12}$$

Integrals of polynomial terms in the area coordinates ξ_1, ξ_2, and ξ_3 can be obtained as follows (2):

$$\int_A \xi_1^a \xi_2^b \xi_3^c \, dA = \frac{a!\, b!\, c!}{(a+b+c+2)!}(2A) \tag{3.2-13}$$

The ease with which such integrations may be carried out is one of the primary advantages of using area coordinates.

Example 3

$$\int_A \xi_1^3 \xi_2 \xi_3^2 \, dA = \frac{3!\,1!\,2!}{8!}(2A) = \frac{A}{1680}$$

Example 4

$$\begin{aligned} \int_A x^2 \, dA &= \int_A (\xi_1 x_1 + \xi_2 x_2 + \xi_3 x_3)^2 \, dA \\ &= \int_A (\xi_1^2 x_1^2 + \xi_2^2 x_2^2 + \xi_3^2 x_3^2 + 2\xi_1 x_1 \xi_2 x_2 + 2\xi_2 x_2 \xi_3 x_3 + 2\xi_1 x_1 \xi_3 x_3)\, dA \end{aligned}$$

$$= \frac{2}{4!}(x_1^2 + x_2^2 + x_3^2 + x_1x_2 + x_2x_3 + x_1x_3)(2A)$$

$$= (x_1^2 + x_2^2 + x_3^2 + x_1x_2 + x_2x_3 + x_1x_3)\frac{A}{6}$$

If the origin is located at the centroid of the triangle, $x_1 + x_2 + x_3 = 0$. Therefore, $(x_1 + x_2 + x_3)^2 = 0$ also, and we have:

$$x_1x_2 + x_2x_3 + x_1x_3 = -\tfrac{1}{2}(x_1^2 + x_2^2 + x_3^2)$$

Substituting this expression into the integral above, we find:

$$\int_A x^2\, dA = (x_1^2 + x_2^2 + x_3^2)\frac{A}{12}$$

which is the same as in Appendix A.

Figure 3.3 shows dimensionless natural coordinates ξ and η for a *quadrilateral element*. Point g is the *geometric center*, where:

$$x_g = \tfrac{1}{4}(x_1 + x_2 + x_3 + x_4) \qquad y_g = \tfrac{1}{4}(y_1 + y_2 + y_3 + y_4) \tag{h}$$

This point is not necessarily the same as the centroid of the element. Note that $\eta = -1$ along edge 1-2, $\xi = 1$ along edge 2-3, $\eta = 1$ along edge 3-4, and $\xi = -1$ along edge 4-1. With linear interpolation in both the ξ and the η directions, the location of a generic point may be expressed as follows:

$$\begin{aligned} x &= \tfrac{1}{4}[(1-\xi)(1-\eta)x_1 + (1+\xi)(1-\eta)x_2 \\ &\qquad + (1+\xi)(1+\eta)x_3 + (1-\xi)(1+\eta)x_4] \\ y &= \tfrac{1}{4}[(1-\xi)(1-\eta)y_1 + (1+\xi)(1-\eta)y_2 \\ &\qquad + (1+\xi)(1+\eta)y_3 + (1-\xi)(1+\eta)y_4] \end{aligned} \tag{3.2-14a}$$

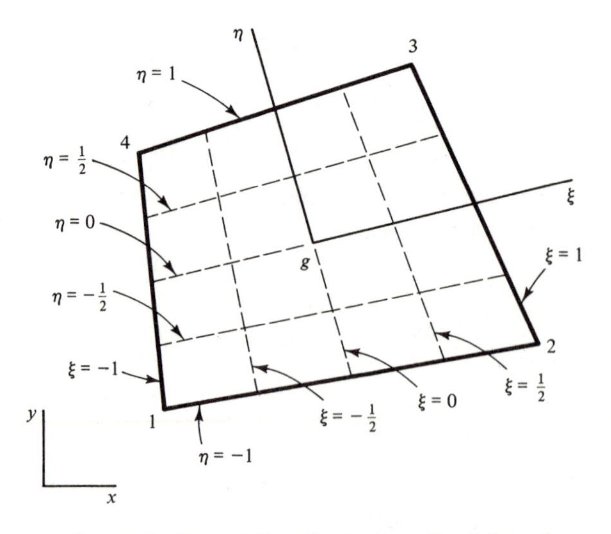

Figure 3.3 Natural Coordinates for a Quadrilateral

Or,

$$x = f_1 x_1 + f_2 x_2 + f_3 x_3 + f_4 x_4 = \sum_{i=1}^{4} f_i x_i$$
$$y = f_1 y_1 + f_2 y_2 + f_3 y_3 + f_4 y_4 = \sum_{i=1}^{4} f_i y_i \qquad (3.2\text{-}14b)$$

in which

$$f_1 = \tfrac{1}{4}(1 - \xi)(1 - \eta) \qquad f_2 = \tfrac{1}{4}(1 + \xi)(1 - \eta)$$
$$f_3 = \tfrac{1}{4}(1 + \xi)(1 + \eta) \qquad f_4 = \tfrac{1}{4}(1 - \xi)(1 + \eta) \qquad (3.2\text{-}15)$$

These functions give the global coordinates in terms of the natural coordinates of the quadrilateral. However, because Eqs. (3.2-14a) are bilinear, the local coordinates ξ and η cannot be expressed in terms of the global coordinates x and y.

The chain rule of differential calculus for differentiation of $f(\xi, \eta)$ with respect to x and y yields:

$$\frac{\partial f}{\partial x} = \frac{\partial f}{\partial \xi}\frac{\partial \xi}{\partial x} + \frac{\partial f}{\partial \eta}\frac{\partial \eta}{\partial x}$$
$$\frac{\partial f}{\partial y} = \frac{\partial f}{\partial \xi}\frac{\partial \xi}{\partial y} + \frac{\partial f}{\partial \eta}\frac{\partial \eta}{\partial y} \qquad \text{(i)}$$

Or,

$$\begin{bmatrix} \dfrac{\partial f}{\partial x} \\[2ex] \dfrac{\partial f}{\partial y} \end{bmatrix} = \begin{bmatrix} \dfrac{\partial \xi}{\partial x} & \dfrac{\partial \eta}{\partial x} \\[2ex] \dfrac{\partial \xi}{\partial y} & \dfrac{\partial \eta}{\partial y} \end{bmatrix} \begin{bmatrix} \dfrac{\partial f}{\partial \xi} \\[2ex] \dfrac{\partial f}{\partial \eta} \end{bmatrix} \qquad \text{(j)}$$

Terms in the coefficient matrix of Eq. (j) are not readily available because we are not able to solve explicitly for ξ and η in terms of x and y. However, if we take the opposite approach and differentiate f with respect to ξ and η, the chain rule produces:

$$\frac{\partial f}{\partial \xi} = \frac{\partial f}{\partial x}\frac{\partial x}{\partial \xi} + \frac{\partial f}{\partial y}\frac{\partial y}{\partial \xi}$$
$$\frac{\partial f}{\partial \eta} = \frac{\partial f}{\partial x}\frac{\partial x}{\partial \eta} + \frac{\partial f}{\partial y}\frac{\partial y}{\partial \eta} \qquad \text{(k)}$$

Or,

$$\begin{bmatrix} \dfrac{\partial f}{\partial \xi} \\[2ex] \dfrac{\partial f}{\partial \eta} \end{bmatrix} = \begin{bmatrix} \dfrac{\partial x}{\partial \xi} & \dfrac{\partial y}{\partial \xi} \\[2ex] \dfrac{\partial x}{\partial \eta} & \dfrac{\partial y}{\partial \eta} \end{bmatrix} \begin{bmatrix} \dfrac{\partial f}{\partial x} \\[2ex] \dfrac{\partial f}{\partial y} \end{bmatrix} \qquad (\ell)$$

For this arrangement the terms in the coefficient matrix are easily obtained by differentiating Eqs. (3.2-14). This array is called the *Jacobian matrix* **J**, which contains the derivatives of the global coordinates with respect to the local coordinates. Thus,

$$\mathbf{J} = \begin{bmatrix} J_{11} & J_{12} \\ J_{21} & J_{22} \end{bmatrix} = \begin{bmatrix} x_{,\xi} & y_{,\xi} \\ x_{,\eta} & y_{,\eta} \end{bmatrix} \tag{3.2-16}$$

Terms in the Jacobian matrix are evaluated as follows:

$$\begin{aligned} J_{11} &= x_{,\xi} = f_{1,\xi}x_1 + f_{2,\xi}x_2 + f_{3,\xi}x_3 + f_{4,\xi}x_4 = \sum_{i=1}^{4} f_{i,\xi}x_i \\ J_{12} &= y_{,\xi} = f_{1,\xi}y_1 + f_{2,\xi}y_2 + f_{3,\xi}y_3 + f_{4,\xi}y_4 = \sum_{i=1}^{4} f_{i,\xi}y_i \\ J_{21} &= x_{,\eta} = f_{1,\eta}x_1 + f_{2,\eta}x_2 + f_{3,\eta}x_3 + f_{4,\eta}x_4 = \sum_{i=1}^{4} f_{i,\eta}x_i \\ J_{22} &= y_{,\eta} = f_{1,\eta}y_1 + f_{2,\eta}y_2 + f_{3,\eta}y_3 + f_{4,\eta}y_4 = \sum_{i=1}^{4} f_{i,\eta}y_i \end{aligned} \tag{3.2-17}$$

Casting these expressions into matrix form gives:

$$\mathbf{J} = \mathbf{D}_L \mathbf{C}_N \tag{3.2-18}$$

The array $\mathbf{D}_L$ in this equation contains derivatives with respect to local coordinates. Thus,

$$\begin{aligned} \mathbf{D}_L &= \begin{bmatrix} f_{1,\xi} & f_{2,\xi} & f_{3,\xi} & f_{4,\xi} \\ f_{1,\eta} & f_{2,\eta} & f_{3,\eta} & f_{4,\eta} \end{bmatrix} \\ &= \frac{1}{4} \begin{bmatrix} -(1-\eta) & (1-\eta) & (1+\eta) & -(1+\eta) \\ -(1-\xi) & -(1+\xi) & (1+\xi) & (1-\xi) \end{bmatrix} \end{aligned} \tag{3.2-19}$$

And the matrix $\mathbf{C}_N$ consists of nodal coordinates, arranged as follows:

$$\mathbf{C}_N = \begin{bmatrix} x_1 & y_1 \\ x_2 & y_2 \\ x_3 & y_3 \\ x_4 & y_4 \end{bmatrix} \tag{3.2-20}$$

By comparing Eqs. (j) and (ℓ), we can see that the coefficient matrix in the former expression is the *inverse of the Jacobian matrix*. Using the formal definition of the inverse, we may obtain $\mathbf{J}^{-1}$ from $\mathbf{J}$ as:

$$\mathbf{J}^{-1} = \frac{\mathbf{J}^a}{|\mathbf{J}|} = \frac{1}{|\mathbf{J}|} \begin{bmatrix} J_{22} & -J_{12} \\ -J_{21} & J_{11} \end{bmatrix} = \frac{1}{|\mathbf{J}|} \begin{bmatrix} y_{,\eta} & -y_{,\xi} \\ -x_{,\eta} & x_{,\xi} \end{bmatrix} \tag{3.2-21}$$

where $\mathbf{J}^a$ denotes the adjoint matrix of $\mathbf{J}$ and $|\mathbf{J}|$ is its determinant. The latter quantity is calculated by:

$$|\mathbf{J}| = J_{11}J_{22} - J_{12}J_{21} = x_{,\xi}y_{,\eta} - x_{,\eta}y_{,\xi} \tag{3.2-22}$$

To determine the derivatives of all the functions with respect to x and y, we can apply Eq. (j) repeatedly. Hence,

$$\begin{bmatrix} f_{i,x} \\ f_{i,y} \end{bmatrix} = \mathbf{J}^{-1} \begin{bmatrix} f_{i,\xi} \\ f_{i,\eta} \end{bmatrix} \qquad (i = 1, 2, 3, 4) \tag{3.2-23}$$

Altogether, we have:

$$\mathbf{D}_G = \mathbf{J}^{-1}\mathbf{D}_L = (\mathbf{D}_L\mathbf{C}_N)^{-1}\mathbf{D}_L \tag{3.2-24}$$

The matrix $\mathbf{D}_G$ given by this expression consists of derivatives of f_i with respect to the global coordinates. That is,

$$\mathbf{D}_G = \begin{bmatrix} f_{1,x} & f_{2,x} & f_{3,x} & f_{4,x} \\ f_{1,y} & f_{2,y} & f_{3,y} & f_{4,y} \end{bmatrix} \tag{3.2-25}$$

Evaluating terms in $\mathbf{D}_G$ yields:

$$\begin{aligned}
D_{G11} &= \frac{1}{4|\mathbf{J}|}[-(1-\eta)J_{22} + (1-\xi)J_{12}] \\
D_{G12} &= \frac{1}{4|\mathbf{J}|}[(1-\eta)J_{22} + (1+\xi)J_{12}] \\
D_{G13} &= \frac{1}{4|\mathbf{J}|}[(1+\eta)J_{22} - (1+\xi)J_{12}] \\
D_{G14} &= \frac{1}{4|\mathbf{J}|}[-(1+\eta)J_{22} - (1-\xi)J_{12}] \\
D_{G21} &= \frac{1}{4|\mathbf{J}|}[(1-\eta)J_{21} - (1-\xi)J_{11}] \\
D_{G22} &= \frac{1}{4|\mathbf{J}|}[-(1-\eta)J_{21} - (1+\xi)J_{11}] \\
D_{G23} &= \frac{1}{4|\mathbf{J}|}[-(1+\eta)J_{21} + (1+\xi)J_{11}] \\
D_{G24} &= \frac{1}{4|\mathbf{J}|}[(1+\eta)J_{21} + (1-\xi)J_{11}]
\end{aligned} \tag{3.2-26}$$

By this approach we can find all the terms in $\mathbf{D}_G$ numerically.

Because of the appearance of the determinant of $\mathbf{J}$ in denominator positions, we usually cannot integrate explicitly to obtain stiffnesses and equivalent nodal loads. Instead, it becomes necessary to use numerical integration.

3.3 NUMERICAL INTEGRATION

With many types of elements, explicit integrations for stiffnesses and equivalent nodal loads are impossible. This is true for most of the isoparametric elements considered in subsequent sections. In such cases it is necessary to use some numerical integration technique. One of the most accurate and convenient methods is *Gaussian quadrature* (3), which is developed in Appendix B. Although its derivation is for one-dimensional integration, this method may be extended to two- and three-dimensional cases as well. However, in this chapter integration will be required only for two-dimensional elements.

For quadrilaterals in Cartesian coordinates, the type of integration to be performed is:

$$I = \int\int f(x, y)\, dx\, dy \tag{a}$$

However, the integral in Eq. (a) is more easily evaluated if it is first transformed to the natural coordinates ξ and η. This can be accomplished by expressing the function f in terms of ξ and η using Eqs. (3.2-14). In addition, the limits of each integration must be changed to become -1 to 1; and the infinitesimal area $dA = dx\, dy$ must be replaced by an appropriate expression in terms of $d\xi$ and $d\eta$. For this purpose Fig. 3.4 shows an infinitesimal area dA in the natural coordinates. Vector $\mathbf{r}$ locates a generic point in the Cartesian coordinates x and y as follows:

$$\mathbf{r} = \mathbf{x} + \mathbf{y} = x\mathbf{i} + y\mathbf{j} \tag{b}$$

The rate of change of $\mathbf{r}$ with respect to ξ is:

$$\frac{\partial \mathbf{r}}{\partial \xi} = \frac{\partial x}{\partial \xi}\mathbf{i} + \frac{\partial y}{\partial \xi}\mathbf{j} \tag{c}$$

Also, the rate of change of $\mathbf{r}$ with respect to η is:

$$\frac{\partial \mathbf{r}}{\partial \eta} = \frac{\partial x}{\partial \eta}\mathbf{i} + \frac{\partial y}{\partial \eta}\mathbf{j} \tag{d}$$

When multiplied by $d\xi$ and $d\eta$, the derivatives in Eqs. (c) and (d) form two adjacent sides of the infinitesimal parallelogram of area dA in the figure. This area may be determined from the following vector triple product:

$$dA = \left(\frac{\partial \mathbf{r}}{\partial \xi} d\xi \times \frac{\partial \mathbf{r}}{\partial \eta} d\eta\right) \cdot \mathbf{k} \tag{e}$$

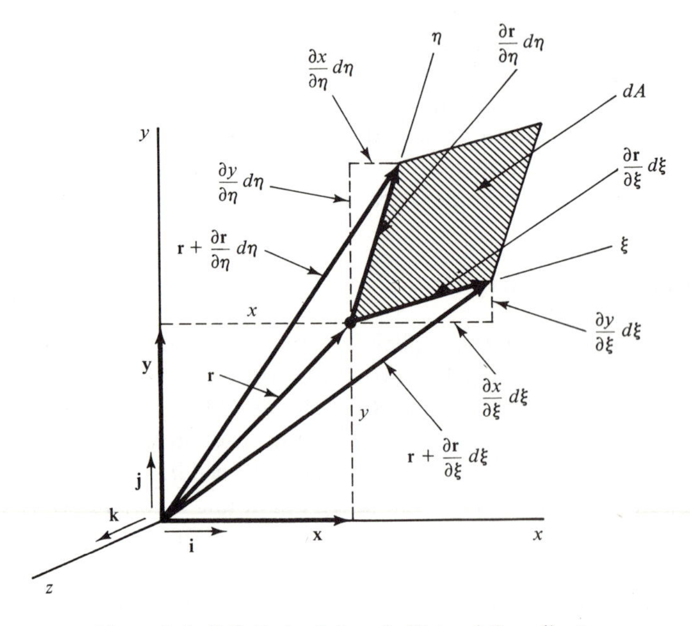

Figure 3.4 Infinitesimal Area in Natural Coordinates

Substitution of Eqs. (c) and (d) into Eq. (e) gives:

$$dA = \left(\frac{\partial x}{\partial \xi}\frac{\partial y}{\partial \eta} - \frac{\partial y}{\partial \xi}\frac{\partial x}{\partial \eta}\right) \partial \xi \, \partial \eta \tag{f}$$

The expression in the parentheses of Eq. (f) may be written as a determinant. That is,

$$dA = \begin{vmatrix} \dfrac{\partial x}{\partial \xi} & \dfrac{\partial y}{\partial \xi} \\ \dfrac{\partial x}{\partial \eta} & \dfrac{\partial y}{\partial \eta} \end{vmatrix} d\xi \, d\eta = |\mathbf{J}| \, d\xi \, d\eta \tag{3.3-1}$$

in which $\mathbf{J}$ is the Jacobian matrix (see Sec. 3.2) and $|\mathbf{J}|$ is its determinant. Thus, the new form of the integral in Eq. (a) becomes:

$$I = \int_{-1}^{1} \int_{-1}^{1} f(\xi, \eta) \, |\mathbf{J}| \, d\xi \, d\eta \tag{3.3-2}$$

Two successive applications of Gaussian quadrature produce:

$$I = \sum_{k=1}^{n} \sum_{j=1}^{n} R_j R_k f(\xi_j, \eta_k) \, |\mathbf{J}(\xi_j, \eta_k)| \tag{3.3-3}$$

where R_j and R_k are weighting factors for the point (ξ_j, η_k). *Integration points* for $n = 1, 2, 3$, and 4 each way on a quadrilateral are shown in Fig. 3.5.

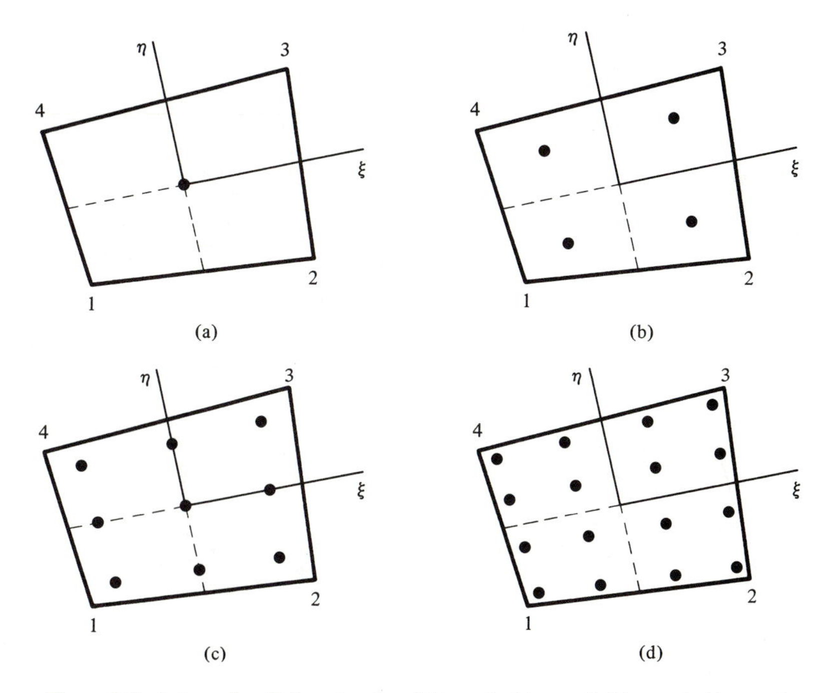

Figure 3.5 Integration Points for Quadrilateral: (a) $n = 1$ (b) $n = 2$ (c) $n = 3$ (d) $n = 4$ (each way)

Example 1

Numerically integrate the function $f(\xi, \eta) = \xi^2\eta^2$ for a rectangular plane stress element of size $2a \times 2b$. In this case the determinant of the Jacobian matrix [see Eq. (3.3-1)] is:

$$|\mathbf{J}| = \begin{vmatrix} a & 0 \\ 0 & b \end{vmatrix} = ab$$

Let $n = 2$; then from Table B.1 in Appendix B we have:

$$\xi_1 = -\xi_2 = \eta_1 = -\eta_2 = -0.577\ldots \qquad R_1 = R_2 = 1$$

Substitution of these quantities into Eq. (3.3-3) gives:

$$\begin{aligned} I &= (1)(1)[(-0.577)^2(-0.577)^2 + (0.577)^2(-0.577)^2 \\ &\quad + (0.577)^2(0.577)^2 + (-0.577)^2(0.577)^2]ab \\ &= 4(0.577)^4 ab \\ &= (0.444\ldots)ab \end{aligned}$$

which is exact.

For triangles in natural coordinates the numerical integration formula is (4):

$$I = A \sum_{j=1}^{n} W_j f(\xi_1, \xi_2, \xi_3)_j \tag{3.3-4}$$

in which W_j is the weighting factor for the jth sampling point. Integration points for $n = 1, 3, 4$, and 6 appear in Fig. 3.6, and their locations and weighting factors are given in Table 3.1.

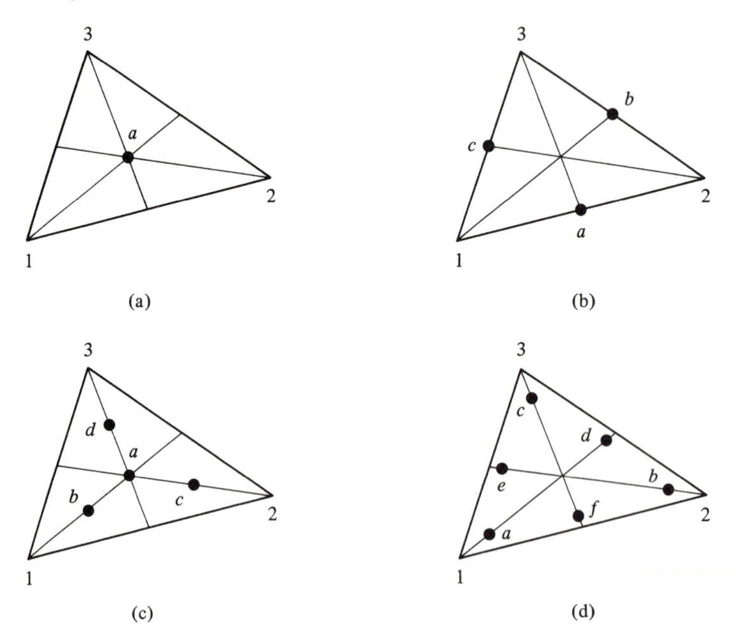

Figure 3.6 Integration Points for Triangle: (a) $n = 1$ (b) $n = 3$ (c) $n = 4$ (d) $n = 6$

TABLE 3.1 Numerical Integration Constants for Triangles

Figure	n	Order	Points	ξ_1	ξ_2	ξ_3	W_j
3.6(a)	1	Linear	a	$\frac{1}{3}$	$\frac{1}{3}$	$\frac{1}{3}$	1
3.6(b)	3	Quadratic	a	$\frac{1}{2}$	$\frac{1}{2}$	0	$\frac{1}{3}$
			b	0	$\frac{1}{2}$	$\frac{1}{2}$	$\frac{1}{3}$
			c	$\frac{1}{2}$	0	$\frac{1}{2}$	$\frac{1}{3}$
3.6(c)	4	Cubic	a	$\frac{1}{3}$	$\frac{1}{3}$	$\frac{1}{3}$	γ_1
			b	0.6	0.2	0.2	γ_2
			c	0.2	0.6	0.2	γ_2
			d	0.2	0.2	0.6	γ_2
3.6(d)	6	Quartic	a	α_1	β_1	β_1	γ_3
			b	β_1	α_1	β_1	γ_3
			c	β_1	β_1	α_1	γ_3
			d	α_2	β_2	β_2	γ_4
			e	β_2	α_2	β_2	γ_4
			f	β_2	β_2	α_2	γ_4

$\alpha_1 = 0.8168475730 \qquad \gamma_1 = -\frac{27}{48}$
$\beta_1 = 0.0915762135 \qquad \gamma_2 = \frac{25}{48}$
$\alpha_2 = 0.1081030182 \qquad \gamma_3 = 0.1099517437$
$\beta_2 = 0.4459484909 \qquad \gamma_4 = 0.2233815897$

Example 2

Repeat Example 4 in Sec. 3.2, using numerical integration with $n = 3$. From Eq. (3.3-4) and Table 3.1 we have:

$$I = A \sum_{j=1}^{3} W_j(\xi_1 x_1 + \xi_2 x_2 + \xi_3 x_3)_j^2$$

$$= \frac{A}{3}\left[\left(\frac{1}{2}x_1 + \frac{1}{2}x_2\right)^2 + \left(\frac{1}{2}x_2 + \frac{1}{2}x_3\right)^2 + \left(\frac{1}{2}x_1 + \frac{1}{2}x_3\right)^2\right]$$

$$= \frac{A}{6}(x_1^2 + x_2^2 + x_3^2 + x_1x_2 + x_2x_3 + x_1x_3)$$

which is the result found previously.

3.4 QUADRILATERAL ELEMENTS

In Sec. 2.3 we studied the bilinear displacement rectangle in detail. That element is the rectangular parent of the *isoparametric quadrilateral* ($Q4$) element depicted in Fig. 3.7(a). The generic displacements indicated in the figure are:

$$\mathbf{u} = \{u, v\} \tag{a}$$

An x and a y translation are shown at each node. Thus, the nodal displacement vector is:

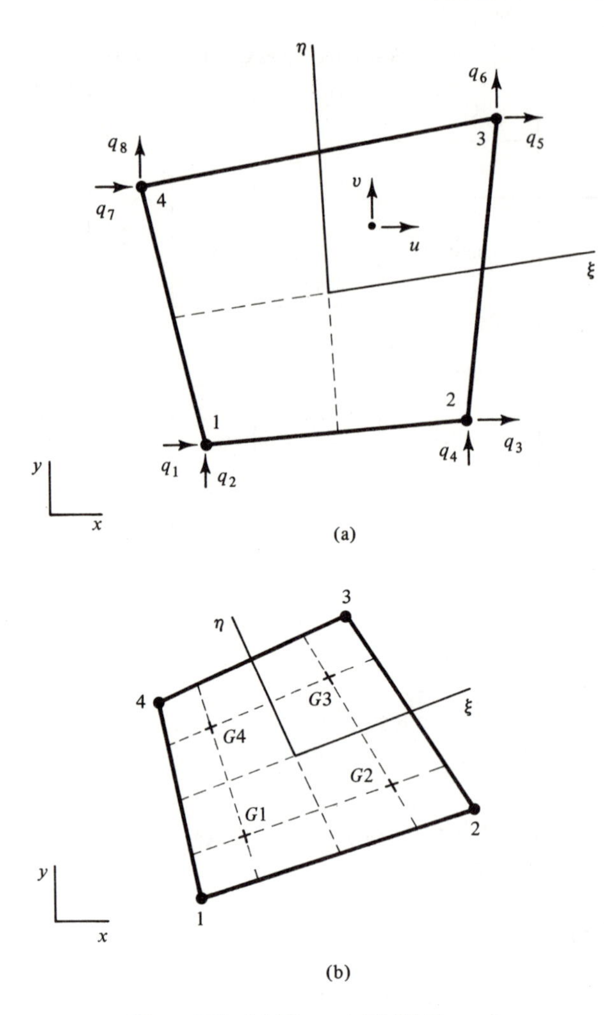

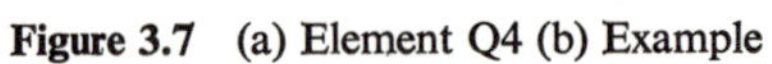
Figure 3.7 (a) Element Q4 (b) Example

$$\mathbf{q} = \{q_1, q_2, \ldots, q_8\} = \{u_1, v_1, \ldots, v_4\} \tag{b}$$

Now let us assume the following displacement shape functions:

$$\begin{aligned} u &= \tfrac{1}{4}[(1-\xi)(1-\eta)u_1 + (1+\xi)(1-\eta)u_2 \\ &\qquad + (1+\xi)(1+\eta)u_3 + (1-\xi)(1+\eta)u_4] \\ v &= \tfrac{1}{4}[(1-\xi)(1-\eta)v_1 + (1+\xi)(1-\eta)v_2 \\ &\qquad + (1+\xi)(1+\eta)v_3 + (1-\xi)(1+\eta)v_4] \end{aligned} \tag{3.4-1a}$$

Or,

$$\begin{aligned} u &= f_1 u_1 + f_2 u_2 + f_3 u_3 + f_4 u_4 = \sum_{i=1}^{4} f_i u_i \\ v &= f_1 v_1 + f_2 v_2 + f_3 v_3 + f_4 v_4 = \sum_{i=1}^{4} f_i v_i \end{aligned} \tag{3.4-1b}$$

In these expressions the functions f_1, f_2, f_3, and f_4 are the same as those in the geometric interpolation formulas [see Eqs. (3.2-14) and (3.2-15)]. Therefore, the Q4 element is isoparametric, and previous statements regarding differentiation and integration of functions hold true. Equations (3.4-1b) can also be written in the matrix form:

$$\mathbf{u}_i = \mathbf{f}_i \mathbf{q}_i \qquad (i = 1, 2, 3, 4) \tag{3.4-2}$$

in which

$$\mathbf{f}_i = \begin{bmatrix} 1 & 0 \\ 0 & 1 \end{bmatrix} f_i \tag{3.4-3}$$

The generic displacements $\mathbf{u}_i$ in Eq. (3.4-2) represent the translations at any point due to the displacements $\mathbf{q}_i$ at node i. As a further efficiency of notation, we can write the functions f_i as:

$$f_i = \tfrac{1}{4}(1 + \xi_0)(1 + \eta_0) \tag{3.4-4}$$

where

$$\xi_0 = \xi_i \xi \qquad \eta_0 = \eta_i \eta \tag{3.4-5}$$

The values of ξ_i and η_i for this element are given in Table 3.2.

TABLE 3.2 Nodal Coordinates for Element Q4

i	1	2	3	4
ξ_i	-1	1	1	-1
η_i	-1	-1	1	1

Similarly, the strain-displacement relationships for the Q4 element can be concisely expressed as:

$$\boldsymbol{\epsilon}_i = \mathbf{B}_i \mathbf{q}_i \qquad (i = 1, 2, 3, 4) \tag{3.4-6}$$

where

$$\mathbf{B}_i = \mathbf{d}\ \mathbf{f}_i = \begin{bmatrix} \dfrac{\partial}{\partial x} & 0 \\ 0 & \dfrac{\partial}{\partial y} \\ \dfrac{\partial}{\partial y} & \dfrac{\partial}{\partial x} \end{bmatrix} \mathbf{f}_i = \begin{bmatrix} f_{i,x} & 0 \\ 0 & f_{i,y} \\ f_{i,y} & f_{i,x} \end{bmatrix} \tag{3.4-7a}$$

Referring to Eqs. (3.2-26), we see that the submatrix $\mathbf{B}_i$ may also be written as follows:

$$\mathbf{B}_i = \begin{bmatrix} D_{G1i} & 0 \\ 0 & D_{G2i} \\ D_{G2i} & D_{G1i} \end{bmatrix} \tag{3.4-7b}$$

Next, we express the stiffness matrix for the Q4 element (with constant thickness t) in Cartesian coordinates. Thus,

$$\mathbf{K} = t \int_A \mathbf{B}^{\mathrm{T}}(x, y)\mathbf{E}\ \mathbf{B}(x, y)\, dx\, dy \tag{c}$$

However, in natural coordinates this formula becomes:

$$\mathbf{K} = t \int_{-1}^{1} \int_{-1}^{1} \mathbf{B}^{\mathrm{T}}(\xi, \eta)\mathbf{E}\ \mathbf{B}(\xi, \eta)\, |\mathbf{J}(\xi, \eta)|\, d\xi\, d\eta \tag{3.4-8}$$

Similarly, equivalent nodal loads due to body forces may be written in natural coordinates as:

$$\mathbf{p}_b = t \int_{-1}^{1} \int_{-1}^{1} \mathbf{f}^{\mathrm{T}}(\xi, \eta)\mathbf{b}(\xi, \eta)\, |\mathbf{J}(\xi, \eta)|\, d\xi\, d\eta \tag{3.4-9}$$

In addition, equivalent nodal loads due to initial strains in natural coordinates are:

$$\mathbf{p}_0 = t \int_{-1}^{1} \int_{-1}^{1} \mathbf{B}^{\mathrm{T}}(\xi, \eta)\mathbf{E}\ \boldsymbol{\epsilon}_0(\xi, \eta)\, |\mathbf{J}(\xi, \eta)|\, d\xi\, d\eta \tag{3.4-10}$$

Except in special cases, the integrals in Eqs. (3.4-8) and (3.4-9) must be performed by numerical integration. However, if the element is rectangular, direct explicit integration may be used, as in Sec. 2.3. Also, line loadings with ξ or η constant may be handled by explicit line integrations. Of course, if the body forces consist of point loads, no integration is required at all. Furthermore, note that the determinant of the Jacobian matrix appears in the denominators of all the terms in matrix $\mathbf{B}$ [see Eqs. (3.2-26) and (3.4-7b)]. Therefore, the determinant $|\mathbf{J}|$ in Eq. (3.4-10) cancels throughout; so the equivalent loads for initial strains may be integrated either explicitly or numerically. If a temperature change ΔT varies bilinearly, it is defined as:

$$\Delta T = \sum_{i=1}^{4} f_i\, \Delta T_i \tag{d}$$

Example

Derive numerically the stiffness term K_{12} for the isoparametric Q4 element in Fig. 3.7(b), using Gaussian integration with $n = 2$ each way. Assume that the thickness t is constant and that the coordinates of nodes 1, 2, 3, and 4 are (7, 1), (20, 5), (14, 14), and (5, 10), respectively. The formula for numerical integration of terms in $\mathbf{K}$ is:

$$\mathbf{K} = t \sum_{k=1}^{n} \sum_{j=1}^{n} R_j R_k \mathbf{B}^{\mathrm{T}}(\xi_j, \eta_k)\mathbf{E}\ \mathbf{B}(\xi_j, \eta_k)\, |\mathbf{J}(\xi_j, \eta_k)| \tag{e}$$

Matrices $\mathbf{B}$ and $\mathbf{J}$ in this expression are functions of the coordinates ξ_j and η_k for the integration points (or Gauss points) $G1$, $G2$, $G3$, and $G4$ shown in the figure. In particular, for K_{12} with $n = 2$, $R_j = R_k = 1$ so that

$$K_{12} = t \sum_{k=1}^{2} \sum_{j=1}^{2} \mathbf{B}_{1,1}^{\mathrm{T}}\mathbf{E}\ \mathbf{B}_{1,2}\, |\mathbf{J}| \tag{f}$$

In this formula the symbol $\mathbf{B}_{1,1}$ denotes the first column of submatrix $\mathbf{B}_1$, and $\mathbf{B}_{1,2}$ is the second column [see Eq. (3.4-7b)]. Substituting these columns into Eq. (f), we obtain:

$$K_{12} = t(E_{12} + E_{33}) \sum_{k=1}^{2} \sum_{j=1}^{2} D_{G11} D_{G21} |\mathbf{J}| \tag{g}$$

in which the terms E_{12} and E_{33} are left in unspecified form.

To evaluate Eq. (g), we first calculate the Jacobian matrix as:

$$\mathbf{J} = \mathbf{D}_L \mathbf{C}_N = \frac{1}{4} \begin{bmatrix} -(1-\eta) & (1-\eta) & (1+\eta) & -(1+\eta) \\ -(1-\xi) & -(1+\xi) & (1+\xi) & (1-\xi) \end{bmatrix} \begin{bmatrix} 7 & 1 \\ 20 & 5 \\ 14 & 14 \\ 5 & 10 \end{bmatrix}$$

$$= \frac{1}{2} \begin{bmatrix} 11 - 2\eta & 4 \\ -4 - 2\xi & 9 \end{bmatrix}$$

Then the determinant of $\mathbf{J}$ is:

$$|\mathbf{J}| = \frac{1}{4}(115 - 18\eta + 8\xi)$$

Terms required from matrix $\mathbf{D}_G$ [see Eqs. (3.2-26)] are:

$$D_{G11} = \frac{1}{4|\mathbf{J}|}[-(1-\eta)J_{22} + (1-\xi)J_{12}] = \frac{-5 + 9\eta - 4\xi}{2(115 - 18\eta + 8\xi)}$$

$$D_{G21} = \frac{1}{4|\mathbf{J}|}[(1-\eta)J_{21} - (1-\xi)J_{11}] = \frac{3(-5 + 2\eta + 3\xi)}{2(115 - 18\eta + 8\xi)}$$

Evaluating the product $D_{G11} D_{G21} |\mathbf{J}|$ at each of the four integration points and summing the results as indicated by Eq. (g) produces:

$$K_{12} = 0.1578t(E_{12} + E_{33})$$

which can be finalized using numerical values of t, E_{12}, and E_{33}.

Figure 3.8(a) shows the rectangular parent of the *isoparametric quadrilateral* (*Q8*) element, which appears in Fig. 3.8(b). In order to understand this higher-order isoparametric element, it is helpful to study first its rectangular parent. Nodal displacements for either element consist of x and y translations at each node. Thus,

$$\mathbf{q} = \{q_1, q_2, \ldots, q_{16}\} = \{u_1, v_1, \ldots, v_8\} \tag{h}$$

The following displacement functions are assumed:

$$\begin{aligned} u &= c_1 + c_2\xi + c_3\eta + c_4\xi^2 + c_5\xi\eta + c_6\eta^2 + c_7\xi^2\eta + c_8\xi\eta^2 \\ v &= c_9 + c_{10}\xi + c_{11}\eta + c_{12}\xi^2 + c_{13}\xi\eta + c_{14}\eta^2 + c_{15}\xi^2\eta + c_{16}\xi\eta^2 \end{aligned} \tag{i}$$

Corresponding displacement shape functions may be written in the form:

$$u = \sum_{i=1}^{8} f_i u_i \qquad v = \sum_{i=1}^{8} f_i v_i \tag{3.4-11}$$

where

$$\begin{aligned} f_i &= \tfrac{1}{4}(1+\xi_0)(1+\eta_0)(\xi_0 + \eta_0 - 1) \qquad && (i = 1, 2, 3, 4) \\ f_i &= \tfrac{1}{2}(1-\xi^2)(1+\eta_0) && (i = 5, 7) \\ f_i &= \tfrac{1}{2}(1+\xi_0)(1-\eta^2) && (i = 6, 8) \end{aligned} \tag{3.4-12}$$

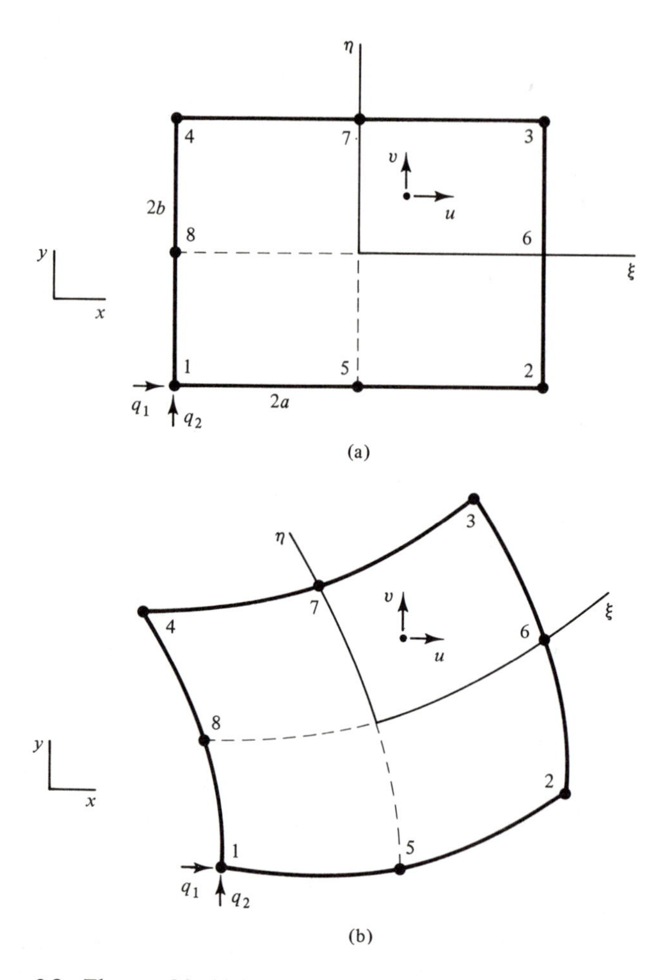

Figure 3.8 Element Q8: (a) Rectangular Parent (b) Isoparametric Counterpart

The values of ξ_i and η_i required in these formulas [see also Eqs. (3.4-5)] are listed in Table 3.3.

TABLE 3.3 Nodal Coordinates for Element Q8

i	1	2	3	4	5	6	7	8
ξ_i	−1	1	1	−1	0	1	0	−1
η_i	−1	−1	1	1	−1	0	1	0

This rectangle is called a *Serendipity element* (1) because nodes appear only on its edges. To understand physically the displacement shape functions, let us consider node 1, where $\xi_1 = \eta_1 = -1$. From Eqs. (3.4-12) we have:

$$f_1 = \tfrac{1}{4}(1 - \xi)(1 - \eta)(-\xi - \eta - 1) \tag{j}$$

This function may be visualized as the combination of three patterns, as shown in Figs. 3.9(a), (b), (c), and (d). Thus,

$$\begin{aligned} f_1 = f_d &= f_a - \frac{1}{2} f_b - \frac{1}{2} f_c \\ &= \tfrac{1}{4}[(1 - \xi)(1 - \eta) - (1 - \xi^2)(1 - \eta) - (1 - \xi)(1 - \eta^2)] \\ &= \tfrac{1}{4}(1 - \xi)(1 - \eta)(-\xi - \eta - 1) \end{aligned}$$

which is the same as Eq. (j). In addition, we can see that $f_5 = f_b$, and so on.

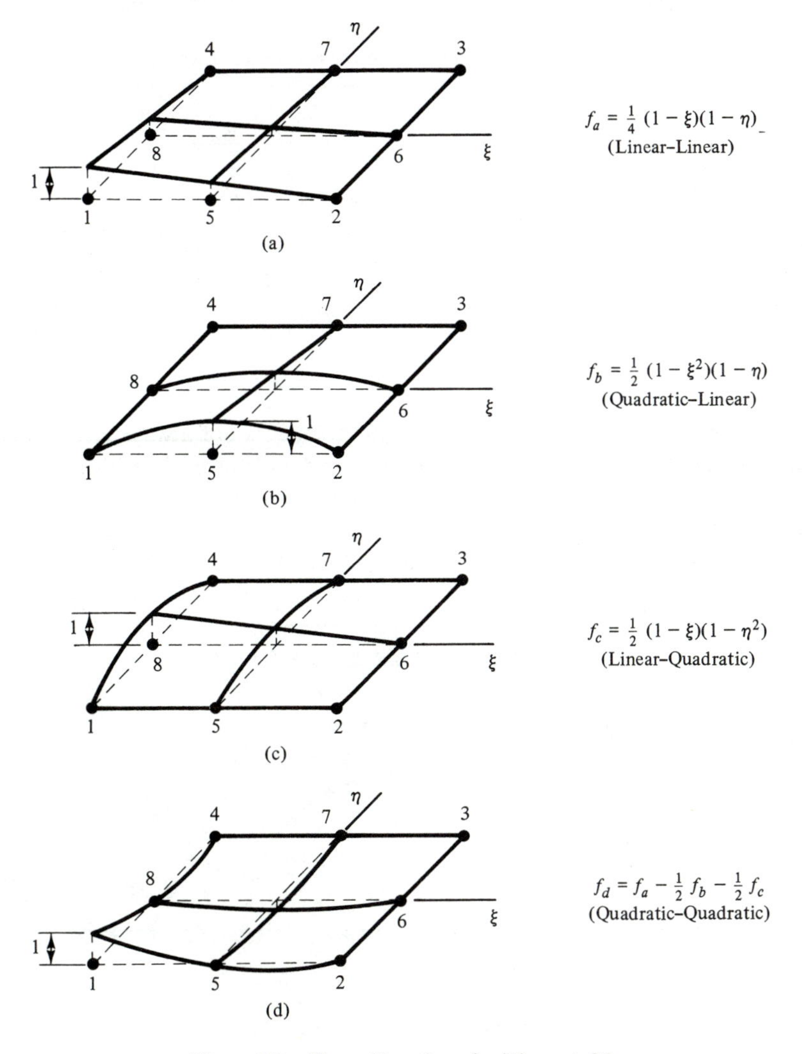

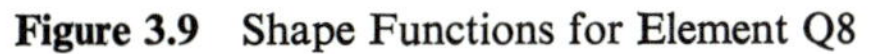

Figure 3.9 Shape Functions for Element Q8

Turning now to the isoparametric Q8 element in Fig. 3.8(b), we take the geometric interpolation functions to be the same as the displacement shape functions in Eqs. (3.4-12). Physically, this means that the natural coordinates ξ and η are curvilinear, and all sides of the element become quadratic curves (6). Thus,

$$x = \sum_{i=1}^{8} f_i x_i \qquad y = \sum_{i=1}^{8} f_i y_i \tag{3.4-13}$$

The formulation of stiffnesses and equivalent nodal loads for element Q8 is very similar to that for element Q4 given earlier. Table 3.4 contains the necessary functions and their derivatives in explicit form. Numerical integration also follows the same pattern as before, even though the local coordinates are curved.

TABLE 3.4 Shape Functions and Derivatives for Element Q8

i	f_i	$f_{i,\xi}$	$f_{i,\eta}$
1	$\frac{1}{4}(1-\xi)(1-\eta)(-\xi-\eta-1)$	$\frac{1}{4}(2\xi+\eta)(1-\eta)$	$\frac{1}{4}(1-\xi)(2\eta+\xi)$
2	$\frac{1}{4}(1+\xi)(1-\eta)(\xi-\eta-1)$	$\frac{1}{4}(2\xi-\eta)(1-\eta)$	$\frac{1}{4}(1+\xi)(2\eta-\xi)$
3	$\frac{1}{4}(1+\xi)(1+\eta)(\xi+\eta-1)$	$\frac{1}{4}(2\xi+\eta)(1+\eta)$	$\frac{1}{4}(1+\xi)(2\eta+\xi)$
4	$\frac{1}{4}(1-\xi)(1+\eta)(-\xi+\eta-1)$	$\frac{1}{4}(2\xi-\eta)(1+\eta)$	$\frac{1}{4}(1-\xi)(2\eta-\xi)$
5	$\frac{1}{2}(1-\xi^2)(1-\eta)$	$-\xi(1-\eta)$	$-\frac{1}{2}(1-\xi^2)$
6	$\frac{1}{2}(1+\xi)(1-\eta^2)$	$\frac{1}{2}(1-\eta^2)$	$-(1+\xi)\eta$
7	$\frac{1}{2}(1-\xi^2)(1+\eta)$	$-\xi(1+\eta)$	$\frac{1}{2}(1-\xi^2)$
8	$\frac{1}{2}(1-\xi)(1-\eta^2)$	$-\frac{1}{2}(1-\eta^2)$	$-(1-\xi)\eta$

When higher-order elements are available, it is not always clear whether to use them or to refine the network of lower-order elements to achieve the same accuracy. Figure 3.10 depicts a two-dimensional continuum discretized into Q4 elements (shown by dashed lines) and a smaller number of Q8 elements (solid lines). In this case it seems clear that using Q8 elements instead of four times as

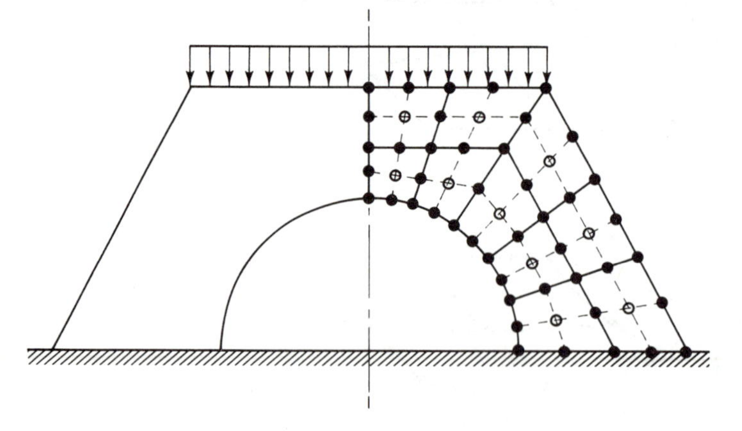

Figure 3.10 Networks for Elements Q4 and Q8

many Q4 elements has advantages. First, we see that the number of nodes is smaller when Q8 elements are used. Second, this type of element is able to model the curved boundary with superior accuracy. Third, we can expect greater numerical accuracy in the results because of quadratic displacement functions in place of linear functions. Of course, the validity of these predictions can only be verified by numerical experimentation with practical applications (see Example 3, Sec. 3.6).

3.5 TRIANGULAR ELEMENTS

In Sec. 2.2 we developed the constant strain triangle. Now the same entity will be referred to as *element T3* (for triangle with three nodes). Because its geometric and displacement shape functions are both linear, this element is seen to be isoparametric. Figure 3.11 shows element T3 with a generic point located by the subareas A_1, A_2, and A_3. As before, displacements for such a point are:

$$\mathbf{u} = \{u, v\} \tag{a}$$

and the nodal displacements are:

$$\mathbf{q} = \{q_1, q_2, \ldots, q_6\} = \{u_1, v_1, \ldots, v_3\} \tag{b}$$

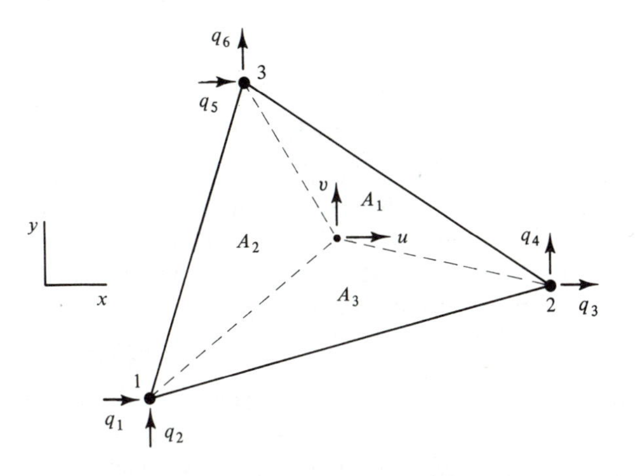

Figure 3.11 Element T3 (CST)

Generic displacements can be written in terms of nodal displacements as follows:

$$\begin{aligned} u &= \sum_{i=1}^{3} \xi_i u_i = \sum_{i=1}^{3} f_i u_i \\ v &= \sum_{i=1}^{3} \xi_i v_i = \sum_{i=1}^{3} f_i v_i \end{aligned} \tag{3.5-1}$$

Comparing these expressions with Eqs. (3.2-8), we see that they have the same form. Thus, the displacement shape functions are given by:

$$
\begin{aligned}
f_1 &= \xi_1 = \frac{1}{2A}(2A_{23} + b_1 x + a_1 y) \\
f_2 &= \xi_2 = \frac{1}{2A}(2A_{31} + b_2 x + a_2 y) \\
f_3 &= \xi_3 = \frac{1}{2A}(2A_{12} + b_3 x + a_3 y)
\end{aligned}
\tag{3.5-2}
$$

which are drawn from Eq. (3.2-10). Derivatives required for strain-displacement relationships are found to be:

$$
\begin{aligned}
f_{1,x} &= \frac{b_1}{2A} \qquad f_{1,y} = \frac{a_1}{2A} \\
f_{2,x} &= \frac{b_2}{2A} \qquad f_{2,y} = \frac{a_2}{2A} \\
f_{3,x} &= \frac{b_3}{2A} \qquad f_{3,y} = \frac{a_3}{2A}
\end{aligned}
\tag{3.5-3}
$$

Then the ith part of the **B** matrix becomes:

$$
\mathbf{B}_i = \begin{bmatrix} f_{i,x} & 0 \\ 0 & f_{i,y} \\ f_{i,y} & f_{i,x} \end{bmatrix} \qquad (i = 1, 2, 3) \tag{3.5-4}
$$

And as in Sec. 2.2, the stiffnesses are obtained from:

$$
\mathbf{K} = \mathbf{B}^{\mathrm{T}}\mathbf{E}\ \mathbf{B}At \tag{3.5-5}
$$

We also looked briefly at the linear strain triangle in Sec. 2.2. That element is shown in Fig. 3.12(a) with its natural coordinates, and its isoparametric counterpart (*element T6*) appears in Fig. 3.12(b). As before, we will approach this isoparametric element by first examining its straight-sided parent. Nodal displacements for either element are:

$$
\mathbf{q} = \{q_1, q_2, \ldots, q_{12}\} = \{u_1, v_1, \ldots, v_6\} \tag{c}
$$

Geometric interpolation functions for the parent element are linear, as is the case for any triangle with straight sides [see Eqs. (3.2-8)]. Quadratic displacement shape functions may be written in natural coordinates as:

$$
u = \sum_{i=1}^{6} f_i u_i \qquad v = \sum_{i=1}^{6} f_i v_i \tag{3.5-6}
$$

where

$$
\begin{aligned}
f_1 &= (2\xi_1 - 1)\xi_1 \qquad f_4 = 4\xi_1\xi_2 \\
f_2 &= (2\xi_2 - 1)\xi_2 \qquad f_5 = 4\xi_2\xi_3 \\
f_3 &= (2\xi_3 - 1)\xi_3 \qquad f_6 = 4\xi_1\xi_3
\end{aligned}
\tag{3.5-7}
$$

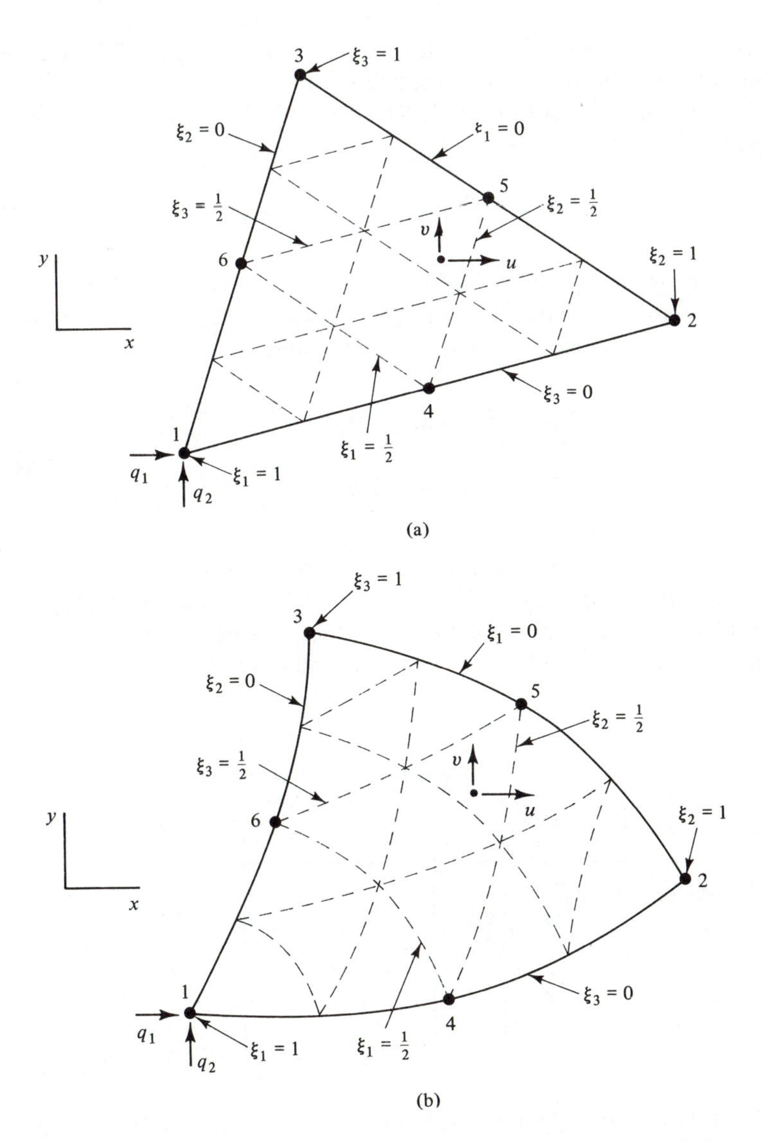

Figure 3.12 Element T6: (a) Triangular Parent (LST) (b) Isoparametric Counterpart

Derivatives for strain-displacement relationships in this subparametric triangle are found by substituting the expressions for ξ_1, ξ_2, and ξ_3 from Eqs. (3.5-2) into the quadratic functions of Eqs. (3.5-7) and differentiating with respect to x and y. Then the stiffness matrix for the element is obtained from:

$$\mathbf{K} = t \int_A \mathbf{B}^{\mathrm{T}}\mathbf{E} \; \mathbf{B} \, dA \tag{3.5-8}$$

Wherever x and y appear in this integral, we substitute Eqs. (3.2-8), and we use area integrals to find the explicit results in Table 2.2.

For the isoparametric triangle T6 in Fig. 3.12(b), we take the geometric interpolation functions to be:

$$x = \sum_{i=1}^{6} f_i x_i \qquad y = \sum_{i=1}^{6} f_i y_i \tag{3.5-9}$$

where f_i are given by Eqs. (3.5-7). Thus, the edges of the element become quadratic curves, as indicated in the figure. Because the natural coordinates are curvilinear, the Jacobian matrix is required. Thus,

$$\begin{aligned} J_{11} &= \frac{\partial x}{\partial \xi_1} = \sum_{i=1}^{6} f_{i,\xi_1} x_i \qquad J_{12} = \frac{\partial y}{\partial \xi_1} = \sum_{i=1}^{6} f_{i,\xi_1} y_i \\ J_{21} &= \frac{\partial x}{\partial \xi_2} = \sum_{i=1}^{6} f_{i,\xi_2} x_i \qquad J_{22} = \frac{\partial y}{\partial \xi_2} = \sum_{i=1}^{6} f_{i,\xi_2} y_i \end{aligned} \tag{3.5-10}$$

Before the differentiations in Eqs. (3.5-10) can be carried out, a dependent coordinate (such as $\xi_3 = 1 - \xi_1 - \xi_2$) must be substituted into the functions in Eqs. (3.5-7). These shape functions and their derivatives with respect to the natural coordinates ξ_1 and ξ_2 are given in Table 3.5 (in terms of ξ_1, ξ_2, and ξ_3).

TABLE 3.5 Shape Functions and Derivatives for Element T6

i	f_i	f_{i,ξ_1}	f_{i,ξ_2}
1	$(2\xi_1 - 1)\xi_1$	$4\xi_1 - 1$	0
2	$(2\xi_2 - 1)\xi_2$	0	$4\xi_2 - 1$
3	$(2\xi_3 - 1)\xi_3$	$1 - 4\xi_3$	$1 - 4\xi_3$
4	$4\xi_1\xi_2$	$4\xi_2$	$4\xi_1$
5	$4\xi_2\xi_3$	$-4\xi_2$	$4(\xi_3 - \xi_2)$
6	$4\xi_1\xi_3$	$4(\xi_3 - \xi_1)$	$-4\xi_1$

Numerical integration is required to find the stiffnesses and equivalent nodal loads for this element. In this case the integration formula is:

$$I = \frac{1}{2} \sum_{j=1}^{n} W_j f(\xi_1, \xi_2, \xi_3)_j \,|\mathbf{J}(\xi_1, \xi_2)_j| \tag{3.5-11}$$

If the triangle were to have straight edges, we would find $|\mathbf{J}| = 2A$; and Eq. (3.5-11) would become the same as Eq. (3.3-4). In any case, the numerical integration constants given in Table 3.1 apply.

3.6 PROGRAMS PSQ4 AND PSQ8 AND APPLICATIONS

We can construct a computer program named PSQ4 that analyzes isotropic continua in plane stress or plane strain using the isoparametric quadrilateral Q4 (see Sec. 3.4). This program is easily conceived by modifying Program PSCST, which was thoroughly described in Sec. 2.5. Moreover, the only subprograms that need to be changed are marked with asterisks in the main program of Appendix D. Of these subprograms, the primary alterations occur in the one named STIFF, for generating the element stiffness matrix. Macro-Flow Chart 3.1 shows the main steps in a new version of STIFF, called STIFQ4. A flow chart of this type gives the computer logic without all of the required statements. The chart implies that the element stiffness matrix is evaluated by summing contributions from four integration points using matrix multiplication. Instead of this "brute-force" approach, it is more efficient to factor matrix **E** by the Cholesky square-root method (see Appendix C) and to use the resulting terms to scale and combine the rows of matrix **B**. Note that only the upper triangular part of the element stiffness matrix need be generated for transferring to the nodal stiffness matrix.

Subprogram STIFQ4 calls upon another subprogram, named BMATQ4, which generates the **B** matrix for element Q4 at every integration point. It also calculates the determinant of the Jacobian matrix that is needed for the evaluation of element stiffnesses. The simple logic of Subprogram BMATQ4 is given in Macro-Flow Chart 3.2. Alternatively, we may code explicit terms directly, instead of using the matrix operations implied in the chart.

Preparation of data for Program PSQ4 is very similar to that for Program PSCST (see Table 2.7 in Sec. 2.5). However, in the structural data [part (d), Element information] each element has four nodes instead of three. Also, in the load data [part (g), Support displacements] every element has eight possible support displacements instead of six.

In Program PSQ4 stresses in the elements are calculated at numerical integration points. Then they are averaged, and one of three printing formats is used, as follows:

IPR = 0: Stresses at integration points are printed (on four lines).

IPR = 1: Average stresses are printed (on one line).

IPR = 2: Both the stresses at integration points and their average values are printed (on five lines).

For this purpose the parameter IPR must be added to the list of load parameters in the data table. Alternatively, stresses could be computed at the geometric

MACRO-FLOW CHART 3.1: SUBPROGRAM STIFQ4 FOR PROGRAM PSQ4

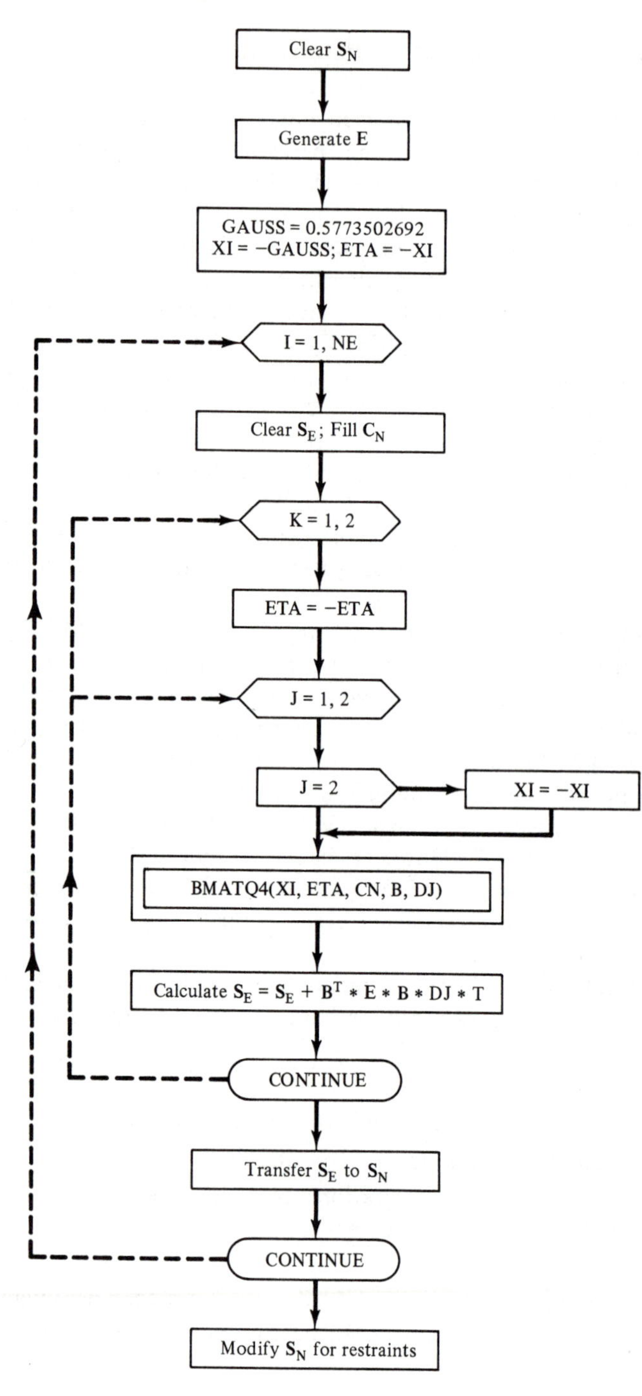

MACRO-FLOW CHART 3.2: SUBPROGRAM BMATQ4 FOR PROGRAM PSQ4

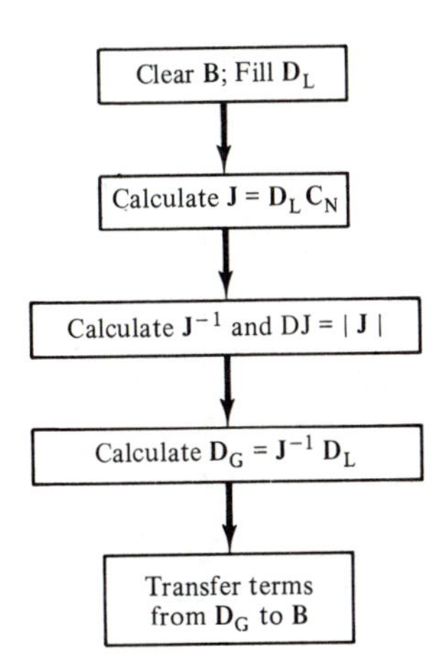

centers of elements or at the nodes. If stresses were computed at the nodes of elements, they should also be averaged for nodes that are common to more than one element.

By extending Program PSQ4, we can also construct a program named PSQ8, which uses the isoparametric quadrilateral Q8 from Sec. 3.4. Of course, the geometric and displacement shape functions must be changed from the bilinear expressions in Eqs. (3.4-1) to the biquadratic formulas in Eqs. (3.4-12). Otherwise, Program PSQ8 is very similar to Program PSQ4. Results from these two programs are compared in Example 3.

Example 1

As a demonstration of Program PSQ4, let us take a single square Q4 element, as shown in Fig. 3.13(a). To make the numbers easy, we use values for the structural parameters as follows:

$$L = 1 \qquad E = 1 \qquad \nu = 0.3 \qquad t = 1$$

where no particular system of units is implied. In addition, if the parameter IPS is set equal to one, the analysis for any in-plane loads will be a plane-strain problem.

Various loading conditions and other influences are depicted in part (b) of Fig. 3.13. For loading 1 we see two nodal forces, each having the value 1.0, applied in the y direction at points 1 and 2. Loading 2 consists of a uniformly distributed force (per unit length) of intensity 1.0 applied in the y direction on edge 1-2. The third loading is a triangular distribution of force (per unit length) of maximum intensity 3.0 on the same edge in the same direction. On the other hand, loading 4 has the same tri-

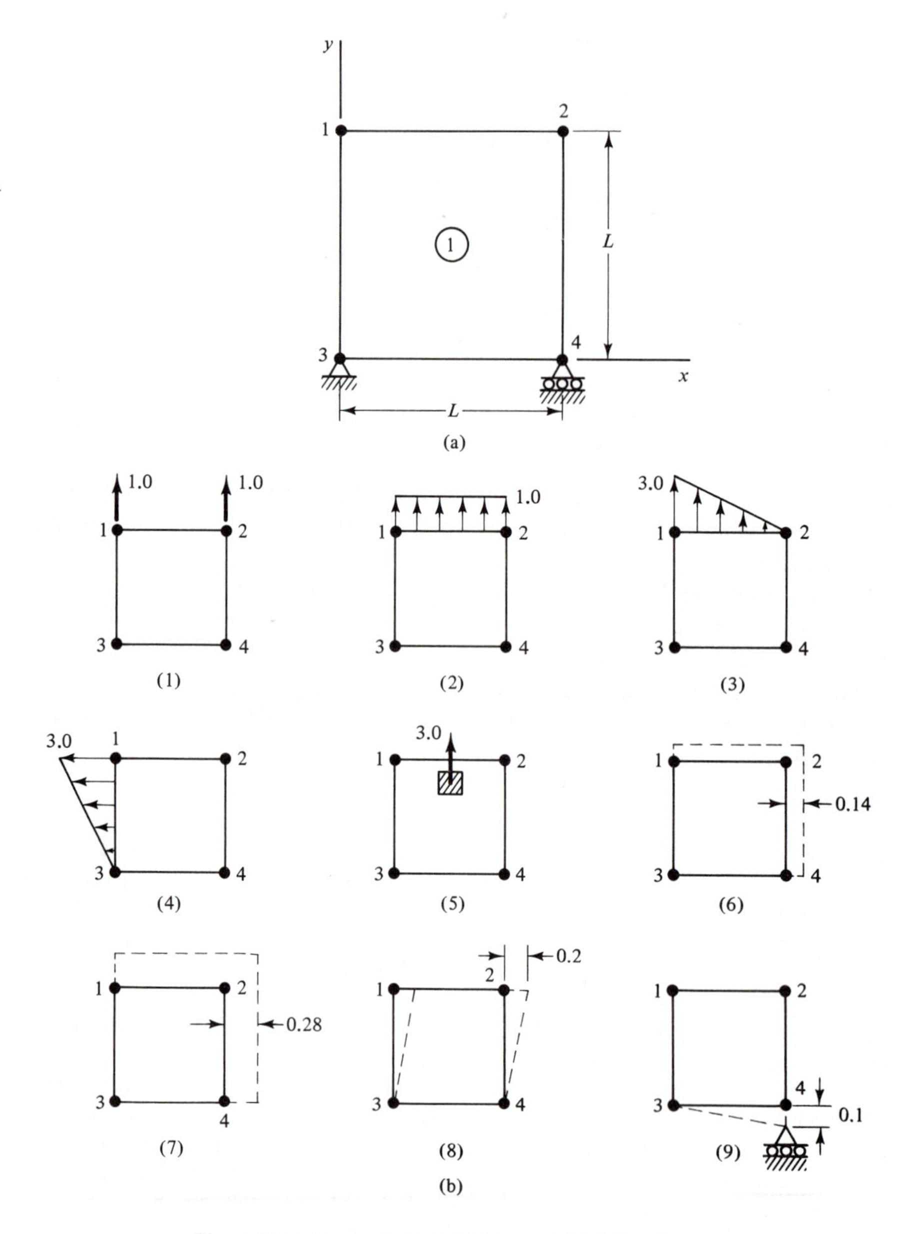

Figure 3.13 Example 1: (a) Element Q4 (b) Loadings

TABLE 3.6 Computer Output for Example 1

```
PROGRAM PSQ4

***  EXAMPLE 1:  ONE SQUARE ELEMENT IN PLANE STRAIN  ***

STRUCTURAL PARAMETERS
   NN   NE  NRN  NLS  IPS           E            PR            T
    4    1    2    9    1  1.0000D 00  3.0000D-01  1.0000D 00

PLANE STRAIN

COORDINATES OF NODES
 NODE          X           Y
    1  0.0000D-01  1.0000D 00
    2  1.0000D 00  1.0000D 00
    3  0.0000D-01  0.0000D-01
    4  1.0000D 00  0.0000D-01

ELEMENT INFORMATION
ELEM.    I    J    K    L
    1    3    4    2    1

NODAL RESTRAINTS
 NODE   R1   R2
    3    1    1
    4    0    1

NUMBER OF DEGREES OF FREEDOM =    5
NUMBER OF NODAL RESTRAINTS   =    3
NUMBER OF TERMS IN SN        =   20

**********  LOADING NUMBER    1  **********
  NLN  NEL  NEV  NET  NEP  NED  IPR
    2    0    0    0    0    0    0

ACTIONS AT NODES
 NODE        AN1         AN2
    1  0.0000D-01  1.0000D 00
    2  0.0000D-01  1.0000D 00

NODAL DISPLACEMENTS
 NODE         DN1         DN2
    1 -6.9389D-17  1.8200D 00
    2 -7.8000D-01  1.8200D 00
    3 -0.0000D-01 -0.0000D-01
    4 -7.8000D-01  0.0000D-01

ELEMENT STRESSES
ELEM. INT.    X STRESS    Y STRESS   XY STRESS    Z STRESS
    1    1 -2.2204D-16  2.0000D 00  5.3376D-18  6.0000D-01
    1    2  0.0000D-01  2.0000D 00  4.8038D-17  6.0000D-01
    1    3  0.0000D-01  2.0000D 00  0.0000D-01  6.0000D-01
    1    4 -2.2204D-16  2.0000D 00  0.0000D-01  6.0000D-01

SUPPORT REACTIONS
 NODE        AR1         AR2
    3  5.1921D-17 -1.0000D 00
    4  0.0000D-01 -1.0000D 00

**********  LOADING NUMBER    2  **********
  NLN  NEL  NEV  NET  NEP  NED  IPR
```

TABLE 3.6 (cont.)

```
     0    1     0     0     0     0     0

LINE LOADS
     I    J           BL1           BL2            BL3           BL4
     1    2   0.0000D-01   0.0000D-01   1.0000D 00   1.0000D 00

NODAL DISPLACEMENTS
 NODE            DN1             DN2
     1 -0.0000D-01   9.1000D-01
     2 -3.9000D-01   9.1000D-01
     3 -0.0000D-01  -0.0000D-01
     4 -3.9000D-01   0.0000D-01

ELEMENT STRESSES
ELEM. INT.      X STRESS     Y STRESS    XY STRESS     Z STRESS
     1     1 -1.3878D-17   1.0000D 00   0.0000D-01   3.0000D-01
     1     2  1.3878D-17   1.0000D 00  -1.6013D-17   3.0000D-01
     1     3 -4.1633D-17   1.0000D 00  -5.3376D-18   3.0000D-01
     1     4 -6.9389D-17   1.0000D 00   1.0675D-17   3.0000D-01

SUPPORT REACTIONS
 NODE           AR1              AR2
     3  4.8886D-18  -5.0000D-01
     4  0.0000D-01  -5.0000D-01

**********  LOADING NUMBER     3  **********
   NLN  NEL   NEV   NET   NEP   NED   IPR
     0    1     0     0     0     0     0

LINE LOADS
     I    J           BL1           BL2            BL3           BL4
     1    2   0.0000D-01   0.0000D-01   3.0000D 00   0.0000D-01

NODAL DISPLACEMENTS
 NODE            DN1             DN2
     1  8.6667D-01   2.2317D 00
     2  2.8167D-01   4.9833D-01
     3 -0.0000D-01  -0.0000D-01
     4 -5.8500D-01   0.0000D-01

ELEMENT STRESSES
ELEM. INT.      X STRESS     Y STRESS    XY STRESS     Z STRESS
     1     1  2.8868D-01   2.1736D 00   1.9245D-01   7.3868D-01
     1     2 -2.8868D-01   8.2642D-01   1.9245D-01   1.6132D-01
     1     3 -2.8868D-01   8.2642D-01  -1.9245D-01   1.6132D-01
     1     4  2.8868D-01   2.1736D 00  -1.9245D-01   7.3868D-01

SUPPORT REACTIONS
 NODE           AR1              AR2
     3  4.9440D-17  -1.0000D 00
     4  0.0000D-01  -5.0000D-01

**********  LOADING NUMBER     4  **********
   NLN  NEL   NEV   NET   NEP   NED   IPR
     0    1     0     0     0     0     0

LINE LOADS
     I    J           BL1           BL2            BL3           BL4
     1    3 -3.0000D 00   0.0000D-01   0.0000D-01   0.0000D-01
```

TABLE 3.6 (cont.)

```
NODAL DISPLACEMENTS
 NODE          DN1          DN2
    1 -6.0667D 00 -1.7333D 00
    2 -4.3333D 00  1.7333D 00
    3 -0.0000D-01 -0.0000D-01
    4 -1.7333D 00  0.0000D-01

ELEMENT STRESSES
ELEM. INT.    X STRESS     Y STRESS    XY STRESS     Z STRESS
    1    1 -1.9245D 00 -1.9245D 00 -1.7698D 00 -1.1547D 00
    1    2 -7.6980D-01  7.6980D-01 -1.0000D 00 -1.7070D-16
    1    3  1.9245D 00  1.9245D 00 -2.3020D-01  1.1547D 00
    1    4  7.6980D-01 -7.6980D-01 -1.0000D 00  1.0408D-16

SUPPORT REACTIONS
 NODE          AR1          AR2
    3  1.5000D 00  1.0000D 00
    4  0.0000D-01 -1.0000D 00

**********  LOADING NUMBER     5 **********
  NLN  NEL  NEV  NET  NEP  NED  IPR
    0    0    1    0    0    0    0

VOLUME LOADS
ELEM.          BV1          BV2
    1  0.0000D-01  3.0000D 00

NODAL DISPLACEMENTS
 NODE          DN1          DN2
    1  6.9389D-17  1.3650D 00
    2 -5.8500D-01  1.3650D 00
    3 -0.0000D-01 -0.0000D-01
    4 -5.8500D-01  0.0000D-01

ELEMENT STRESSES
ELEM. INT.    X STRESS     Y STRESS    XY STRESS     Z STRESS
    1    1 -1.9429D-16  1.5000D 00  2.6688D-17  4.5000D-01
    1    2 -1.9429D-16  1.5000D 00  5.8714D-17  4.5000D-01
    1    3 -8.3267D-17  1.5000D 00  0.0000D-01  4.5000D-01
    1    4 -8.3267D-17  1.5000D 00  0.0000D-01  4.5000D-01

SUPPORT REACTIONS
 NODE          AR1          AR2
    3  7.7050D-17 -1.5000D 00
    4  0.0000D-01 -1.5000D 00

**********  LOADING NUMBER     6 **********
  NLN  NEL  NEV  NET  NEP  NED  IPR
    0    0    0    1    0    0    0

TEMPERATURE STRAINS
ELEM.          TS
    1  1.4000D-01

NODAL DISPLACEMENTS
 NODE          DN1          DN2
    1 -1.3878D-17  1.4000D-01
    2  1.4000D-01  1.4000D-01
    3 -0.0000D-01 -0.0000D-01
    4  1.4000D-01  0.0000D-01
```

TABLE 3.6 (cont.)

```
ELEMENT STRESSES
ELEM. INT.    X STRESS     Y STRESS     XY STRESS     Z STRESS
    1    1 -1.3344D-17 -6.6720D-17 -4.0032D-18 -2.4019D-17
    1    2  2.6688D-18 -2.9357D-17 -9.6744D-18 -8.0064D-18
    1    3 -1.6013D-17 -3.7363D-17 -5.3376D-18 -1.6013D-17
    1    4 -3.2026D-17 -7.4727D-17  0.0000D-01 -3.2026D-17

SUPPORT REACTIONS
 NODE         AR1          AR2
    3  6.2252D-18  3.4393D-17
    4  0.0000D-01  1.7649D-17

**********  LOADING NUMBER     7  **********
  NLN  NEL  NEV  NET  NEP  NED  IPR
    0    0    0    0    1    0    0

PRESTRAINS
ELEM.         PS1          PS2          PS3
    1  2.8000D-01  2.8000D-01  0.0000D-01

NODAL DISPLACEMENTS
 NODE         DN1          DN2
    1 -6.9389D-17  2.8000D-01
    2  2.8000D-01  2.8000D-01
    3 -0.0000D-01 -0.0000D-01
    4  2.8000D-01  0.0000D-01

ELEMENT STRESSES
ELEM. INT.    X STRESS     Y STRESS     XY STRESS     Z STRESS
    1    1 -2.4019D-17 -5.6045D-17 -1.7014D-17 -2.4019D-17
    1    2 -1.6013D-17 -3.7363D-17  6.0048D-18 -1.6013D-17
    1    3  4.0032D-17 -1.3344D-17  1.0675D-17  8.0064D-18
    1    4  3.2026D-17 -3.2026D-17 -1.6013D-17  0.0000D-01

SUPPORT REACTIONS
 NODE         AR1          AR2
    3  9.7167D-18  2.2496D-17
    4  0.0000D-01  1.2198D-17

**********  LOADING NUMBER     8  **********
  NLN  NEL  NEV  NET  NEP  NED  IPR
    0    0    0    0    1    0    0

PRESTRAINS
ELEM.         PS1          PS2          PS3
    1  0.0000D-01  0.0000D-01  2.0000D-01

NODAL DISPLACEMENTS
 NODE         DN1          DN2
    1  2.0000D-01  6.3452D-18
    2  2.0000D-01 -7.4795D-18
    3 -0.0000D-01 -0.0000D-01
    4 -1.0111D-17  0.0000D-01

ELEMENT STRESSES
ELEM. INT.    X STRESS     Y STRESS     XY STRESS     Z STRESS
    1    1 -5.0304D-18  1.6064D-18  0.0000D-01 -1.0272D-18
    1    2 -9.6353D-18 -9.1382D-18 -5.3376D-18 -5.6320D-18
    1    3 -2.6296D-18 -6.1358D-18 -5.3376D-18 -2.6296D-18
```

TABLE 3.6 (cont.)

```
    1     4   1.9752D-18   4.6088D-18  -5.3376D-18   1.9752D-18

SUPPORT REACTIONS
 NODE          AR1            AR2
    3   4.5426D-18   1.1979D-18
    4   0.0000D-01   1.0668D-18

**********  LOADING NUMBER     9  **********
  NLN  NEL  NEV  NET  NEP  NED  IPR
    0    0    0    0    0    1    0

SUPPORT DISPLACEMENTS
ELEM.        DR1,5         DR2,6          DR3,7          DR4,8
    1   0.0000D-01   0.0000D-01   0.0000D-01  -1.0000D-01
        0.0000D-01   0.0000D-01   0.0000D-01   0.0000D-01

NODAL DISPLACEMENTS
 NODE          DN1           DN2
    1   1.0000D-01  -5.7829D-18
    2   1.0000D-01  -1.0000D-01
    3  -0.0000D-01  -0.0000D-01
    4  -4.5855D-18  -1.0000D-01

ELEMENT STRESSES
ELEM. INT.     X STRESS     Y STRESS    XY STRESS     Z STRESS
    1    1  -3.7988D-18  -6.6400D-18  -4.8676D-18  -3.1316D-18
    1    2  -1.8726D-18  -2.1455D-18  -1.8652D-18  -1.2054D-18
    1    3  -7.0504D-19  -1.6451D-18  -4.2506D-18  -7.0504D-19
    1    4  -2.6312D-18  -6.1396D-18  -8.9211D-18  -2.6312D-18

SUPPORT REACTIONS
 NODE          AR1            AR2
    3   4.3363D-18   4.7434D-18
    4   0.0000D-01  -6.0082D-19
```

angular distribution of force applied in the negative x direction on edge 1-3. For loading 5 we have a uniformly distributed force (per unit volume) of intensity 3.0 acting in the y direction. Loading 6 involves a uniform increase of temperature ΔT that produces a uniform normal prestrain $\alpha\,\Delta T(1 + \nu)$ of 0.14 in all directions. Similarly, loading 7 is caused by a uniform normal prestrain of 0.28 in both the x and y directions. Also, loading 8 shows a uniform shearing prestrain of 0.2. Last, loading 9 is due to a support displacement of 0.1 in the negative y direction at node 4.

The computer output for this example is given in Table 3.6. As before, the first page of the table contains a print of the structural data. Appearing next are the results of calculations for each of the nine loading cases. We see from these results that the program functions correctly for plane strain, and it works equally well for plane stress.

Example 2

Shown in Fig. 3.14(a) is the thin, square plate with a circular hole; this problem was used previously as Example 2 in Sec. 2.5. We take the same physical parameters as before and discretize the shaded quarter into a network of Q4 elements, as indicated in Fig. 3.14(b).

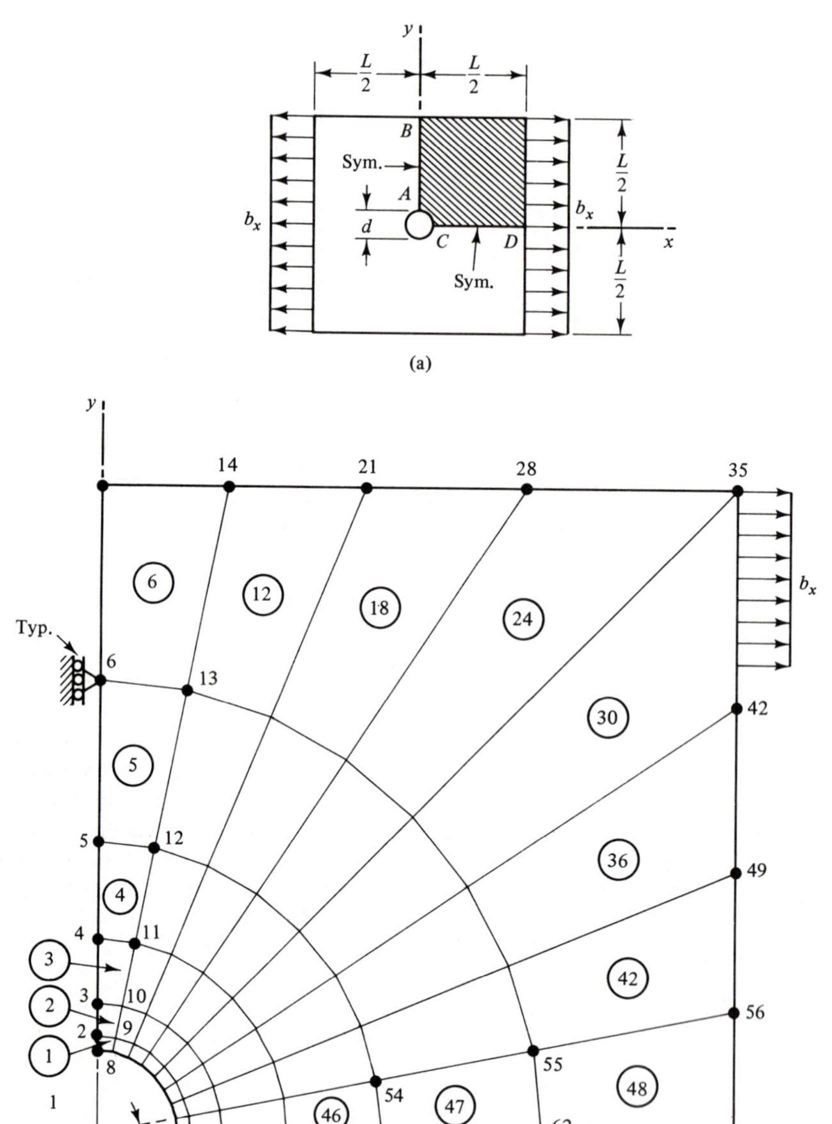

Figure 3.14 Example 2: (a) Problem (b) Q4 Network

Figure 3.15 shows graphs of the stress ratios $\sigma_x t/b_x$ and $\sigma_y t/b_x$ near lines AB and CD. They were obtained using Program PSQ4, and the plotted points represent average stress ratios for elements 1 through 6 (near line AB) and for elements 43 through 48 (near line CD). These graphs are very similar to those in Fig. 2.17 (see Sec. 2.5). However, the absolute values of the ordinates in Fig. 3.15 are smaller than those in Fig. 2.17 because the centers of the Q4 elements do not lie on the lines AB and CD. On the other hand, if the values for the two triangles within each quadrilateral were averaged [compare Figs. 2.16(b) and 3.14(b)], the resulting graphs of stress ratios would be indistinguishable from those in Fig. 3.15.

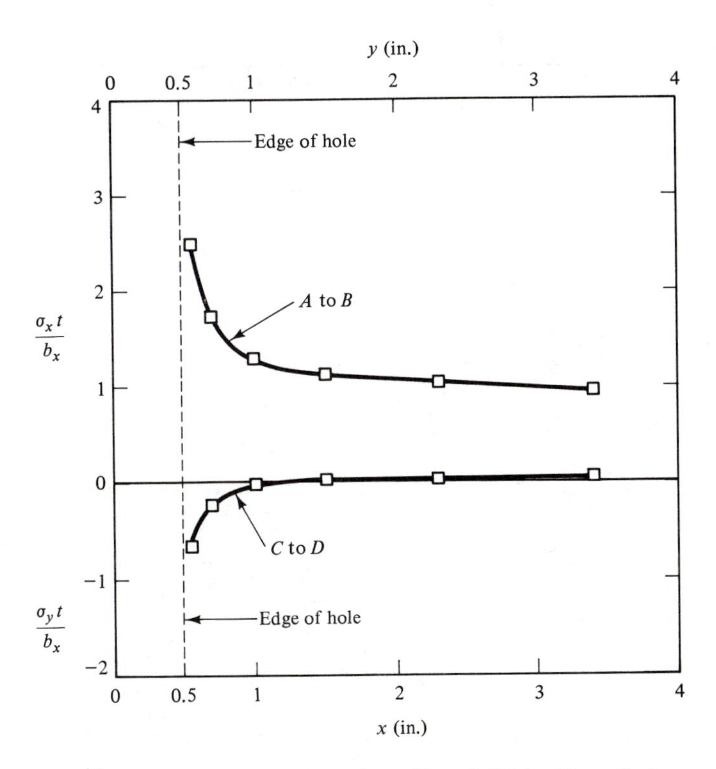

Figure 3.15 Stress Ratios Near Lines AB and CD for Example 2

Example 3

Figure 3.16(a) shows again the concrete culvert analyzed before as Example 3 in Sec. 2.5. We assume the same physical parameters given previously and discretize the right-hand half of the problem into a network of 48 Q4 elements, as given in Fig. 3.16(b). Alternatively, we can use the network of only twelve Q8 elements that is depicted in Fig. 3.16(c).

Graphs of the stress ratios σ_x/b_y (near line AB), $\sigma_{y'}/b_y$ (on line CD), and σ_y/b_y

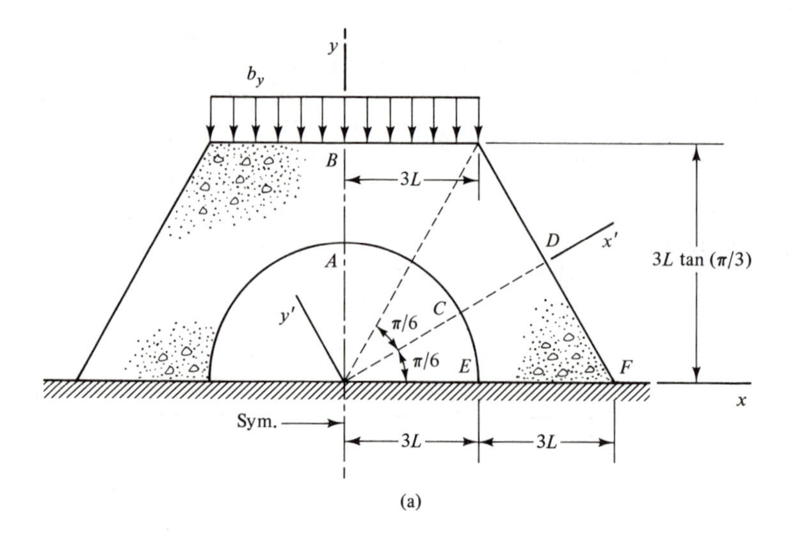

(a)

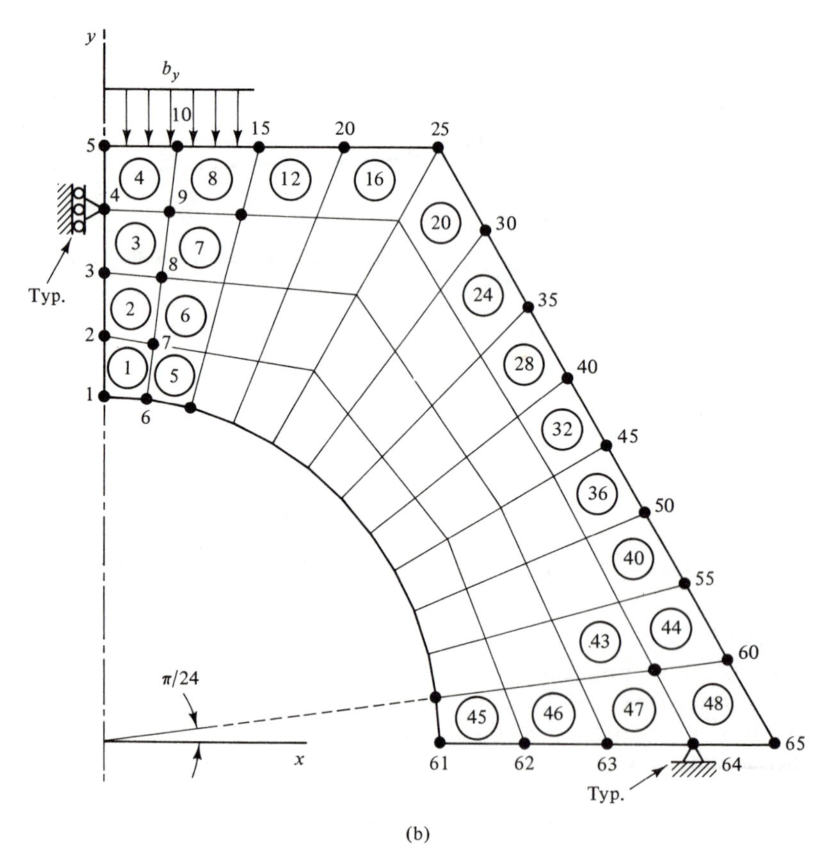

(b)

Figure 3.16 Example 3: (a) Culvert (b) Q4 Network (c) Q8 Network

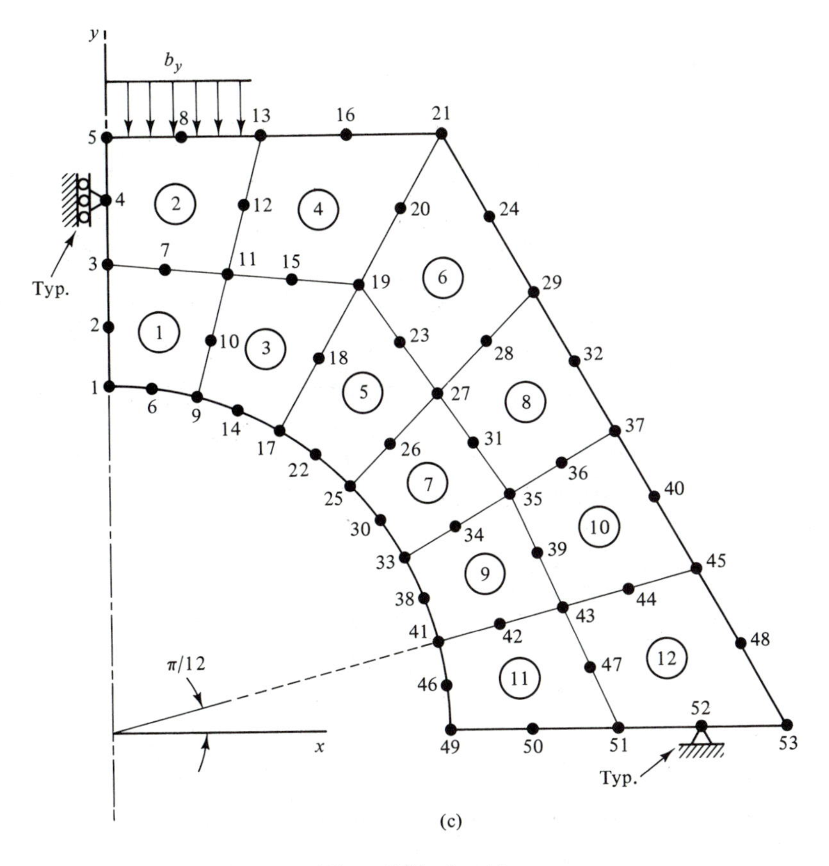

Figure 3.16 (cont.)

(near line *EF*) are plotted in Fig. 3.17. Each graph shows the results from Programs PSCST, PSQ4, and PSQ8. In the first and third graphs the plotted points represent the following values of stress ratios: CST—average from two triangles within a Q4 element; Q4—average of four integration points; Q8—value from the nearest integration point. On the other hand, the stress ratios plotted in the second graph (for line *CD*) have the values: CST—average from four triangles (two on each side of *CD*); Q4—average from two quadrilaterals (one on each side of *CD*); Q8—average from two integration points (one on each side of *CD*). We see from these graphs that only for the first plotted point is there a significant difference in the values. Because the Q8 element is the most accurate of the three types used, the dot-dash curves in Fig. 3.17 presumably represent the best results. For this example, the execution times on the Stanford IBM 3081 computer were about equal for Programs PSCST and PSQ8, but Program PSQ4 required about 50 percent more time. Therefore, we conclude that Program PSQ8 not only is the most accurate but also is the most efficient.

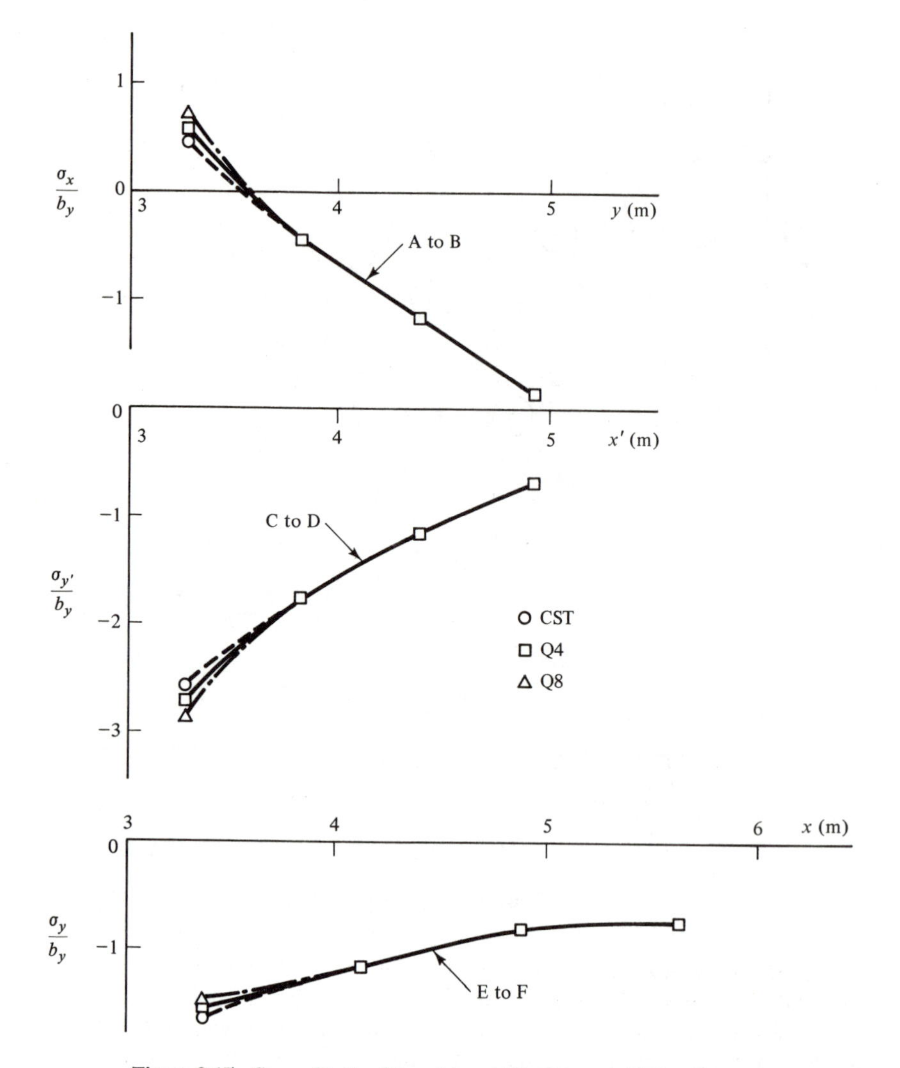

Figure 3.17 Stress Ratios Near Lines *AB*, *CD*, and *EF* for Example 3

REFERENCES

1. Zienkiewicz, O. C., *The Finite Element Method*, 3d ed., McGraw-Hill Ltd., London, 1977.
2. Eisenberg, M. A., and Malvern, L. E., "On Finite Element Integration in Natural Coordinates," *Int. Jour. Num. Meth. Eng.*, Vol. 7, No. 4, 1973, pp. 574–575.
3. Scarborough, J. B., *Numerical Mathematical Analysis*, 6th ed., Johns Hopkins Press, Baltimore, 1966.

4. Cowper, G. R., "Gaussian Quadrature Formulas for Triangles," *Int. Jour. Num. Meth. Eng.*, Vol. 7, No. 3, 1973, pp. 405–408.
5. Irons, B. M., "Engineering Applications of Numerical Integration in Stiffness Methods," *AIAA Jour.*, Vol. 4, No. 11, 1966, pp. 2035–2037.
6. Ergatoudis, B., Irons, B. M., and Zienkiewicz, O. C., "Curved Isoparametric 'Quadrilateral' Elements for Finite Element Analysis," *Int. Jour. Solids Struct.*, Vol. 4, No. 1, 1968, pp. 31–42.

PROBLEMS

3.2-1. Using natural coordinates for a line, derive the result for $\int_L x^4\, dL$ given in Appendix A.

3.2-2. Using area coordinates for a triangle, derive the result for $\int_A xy\, dA$ given in Appendix A.

3.2-3. Using area coordinates for a triangle, derive the result for $\int_A x^3\, dA$ given in Appendix A.

3.2-4. Using area coordinates for a triangle, derive the result for $\int_A x^2y\, dA$ given in Appendix A.

3.2-5. Using area coordinates for a triangle, derive the result for $\int_A x^4\, dA$ given in Appendix A.

3.2-6. Using area coordinates for a triangle, derive the result for $\int_A x^3y\, dA$ given in Appendix A.

3.2-7. Using area coordinates for a triangle, derive the result for $\int_A x^2y^2\, dA$ given in Appendix A.

3.3-1. Derive the values of ξ_i and R_i in Table B.1 for $n = 1$.

3.3-2. Derive the values of ξ_i and R_i in Table B.1 for $n = 2$.

3.3-3. Derive the values of ξ_i and R_i in Table B.1 for $n = 3$.

3.3-4. Using Gaussian numerical integration with $n = 2$, derive the equivalent nodal loads for the axial element (see Sec. 1.5) due to a linearly varying axial load (Fig. 1.9).

3.3-5. For the torsional element (see Sec. 1.5) find equivalent nodal loads due to a parabolically distributed axial torque [Fig. 1.15(b)]. Use Gaussian numerical integration with $n = 2$.

3.3-6. Using Gaussian numerical integration with $n = 2$ each way, derive the stiffness term K_{11} in Table 2.3 for the bilinear-displacement rectangle. Note that $K_{11} = (\mathbf{K}_1)_{11} + (\mathbf{K}_2)_{11}$ in the table.

3.3-7. Repeat Prob. 2.3-6, using Gaussian numerical integration with $n = 2$ each way.

3.3-8. Repeat Prob. 2.3-8, using Gaussian numerical integration with $n = 2$ each way.

3.3-9. Repeat Prob. 3.2-1, using numerical integration with $n = 3$ (see Table B.1).

3.3-10. Repeat Prob. 3.2-2, using numerical integration with $n = 3$ (see Table 3.1).

3.3-11. Repeat Prob. 3.2-3, using numerical integration with $n = 4$ (see Table 3.1).

3.3-12. Repeat Prob. 3.2-4, using numerical integration with $n = 4$ (see Table 3.1).

3.3-13. Repeat Prob. 3.2-5, using numerical integration with $n = 6$ (see Table 3.1).

3.3-14. Repeat Prob. 3.2-6, using numerical integration with $n = 6$ (see Table 3.1).

3.3-15. Repeat Prob. 3.2-7, using numerical integration with $n = 6$ (see Table 3.1).

3.4-1. Repeat the example problem in Sec. 3.4, but find K_{11} instead. Give the numerical results to four significant figures.

3.4-2. Repeat the example problem in Sec. 3.4, but find K_{22} instead. Give the numerical results to four significant figures.

3.4-3. Derive the stiffness term K_{56} for the isoparametric Q4 element in the figure, using Gaussian numerical integration with $n = 2$ each way. Assume that the thickness t is constant and that the coordinates for nodes 1, 2, 3, and 4 are (2, 1), (8, 1), (7, 6), and (1, 5), respectively. Give the numerical results to four significant figures.

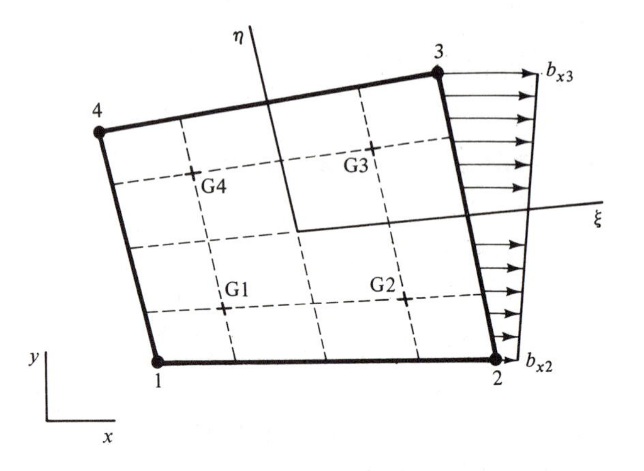

Problems 3.4-3 through 3.4-10

3.4-4. Repeat Prob. 3.4-3, but find K_{55} instead. Give the numerical results to four significant figures.

3.4-5. Repeat Prob. 3.4-3, but find K_{66} instead. Give the numerical results to four significant figures.

3.4-6. Let the Q4 element in the figure be subjected to a uniform gravitational loading $-b_y$ (force per unit volume). Determine the equivalent nodal loads at node 1, using Gaussian numerical integration with $n = 2$ each way.

3.4-7. The figure shows a linearly varying load b_x (force per unit length) applied in the x direction along side 2-3 of the Q4 element. By explicit integration derive the equivalent nodal loads due to such an edge loading. For this purpose, let L_{23} be the length of edge 2-3 (where $\xi = 1$), and express the intensity of b_x in terms of η before integrating.

3.4-8. Repeat Prob. 3.4-7, using Gaussian numerical integration with $n = 2$.

3.4-9. Suppose that the Q4 element in the figure is subjected to a uniform temperature change ΔT. Find the equivalent loads at node 2 due to this influence, using Gaussian numerical integration with $n = 2$ each way. For this purpose, assume that the material is isotropic and in a state of plane stress.

3.4-10. Repeat Prob. 3.4-9 using exact integration.

3.4-11. For the rectangular parent of the Q8 element [see Fig. 3.8(a)], integrate explicitly to find equivalent loads at nodes 1, 2, and 5 due to a uniformly distributed body force b_x (force per unit volume) acting in the x direction. Assume that the thickness t is constant throughout the element.

3.4-12. Repeat Prob. 3.4-11, using Gaussian numerical integration with $n = 2$ each way.

3.4-13. For the isoparametric Q8 element shown in the figure, express the global coordinates of the geometric center g in terms of the global coordinates of nodes 1 through 8.

3.4-14. The figure shows a Q8 element with the x and y components P_x and P_y of a concentrated force applied at the location $(\xi, \eta) = (\frac{1}{4}, \frac{1}{2})$. Find the equivalent loads at nodes 3, 4, and 7 due to these forces.

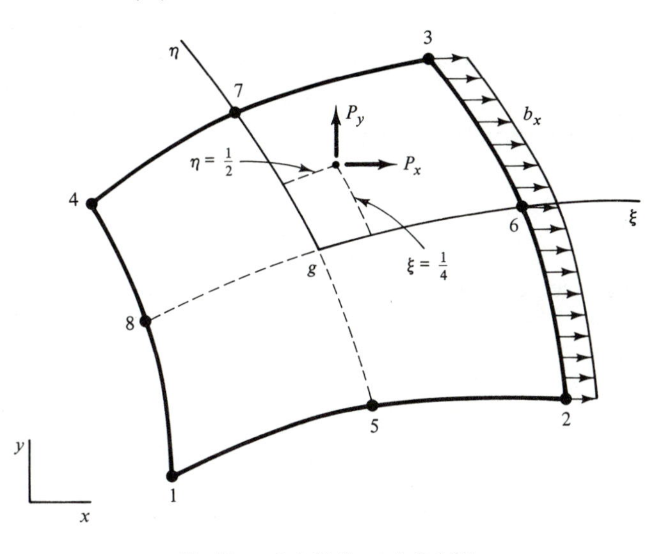

Problems 3.4-13 through 3.4-15

3.4-15. The Q8 element in the figure has a uniformly distributed force b_x (per unit length) applied in the x direction on edge 2-3. Find the equivalent loads at nodes 2, 3, and 6 due to this influence.

CHAPTER

4

General Solids

4.1 STRESSES AND STRAINS IN SOLIDS

In Sec. 1.2 we defined *strain-displacement relationships* for a solid in Cartesian coordinates. Now these relationships will be expressed in matrix form as:

$$\boldsymbol{\epsilon} = \mathbf{d}\ \mathbf{u} \tag{a}$$

in which the displacement vector is:

$$\mathbf{u} = \{u, v, w\} \tag{b}$$

and the strain vector is:

$$\boldsymbol{\epsilon} = \{\epsilon_x, \epsilon_y, \epsilon_z, \gamma_{xy}, \gamma_{yz}, \gamma_{zx}\} \tag{c}$$

The strain-displacement operator **d** in Eq. (a) has the form:

$$\mathbf{d} = \begin{bmatrix} \frac{\partial}{\partial x} & 0 & 0 \\ 0 & \frac{\partial}{\partial y} & 0 \\ 0 & 0 & \frac{\partial}{\partial z} \\ \frac{\partial}{\partial y} & \frac{\partial}{\partial x} & 0 \\ 0 & \frac{\partial}{\partial z} & \frac{\partial}{\partial y} \\ \frac{\partial}{\partial z} & 0 & \frac{\partial}{\partial x} \end{bmatrix} \tag{4.1-1}$$

Stress-strain relationships were also developed in Sec. 1.2 for an *isotropic material* [see Eqs. (1.2-3) and (1.2-4)]. In general, we can write these relationships as:

$$\boldsymbol{\sigma} = \mathbf{E}\,\boldsymbol{\epsilon} \tag{d}$$

where the stress vector is:

$$\boldsymbol{\sigma} = \{\sigma_x, \sigma_y, \sigma_z, \tau_{xy}, \tau_{yz}, \tau_{zx}\} \tag{e}$$

If the material happens to be *anisotropic* (1), the stress-strain operator **E** in Eq. (d) may be expressed as follows:

$$\mathbf{E} = \begin{bmatrix} E_{11} & E_{12} & E_{13} & E_{14} & E_{15} & E_{16} \\ & E_{22} & E_{23} & E_{24} & E_{25} & E_{26} \\ & & E_{33} & E_{34} & E_{35} & E_{36} \\ & & & E_{44} & E_{45} & E_{46} \\ & \text{Sym.} & & & E_{55} & E_{56} \\ & & & & & E_{66} \end{bmatrix} \tag{4.1-2}$$

This matrix has 21 independent constants that define the stress-strain properties. On the other hand, if the material is *orthotropic*, the matrix **E** becomes:

$$\mathbf{E} = \begin{bmatrix} E_{11} & E_{12} & E_{13} & 0 & 0 & 0 \\ & E_{22} & E_{23} & 0 & 0 & 0 \\ & & E_{33} & 0 & 0 & 0 \\ & & & E_{44} & 0 & 0 \\ & \text{Sym.} & & & E_{55} & 0 \\ & & & & & E_{66} \end{bmatrix} \tag{4.1-3}$$

which has nine independent constants. The most simple form of **E** occurs when the material is *isotropic*, in which case we have:

$$\mathbf{E} = \frac{E}{(1+\nu)e_2} \begin{bmatrix} e_1 & \nu & \nu & 0 & 0 & 0 \\ & e_1 & \nu & 0 & 0 & 0 \\ & & e_1 & 0 & 0 & 0 \\ & & & e_3 & 0 & 0 \\ & \text{Sym.} & & & e_3 & 0 \\ & & & & & e_3 \end{bmatrix} \tag{4.1-4}$$

Equation (4.1-4) is the same as Eq. (1.2-4), except for slightly different notation. With this type of material we have only two independent constants, E and ν. Also, the parameters e_1, e_2, and e_3 have the same definitions as those for plane strain (see Sec. 2.2). That is,

$$e_1 = 1 - \nu \qquad e_2 = 1 - 2\nu \qquad e_3 = \frac{e_2}{2}$$

For convenience in *rotation of axes*, the stress vector [Eq. (e)] may be recast into the form of a symmetric 3 × 3 matrix, as follows:

$$\boldsymbol{\sigma} = \begin{bmatrix} \sigma_x & \tau_{xy} & \tau_{xz} \\ \tau_{yx} & \sigma_y & \tau_{yz} \\ \tau_{zx} & \tau_{zy} & \sigma_z \end{bmatrix} \tag{4.1-5}$$

Then the rotation-of-axes transformation for stresses can be stated as:

$$\boldsymbol{\sigma}' = \mathbf{R}\ \boldsymbol{\sigma}\ \mathbf{R}^{\mathrm{T}} \tag{4.1-6}$$

in which $\boldsymbol{\sigma}'$ is similar to $\boldsymbol{\sigma}$, but for primed axes [see Fig. 4.1(a)]. The rotation matrix $\mathbf{R}$ in Eq. (4.1-6) has the form:

$$\mathbf{R} = \begin{bmatrix} \lambda_{11} & \lambda_{12} & \lambda_{13} \\ \lambda_{21} & \lambda_{22} & \lambda_{23} \\ \lambda_{31} & \lambda_{32} & \lambda_{33} \end{bmatrix} = \begin{bmatrix} \ell_1 & m_1 & n_1 \\ \ell_2 & m_2 & n_2 \\ \ell_3 & m_3 & n_3 \end{bmatrix} \tag{41.-7}$$

In this matrix the terms ℓ_1, m_1, and so on, are slightly more efficient symbols for the direction cosines $\lambda_{11}, \lambda_{12}, \ldots$, which were explained in Sec. 1.7.

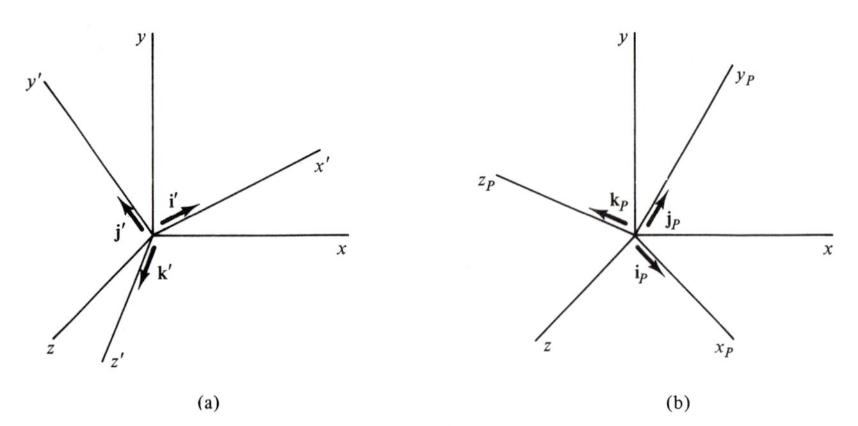

Figure 4.1 Orthogonal Axes: (a) Inclined (b) Principal

Similarly, the strain vector [Eq. (c)] may be recast as the symmetric 3 × 3 matrix:

$$\boldsymbol{\epsilon} = \begin{bmatrix} \epsilon_x & \gamma_{xy} & \gamma_{xz} \\ \gamma_{yx} & \epsilon_y & \gamma_{yz} \\ \gamma_{zx} & \gamma_{zy} & \epsilon_z \end{bmatrix} \tag{4.1-8}$$

for which the rotation transformation is:

$$\boldsymbol{\epsilon}' = \mathbf{R}\ \boldsymbol{\epsilon}\ \mathbf{R}^{\mathrm{T}} \tag{4.1-9}$$

Again, the matrix $\boldsymbol{\epsilon}'$ is similar to $\boldsymbol{\epsilon}$, but for primed axes.

It is possible to calculate *principal normal stresses* and their directions as the solution of an eigenvalue problem (2). By that approach we have:

$$\boldsymbol{\sigma}_P = \mathbf{R}_P \boldsymbol{\sigma} \ \mathbf{R}_P^{\mathrm{T}} \tag{4.1-10}$$

In this equation the symbol $\boldsymbol{\sigma}_P$ represents a diagonal matrix of principal stresses (or *spectral matrix*). Thus,

$$\boldsymbol{\sigma}_P = \begin{bmatrix} \sigma_{P1} & 0 & 0 \\ 0 & \sigma_{P2} & 0 \\ 0 & 0 & \sigma_{P3} \end{bmatrix} \tag{4.1-11}$$

In addition, the symbol $\mathbf{R}_P$ in Eq. (4.1-10) denotes the *rotation matrix for principal axes* [see Fig. 4.1(b)]. This matrix is obtained as the transpose of the *normalized modal matrix*. That is,

$$\mathbf{R}_P = \boldsymbol{\Phi}_N^{\mathrm{T}} \tag{4.1-12}$$

in which the rows are normalized to unit lengths.

Similarly, *principal normal strains* may be calculated as:

$$\boldsymbol{\epsilon}_P = \mathbf{R}_P \boldsymbol{\epsilon} \ \mathbf{R}_P^{\mathrm{T}} \tag{4.1-13}$$

in which the symbol $\boldsymbol{\epsilon}_P$ stands for a diagonal matrix of strains corresponding to $\boldsymbol{\sigma}_P$.

Example

Suppose that in a two-dimensional continuum we have a state of stress at a point exemplified by:

$$\boldsymbol{\sigma} = \begin{bmatrix} \sigma_x & \tau_{xy} \\ \tau_{yx} & \sigma_y \end{bmatrix} \tag{f}$$

Then the eigenvalue problem to be solved is:

$$(\boldsymbol{\sigma} - \lambda_i \mathbf{I})\boldsymbol{\Phi}_i = \mathbf{O} \tag{g}$$

In this expression the symbol λ_i denotes the ith eigenvalue of $\boldsymbol{\sigma}$, and $\boldsymbol{\Phi}_i$ represents the corresponding eigenvector. To find λ_i we set the determinant of the coefficient matrix equal to zero, as follows:

$$|\boldsymbol{\sigma} - \lambda_i \mathbf{I}| = 0 \tag{h}$$

Or,

$$\begin{vmatrix} \sigma_x - \lambda_i & \tau_{xy} \\ \tau_{yx} & \sigma_y - \lambda_i \end{vmatrix} = 0 \tag{i}$$

Expanding this determinant yields:

$$\lambda_i^2 - (\sigma_x + \sigma_y)\lambda_i + \sigma_x \sigma_y - \tau_{xy}^2 = 0 \tag{j}$$

From this expression we obtain the solution:

$$\lambda_i = \frac{\sigma_x + \sigma_y}{2} \pm \sqrt{\left(\frac{\sigma_x - \sigma_y}{2}\right)^2 + \tau_{xy}^2} \tag{k}$$

This formula is the same as Eqs. (2.1-5) and (2.1-6) found previously for principal normal stresses in two dimensions.

When the material is not isotropic, it may be necessary to transform the stress-strain relationships from one set of coordinates to another by rotation of axes. For this purpose, we rewrite the expanded results of Eq. (4.1-9) as:

$$\boldsymbol{\epsilon}' = \mathbf{T}_\epsilon \boldsymbol{\epsilon} \tag{4.1-14}$$

In this equation $\boldsymbol{\epsilon}$ and $\boldsymbol{\epsilon}'$ are in the form of Eq. (c) instead of Eq. (4.1-8). The 6×6 *strain transformation matrix* $\mathbf{T}_\epsilon$ in Eq. (4.1-14) is as follows:

$$\mathbf{T}_\epsilon = \left[\begin{array}{ccc|ccc} \ell_1^2 & m_1^2 & n_1^2 & \ell_1 m_1 & m_1 n_1 & n_1 \ell_1 \\ \ell_2^2 & m_2^2 & n_2^2 & \ell_2 m_2 & m_2 n_2 & n_2 \ell_2 \\ \ell_3^2 & m_3^2 & n_3^2 & \ell_3 m_3 & m_3 n_3 & n_3 \ell_3 \\ \hline 2\ell_1\ell_2 & 2m_1m_2 & 2n_1n_2 & \ell_1 m_2 + \ell_2 m_1 & m_1 n_2 + m_2 n_1 & n_1 \ell_2 + n_2 \ell_1 \\ 2\ell_2\ell_3 & 2m_2m_3 & 2n_2n_3 & \ell_2 m_3 + \ell_3 m_2 & m_2 n_3 + m_3 n_2 & n_2 \ell_3 + n_3 \ell_2 \\ 2\ell_3\ell_1 & 2m_3m_1 & 2n_3n_1 & \ell_3 m_1 + \ell_1 m_3 & m_3 n_1 + m_1 n_3 & n_3 \ell_1 + n_1 \ell_3 \end{array}\right] \tag{4.1-15a}$$

The partition lines in matrix $\mathbf{T}_\epsilon$ separate terms pertaining to normal and shearing strains. Thus,

$$\mathbf{T}_\epsilon = \begin{bmatrix} \mathbf{T}_{\epsilon 11} & \mathbf{T}_{\epsilon 12} \\ \mathbf{T}_{\epsilon 21} & \mathbf{T}_{\epsilon 22} \end{bmatrix} \tag{4.1-15b}$$

in which the subscripts 1 and 2 denote normal and shearing strains, respectively.

In order to discover the form of the *stress transformation matrix*, we equate the *virtual strain energy densities* for the primed and unprimed axes [see Fig. 4.1(a)]:

$$(\delta\boldsymbol{\epsilon}')^{\mathrm{T}}\boldsymbol{\sigma}' = \delta\boldsymbol{\epsilon}^{\mathrm{T}}\boldsymbol{\sigma} \tag{ℓ}$$

Then we substitute the transposed incremental form of Eq. (4.1-14) into Eq. (ℓ) to obtain:

$$\delta\boldsymbol{\epsilon}^{\mathrm{T}}\mathbf{T}_\epsilon^{\mathrm{T}}\boldsymbol{\sigma}' = \delta\boldsymbol{\epsilon}^{\mathrm{T}}\boldsymbol{\sigma} \tag{m}$$

Hence, we conclude that

$$\boldsymbol{\sigma}' = \mathbf{T}_\sigma \boldsymbol{\sigma} \tag{4.1-16}$$

where

$$\mathbf{T}_\sigma = \mathbf{T}_\epsilon^{-\mathrm{T}} \tag{4.1-17}$$

Thus, the stress transformation matrix $\mathbf{T}_\sigma$ is proven to be the transposed inverse of the strain transformation matrix $\mathbf{T}_\epsilon$. The inversion implied by Eq. (4.1-17) is not actually necessary, because expansion of Eq. (4.1-6) shows that $\mathbf{T}_\sigma$ bears the following relationship to $\mathbf{T}_\epsilon$:

$$\mathbf{T}_\sigma = \begin{bmatrix} \mathbf{T}_{\sigma 11} & \mathbf{T}_{\sigma 12} \\ \mathbf{T}_{\sigma 21} & \mathbf{T}_{\sigma 22} \end{bmatrix} = \begin{bmatrix} \mathbf{T}_{\epsilon 11} & 2\mathbf{T}_{\epsilon 12} \\ \frac{1}{2}\mathbf{T}_{\epsilon 21} & \mathbf{T}_{\epsilon 22} \end{bmatrix} \tag{4.1-18}$$

Again, the subscripts 1 and 2 on the submatrices of $\mathbf{T}_\sigma$ refer to normal and shearing stresses, respectively.

Now the transformation of $\mathbf{E}'$ to $\mathbf{E}$ can be implemented by first writing the stress-strain relationships in the primed coordinates as:

$$\boldsymbol{\sigma}' = \mathbf{E}'\boldsymbol{\epsilon}' \tag{n}$$

Next, we substitute Eqs. (4.1-14) and (4.1-16) into Eq. (n) to obtain:

$$\mathbf{T}_\sigma \boldsymbol{\sigma} = \mathbf{E}'\mathbf{T}_\epsilon \boldsymbol{\epsilon} \tag{o}$$

Then premultiply Eq. (o) by $\mathbf{T}_\sigma^{-1}$, and use Eq. (4.1-17) to find:

$$\boldsymbol{\sigma} = \mathbf{T}_\epsilon^{\mathrm{T}}\mathbf{E}'\mathbf{T}_\epsilon \boldsymbol{\epsilon} \tag{p}$$

Thus, we see that

$$\mathbf{E} = \mathbf{T}_\epsilon^{\mathrm{T}}\mathbf{E}'\mathbf{T}_\epsilon \tag{4.1-19}$$

which represents the transformation of $\mathbf{E}'$ to $\mathbf{E}$. The reverse transformation is:

$$\mathbf{E}' = \mathbf{T}_\sigma \mathbf{E} \mathbf{T}_\sigma^{\mathrm{T}} \tag{4.1-20}$$

For stresses and strains in two-dimensional continua (see Sec. 2.1), we have:

$$\begin{aligned} \ell_1 &= \cos\theta & m_1 &= \sin\theta & n_1 &= 0 \\ \ell_2 &= -\sin\theta & m_2 &= \cos\theta & n_2 &= 0 \\ \ell_3 &= 0 & m_3 &= 0 & n_3 &= 1 \end{aligned} \tag{q}$$

Then the stress transformation matrix simplifies to:

$$\mathbf{T}_\sigma = \mathbf{T}_\theta = \begin{bmatrix} \ell_1^2 & m_1^2 & 2\ell_1 m_1 \\ \ell_2^2 & m_2^2 & 2\ell_2 m_2 \\ \ell_1\ell_2 & m_1 m_2 & \ell_1 m_2 + \ell_2 m_1 \end{bmatrix} \tag{4.1-21}$$

which is the same as Eq. (2.1-2). Also, the strain transformation matrix becomes:

$$\mathbf{T}_\epsilon = \mathbf{T}_\theta^{-\mathrm{T}} = \begin{bmatrix} \ell_1^2 & m_1^2 & \ell_1 m_1 \\ \ell_2^2 & m_2^2 & \ell_2 m_2 \\ 2\ell_1\ell_2 & 2m_1 m_2 & \ell_1 m_2 + \ell_2 m_1 \end{bmatrix} \tag{4.1-22}$$

which is the same as Eq. (2.1-14).

4.2 NATURAL COORDINATES

Figure 4.2 shows dimensionless natural coordinates ξ, η, and ζ for a *hexahedron*. At the geometric center (point g) the coordinates are:

$$x_g = \frac{1}{8}\sum_{i=1}^{8} x_i \qquad y_g = \frac{1}{8}\sum_{i=1}^{8} y_i \qquad z_g = \frac{1}{8}\sum_{i=1}^{8} z_i \tag{a}$$

where x_i, y_i, and z_i are the Cartesian coordinates of the corners. Note that $\xi = 1$ on face 2-3-7-6, $\eta = 1$ on face 3-4-8-7, and so on. With linear interpolation in the ξ, η, and ζ directions, the location of any point in the hexahedron may be written as:

$$x = \sum_{i=1}^{8} f_i x_i \qquad y = \sum_{i=1}^{8} f_i y_i \qquad z = \sum_{i=1}^{8} f_i z_i \tag{4.2-1}$$

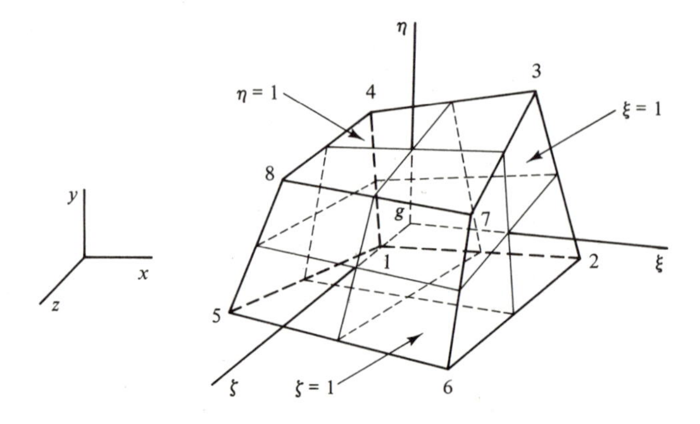

Figure 4.2 Natural Coordinates for a Hexahedron

in which

$$
\begin{aligned}
f_1 &= \tfrac{1}{8}(1-\xi)(1-\eta)(1-\zeta) & f_2 &= \tfrac{1}{8}(1+\xi)(1-\eta)(1-\zeta) \\
f_3 &= \tfrac{1}{8}(1+\xi)(1+\eta)(1-\zeta) & f_4 &= \tfrac{1}{8}(1-\xi)(1+\eta)(1-\zeta) \\
f_5 &= \tfrac{1}{8}(1-\xi)(1-\eta)(1+\zeta) & f_6 &= \tfrac{1}{8}(1+\xi)(1-\eta)(1+\zeta) \\
f_7 &= \tfrac{1}{8}(1+\xi)(1+\eta)(1+\zeta) & f_8 &= \tfrac{1}{8}(1-\xi)(1+\eta)(1+\zeta)
\end{aligned}
\tag{4.2-2}
$$

Because these interpolation formulas are trilinear, the local coordinates ξ, η, and ζ cannot be expressed in terms of the global coordinates x, y, and z.

In three dimensions the chain rule for differentiation with respect to the natural coordinates for a hexahedron leads to the following 3×3 *Jacobian matrix*:

$$
\mathbf{J} = \begin{bmatrix} J_{11} & J_{12} & J_{13} \\ J_{21} & J_{22} & J_{23} \\ J_{31} & J_{32} & J_{33} \end{bmatrix} = \begin{bmatrix} x_{,\xi} & y_{,\xi} & z_{,\xi} \\ x_{,\eta} & y_{,\eta} & z_{,\eta} \\ x_{,\zeta} & y_{,\zeta} & z_{,\zeta} \end{bmatrix}
\tag{4.2-3}
$$

Terms in this matrix are found by the differentiations indicated. Thus

$$
\begin{aligned}
J_{11} &= \sum_{i=1}^{8} f_{i,\xi} x_i & J_{12} &= \sum_{i=1}^{8} f_{i,\xi} y_i & J_{13} &= \sum_{i=1}^{8} f_{i,\xi} z_i \\
J_{21} &= \sum_{i=1}^{8} f_{i,\eta} x_i & J_{22} &= \sum_{i=1}^{8} f_{i,\eta} y_i & J_{23} &= \sum_{i=1}^{8} f_{i,\eta} z_i \\
J_{31} &= \sum_{i=1}^{8} f_{i,\zeta} x_i & J_{32} &= \sum_{i=1}^{8} f_{i,\zeta} y_i & J_{33} &= \sum_{i=1}^{8} f_{i,\zeta} z_i
\end{aligned}
\tag{4.2-4}
$$

As before, these calculations may be arranged in the matrix format:

$$
\mathbf{J} = \mathbf{D}_L \mathbf{C}_N \tag{3.2-18}
$$

repeated

In this instance, the matrix $\mathbf{D}_L$ is the following 3×8 array of derivatives with respect to local coordinates:

$$\mathbf{D}_L = \begin{bmatrix} f_{1,\xi} & f_{2,\xi} & f_{3,\xi} & f_{4,\xi} & f_{5,\xi} & f_{6,\xi} & f_{7,\xi} & f_{8,\xi} \\ f_{1,\eta} & f_{2,\eta} & f_{3,\eta} & f_{4,\eta} & f_{5,\eta} & f_{6,\eta} & f_{7,\eta} & f_{8,\eta} \\ f_{1,\zeta} & f_{2,\zeta} & f_{3,\zeta} & f_{4,\zeta} & f_{5,\zeta} & f_{6,\zeta} & f_{7,\zeta} & f_{8,\zeta} \end{bmatrix}$$

$$= \frac{1}{8}\begin{bmatrix} -(1-\eta)(1-\zeta) & (1-\eta)(1-\zeta) & (1+\eta)(1-\zeta) & -(1+\eta)(1-\zeta) \\ -(1-\xi)(1-\zeta) & -(1+\xi)(1-\zeta) & (1+\xi)(1-\zeta) & (1-\xi)(1-\zeta) \\ -(1-\xi)(1-\eta) & -(1+\xi)(1-\eta) & -(1+\xi)(1+\eta) & -(1-\xi)(1+\eta) \end{bmatrix.$$

$$\begin{matrix} -(1-\eta)(1+\zeta) & (1-\eta)(1+\zeta) & (1+\eta)(1+\zeta) & -(1+\eta)(1+\zeta) \\ -(1-\xi)(1+\zeta) & -(1+\xi)(1+\zeta) & (1+\xi)(1+\zeta) & (1-\xi)(1+\zeta) \\ (1-\xi)(1-\eta) & (1+\xi)(1-\eta) & (1+\xi)(1+\eta) & (1-\xi)(1+\eta) \end{bmatrix} \quad (4.2\text{-}5)$$

Also, the matrix $\mathbf{C}_N$ becomes an 8×3 array of nodal coordinates. Thus,

$$\mathbf{C}_N = \begin{bmatrix} x_1 & y_1 & z_1 \\ x_2 & y_2 & z_2 \\ \cdots & \cdots & \cdots \\ x_8 & y_8 & z_8 \end{bmatrix} \quad (4.2\text{-}6)$$

The *inverse of the Jacobian matrix* may be expressed as:

$$\mathbf{J}^{-1} = \frac{\mathbf{J}^a}{|\mathbf{J}|} = \frac{1}{|\mathbf{J}|}\begin{bmatrix} J^a_{11} & J^a_{12} & J^a_{13} \\ J^a_{21} & J^a_{22} & J^a_{23} \\ J^a_{31} & J^a_{32} & J^a_{33} \end{bmatrix} \quad (4.2\text{-}7)$$

where the symbol $\mathbf{J}^a$ represents the adjoint matrix of $\mathbf{J}$, and $|\mathbf{J}|$ is its determinant. To find the derivatives of all the functions with respect to global coordinates, we have:

$$\mathbf{D}_G = \mathbf{J}^{-1}\mathbf{D}_L \quad (3.2\text{-}24) \text{ repeated}$$

In this case, the matrix $\mathbf{D}_G$ consists of the following terms:

$$\mathbf{D}_G = \begin{bmatrix} f_{1,x} & f_{2,x} & f_{3,x} & f_{4,x} & f_{5,x} & f_{6,x} & f_{7,x} & f_{8,x} \\ f_{1,y} & f_{2,y} & f_{3,y} & f_{4,y} & f_{5,y} & f_{6,y} & f_{7,y} & f_{8,y} \\ f_{1,z} & f_{2,z} & f_{3,z} & f_{4,z} & f_{5,z} & f_{6,z} & f_{7,z} & f_{8,z} \end{bmatrix} \quad (4.2\text{-}8)$$

Explicit evaluation of the 24 items in this array is straightforward but tedious, so the work should be relegated to a digital computer.

A *tetrahedron* of volume V is shown in Fig. 4.3. The corners 1, 2, and 3 are numbered counterclockwise when viewed from corner 4. To locate any point in the tetrahedron, we imagine four subtetrahedra having that point as a common vertex. If those subtetrahedra have volumes V_1, V_2, V_3, and V_4, we can define dimensionless *volume coordinates* as:

$$\xi_1 = \frac{V_1}{V} \qquad \xi_2 = \frac{V_2}{V} \qquad \xi_3 = \frac{V_3}{V} \qquad \xi_4 = \frac{V_4}{V} \quad \text{(b)}$$

Since we know that:

$$V_1 + V_2 + V_3 + V_4 = V \quad \text{(c)}$$

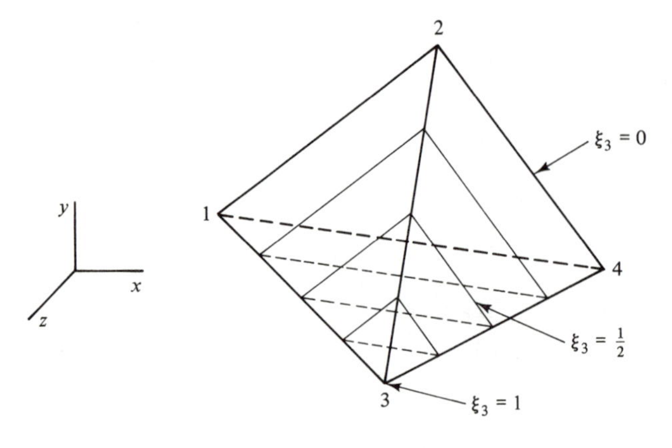

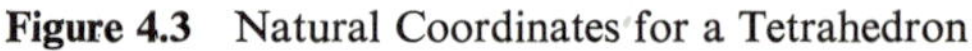
Figure 4.3 Natural Coordinates for a Tetrahedron

then

$$\xi_1 + \xi_2 + \xi_3 + \xi_4 = 1 \tag{4.2-9}$$

This expression shows that ξ_1, ξ_2, ξ_3, and ξ_4 are interdependent. Figure 4.3 indicates that $\xi_3 = 1$ at point 3 and that $\xi_3 = 0$ on the face 1-4-2, which is opposite point 3. Also depicted is a linear variation of ξ_3 from point 3 to the opposite face, and similarly for ξ_1, ξ_2, and ξ_4. When global coordinates x, y, and z are written in terms of the local coordinates, we have:

$$\begin{aligned} x &= \xi_1 x_1 + \xi_2 x_2 + \xi_3 x_3 + \xi_4 x_4 \\ y &= \xi_1 y_1 + \xi_2 y_2 + \xi_3 y_3 + \xi_4 y_4 \\ z &= \xi_1 z_1 + \xi_2 z_2 + \xi_3 z_3 + \xi_4 z_4 \end{aligned} \tag{4.2-10}$$

Conversely, we can express the volume coordinates in terms of x, y, and z by first combining Eqs. (4.2-9) and (4.2-10) into the matrix format:

$$\begin{bmatrix} 1 \\ x \\ y \\ z \end{bmatrix} = \begin{bmatrix} 1 & 1 & 1 & 1 \\ x_1 & x_2 & x_3 & x_4 \\ y_1 & y_2 & y_3 & y_4 \\ z_1 & z_2 & z_3 & z_4 \end{bmatrix} \begin{bmatrix} \xi_1 \\ \xi_2 \\ \xi_3 \\ \xi_4 \end{bmatrix} \tag{4.2-11a}$$

Or,

$$\mathbf{x} = \mathbf{H}\ \boldsymbol{\xi} \tag{4.2-11b}$$

In the latter equation the vectors $\mathbf{x}$ and $\boldsymbol{\xi}$ contain the global and local coordinates appearing in Eq. (4.2-11a), and $\mathbf{H}$ is the coefficient matrix. Next, we solve for the vector $\boldsymbol{\xi}$ in determinantal form:

$$\boldsymbol{\xi} = \mathbf{H}^{-1}\mathbf{x} = \frac{\mathbf{H}^a}{|\mathbf{H}|}\mathbf{x} = \frac{1}{|\mathbf{H}|} \begin{bmatrix} H^a_{11} & H^a_{12} & H^a_{13} & H^a_{14} \\ H^a_{21} & H^a_{22} & H^a_{23} & H^a_{24} \\ H^a_{31} & H^a_{32} & H^a_{33} & H^a_{34} \\ H^a_{41} & H^a_{42} & H^a_{43} & H^a_{44} \end{bmatrix} \begin{bmatrix} 1 \\ x \\ y \\ z \end{bmatrix} \tag{4.2-12}$$

where $\mathbf{H}^a$ is the adjoint matrix of $\mathbf{H}$, and $|\mathbf{H}|$ is its determinant. It can be shown that

$$|\mathbf{H}| = 6V \tag{4.2-13}$$

If we have a function $f(\xi_1, \xi_2, \xi_3, \xi_4)$ to be differentiated with respect to x, y, and z, the chain rule gives:

$$\begin{aligned} \frac{\partial f}{\partial x} &= \sum_{i=1}^{4} \frac{\partial f}{\partial \xi_i} \frac{\partial \xi_i}{\partial x} \\ \frac{\partial f}{\partial y} &= \sum_{i=1}^{4} \frac{\partial f}{\partial \xi_i} \frac{\partial \xi_i}{\partial y} \\ \frac{\partial f}{\partial z} &= \sum_{i=1}^{4} \frac{\partial f}{\partial \xi_i} \frac{\partial \xi_i}{\partial z} \end{aligned} \tag{4.2-14}$$

From Eqs. (4.2-12) and (4.2.13) we see that:

$$\frac{\partial \xi_i}{\partial x} = \frac{H^a_{i2}}{6V} \qquad \frac{\partial \xi_i}{\partial y} = \frac{H^a_{i3}}{6V} \qquad \frac{\partial \xi_i}{\partial z} = \frac{H^a_{i4}}{6V} \tag{d}$$

Thus, Eqs. (4.2-14) become:

$$\begin{aligned} \frac{\partial f}{\partial x} &= \frac{1}{6V} \sum_{i=1}^{4} H^a_{i2} \frac{\partial f}{\partial \xi_i} \\ \frac{\partial f}{\partial y} &= \frac{1}{6V} \sum_{i=1}^{4} H^a_{i3} \frac{\partial f}{\partial \xi_i} \\ \frac{\partial f}{\partial z} &= \frac{1}{6V} \sum_{i=1}^{4} H^a_{i4} \frac{\partial f}{\partial \xi_i} \end{aligned} \tag{4.2-15}$$

Integrals of polynomial terms in the volume coordinates ξ_1, ξ_2, ξ_3, and ξ_4 can be obtained as follows (3):

$$\int_V \xi_1^a \xi_2^b \xi_3^c \xi_4^d \, dV = \frac{a!\, b!\, c!\, d!}{(a+b+c+d+3)!}(6V) \tag{4.2-16}$$

Such integrations are easily performed, and they provide a distinct advantage to the use of volume coordinates.

Example 1

$$\int_V \xi_1^2 \xi_3 \, dV = \frac{2!\, 1!}{6!}(6V) = \frac{V}{60}$$

Example 2

$$\begin{aligned} \int_V z^2 \, dV &= \int_V (\xi_1 z_1 + \xi_2 z_2 + \xi_3 z_3 + \xi_4 z_4)^2 \, dV \\ &= \int_V (\xi_1^2 z_1^2 + \xi_2^2 z_2^2 + \xi_3^2 z_3^2 + \xi_4^2 z_4^2 + 2\xi_1 z_1 \xi_2 z_2 + 2\xi_1 z_1 \xi_3 z_3 \\ &\qquad + 2\xi_1 z_1 \xi_4 z_4 + 2\xi_2 z_2 \xi_3 z_3 + 2\xi_2 z_2 \xi_4 z_4 + 2\xi_3 z_3 \xi_4 z_4) \, dV \\ &= \frac{2}{5!}(z_1^2 + z_2^2 + z_3^2 + z_4^2 + z_1 z_2 + z_1 z_3 + z_1 z_4 + z_2 z_3 + z_2 z_4 + z_3 z_4)(6V) \\ &= (z_1^2 + z_2^2 + z_3^2 + z_4^2 + z_1 z_2 + z_1 z_3 + z_1 z_4 + z_2 z_3 + z_2 z_4 + z_3 z_4)\frac{V}{10} \end{aligned}$$

If the origin is located at the centroid of the tetrahedron, $z_1 + z_2 + z_3 + z_4 = 0$. Thus, $(z_1 + z_2 + z_3 + z_4)^2 = 0$ becomes:

$$z_1 z_2 + z_1 z_3 + z_1 z_4 + z_2 z_3 + z_2 z_4 + z_3 z_4 = -\tfrac{1}{2}(z_1^2 + z_2^2 + z_3^2 + z_4^2)$$

When this expression is substituted into the integral above, we obtain:

$$\int_V z^2 \, dV = (z_1^2 + z_2^2 + z_3^2 + z_4^2)\frac{V}{20}$$

This result is the same as the formula listed in Appendix A.

4.3 NUMERICAL INTEGRATION

In Sec. 3.3 we discussed numerical integration for two-dimensional elements. With solids the geometric complexities are worse, and it is usually necessary to integrate numerically.

For hexahedra in Cartesian coordinates, the type of integral to be evaluated has the form:

$$I = \int\int\int f(x, y, z) \, dx \, dy \, dz \tag{a}$$

Before integrating, we rewrite the function in terms of the natural coordinates ξ, η, and ζ using Eqs. (4.2-1) and (4.2-2). We also change the limits of each integration to -1 and 1; and we must replace the infinitesimal volume $dV = dx \, dy \, dz$ by an expression involving $d\xi$, $d\eta$, and $d\zeta$. Toward this end, Fig.

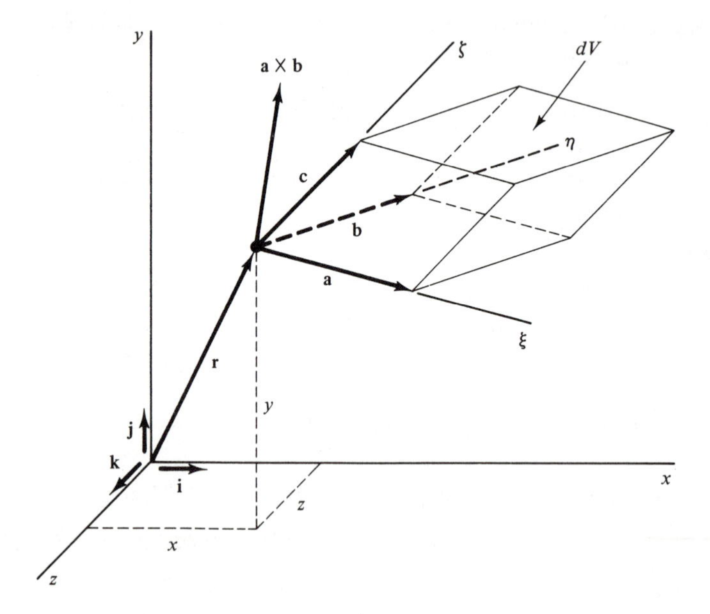

Figure 4.4 Infinitesimal Volume in Natural Coordinates

4.4 shows an infinitesimal volume dV in the natural coordinates. Also shown is a vector $\mathbf{r}$, which locates a generic point in the Cartesian coordinates. Thus,

$$\mathbf{r} = x\mathbf{i} + y\mathbf{j} + z\mathbf{k} \tag{b}$$

The rates of change of $\mathbf{r}$ with respect to ξ, η, and ζ are:

$$\frac{\partial \mathbf{r}}{\partial \xi} = \frac{\partial x}{\partial \xi}\mathbf{i} + \frac{\partial y}{\partial \xi}\mathbf{j} + \frac{\partial z}{\partial \xi}\mathbf{k} \tag{c}$$

$$\frac{\partial \mathbf{r}}{\partial \eta} = \frac{\partial x}{\partial \eta}\mathbf{i} + \frac{\partial y}{\partial \eta}\mathbf{j} + \frac{\partial z}{\partial \eta}\mathbf{k} \tag{d}$$

$$\frac{\partial \mathbf{r}}{\partial \zeta} = \frac{\partial x}{\partial \zeta}\mathbf{i} + \frac{\partial y}{\partial \zeta}\mathbf{j} + \frac{\partial z}{\partial \zeta}\mathbf{k} \tag{e}$$

Let

$$\mathbf{a} = \frac{\partial \mathbf{r}}{\partial \xi}\,d\xi \qquad \mathbf{b} = \frac{\partial \mathbf{r}}{\partial \eta}\,d\eta \qquad \mathbf{c} = \frac{\partial \mathbf{r}}{\partial \zeta}\,d\zeta \tag{f}$$

These vectors are shown in Fig. 4.4 as the edges of the infinitesimal parallelopiped of volume dV. This volume may be determined from the following vector triple product:

$$dV = (\mathbf{a} \times \mathbf{b}) \cdot \mathbf{c} = |\mathbf{J}|\,d\xi\,d\eta\,d\zeta$$
$$= \begin{vmatrix} \dfrac{\partial x}{\partial \xi} & \dfrac{\partial y}{\partial \xi} & \dfrac{\partial z}{\partial \xi} \\ \dfrac{\partial x}{\partial \eta} & \dfrac{\partial y}{\partial \eta} & \dfrac{\partial z}{\partial \eta} \\ \dfrac{\partial x}{\partial \zeta} & \dfrac{\partial y}{\partial \zeta} & \dfrac{\partial z}{\partial \zeta} \end{vmatrix} d\xi\,d\eta\,d\zeta \tag{4.3-1}$$

in which $\mathbf{J}$ is the 3×3 Jacobian matrix, and $|\mathbf{J}|$ is its determinant. Hence, the revised form of the integral in Eq. (a) becomes:

$$I = \int_{-1}^{1}\int_{-1}^{1}\int_{-1}^{1} f(\xi, \eta, \zeta)\,|\mathbf{J}|\,d\xi\,d\eta\,d\zeta \tag{4.3-2}$$

Three successive applications of Gaussian quadrature yield:

$$I = \sum_{\ell=1}^{n}\sum_{k=1}^{n}\sum_{j=1}^{n} R_j R_k R_\ell f(\xi_j, \eta_k, \zeta_\ell)\,|\mathbf{J}(\xi_j, \eta_k, \zeta_\ell)| \tag{4.3-3}$$

Integration points for $n = 1, 2, 3$, and 4 each way number 1, 8, 27, and 64, respectively.

For tetrahedra in natural coordinates the numerical integration formula is (4):

$$I = V\sum_{j=1}^{n} W_j f(\xi_1, \xi_2, \xi_3, \xi_4)_j \tag{4.3-4}$$

where W_j is the weighting factor for the jth sampling point. Integration points for $n = 1, 4$, and 5 appear in Fig. 4.5, and their locations and weighting factors W_j are given in Table 4.1.

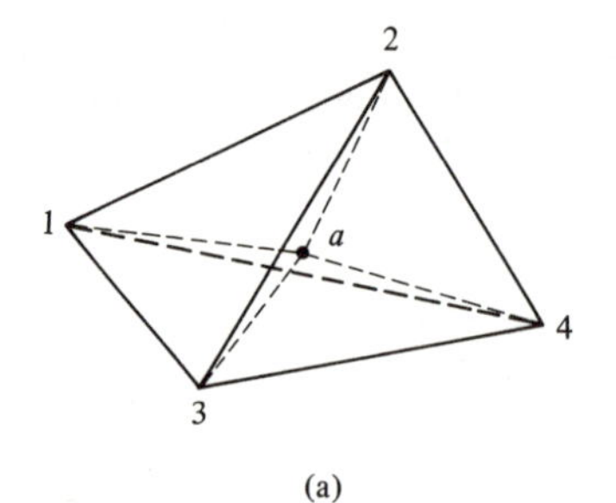

(a)

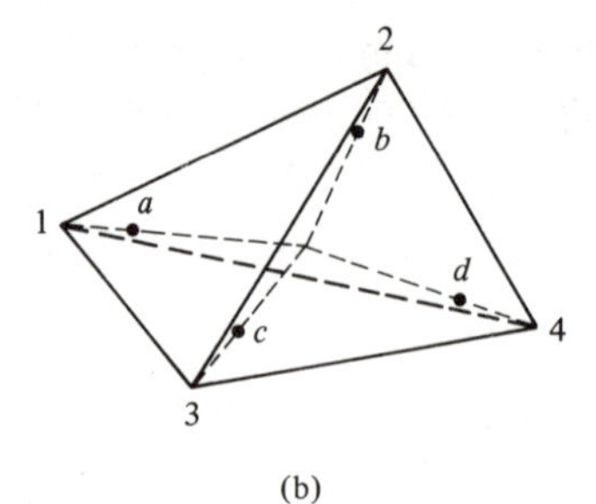

(b)

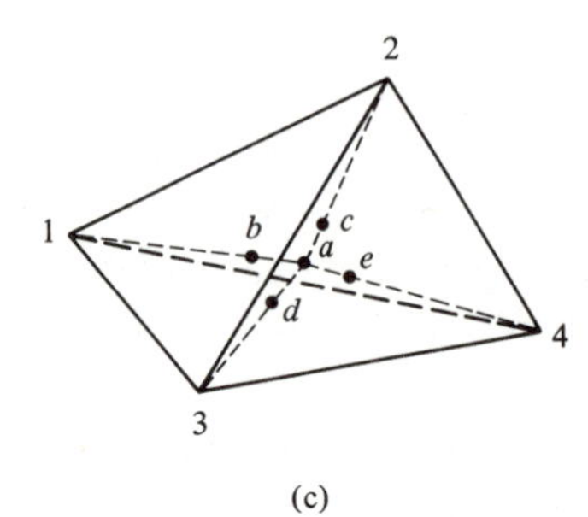

(c)

Figure 4.5 Integration Points for Tetrahedron: (a) $n = 1$ (b) $n = 4$ (c) $n = 5$

TABLE 4.1 Numerical Integration Constants for Tetrahedra

Figure	n	Order	Points	ξ_1	ξ_2	ξ_3	ξ_4	W_j
4.5(a)	1	Linear	a	$\frac{1}{4}$	$\frac{1}{4}$	$\frac{1}{4}$	$\frac{1}{4}$	1
4.5(b)	4	Quadratic	a	α	β	β	β	$\frac{1}{4}$
			b	β	α	β	β	$\frac{1}{4}$
			c	β	β	α	β	$\frac{1}{4}$
			d	β	β	β	α	$\frac{1}{4}$
4.5(c)	5	Cubic	a	$\frac{1}{4}$	$\frac{1}{4}$	$\frac{1}{4}$	$\frac{1}{4}$	γ
			b	$\frac{1}{3}$	$\frac{1}{6}$	$\frac{1}{6}$	$\frac{1}{6}$	δ
			c	$\frac{1}{6}$	$\frac{1}{3}$	$\frac{1}{6}$	$\frac{1}{6}$	δ
			d	$\frac{1}{6}$	$\frac{1}{6}$	$\frac{1}{3}$	$\frac{1}{6}$	δ
			e	$\frac{1}{6}$	$\frac{1}{6}$	$\frac{1}{6}$	$\frac{1}{3}$	δ

$\alpha = 0.58541020 \quad \beta = 0.13819660 \quad \gamma = -\frac{4}{5} \quad \delta = \frac{9}{20}$

4.4 HEXAHEDRAL ELEMENTS

Figure 4.6(a) shows the rectangular solid (*element RS8*) that is the parent of the *isoparametric hexahedron* (*element H8*) in Fig. 4.6(b). To help understand this isoparametric element, it is useful to study first its rectangular parent. For either

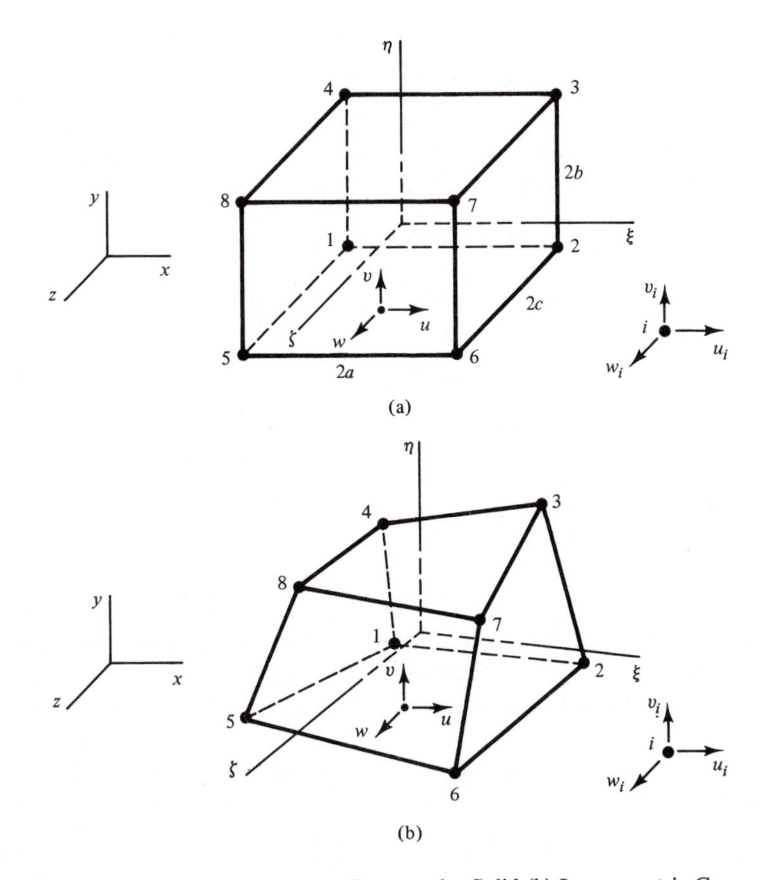

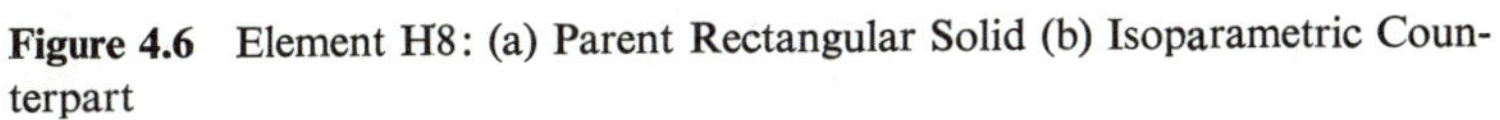

Figure 4.6 Element H8: (a) Parent Rectangular Solid (b) Isoparametric Counterpart

element the generic displacements are:

$$\mathbf{u} = \{u, v, w\} \tag{a}$$

Nodal displacements consist of x, y, and z translations at each corner node. Thus,

$$\mathbf{q} = \{q_1, q_2, q_3, \ldots, q_{24}\} = \{u_1, v_1, w_1, \ldots, w_8\} \tag{b}$$

Assumed displacement functions are trilinear, as follows:

$$u = c_1 + c_2\xi + c_3\eta + c_4\zeta + c_5\xi\eta + c_6\eta\zeta + c_7\zeta\xi + c_8\xi\eta\zeta \tag{c}$$

(v and w similar)

The corresponding displacement shape functions may be expressed as:

$$u = \sum_{i=1}^{8} f_i u_i \qquad v = \sum_{i=1}^{8} f_i v_i \qquad w = \sum_{i=1}^{8} f_i w_i \tag{4.4-1}$$

where

$$f_i = \tfrac{1}{8}(1 + \xi_0)(1 + \eta_0)(1 + \zeta_0) \tag{4.4-2}$$

and

$$\xi_0 = \xi_i \xi \qquad \eta_0 = \eta_i \eta \qquad \zeta_0 = \zeta_i \zeta \tag{4.4-3}$$

The values of ξ_i, η_i, and ζ_i required in these formulas are given in Table 4.2. For the subparametric parent element explicit integrations are feasible, and stiffnesses for an orthotropic material were presented by Melosh (5).

TABLE 4.2 Nodal Coordinates for Element H8

i	ξ_i	η_i	ζ_i
1	−1	−1	−1
2	1	−1	−1
3	1	1	−1
4	−1	1	−1
5	−1	−1	1
6	1	−1	1
7	1	1	1
8	−1	1	1

Considering now the isoparametric H8 element (6) in Fig. 4.6(b), we take the geometric interpolation functions to be those that were given earlier as Eqs. (4.2-1) and (4.2-2). Because those formulas are the same as the displacement shape functions [see Eqs. (4.4-1), (4.4-2), and (4.4-3)], the H8 element is verified as isoparametric. Equations (4.4-1) can also be stated as a matrix expression. That is,

$$\mathbf{u}_i = \mathbf{f}_i \mathbf{q}_i \qquad (i = 1, 2, \ldots, 8) \tag{4.4-4}$$

in which

$$\mathbf{f}_i = \begin{bmatrix} 1 & 0 & 0 \\ 0 & 1 & 0 \\ 0 & 0 & 1 \end{bmatrix} f_i \tag{4.4-5}$$

As before, the generic displacements $\mathbf{u}_i$ in Eq. (4.4-4) denote the translations at any point due to the displacements $\mathbf{q}_i$ at node i.

Strain-displacement relationships can also be expressed efficiently as:

$$\boldsymbol{\epsilon}_i = \mathbf{B}_i \mathbf{q}_i \qquad (i = 1, 2, \ldots, 8) \tag{4.4-6}$$

where

$$\mathbf{B}_i = \mathbf{d}\ \mathbf{f}_i = \begin{bmatrix} f_{i,x} & 0 & 0 \\ 0 & f_{i,y} & 0 \\ 0 & 0 & f_{i,z} \\ f_{i,y} & f_{i,x} & 0 \\ 0 & f_{i,z} & f_{i,y} \\ f_{i,z} & 0 & f_{i,x} \end{bmatrix} = \begin{bmatrix} D_{G1i} & 0 & 0 \\ 0 & D_{G2i} & 0 \\ 0 & 0 & D_{G3i} \\ D_{G2i} & D_{G1i} & 0 \\ 0 & D_{G3i} & D_{G2i} \\ D_{G3i} & 0 & D_{G1i} \end{bmatrix} \tag{4.4-7}$$

Then the stiffness matrix for element H8 may be written in Cartesian coordinates as:

$$\mathbf{K} = \int_V \mathbf{B}^{\mathrm{T}}(x, y, z)\mathbf{E}\ \mathbf{B}(x, y, z)\, dx\, dy\, dz \tag{d}$$

In natural coordinates Eq. (d) becomes:

$$\mathbf{K} = \int_{-1}^{1}\int_{-1}^{1}\int_{-1}^{1} \mathbf{B}^{\mathrm{T}}(\xi, \eta, \zeta)\mathbf{E}\ \mathbf{B}(\xi, \eta, \zeta)\,|\,\mathbf{J}(\xi, \eta, \zeta)\,|\, d\xi\, d\eta\, d\zeta \tag{4.4-8}$$

We can also express equivalent nodal loads due to body forces in natural coordinates, as follows:

$$\mathbf{p}_b = \int_{-1}^{1}\int_{-1}^{1}\int_{-1}^{1} \mathbf{f}^{\mathrm{T}}(\xi, \eta, \zeta)\mathbf{b}(\xi, \eta, \zeta)\,|\,\mathbf{J}(\xi, \eta, \zeta)\,|\, d\xi\, d\eta\, d\zeta \tag{4.4-9}$$

Furthermore, equivalent nodal loads due to initial strains in natural coordinates are:

$$\mathbf{p}_0 = \int_{-1}^{1}\int_{-1}^{1}\int_{-1}^{1} \mathbf{B}^{\mathrm{T}}(\xi, \eta, \zeta)\mathbf{E}\ \boldsymbol{\epsilon}_0(\xi, \eta, \zeta)\,|\,\mathbf{J}(\xi, \eta, \zeta)\,|\, d\xi\, d\eta\, d\zeta \tag{4.4-10}$$

The integrals in Eqs. (4.4-8) and (4.4-9) must be evaluated by numerical integration, except in special cases. For convenience, the integrals in Eq. (4.4-10) may also be found numerically, even though it is known that the Jacobian matrix cancels.

Example

For the rectangular solid element RS8, find the stiffness term K_{11}, assuming isotropic material. From Eq. (4.4-7) the first column of matrix $\mathbf{B}$ is:

$$\begin{aligned}\mathbf{B}_{1,1} &= \{f_{1,x}, 0, 0, f_{1,y}, 0, f_{1,z}\} \\ &= -\frac{1}{8}\left\{\frac{1}{a}(1-\eta)(1-\zeta), 0, 0, \frac{1}{b}(1-\xi)(1-\zeta), 0, \frac{1}{c}(1-\xi)(1-\eta)\right\}\end{aligned}$$

Then Eq. (4.4-8) yields:

$$K_{11} = abc\int_{-1}^{1}\int_{-1}^{1}\int_{-1}^{1} \mathbf{B}_{1,1}^{\mathrm{T}}\mathbf{E}\ \mathbf{B}_{1,1}\, d\xi\, d\eta\, d\zeta$$

in which the matrix $\mathbf{E}$ is given by Eq. (4.1-4). Substitution of $\mathbf{B}_{1,1}$ and $\mathbf{E}$ into the expression for K_{11} produces:

$$K_{11} = \frac{2abcE}{9(1+\nu)e_2}\left(\frac{e_1}{a^2} + \frac{e_3}{b^2} + \frac{e_3}{c^2}\right)$$

Next, we shall examine a higher-order hexahedral element with additional nodes and curved surfaces. The parent rectangular solid (*element RS20*) is illustrated in Fig. 4.7(a), and its isoparametric counterpart (*element H20*) appears in Fig. 4.7(b). Each element has the generic displacements given in Eq. (a), and the figures show x, y, and z translations at each of the twenty nodes. Hence, the nodal displacements are:

$$\mathbf{q} = \{q_1, q_2, q_3, \ldots, q_{60}\} = \{u_1, v_1, w_1, \ldots, w_{20}\} \tag{e}$$

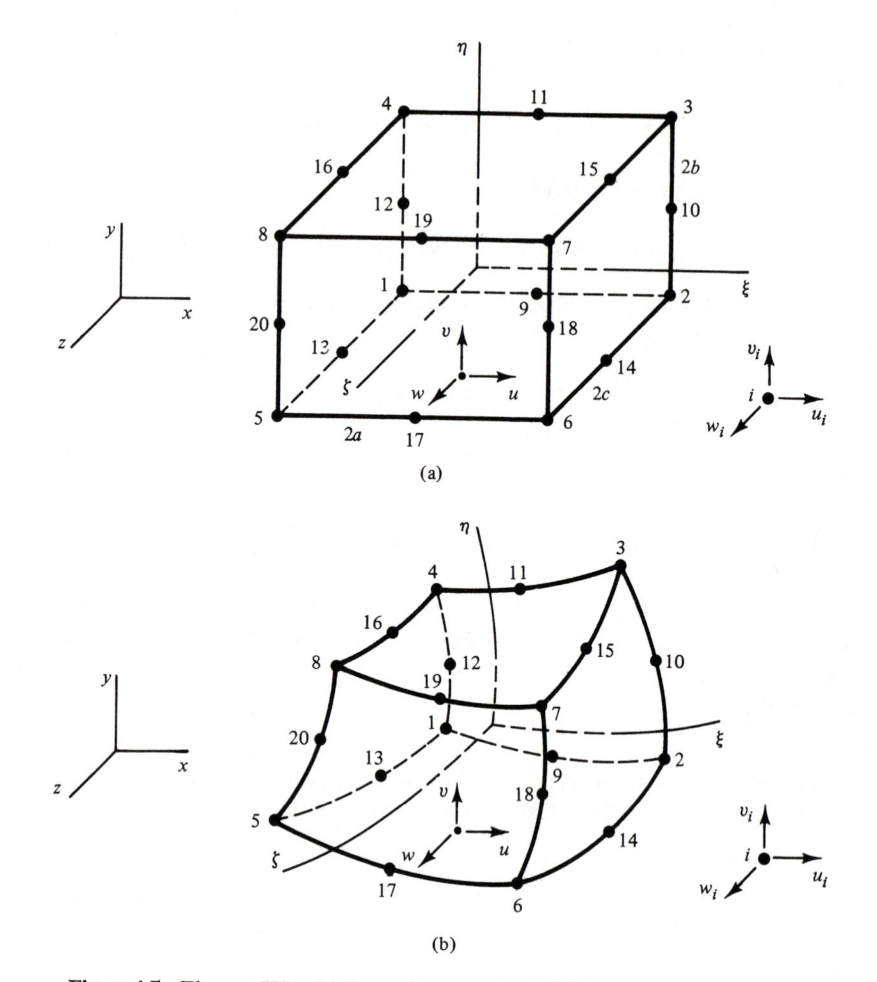

Figure 4.7 Element H20: (a) Parent Rectangular Solid (b) Isoparametric Counterpart

For these elements we assume the displacement shape functions to be:

$$u = \sum_{i=1}^{20} f_i u_i \qquad v = \sum_{i=1}^{20} f_i v_i \qquad w = \sum_{i=1}^{20} f_i w_i \tag{4.4-11}$$

where

$$\begin{aligned}
f_i &= \tfrac{1}{8}(1 + \xi_0)(1 + \eta_0)(1 + \zeta_0)(\xi_0 + \eta_0 + \zeta_0 - 2) && (i = 1, 2, \ldots, 8) \\
f_i &= \tfrac{1}{4}(1 - \xi^2)(1 + \eta_0)(1 + \zeta_0) && (i = 9, 11, 17, 19) \\
f_i &= \tfrac{1}{4}(1 - \eta^2)(1 + \zeta_0)(1 + \xi_0) && (i = 10, 12, 18, 20) \\
f_i &= \tfrac{1}{4}(1 - \zeta^2)(1 + \xi_0)(1 + \eta_0) && (i = 13, 14, 15, 16)
\end{aligned} \tag{4.4-12}$$

Values of ξ_i, η_i, and ζ_i for these quadratic formulas are given in Table 4.3. Explicit integrations are possible for the subparametric parent element.

TABLE 4.3 Nodal Coordinates for Element H20

i	ξ_i	η_i	ζ_i	i	ξ_i	η_i	ζ_i
1	−1	−1	−1	11	0	1	−1
2	1	−1	−1	12	−1	0	−1
3	1	1	−1	13	−1	−1	0
4	−1	1	−1	14	1	−1	0
5	−1	−1	1	15	1	1	0
6	1	−1	1	16	−1	1	0
7	1	1	1	17	0	−1	1
8	−1	1	1	18	1	0	1
9	0	−1	−1	19	0	1	1
10	1	0	−1	20	−1	0	1

For the isoparametric H20 element (7) in Fig. 4.7(b), we use geometric interpolation functions that are the same as the displacement shape functions in Eqs. (4.4-12). Thus,

$$x = \sum_{i=1}^{20} f_i x_i \qquad y = \sum_{i=1}^{20} f_i y_i \qquad z = \sum_{i=1}^{20} f_i z_i \tag{4.4-13}$$

In this instance, the faces and edges of the element are quadratic surfaces and curves, as indicated in the figure.

Terms in the Jacobian matrix for element H20 are the same as those given in Eqs. (4.2-4), but with the upper index 8 changed to 20. Furthermore, Eqs. (4.4-4) through (4.4-10) for element H8 will pertain to element H20 if the number 8 is changed to 20 in appropriate locations. Derivatives $f_{i,\xi}$, and so on, required for the development of element H20 are easily obtained and need not be tabulated. For example,

$$f_{1,\xi} = (1 + 2\xi + \eta + \zeta)(1 - \eta)(1 - \zeta) \tag{f}$$

Of course, numerical integration is required for this element.

4.5 TETRAHEDRAL ELEMENTS

In this section we discuss briefly the three-dimensional counterparts of the triangular elements T3 and T6 covered in Sec. 3.5. The new elements are tetrahedra and will be referred to as Tet-4 and Tet-10 to indicate the number of nodes involved in each case.

Figure 4.8(a) shows *element Tet-4*, which has nodes at the four vertices. For this tetrahedron the generic displacements are:

$$\mathbf{u} = \{u, v, w\} \tag{a}$$

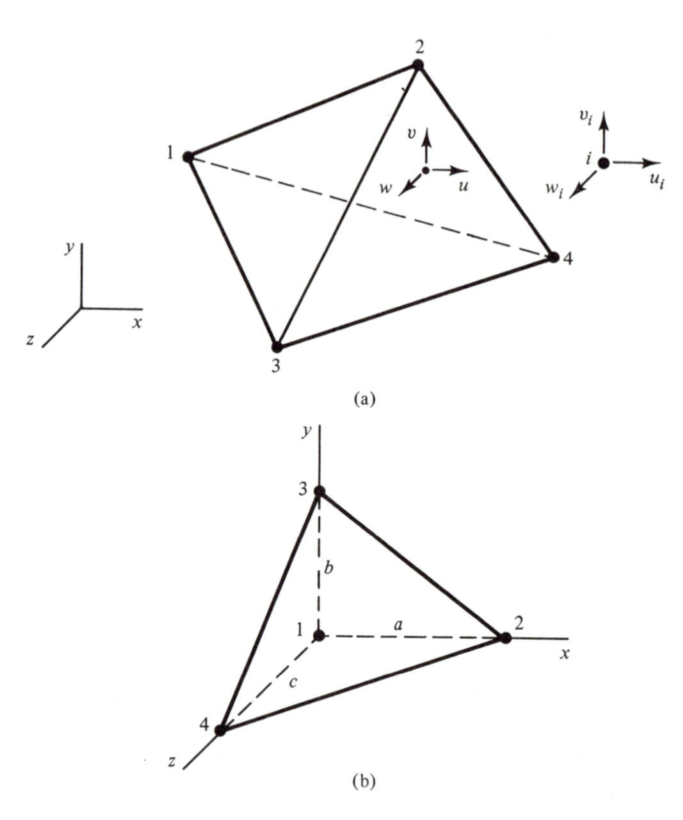

Figure 4.8 (a) Element Tet-4 (b) Right Tetrahedron

At each node there are translations in the x, y, and z directions, so the nodal displacements can be written as:

$$\mathbf{q} = \{q_1, q_2, q_3, \ldots, q_{12}\} = \{u_1, v_1, w_1, \ldots, w_4\} \tag{b}$$

Displacement functions assumed for this element are linear, as follows:

$$\begin{aligned} u &= c_1 + c_2 x + c_3 y + c_4 z \\ v &= c_5 + c_6 x + c_7 y + c_8 z \\ w &= c_9 + c_{10} x + c_{11} y + c_{12} z \end{aligned} \tag{c}$$

Corresponding displacement shape functions are found to be:

$$u = \sum_{i=1}^{4} f_i u_i \qquad v = \sum_{i=1}^{4} f_i v_i \qquad w = \sum_{i=1}^{4} f_i w_i \tag{4.5-1}$$

In these expressions the functions f_i are the same as the natural coordinates [see Eq. (4.2-12)]. Thus, the element is isoparametric, and we have:

$$f_i = \xi_i \qquad (i = 1, 2, 3, 4) \tag{4.5-2}$$

Derivatives $f_{i,x}$, and so on, for strain-displacement relationships [see Eq. (4.4-7)] are readily obtained and are all constants. For this reason the Tet-4 element is sometimes called the *constant strain tetrahedron*, which is analogous to the constant strain triangle. Stiffnesses for the Tet-4 element are given by Melosh (5) for an orthotropic material.

Example

Determine the stiffness K_{22} of the right tetrahedron shown in Fig. 4.8(b), assuming isotropic material. For this special geometry Eq. (4.2-11a) simplifies to:

$$\begin{bmatrix} 1 \\ x \\ y \\ z \end{bmatrix} = \begin{bmatrix} 1 & 1 & 1 & 1 \\ 0 & a & 0 & 0 \\ 0 & 0 & b & 0 \\ 0 & 0 & 0 & c \end{bmatrix} \begin{bmatrix} \xi_1 \\ \xi_2 \\ \xi_3 \\ \xi_4 \end{bmatrix} \tag{d}$$

And Eq. (4.2-12) becomes:

$$\begin{bmatrix} \xi_1 \\ \xi_2 \\ \xi_3 \\ \xi_4 \end{bmatrix} = \frac{1}{abc} \begin{bmatrix} abc & -bc & -ac & -ab \\ 0 & bc & 0 & 0 \\ 0 & 0 & ac & 0 \\ 0 & 0 & 0 & ab \end{bmatrix} \begin{bmatrix} 1 \\ x \\ y \\ z \end{bmatrix} \tag{e}$$

Hence,

$$f_1 = 1 - \frac{x}{a} - \frac{y}{b} - \frac{z}{c} \qquad f_2 = \frac{x}{a} \qquad f_3 = \frac{y}{b} \qquad f_4 = \frac{z}{c} \tag{f}$$

The second column of matrix **B** [see Eq. (4.4-7)] is:

$$\mathbf{B}_{1,2} = \{0, f_{1,y}, 0, f_{1,x}, f_{1,z}, 0\} = \left\{0, -\frac{1}{b}, 0, -\frac{1}{a}, -\frac{1}{c}, 0\right\} \tag{g}$$

Then

$$K_{22} = \mathbf{B}_{1,2}^{\mathrm{T}} \mathbf{E}\, \mathbf{B}_{1,2} V = \frac{EV}{(1+\nu)e_2}\left(\frac{e_1}{b^2} + \frac{e_3}{a^2} + \frac{e_3}{c^2}\right) \tag{h}$$

Appearing in Fig. 4.9(a) is a tetrahedron with four corner nodes and six midedge nodes. This element is the three-dimensional counterpart of the linear strain triangle studied in Secs. 2.2 and 3.5. For this reason it is sometimes called the *linear strain tetrahedron*. It is also the tetrahedral parent of *element Tet-10*, which is depicted in Fig. 4.9(b). For either element the generic displacements are given by Eq. (a), and the nodal displacements are:

$$\mathbf{q} = \{q_1, q_2, q_3, \ldots, q_{30}\} = \{u_1, v_1, w_1, \ldots, w_{10}\} \tag{i}$$

Assumed displacement functions are complete quadratics, as follows:

$$u = c_1 + c_2 x + c_3 y + c_4 z + c_5 xy + c_6 yz + c_7 xz + c_8 x^2 + c_9 y^2 + c_{10} z^2 \tag{j}$$

(v and w similar)

Corresponding displacement shape functions are found to be:

$$u = \sum_{i=1}^{10} f_i u_i \qquad v = \sum_{i=1}^{10} f_i v_i \qquad w = \sum_{i=1}^{10} f_i w_i \tag{4.5-3}$$

where the quadratic functions in natural coordinates have the following formulas:

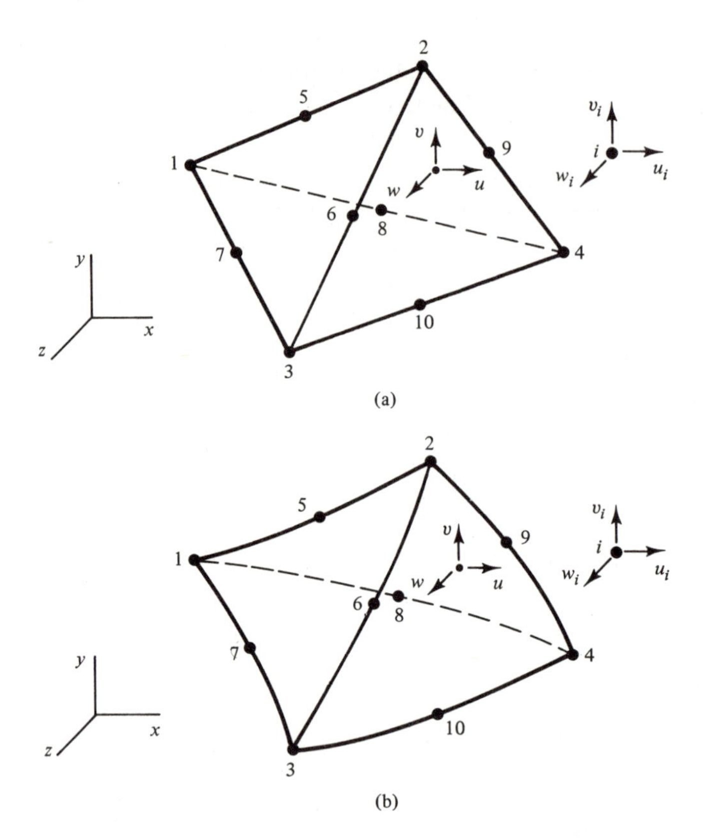

Figure 4.9 Element Tet-10: (a) Tetrahedral Parent (b) Isoparametric Counterpart

$$
\begin{aligned}
f_1 &= (2\xi_1 - 1)\xi_1 \qquad f_2 = (2\xi_2 - 1)\xi_2 \\
f_3 &= (2\xi_3 - 1)\xi_3 \qquad f_4 = (2\xi_4 - 1)\xi_4 \\
f_5 &= 4\xi_1\xi_2 \qquad f_6 = 4\xi_2\xi_3 \qquad f_7 = 4\xi_1\xi_3 \\
f_8 &= 4\xi_1\xi_4 \qquad f_9 = 4\xi_2\xi_4 \qquad f_{10} = 4\xi_3\xi_4
\end{aligned} \tag{4.5-4}
$$

To find derivatives for strain-displacement relationships in the tetrahedral parent, we must substitute expressions for the natural coordinates from Eq. (4.2-12) into Eqs. (4.5-4) and differentiate with respect to x, y, and z. For this element explicit integrations are possible (8).

Proceeding now to the isoparametric element Tet-10 in Fig. 4.9(b), we take the geometric interpolation functions to be:

$$
x = \sum_{i=1}^{10} f_i x_i \qquad y = \sum_{i=1}^{10} f_i y_i \qquad z = \sum_{i=1}^{10} f_i z_i \tag{4.5-5}
$$

where f_i are the quadratic formulas given by Eqs. (4.5-4). We see that the faces and edges of the element become quadratic surfaces and curves. Since the natural coordinates are now curvilinear, the Jacobian matrix is needed. Hence,

$$
\begin{aligned}
J_{11} &= \sum_{i=1}^{10} f_{i,\xi_1} x_i \qquad & J_{12} &= \sum_{i=1}^{10} f_{i,\xi_1} y_i \qquad & J_{13} &= \sum_{i=1}^{10} f_{i,\xi_1} z_i \\
J_{21} &= \sum_{i=1}^{10} f_{i,\xi_2} x_i \qquad & J_{22} &= \sum_{i=1}^{10} f_{i,\xi_2} y_i \qquad & J_{23} &= \sum_{i=1}^{10} f_{i,\xi_2} z_i \\
J_{31} &= \sum_{i=1}^{10} f_{i,\xi_3} x_i \qquad & J_{32} &= \sum_{i=1}^{10} f_{i,\xi_3} y_i \qquad & J_{33} &= \sum_{i=1}^{10} f_{i,\xi_3} z_i
\end{aligned}
\tag{4.5-6}
$$

However, before the differentiations indicated in Eqs. (4.5-6) can be made, it is necessary to substitute a dependent coordinate (such as $\xi_4 = 1 - \xi_1 - \xi_2 - \xi_3$) into the quadratic functions f_i in Eqs. (4.5-4). These shape functions and their derivatives with respect to the natural coordinates ξ_1, ξ_2, and ξ_3 are listed in Table 4.4 (in terms of ξ_1, ξ_2, ξ_3, and ξ_4).

TABLE 4.4 Shape Functions and Derivatives for Element Tet-10

i	f_i	f_{i,ξ_1}	f_{i,ξ_2}	f_{i,ξ_3}
1	$(2\xi_1 - 1)\xi_1$	$4\xi_1 - 1$	0	0
2	$(2\xi_2 - 1)\xi_2$	0	$4\xi_2 - 1$	0
3	$(2\xi_3 - 1)\xi_3$	0	0	$4\xi_3 - 1$
4	$(2\xi_4 - 1)\xi_4$	$1 - 4\xi_4$	$1 - 4\xi_4$	$1 - 4\xi_4$
5	$4\xi_1\xi_2$	$4\xi_2$	$4\xi_1$	0
6	$4\xi_2\xi_3$	0	$4\xi_3$	$4\xi_2$
7	$4\xi_1\xi_3$	$4\xi_3$	0	$4\xi_1$
8	$4\xi_1\xi_4$	$4(\xi_4 - \xi_1)$	$-4\xi_1$	$-4\xi_1$
9	$4\xi_2\xi_4$	$-4\xi_2$	$4(\xi_4 - \xi_2)$	$-4\xi_2$
10	$4\xi_3\xi_4$	$-4\xi_3$	$-4\xi_3$	$4(\xi_4 - \xi_3)$

To find the stiffnesses and equivalent nodal loads for this element, numerical integration is required. The formula is:

$$
I = \frac{1}{6} \sum_{j=1}^{n} W_j f(\xi_1, \xi_2, \xi_3, \xi_4)_j \, |\mathbf{J}(\xi_1, \xi_2, \xi_3)_j| \tag{4.5-7}
$$

If the tetrahedron had straight edges, we would find $|\mathbf{J}| = 6V$; and Eq. (4.5-7) would become the same as Eq. (4.3-4). In any case, the numerical integration constants given in Table 4.1 would apply.

4.6 PROGRAMS SOH8 AND SOH20 AND APPLICATIONS

In this section we briefly discuss two computer programs named SOH8 and SOH20 for analysis of solids using the isoparametric hexahedra H8 and H20 (see Sec. 4.4). These programs can be constructed by modifying Programs PSQ4 and PSQ8, which were described in Sec. 3.6.

Table 4.5 shows how data must be prepared for Program SOH8. Because the material is assumed to be isotropic, the only material constants required in the structural parameters are Young's modulus (E) and Poisson's ratio (PR). There are three coordinates for each node, and the element information must contain eight node numbers. In addition, there are three possible conditions of restraint at each restrained node.

The load data in Table 4.5 includes various items that deserve mention. Nodal loads consist of three possible components of force applied at each loaded node. The line loads along edge ij are defined by three pairs of numbers (b_{L1} through b_{L6}) that represent intensities of force (per unit length) at nodes i and j in the x, y, and z directions, respectively. Area loads on face $ijk\ell$ consist of twelve numbers, of which the first four (b_{A1} through b_{A4}) pertain to force (per unit area) in the x direction, as indicated in Fig. 4.10. The next four (b_{A5} through b_{A8}) apply

TABLE 4.5 Preparation of Data for Program SOH8

Data	No. of Lines	Items on Data Lines
STRUCTURAL DATA		
(a) Problem identification	1	Descriptive title
(b) Structural parameters	1	NN,NE,NRN,NLS,E,PR
(c) Nodal coordinates	NN	J,X(J),Y(J),Z(J)
(d) Element information*	NE	I,IN(I,1),IN(I,2), . . . ,IN(I,8)
(e) Nodal restraint list	NRN	K,NRL(3K−2),NRL(3K−1),NRL(3K)
LOAD DATA		
(a) Load parameters	1	NLN,NEL,NEA,NEV,IPR
(b) Nodal loads	NLN	K,AN(3K−2),AN(3K−1),AN(3K)
(c) Line loads†	NEL	I,J,BL1,BL2, . . . ,BL6
(d) Area loads†	NEA	I,J,K,L,BA1,BA2, . . . ,BA12
(e) Volume loads†	NEV	I,BV1,BV2,BV3

*See Fig. 4.6(b) for sequence of node numbers.

†Optional supplementary influences.

to the y direction, and the last four (b_{A9} through b_{A12}) refer to the z direction. It is assumed that each component of loading has a bilinear variation over the surface $ijk\ell$. On the other hand, volume loads b_{V1}, b_{V2}, and b_{V3} simply consist of uniform intensities of force (per unit volume) in the x, y, and z directions.

In Program SOH8 the second subprogram called by the main program is named STIFH8, which is very similar to STIFQ4 in Sec. 3.6. However, an additional loop for ZETA is required to account for the fact that the number of integration points is eight instead of four. Within the inner loop another subprogram named BMATH8 is called for the purpose of generating the **B** matrix and the determinant of the Jacobian matrix for a particular integration

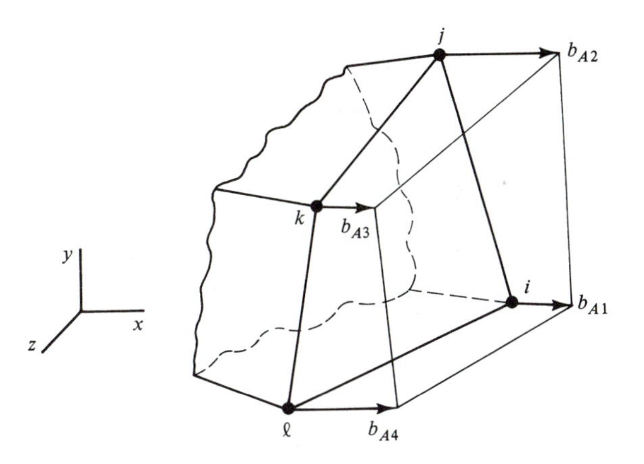

Figure 4.10 Area Loads for Element H8

point. This subprogram has the same logic as that in BMATQ4, which was described in Sec. 3.6.

The data and logic for Program SOH20 are very similar to those for Program SOH8. Certain differences arise, however, when exchanging an element with eight nodes for one with twenty nodes. Most importantly, the trilinear geometric and displacement shape functions given by Eq. (4.4-2) must be replaced by the triquadratic expressions in Eqs. (4.4-12). Also, the summations from 1 to 8 are replaced by those from 1 to 20.

Example 1

To show how Program SOH8 works, we consider the single cubic element given in Fig. 4.11(a). Assume the following values for the structural parameters:

$$L = 1 \qquad E = 1 \qquad \nu = 0.3$$

which are not referred to any particular system of units. Support restraints are indicated in the figure (at points 1, 2, 5, and 6) by arrows with small cross lines on their shafts.

Certain loading conditions appear in part (b) of Fig. 4.11. The first loading has four nodal forces of equal magnitudes 1.0 applied in the y direction at points 3, 4, 7, and 8. The second loading consists of a uniformly distributed force (per unit length) of intensity 2.0, applied in the x direction on edge 3-7. Loading 3 illustrates a wedge-shaped load (force per unit area) of maximum intensity 3.0, applied in the z direction on surface 5-6-7-8. This loading varies linearly in the y direction but is constant in the x direction. Loading 4 shows a uniformly distributed force (per unit volume) of intensity 1.0, applied in the negative y direction.

Table 4.6 contains the computer output for this example. The print of structural data on the first page of the table is followed by the results of calculations for each of the four loading cases. By checking the displacements, stresses, and reactions due to these loads, we see that the program is working properly.

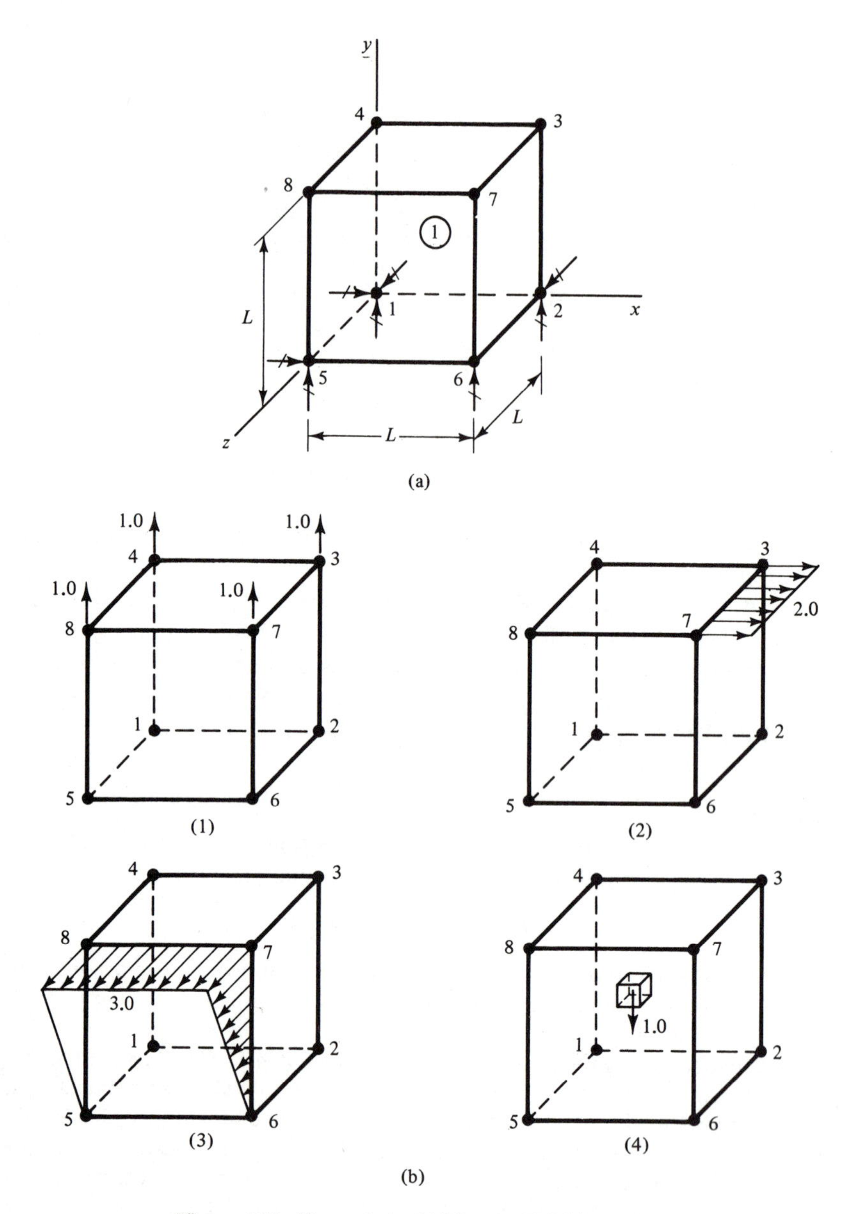

Figure 4.11 Example 1: (a) Element H8 (b) Loadings

Example 2

Let us now apply Program SOH8 to the short beam illustrated in Fig. 4.12(a). This beam is so stout that it acts more like a general solid than a flexural member. Note that the problem is symmetric with respect to both the x-y and the $\bar{y}$-$\bar{z}$ planes. Loads applied to the beam are all in the negative y direction. They consist of a concentrated force P at the center of the upper surface, a uniformly distributed pressure b_A (force

TABLE 4.6 Computer Output for Example 1

```
PROGRAM SOH8

***  EXAMPLE 1:  ONE CUBIC ELEMENT  ***

STRUCTURAL PARAMETERS
   NN   NE  NRN  NLS             E            PR
    8    1    4    4  1.0000D 00   3.0000D-01

COORDINATES OF NODES
 NODE           X            Y            Z
    1  0.0000D-01  0.0000D-01  0.0000D-01
    2  1.0000D 00  0.0000D-01  0.0000D-01
    3  1.0000D 00  1.0000D 00  0.0000D-01
    4  0.0000D-01  1.0000D 00  0.0000D-01
    5  0.0000D-01  0.0000D-01  1.0000D 00
    6  1.0000D 00  0.0000D-01  1.0000D 00
    7  1.0000D 00  1.0000D 00  1.0000D 00
    8  0.0000D-01  1.0000D 00  1.0000D 00

ELEMENT INFORMATION
ELEM.   N1   N2   N3   N4   N5   N6   N7   N8
    1    1    2    3    4    5    6    7    8

NODAL RESTRAINTS
 NODE   R1   R2   R3
    1    1    1    1
    2    0    1    1
    5    1    1    0
    6    0    1    0

NUMBER OF DEGREES OF FREEDOM =   16
NUMBER OF NODAL RESTRAINTS   =    8
NUMBER OF TERMS IN SN        =  199

**********  LOADING NUMBER    1  **********
  NLN  NEL  NEA  NEV  IPR
    4    0    0    0    0

ACTIONS AT NODES
 NODE         AN1          AN2          AN3
    3  0.0000D-01  1.0000D 00  0.0000D-01
    4  0.0000D-01  1.0000D 00  0.0000D-01
    7  0.0000D-01  1.0000D 00  0.0000D-01
    8  0.0000D-01  1.0000D 00  0.0000D-01

NODAL DISPLACEMENTS
 NODE         DN1          DN2          DN3
    1  0.0000D-01  0.0000D-01  0.0000D-01
    2 -1.2000D 00 -0.0000D-01 -0.0000D-01
    3 -1.2000D 00  4.0000D 00  4.3021D-16
    4  1.0825D-15  4.0000D 00  2.7062D-15
    5 -0.0000D-01 -0.0000D-01 -1.2000D 00
    6 -1.2000D 00 -0.0000D-01 -1.2000D 00
    7 -1.2000D 00  4.0000D 00 -1.2000D 00
    8  1.3323D-15  4.0000D 00 -1.2000D 00

ELEMENT STRESSES
ELEM. INT.    X STRESS     Y STRESS     Z STRESS    XY STRESS    XZ STRESS    YZ STRESS
    1    1  8.4655D-16  4.0000D 00  2.2204D-16  2.9357D-16  6.4051D-16 -3.9698D-17
    1    2  6.5226D-16  4.0000D 00  4.4409D-16  2.9357D-16  3.1792D-16  1.4278D-16
```

TABLE 4.6 (cont.)

```
    1    3  7.9103D-16  4.0000D 00  6.6613D-16 -1.3344D-16  2.4086D-16 -2.6688D-16
    1    4  9.7145D-16  4.0000D 00  6.6613D-16 -1.2277D-16  4.2167D-16 -4.5903D-16
    1    5  1.1935D-15  4.0000D 00  4.4409D-16  4.5370D-16  6.5119D-16  6.4051D-17
    1    6  1.4572D-15  4.0000D 00  8.8818D-16  4.2167D-16  4.5903D-16  2.6688D-16
    1    7  1.4572D-15  4.0000D 00  8.8818D-16  4.2701D-16  4.5370D-16 -1.0675D-16
    1    8  1.1935D-15  4.0000D 00  4.4409D-16  4.2701D-16  1.5479D-16 -2.4019D-16

SUPPORT REACTIONS
 NODE          AR1         AR2         AR3
    1 -2.5675D-16 -1.0000D 00 -2.2532D-16
    2  0.0000D-01 -1.0000D 00 -2.4728D-16
    5 -3.9921D-16 -1.0000D 00  0.0000D-01
    6  0.0000D-01 -1.0000D 00  0.0000D-01

**********  LOADING NUMBER    2  **********
  NLN  NEL  NEA  NEV  IPR
    0    1    0    0    0

LINE LOADS
    I    J         BL1         BL2         BL3         BL4         BL5         BL6
    3    7  2.0000D 00  2.0000D 00  0.0000D-01  0.0000D-01  0.0000D-01  0.0000D-01

NODAL DISPLACEMENTS
 NODE          DN1         DN2         DN3
    1  0.0000D-01  0.0000D-01  0.0000D-01
    2  2.3786D 00 -0.0000D-01 -0.0000D-01
    3  1.1371D 01 -4.6002D 00 -4.9134D-01
    4  9.5120D 00  3.3209D 00  4.1201D-01
    5 -0.0000D-01 -0.0000D-01 -1.5214D 00
    6  1.6214D 00 -0.0000D-01  3.2139D-01
    7  1.0749D 01 -4.3205D 00  3.4928D-01
    8  8.6087D 00  3.1998D 00 -1.6286D 00

ELEMENT STRESSES
ELEM. INT.    X STRESS    Y STRESS    Z STRESS   XY STRESS   XZ STRESS   YZ STRESS
    1    1  3.1574D 00  2.7605D 00  5.8006D-01  2.9237D 00  5.7721D-02 -2.2941D-02
    1    2  1.2341D 00 -2.6430D 00 -4.2736D-01  2.8572D 00 -7.5336D-02 -1.4236D-01
    1    3  1.1010D 00 -2.6430D 00 -2.9422D-01  1.1170D 00 -3.2027D-02 -3.1295D-01
    1    4  2.8246D 00  2.5608D 00  2.4721D-01  1.1836D 00  4.9643D-02 -3.2667D-01
    1    5  2.7273D 00  2.5531D 00  3.8882D-01  2.7907D 00 -8.8488D-03  4.3499D-01
    1    6  8.8113D-01 -2.6706D 00 -5.4152D-01  2.8573D 00 -8.7667D-03  3.1558D-01
    1    7  1.2140D 00 -2.4709D 00 -2.0867D-01  1.1685D 00  3.4542D-02  2.7812D-01
    1    8  2.8605D 00  2.5531D 00  2.5568D-01  1.1019D 00 -1.6927D-02  2.6440D-01

SUPPORT REACTIONS
 NODE          AR1         AR2         AR3
    1 -1.0610D 00 -1.0118D 00 -6.1022D-02
    2  0.0000D-01  1.0118D 00  6.1022D-02
    5 -9.3898D-01 -9.8822D-01  0.0000D-01
    6  0.0000D-01  9.8822D-01  0.0000D-01

**********  LOADING NUMBER    3  **********
  NLN  NEL  NEA  NEV  IPR
    0    0    1    0    0

AREA LOADS
    I    J    K    L    BA1,5,9    BA2,6,10    BA3,7,11    BA4,8,12
    5    6    7    8 0.0000D-01  0.0000D-01  0.0000D-01  0.0000D-01
                     0.0000D-01  0.0000D-01  0.0000D-01  0.0000D-01
                     0.0000D-01  0.0000D-01  3.0000D 00  3.0000D 00
```

TABLE 4.6 (cont.)

```
NODAL DISPLACEMENTS
 NODE        DN1          DN2          DN3
    1  0.0000D-01  0.0000D-01  0.0000D-01
    2 -1.2357D 00 -0.0000D-01 -0.0000D-01
    3 -9.6431D-01  1.4499D 00  5.2793D 00
    4 -1.1899D-01  1.5104D 00  5.7310D 00
    5 -0.0000D-01 -0.0000D-01  2.6643D 00
    6 -3.1431D-01 -0.0000D-01  2.2857D 00
    7  2.4642D-02 -2.3102D 00  5.8746D 00
    8 -5.7067D-01 -2.4501D 00  6.1857D 00

ELEMENT STRESSES
ELEM. INT.    X STRESS     Y STRESS     Z STRESS    XY STRESS    XZ STRESS    YZ STRESS
    1    1  3.6220D-01  1.5968D 00  2.7282D 00 -4.3309D-02  1.6784D 00 -1.1471D-02
    1    2  2.6658D-01  1.4931D 00  2.5132D 00  6.7744D-02  1.6119D 00  2.1750D-01
    1    3  5.5672D-02  1.0601D 00  1.2807D 00  6.3705D-02  7.6748D-01  1.3220D-01
    1    4  5.1439D-02  1.0639D 00  1.2628D 00 -4.7347D-02  8.0831D-01 -1.6334D-01
    1    5 -1.4151D-01 -1.1050D 00  1.7666D 00 -1.0984D-01  1.2121D 00 -7.1179D-02
    1    6 -1.9859D-01 -1.1188D 00  1.5901D 00  6.7785D-02  1.2121D 00  1.5779D-01
    1    7 -1.7651D-01 -1.4519D 00  4.5747D-01  8.9440D-02  3.6775D-01  1.3906D-01
    1    8 -2.1928D-01 -1.5380D 00  4.0097D-01 -8.8182D-02  3.4202D-01 -1.5647D-01

SUPPORT REACTIONS
 NODE        AR1          AR2          AR3
    1 -3.0511D-02 -5.0589D-01 -7.8051D-01
    2  0.0000D-01 -4.9411D-01 -7.1949D-01
    5  3.0511D-02  5.0589D-01  0.0000D-01
    6  0.0000D-01  4.9411D-01  0.0000D-01

**********  LOADING NUMBER     4  **********
  NLN  NEL  NEA  NEV  IPR
    0    0    0    1    1

VOLUME LOADS
ELEM.        BV1          BV2          BV3
    1  0.0000D-01 -1.0000D 00  0.0000D-01

NODAL DISPLACEMENTS
 NODE        DN1          DN2          DN3
    1  0.0000D-01  0.0000D-01  0.0000D-01
    2  1.5000D-01 -0.0000D-01 -0.0000D-01
    3  1.5000D-01 -5.0000D-01 -1.9169D-16
    4 -1.1623D-16 -5.0000D-01 -4.0246D-16
    5 -0.0000D-01 -0.0000D-01  1.5000D-01
    6  1.5000D-01 -0.0000D-01  1.5000D-01
    7  1.5000D-01 -5.0000D-01  1.5000D-01
    8 -8.5001D-17 -5.0000D-01  1.5000D-01

ELEMENT STRESSES
ELEM. INT.    X STRESS     Y STRESS     Z STRESS    XY STRESS    XZ STRESS    YZ STRESS
    1 AVG. -1.4051D-16 -5.0000D-01 -1.0235D-16 -1.5179D-17 -5.9131D-17  1.7097D-17

SUPPORT REACTIONS
 NODE        AR1          AR2          AR3
    1  3.9559D-17  2.5000D-01  3.9171D-17
    2  0.0000D-01  2.5000D-01  3.6898D-17
    5  4.9568D-17  2.5000D-01  0.0000D-01
    6  0.0000D-01  2.5000D-01  0.0000D-01
```

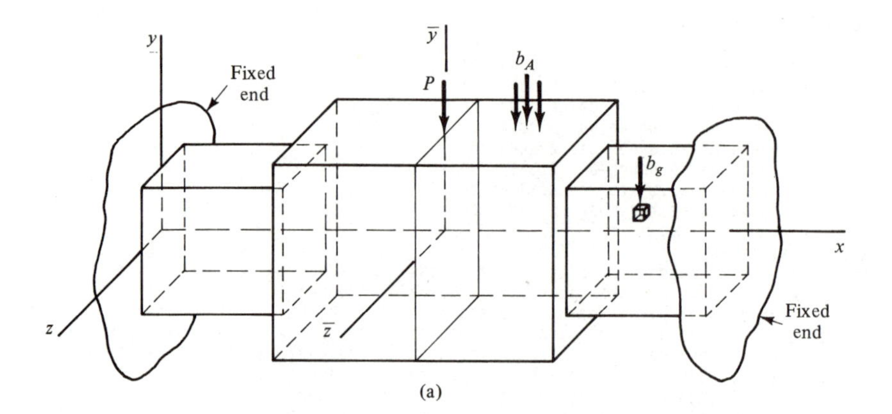

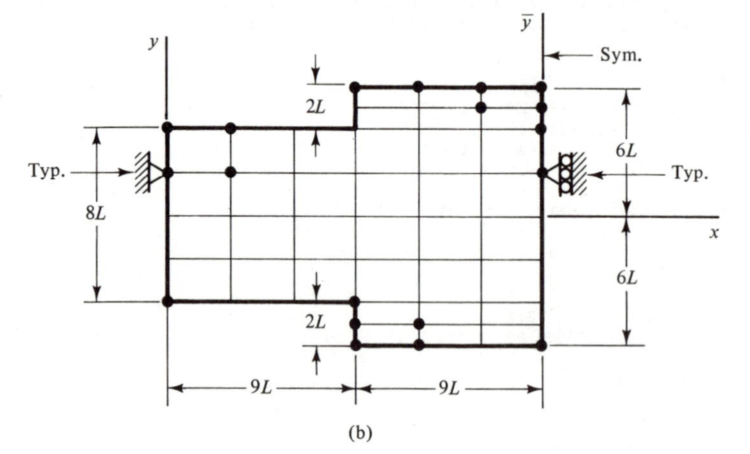

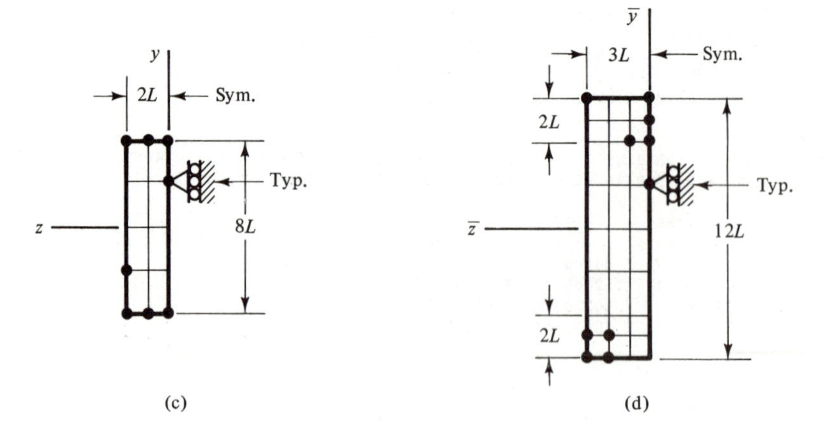

Figure 4.12 Example 2: (a) Short Beam (b) H8 Network at x-y Plane, (c) H8 Network at Small Section (d) H8 Network at Large Section

per unit area) applied to the upper surface, and a uniformly distributed force b_g (per unit volume) due to gravity. Values of these loads and other parameters are:

$$L = 3 \text{ in.} \qquad E = 3 \times 10^4 \text{ k/in.}^2 \qquad \nu = 0.3$$
$$P = 100 \text{ k} \qquad b_A = 1 \times 10^{-2} \text{ k/in.}^2 \qquad b_g = 2.83 \times 10^{-4} \text{ k/in.}^3$$

where the material is steel and U.S. units are used.

For the purpose of approximate analysis, we divide a quarter of the beam into a network of H8 elements, as indicated in Figs. 4.12(b), (c), and (d). Because of symmetry, restraints at nodes on planes x-y and $\bar{y}$-$\bar{z}$ must be rollers that prevent translations across those planes. In addition, pinned supports are required at nodes on the y-z plane to fix the end points.

In this problem we will calculate the deflections of points on the x axis and compare them with values obtained from beam theory. Figure 4.13(a) shows the

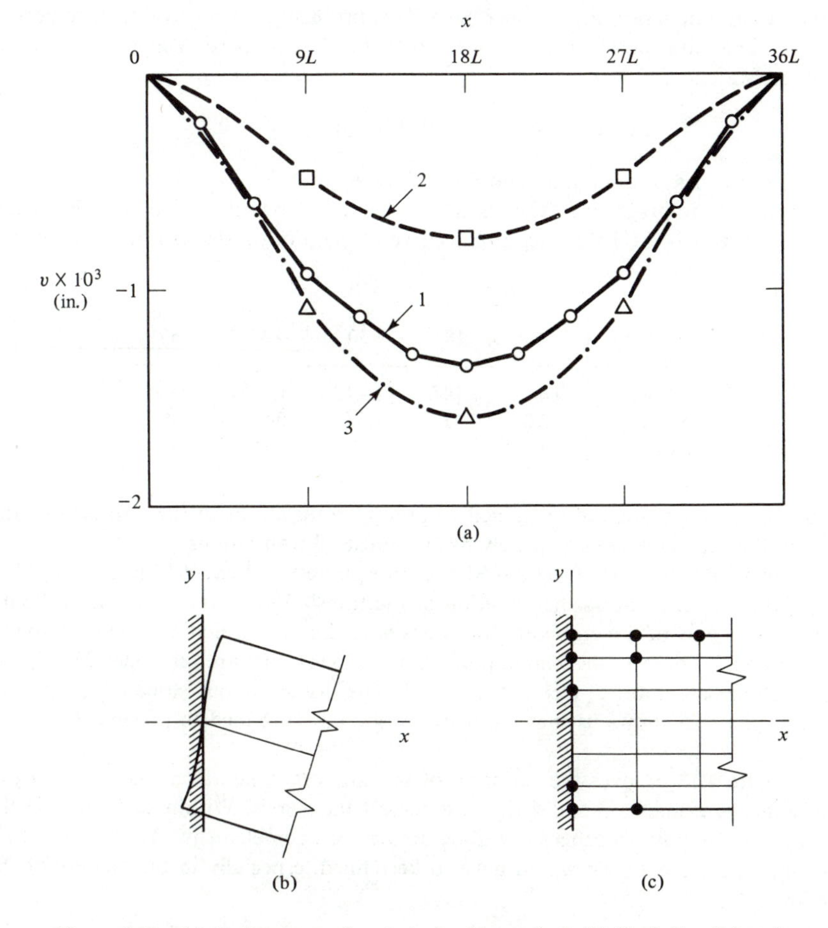

Figure 4.13 (a) Deflections of x Axis for Example 2 (b) End Condition for Beam (c) End Conditon for H8 Network

results of running Program SOH8 with the network given. In the figure, graph 1 represents output from the program; graph 2 is due to flexural deformations only; and graph 3 has both flexural and shearing deformations from beam theory. We see that plots 1 and 3 show much greater deflections than plot 2, demonstrating that shearing deformations are comparable to flexural deformations. Graph 1 is recognized to be better than 3 because its analytical model is more refined. Notice that the restraints at the fixed support are not the same for the beam and the finite-element network. Figure 4.13(b) shows the theoretical restraint condition at the fixed end of sa beam with shearing deformations (9). On the other hand, the fixed end of a discretized beam is depicted in Fig. 4.13(c), where no deformation of the cross section can occur.

Example 3

Figure 4.14(a) shows a gear tooth that is subjected to a line load b_x of 10 kN/cm, acting in the x direction at its upper edge. This problem is symmetric with respect to the x-y plane, and the tooth is assumed to be fixed at its base. Values of structural parameters are as follows:

$$L = 1 \text{ cm} \qquad E = 7 \times 10^3 \text{ kN/cm}^2 \qquad \nu = 0.3$$

where the material is aluminum and S.I. units are given.

Half of the problem is discretized into a crude network of four H20 elements, as seen in Fig. 4.14(b). In this figure the curve of the tooth is defined by the following nodal coordinates:

Point	11	18	30	37	49
x Coord.	$0.6L$	$1.14L$	$1.49L$	$1.78L$	$2.0L$
y Coord.	$3.2L$	$2.4L$	$1.6L$	$0.8L$	0
z Coord.	0	0	0	0	0

Nodes in the x-y plane are restrained to prevent translations in the z direction, and those in the x-z plane are completely fixed against all translations.

Partial results from Program SOH20 are plotted in Figs. 4.15(a) and (b). The first of these figures shows the variation of the translation u along the y axis. Points on the upper part of the graph are for nodes 6, 16, and 25, which are common to elements 1 and 2. On the other hand, points on the lower part are for nodes 25, 35, and 44, which are common to elements 3 and 4. The discontinuous slope of the plotted curve at point 25 shows us that the network should be refined to produce smoother results.

Figure 4.15(b) gives the variation of the stress σ_y taken from the four integration points in elements 3 and 4 that are nearest the x axis. Within each element the stress varies linearly, and there is a discontinuity between elements. To improve these stress results, the network would have to be refined, especially in the vicinity of the line loading.

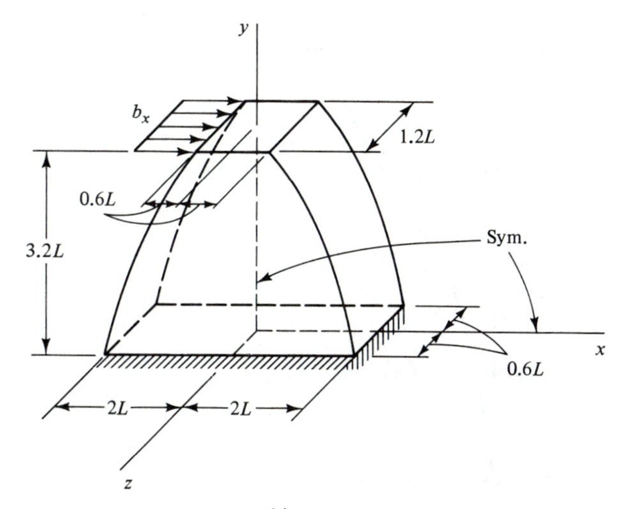

(a)

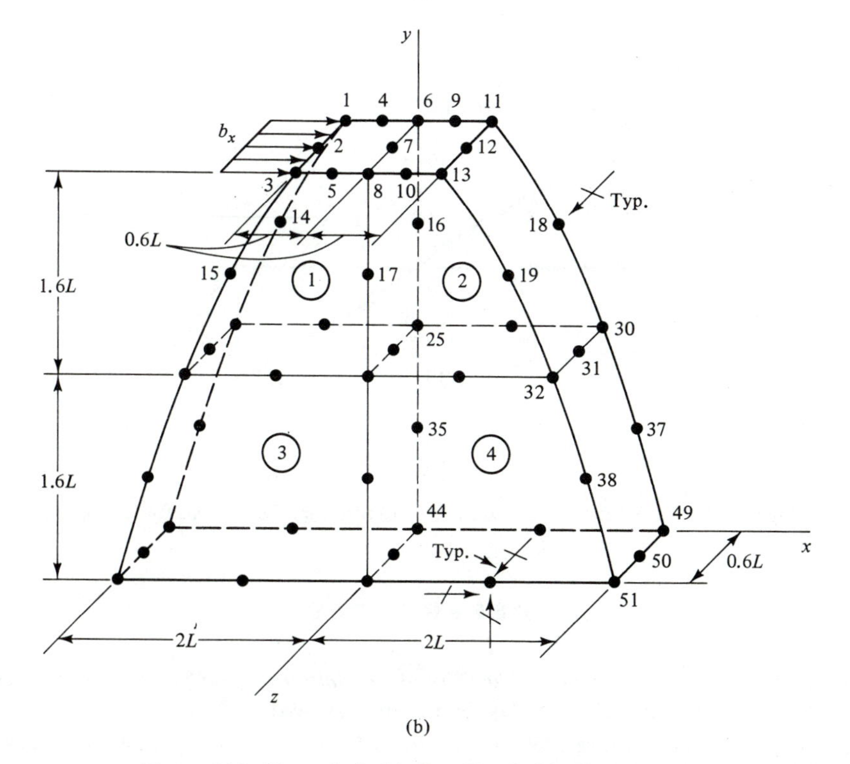

(b)

Figure 4.14 Example 3: (a) Gear Tooth (b) H20 Network

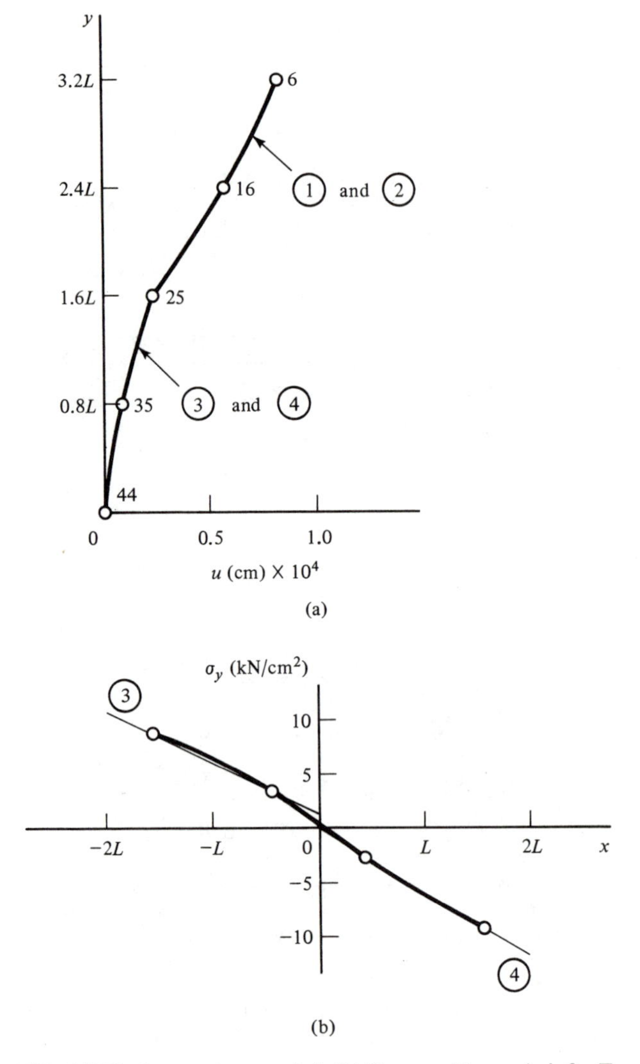

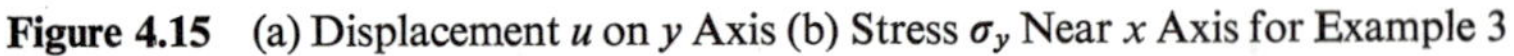
Figure 4.15 (a) Displacement u on y Axis (b) Stress σ_y Near x Axis for Example 3

REFERENCES

1. Lekhnitskii, S. G., *Theory of Elasticity of an Anisotropic Body*, Translation from Russian by P. Fern, Holden-Day, San Francisco, 1963.
2. Gere, J. M., and Weaver, W., Jr., *Matrix Algebra for Engineers*, 2d ed., Brooks-Cole, Monterey, 1983.
3. Eisenberg, M. A., and Malvern, L. E., "On Finite Element Integration in Natural Coordinates," *Int. Jour. Num. Meth. Eng.*, Vol. 7, No. 4, 1973, pp. 574–575.

4. Hammer, P. C., Marlowe, O. P., and Stroud, A. H., "Numerical Integration Over Simplexes and Cones," *Math. Tables Aids Comp.*, Vol. 10, 1956, pp. 130–137.
5. Melosh, R. J., "Structural Analysis of Solids," *Jour. Struc. Div.*, ASCE, Vol. 89, No. ST4, August 1963, pp. 205–223.
6. Irons, B. M., "Engineering Applications of Numerical Integration in Stiffness Methods," *AIAA Jour.*, Vol. 4, No. 11, 1966, pp. 2035–2037.
7. Ergatoudis, J., Irons, B. M., and Zienkiewicz, O. C., "Three-Dimensional Stress Analysis of Arch Dams and Their Foundations," *Proc. Symp. Arch Dams*, Inst. Civ. Eng., London, 1968, pp. 37–50.
8. Argyris, J. H., "Matrix Analysis of Three-Dimensional Elastic Media," *AIAA Jour.*, Vol. 3, No. 1, January 1965, pp. 45–51.
9. Timoshenko, S. P., and Gere, J. M., *Mechanics of Materials*, Van Nostrand-Reinhold, New York, 1972.

PROBLEMS

4.1-1. Determine the magnitudes and directions of principal normal stresses associated with the following stresses: $\sigma_x = 12$, $\sigma_y = -10$, $\sigma_z = 15$, $\tau_{xy} = 18$, $\tau_{xz} = \tau_{yz} = 0$.

4.1-2. Determine the magnitudes and directions of principal normal stresses associated with the following stresses: $\sigma_x = 8$, $\sigma_y = 13$, $\sigma_z = -9$, $\tau_{xy} = \tau_{xz} = 0$, $\tau_{yz} = -12$.

4.1-3. Determine the magnitudes and directions of principal normal stresses associated with the following stresses: $\sigma_x = -16$, $\sigma_y = -11$, $\sigma_z = 0$, $\tau_{xy} = \tau_{yz} = 0$, $\tau_{xz} = 15$.

4.1-4. Determine the magnitudes and directions of principal normal stresses associated with the following stresses: $\sigma_x = 10$, $\sigma_y = 20$, $\sigma_z = -14$, $\tau_{xy} = \tau_{xz} = 5$, $\tau_{yz} = 0$.

4.1-5. Determine the magnitudes and directions of principal normal stresses associated with the following stresses: $\sigma_x = -17$, $\sigma_y = 0$, $\sigma_z = 22$, $\tau_{xy} = 7$, $\tau_{xz} = -10$, $\tau_{yz} = -8$.

4.1-6. Determine the magnitudes and directions of principal normal stresses associated with the following stresses: $\sigma_x = 18$, $\sigma_y = 2$, $\sigma_z = -6$, $\tau_{xy} = -12$, $\tau_{xz} = 0$, $\tau_{yz} = 5$.

4.2-1. Using volume coordinates for a tetrahedron, derive the result for $\int_V x^2\, dV$ given in Appendix A.

4.2-2. Using volume coordinates for a tetrahedron, derive the result for $\int_V y^2\, dV$ given in Appendix A.

4.2-3. Using volume coordinates for a tetrahedron, derive the result for $\int_V xy\, dV$ given in Appendix A.

4.2-4. Using volume coordinates for a tetrahedron, derive the result for $\int_V yz\, dV$ given in Appendix A.

4.2-5. Using volume coordinates for a tetrahedron, derive the result for $\int_V zx\, dV$ given in Appendix A.

4.4-1. For the eight-node rectangular solid element (RS8) shown in the figure, determine the stiffness K_{12}, assuming isotropic material.

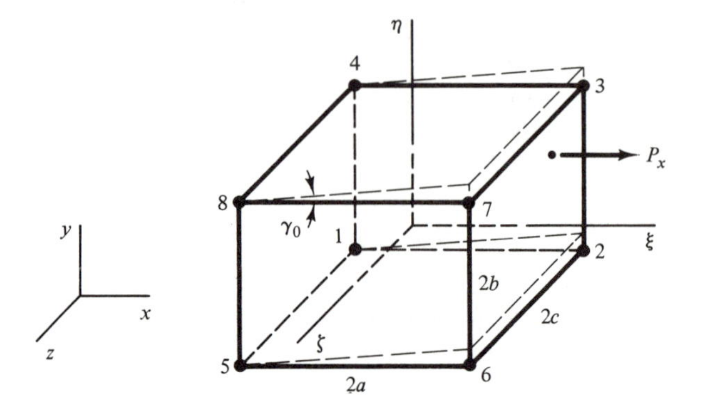

Problems 4.4-1 through 4.4-8

4.4-2. Find the stiffness K_{69} for the RS8 element if the material is isotropic.

4.4-3. Derive equivalent nodal loads for a concentrated force P_x in the positive x direction, applied anywhere on the positive x face of the RS8 element (see figure).

4.4-4. Suppose that the RS8 element is subjected to a temperature variation $\Delta T = \xi\eta\zeta\, \Delta T_0$. Find the initial stresses in the element due to this condition, assuming isotropic material.

4.4-5. Develop explicit terms for the third column of the stress-displacement matrix **EB** for the RS8 element if the material is isotropic.

4.4-6. Derive equivalent nodal loads for the RS8 element due to a uniform gravitational force b_g (per unit volume) acting in the negative y direction.

4.4-7. Suppose that the RS8 element has a uniform shearing prestrain of $\gamma_{xy} = \gamma_0$ (see figure). Find the equivalent nodal loads due to this initial condition, assuming that the material is isotropic.

4.4-8. Let the RS8 element be subjected to a temperature variation $\Delta T = \zeta\, \Delta T_0$. Derive the equivalent nodal loads for this influence with isotropic material.

4.5-1. Find the stiffness K_{11} of the right tetrahedral element in Fig. 4.8(b), assuming isotropic material.

4.5-2. For the right tetrahedron in Fig. 4.8(b), determine the stiffness K_{12}, assuming isotropic material.

4.5-3. Determine the stiffness K_{13} for the right tetrahedron in Fig. 4.8(b), assuming isotropic material.

4.5-4. For the right tetrahedral element in Fig. 4.8(b), find the stiffness K_{33}, assuming isotropic material.

4.5-5. Find the stiffness K_{23} of the right tetrahedron in Fig. 4.8(b), assuming isotropic material.

4.5-6. Let the Tet-4 element be subjected to a concentrated force P_x in the positive x direction (see figure). If the coordinates of the point of application are $(\xi_1, \xi_2, \xi_3, \xi_4) = (0, \frac{2}{3}, \frac{1}{6}, \frac{1}{6})$, find the equivalent nodal loads.

4.5-7. Derive equivalent nodal loads for the Tet-4 element due to a uniformly distributed body force b_z (force per unit volume) acting in the positive z direction.

4.5-8. Suppose that the Tet-4 element is subjected to a uniform temperature change ΔT. Find the initial stresses in the element due to this condition, assuming isotropic material.

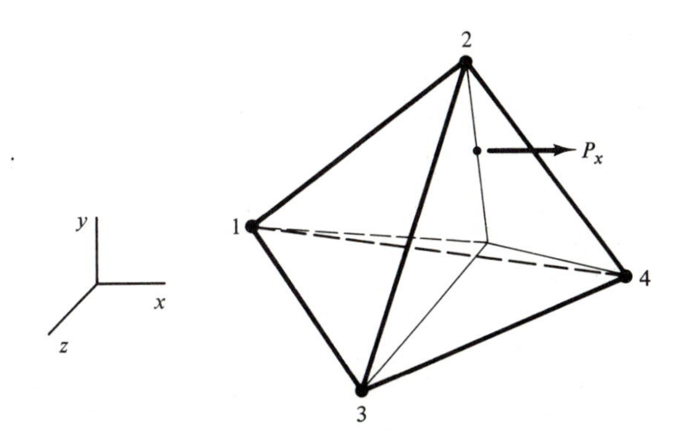

Problems 4.5-6 through 4.5-8

CHAPTER

5

Axisymmetric Solids

5.1 AXISYMMETRIC STRESSES AND STRAINS

An axisymmetric solid is defined as a three-dimensional body that is developed by rotation of a planar section about an axis. Figure 1.3(b) shows such a solid with a portion cut away to expose the section. This type of body is sometimes called a *solid of revolution*. Cylindrical coordinates r, z, and θ provide a suitable reference frame, as indicated in Fig. 5.1. We assume that the body is axisymmetric with respect to the z axis and that a typical finite element is a circular ring. This element may have various cross-sectional shapes, but if the loads are axisymmetric, we may analyze the ring using only a representative two-dimensional section. While nodes are indicated by dots on the cross section, they are actually nodal circles. In this section stresses and strains for axisymmetric loading will be discussed (1); the theory for nonaxisymmetric loading is covered in Sec. 5.4.

For any point on the cross section of an axisymmetrically loaded ring element the generic displacements are:

$$\mathbf{u} = \{u, v\} \tag{a}$$

Translations u and v occur in the r and z directions, respectively, as indicated in Fig. 5.1. For this case the translation w in the θ direction is zero. In addition, the shearing strains $\gamma_{r\theta}$ and $\gamma_{z\theta}$ are also zero. On the other hand, the figure shows four types of strains that are nonzero, as follows:

$$\boldsymbol{\epsilon} = \{\epsilon_r, \epsilon_z, \epsilon_\theta, \gamma_{rz}\} \tag{b}$$

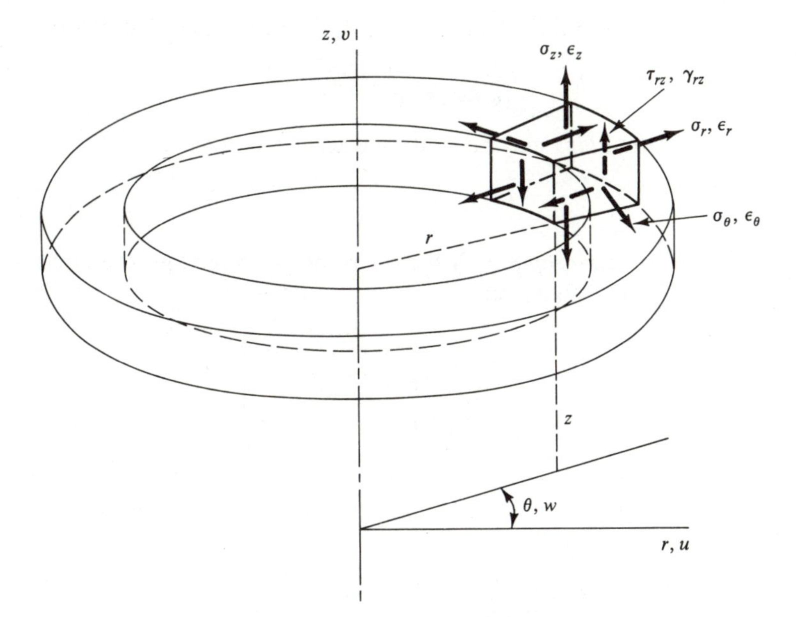

Figure 5.1 Axisymmetric Ring Element

Relationships between these strains and the generic displacements in Eq. (a) are seen to be:

$$
\begin{aligned}
\epsilon_r &= \frac{\partial u}{\partial r} \qquad \epsilon_\theta = \frac{2\pi(r+u) - 2\pi r}{2\pi r} = \frac{u}{r} \\
\epsilon_z &= \frac{\partial v}{\partial z} \qquad \gamma_{rz} = \frac{\partial u}{\partial z} + \frac{\partial v}{\partial r}
\end{aligned}
\tag{5.1-1}
$$

Casting such *strain-displacement relationships* into matrix form gives:

$$
\boldsymbol{\epsilon} = \mathbf{du} \tag{1.3-6}
$$

repeated

where

$$
\mathbf{d} = \begin{bmatrix} \frac{\partial}{\partial r} & 0 \\ 0 & \frac{\partial}{\partial z} \\ \frac{1}{r} & 0 \\ \frac{\partial}{\partial z} & \frac{\partial}{\partial r} \end{bmatrix} \tag{5.1-2}
$$

In this instance the nonzero term $1/r$ in the third row of matrix $\mathbf{d}$ is a multiplier of u, not a derivative.

Corresponding to the strains in Eq. (b), the four types of stresses depicted in Fig. 5.1 are:

$$\boldsymbol{\sigma} = \{\sigma_r, \sigma_z, \sigma_\theta, \tau_{rz}\} \tag{c}$$

Stress-strain relationships may be expressed in matrix form as:

$$\boldsymbol{\sigma} = \mathbf{E}\,\boldsymbol{\epsilon} \tag{1.3-9}$$

repeated

If the material is *orthotropic* with θ as a principal direction, the stress-strain operator $\mathbf{E}$ in Eq. (1.3-9) becomes:

$$\mathbf{E} = \begin{bmatrix} E_{11} & E_{12} & E_{13} & 0 \\ E_{21} & E_{22} & E_{23} & 0 \\ E_{31} & E_{32} & E_{33} & 0 \\ 0 & 0 & 0 & E_{44} \end{bmatrix} \tag{5.1-3}$$

Values of E_{11}, and so on, must be obtained from the inversion of a 4×4 matrix $\mathbf{C}$, where

$$C_{11} = \frac{1}{E_r} \qquad C_{12} = -\frac{\nu_{rz}}{E_z} \qquad \cdots$$

We assume that r and z are also principal directions for the orthotropic material. Otherwise, a rotation-of-axes transformation would be required.

For an *isotropic* material, the matrix $\mathbf{E}$ simplifies to:

$$\mathbf{E} = \frac{E}{(1+\nu)e_2} \begin{bmatrix} e_1 & \nu & \nu & 0 \\ \nu & e_1 & \nu & 0 \\ \nu & \nu & e_1 & 0 \\ 0 & 0 & 0 & e_3 \end{bmatrix} \tag{5.1-4}$$

in which

$$e_1 = 1 - \nu \qquad e_2 = 1 - 2\nu \qquad e_3 = \frac{e_2}{2}$$

These parameters are the same as those for plane strain in Sec. 2.2.

5.2 TRIANGULAR AND RECTANGULAR SECTIONS

The triangular cross section of a ring element (2) in an axisymmetric solid (with axisymmetric loads) appears in Fig. 5.2. Three nodal circles are represented by the dots at the apexes (points 1, 2, and 3). At each node we have translations in the r and z directions, so the vector of nodal displacements becomes:

$$\mathbf{q} = \{q_1, q_2, \ldots, q_6\} = \{u_1, v_1, \ldots, v_3\} \tag{a}$$

As in the constant strain triangle (Sec. 2.2), the assumed displacement functions

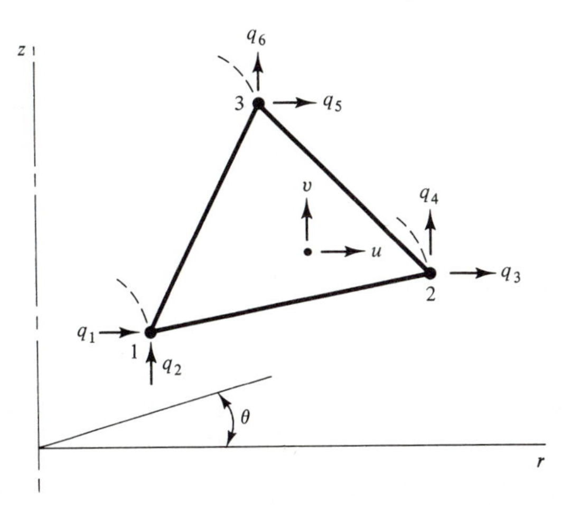

Figure 5.2 Triangular Section

are linear. Thus,

$$\mathbf{g} = \begin{bmatrix} 1 & r & z & 0 & 0 & 0 \\ 0 & 0 & 0 & 1 & r & z \end{bmatrix} \tag{5.2-1}$$

Applying the operator $\mathbf{d}$ in Eq. (5.1-2) to matrix $\mathbf{g}$ in Eq. (5.2-1), we obtain matrix $\mathbf{B}_c$ as follows:

$$\mathbf{B}_c = \mathbf{d}\ \mathbf{g} = \begin{bmatrix} 0 & 1 & 0 & 0 & 0 & 0 \\ 0 & 0 & 0 & 0 & 0 & 1 \\ \frac{1}{r} & 1 & \frac{z}{r} & 0 & 0 & 0 \\ 0 & 0 & 1 & 0 & 1 & 0 \end{bmatrix} \tag{5.2-2}$$

where the subscript c implies the use of generalized displacements (see Sec. 1.4). Premultiplication of matrix $\mathbf{B}_c$ with matrix $\mathbf{E}$ from Eq. (5.1-3) for an orthotropic material yields:

$$\mathbf{E}\ \mathbf{B}_c = \begin{bmatrix} \frac{E_{13}}{r} & E_{11} + E_{13} & \frac{E_{13}z}{r} & 0 & 0 & E_{12} \\ \frac{E_{23}}{r} & E_{21} + E_{23} & \frac{E_{23}z}{r} & 0 & 0 & E_{22} \\ \frac{E_{33}}{r} & E_{31} + E_{33} & \frac{E_{33}z}{r} & 0 & 0 & E_{32} \\ 0 & 0 & E_{44} & 0 & E_{44} & 0 \end{bmatrix} \tag{5.2-3}$$

which can be used for the calculation of stresses. Then premultiplication of Eq. (5.2-3) with $\mathbf{B}_c^{\mathrm{T}}$ produces:

$$\mathbf{B}_c^{\mathrm{T}}\mathbf{E}\ \mathbf{B}_c = \begin{bmatrix} \frac{E_{33}}{r^2} & \frac{E_{13}+E_{33}}{r} & \frac{E_{33}z}{r^2} & 0 & 0 & \frac{E_{23}}{r} \\ & E_{11}+2E_{13}+E_{33} & \frac{(E_{13}+E_{33})z}{r} & 0 & 0 & E_{12}+E_{23} \\ & & \frac{E_{44}+E_{33}z^2}{r^2} & 0 & E_{44} & \frac{E_{23}z}{r} \\ & & & 0 & 0 & 0 \\ & \text{Sym.} & & & E_{44} & 0 \\ & & & & & E_{22} \end{bmatrix} \tag{5.2-4}$$

Generalized element stiffnesses $\mathbf{K}_c$ are obtained by integrating the matrix in Eq. (5.2-4) with respect to θ, r, and z. That is,

$$\mathbf{K}_c = \int_A \int_0^{2\pi} \mathbf{B}_c^{\mathrm{T}}\mathbf{E}\ \mathbf{B}_c r\ d\theta\ dA = 2\pi \int_A \mathbf{B}_c^{\mathrm{T}}\mathbf{E}\ \mathbf{B}_c r\ dA \tag{5.2-5}$$

in which A is the area of the cross section of the ring. Because integrals in Eq. (5.2-5) are logarithmic, numerical integration is preferred.

Proceeding in a similar manner, we can also develop generalized equivalent nodal loads due to body forces as:

$$\mathbf{p}_{bc} = 2\pi \int_A \mathbf{g}^{\mathrm{T}}\mathbf{b} r\ dA \tag{5.2-6}$$

And those caused by initial strains are:

$$\mathbf{p}_{0c} = 2\pi \int_A \mathbf{B}_c^{\mathrm{T}}\mathbf{E}\ \boldsymbol{\epsilon}_0 r\ dA \tag{5.2-7}$$

Because the radius r does not appear in denominator positions, Eqs. (5.2-6) and (5.2-7) can be integrated directly. For any of the Eqs. (5.2-5) through (5.2-7), actions per radian may be obtained by deleting the factor 2π.

Figure 5.3 shows a rectangular cross section for a ring element (3). In

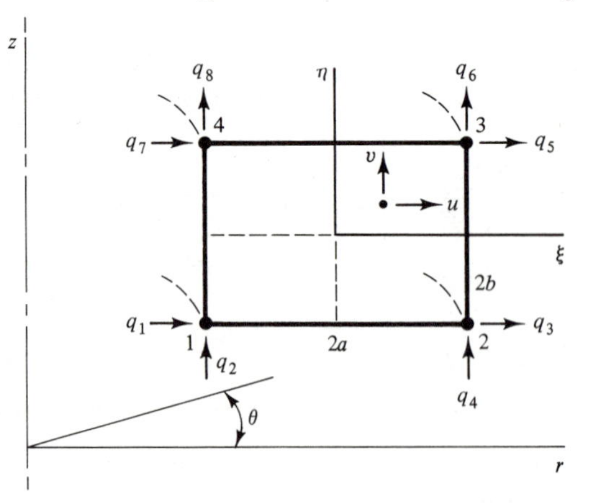

Figure 5.3 Rectangular Section

this case there are four nodal circles represented by dots at the corners. Thus, the nodal displacement vector consists of:

$$\mathbf{q} = \{q_1, q_2, \ldots, q_8\} = \{u_1, v_1, \ldots, v_4\} \tag{b}$$

The assumed displacement functions are bilinear, as in the first rectangle described in Sec. 2.3. Therefore, matrix $\mathbf{g}$ becomes:

$$\mathbf{g} = \begin{bmatrix} 1 & \xi & \eta & \xi\eta & 0 & 0 & 0 & 0 \\ 0 & 0 & 0 & 0 & 1 & \xi & \eta & \xi\eta \end{bmatrix} \tag{5.2-8}$$

For the dimensionless coordinates ξ and η the operator $\mathbf{d}$ is:

$$\mathbf{d} = \begin{bmatrix} \frac{1}{a}\frac{\partial}{\partial\xi} & 0 \\ 0 & \frac{1}{b}\frac{\partial}{\partial\eta} \\ \frac{1}{r} & 0 \\ \frac{1}{b}\frac{\partial}{\partial\eta} & \frac{1}{a}\frac{\partial}{\partial\xi} \end{bmatrix} \tag{5.2-9}$$

Using this matrix, we find:

$$\mathbf{B}_c = \mathbf{d}\,\mathbf{g} = \begin{bmatrix} 0 & \frac{1}{a} & 0 & \frac{\eta}{a} & 0 & 0 & 0 & 0 \\ 0 & 0 & 0 & 0 & 0 & 0 & \frac{1}{b} & \frac{\xi}{b} \\ \frac{1}{r} & \frac{\xi}{r} & \frac{\eta}{r} & \frac{\xi\eta}{r} & 0 & 0 & 0 & 0 \\ 0 & 0 & \frac{1}{b} & \frac{\xi}{b} & 0 & \frac{1}{a} & 0 & \frac{\eta}{a} \end{bmatrix} \tag{5.2-10}$$

Then, for an orthotropic material, premultiplication of $\mathbf{B}_c$ with $\mathbf{E}$ gives:

$$\mathbf{E}\,\mathbf{B}_c = \begin{bmatrix} \frac{E_{13}}{r} & \frac{E_{11}}{a} + \frac{E_{13}\xi}{r} & \frac{E_{13}\eta}{r} & \frac{E_{11}\eta}{a} + \frac{E_{13}\xi\eta}{r} & 0 & 0 & \frac{E_{12}}{b} & \frac{E_{12}\xi}{b} \\ \frac{E_{23}}{r} & \frac{E_{21}}{a} + \frac{E_{23}\xi}{r} & \frac{E_{23}\eta}{r} & \frac{E_{21}\eta}{a} + \frac{E_{23}\xi\eta}{r} & 0 & 0 & \frac{E_{22}}{b} & \frac{E_{22}\xi}{b} \\ \frac{E_{33}}{r} & \frac{E_{31}}{a} + \frac{E_{33}\xi}{r} & \frac{E_{33}\eta}{r} & \frac{E_{31}\eta}{a} + \frac{E_{33}\xi\eta}{r} & 0 & 0 & \frac{E_{32}}{b} & \frac{E_{32}\xi}{b} \\ 0 & 0 & \frac{E_{44}}{b} & \frac{E_{44}\xi}{b} & 0 & \frac{E_{44}}{a} & 0 & \frac{E_{44}\eta}{a} \end{bmatrix} \tag{5.2-11}$$

When this matrix is premultiplied by $\mathbf{B}_c^{\mathrm{T}}$, we obtain an 8×8 matrix similar to the 6×6 array in Eq. (5.2-4). Integration formulas for $\mathbf{K}_c$, $\mathbf{p}_{bc}$, and $\mathbf{p}_{0c}$ for the rectangular section are the same as those for the triangular section [see Eqs. (5.2-5) through (5.2-7)], except that the coordinates are dimensionless.

Example

For the ring element with a rectangular cross section (see Fig. 5.3), determine equivalent nodal loads due to a uniform prestrain ϵ_{z0}, assuming isotropic material. For this purpose, let the coordinates of points 1 through 4 be (1, 1), (5, 1), (5, 3), and (1, 3). For vector $\mathbf{p}_{0c}$ we have:

$$\mathbf{p}_{0c} = 2\pi ab \int_{-1}^{1} \int_{-1}^{1} \mathbf{B}_c^{\mathrm{T}} \mathbf{E}\ \boldsymbol{\epsilon}_0 r\, d\xi\, d\eta$$

Substitution of $\boldsymbol{\epsilon}_0 = \{0, \epsilon_{z0}, 0, 0\}$ along with Eq. (5.1-4) for $\mathbf{E}$ and Eq. (5.2-10) for $\mathbf{B}_c$ yields:

$$\mathbf{p}_{0c} = \frac{2\pi ab E \epsilon_{z0}}{(1+\nu)e_2} \int_{-1}^{1} \int_{-1}^{1} \left\{\nu, \frac{\nu r}{a} + \nu\xi, \nu\eta, \frac{\nu\eta r}{a} + \nu\xi\eta, 0, 0, \frac{e_1 r}{b}, \frac{e_1 \xi r}{b}\right\} d\xi\, d\eta$$

where

$$a = 2 \qquad b = 1 \qquad r = 3 + 2\xi$$

Integration of the terms in $\mathbf{p}_{0c}$ gives:

$$\mathbf{p}_{0c} = \frac{2\pi(2)E\epsilon_{z0}}{(1+\nu)e_2}\left\{4\nu, 6\nu, 0, 0, 0, 0, 12e_1, \frac{8}{3}e_1\right\}$$

In this case there is no barrier against explicit integration. Premultiplication of $\mathbf{p}_{0c}$ by $\mathbf{h}^{-\mathrm{T}}$ [see Eq. (j) of Sec. 2.3] produces:

$$\mathbf{p}_0 = \mathbf{h}^{-\mathrm{T}}\mathbf{p}_{0c} = \frac{2\pi E\epsilon_{z0}}{3(1+\nu)e_2}\{-3\nu, -14e_1, 15\nu, -22e_1, 15\nu, 22e_1, -3\nu, 14e_1\}$$

which is the desired solution.

5.3 QUADRILATERAL SECTIONS

In this section we shall examine ring elements having cross sections that are isoparametric quadrilaterals. Previously (see Sec. 3.4) we discussed the isoparametric elements Q4 and Q8 for analysis of plane stress and plane strain. Comments in this section will depend heavily upon that earlier work.

Figure 5.4(a) shows *element AXQ4*, which is an axisymmetric ring element with a cross section in the shape of element Q4. Thus, the nodal displacement vector is:

$$\mathbf{q} = \{q_1, q_2, \ldots, q_8\} = \{u_1, v_1, \ldots, v_4\} \tag{a}$$

Bilinear displacement shape functions in matrix $\mathbf{f}$ are the same as for element Q4, and the strain-displacement matrix becomes:

$$\mathbf{B}_i = \begin{bmatrix} f_{i,r} & 0 \\ 0 & f_{i,z} \\ \dfrac{f_i}{r} & 0 \\ f_{i,z} & f_{i,r} \end{bmatrix} \qquad (i = 1, 2, 3, 4) \tag{5.3-1}$$

which is obtained using Eq. (5.1-2). The radius r in Eq. (5.3-1) is found as:

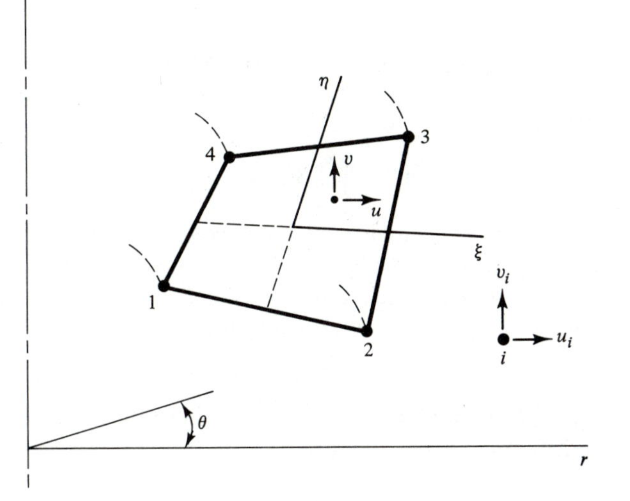

(a)

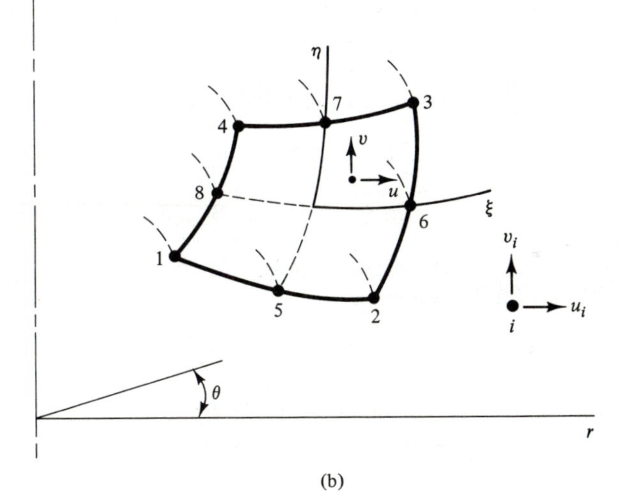

(b)

Figure 5.4 Quadrilateral Sections: (a) Element AXQ4 (b) Element AXQ8

$$r = \sum_{i=1}^{4} f_i r_i \tag{b}$$

In addition, the derivatives $f_{i,r} = D_{G1i}$, and so on, are given by Eqs. (3.2-26), except that r and z replace the coordinates x and y.

Stiffnesses for element AXQ4 may be calculated from:

$$\mathbf{K} = 2\pi \int_{-1}^{1} \int_{-1}^{1} \mathbf{B}^{\mathrm{T}}\mathbf{E}\ \mathbf{B}|\mathbf{J}|\, r\, d\xi\, d\eta \tag{5.3-2}$$

Also, equivalent nodal loads due to body forces are:

$$\mathbf{p}_b = 2\pi \int_{-1}^{1} \int_{-1}^{1} \mathbf{f}^{\mathrm{T}}\mathbf{b}\,|\mathbf{J}|\, r\, d\xi\, d\eta \tag{5.3-3}$$

And those for initial strains become:

$$\mathbf{p}_0 = 2\pi \int_{-1}^{1} \int_{-1}^{1} \mathbf{B}^{\mathrm{T}}\mathbf{E}\ \boldsymbol{\epsilon}_0\,|\mathbf{J}|\, r\, d\xi\, d\eta \tag{5.3-4}$$

Numerical integration is required to evaluate Eqs. (5.3-2) through (5.3-4).

Element AXQ8 appears in Fig. 5.4(b). Its cross section has the same shape as element Q8, and its quadratic interpolation functions were given as Eqs. (3.4-12). The nodal displacement vector for this element is:

$$\mathbf{q} = \{q_1, q_2, \ldots, q_{16}\} = \{u_1, v_1, \ldots, v_8\} \tag{c}$$

Expressions for $\mathbf{B}_i$, $\mathbf{K}$, $\mathbf{p}_b$, and $\mathbf{p}_0$ are the same as those for element AXQ4 shown in Eqs. (5.3-1) through (5.3-4), except that $i = 1, 2, \ldots, 8$.

Example

Find the stiffness term K_{22} for element AXQ4 in Fig. 5.4(a), using Gaussian numerical integration with $n = 2$ each way. Assume that the coordinates of nodes 1, 2, 3, and 4 in the r-z plane are (13, 15), (32, 11), (36, 29), and (19, 27), respectively. For numerical integration of terms in matrix $\mathbf{K}$, the formula is:

$$\mathbf{K} = 2\pi \sum_{k=1}^{n} \sum_{j=1}^{n} R_j R_k \mathbf{B}^{\mathrm{T}}(\xi_j, \eta_k)\mathbf{E}\ \mathbf{B}(\xi_j, \eta_k)\,|\mathbf{J}(\xi_j, \eta_k)|\, r(\xi_j, \eta_k) \tag{d}$$

In particular, for the term K_{22} with $n = 2$, $R_j = R_k = 1$. Therefore,

$$K_{22} = 2\pi \sum_{k=1}^{2} \sum_{j=1}^{2} \mathbf{B}_{1,2}^{\mathrm{T}}\mathbf{E}\ \mathbf{B}_{1,2}\,|\mathbf{J}|\, r \tag{e}$$

in which $\mathbf{B}_{1,2}$ is the second column of submatrix $\mathbf{B}_1$. Substituting this column of $\mathbf{B}$ from Eq. (5.3-1) into Eq. (e), we find:

$$K_{22} = 2\pi \sum_{k=1}^{2} \sum_{j=1}^{2} (E_{22}D_{G21}^2 + E_{44}D_{G11}^2)\,|\mathbf{J}|\, r \tag{f}$$

where the terms E_{22} and E_{44} remain in unspecified form.

Now we can evaluate Eq. (f) using Eq. (b) to find r at each integration point. In this case the Jacobian matrix becomes:

$$\mathbf{J} = \mathbf{D}_L\mathbf{C}_N = \frac{1}{4}\begin{bmatrix} -(1-\eta) & (1-\eta) & (1+\eta) & -(1+\eta) \\ -(1-\xi) & -(1+\xi) & (1+\xi) & (1-\xi) \end{bmatrix} \begin{bmatrix} 13 & 15 \\ 32 & 11 \\ 36 & 29 \\ 19 & 27 \end{bmatrix}$$

$$= \frac{1}{2}\begin{bmatrix} 18-\eta & -1+3\eta \\ 5-\xi & 15+3\xi \end{bmatrix}$$

And its determinant is:

$$|\mathbf{J}| = \frac{1}{4}(275 + 53\xi - 30\eta)$$

From matrix $\mathbf{D}_G$ [see Eqs. (3.2-26)] we find:

$$D_{G11} = \frac{1}{4|\mathbf{J}|}[-(1-\eta)J_{22} + (1-\xi)J_{12}] = \frac{-8-\xi+9\eta}{275+53\xi}$$

$$D_{G21} = \frac{1}{4|\mathbf{J}|}[(1-\eta)J_{21} - (1-\xi)J_{11}] = \frac{-13+17\xi-4\eta}{2(275+53\xi)}$$

Substitution of all required terms into Eq. (f) at each of the four integration points yields:

$$K_{22} = 2\pi(5.758E_{22} + 7.444E_{44})$$

5.4 NONAXISYMMETRIC LOADS

In many cases of axisymmetric solids the loads are not distributed axisymmetrically. For example, wind and earthquake forces on chimneys and other axisymmetric structures do not have axisymmetric patterns. However, any set of loads on a solid of revolution may be decomposed into two sets, which are symmetric and antisymmetric with respect to a plane of symmetry containing the axis of revolution (2, 4). *Fourier decomposition* of the symmetric loads (5) for m harmonic terms produces:

$$b_r = \sum_{j=0}^{m} b_{rj}\cos j\theta \qquad b_z = \sum_{j=0}^{m} b_{zj}\cos j\theta$$
$$b_\theta = \sum_{j=0}^{m} b_{\theta j}\sin j\theta \tag{5.4-1}$$

where b_{rj}, b_{zj}, and $b_{\theta j}$ are functions of r and z only. When $j = 0$, we have $b_\theta = 0$; and Eqs. (5.4-1) become the case of axisymmetric loads. Otherwise, $j = 1, 2, \ldots, m$ represent cases of nonaxisymmetric loads that are symmetric with respect to a plane through the z axis. Figures 5.5(a), (b), and (c) show the first harmonic loads for the r, z, and θ directions, respectively. If the loads were antisymmetric with respect to the plane of symmetry, the functions $\sin j\theta$ and $\cos j\theta$ in Eqs. (5.4-1) would be interchanged.

Generic displacements for nonaxisymmetric loads must include the translation w in the θ direction. Thus,

$$\mathbf{u} = \{u, v, w\} \tag{a}$$

And the strain vector must contain $\gamma_{z\theta}$ and $\gamma_{r\theta}$, as follows:

$$\boldsymbol{\epsilon} = \{\epsilon_r, \epsilon_z, \epsilon_\theta, \gamma_{rz}, \gamma_{z\theta}, \gamma_{r\theta}\} \tag{b}$$

Strain-displacement relationships developed by Love (6) are:

$$\epsilon_r = \frac{\partial u}{\partial r} \qquad \epsilon_z = \frac{\partial v}{\partial z} \qquad \epsilon_\theta = \frac{u}{r} + \frac{1}{r}\frac{\partial w}{\partial \theta}$$

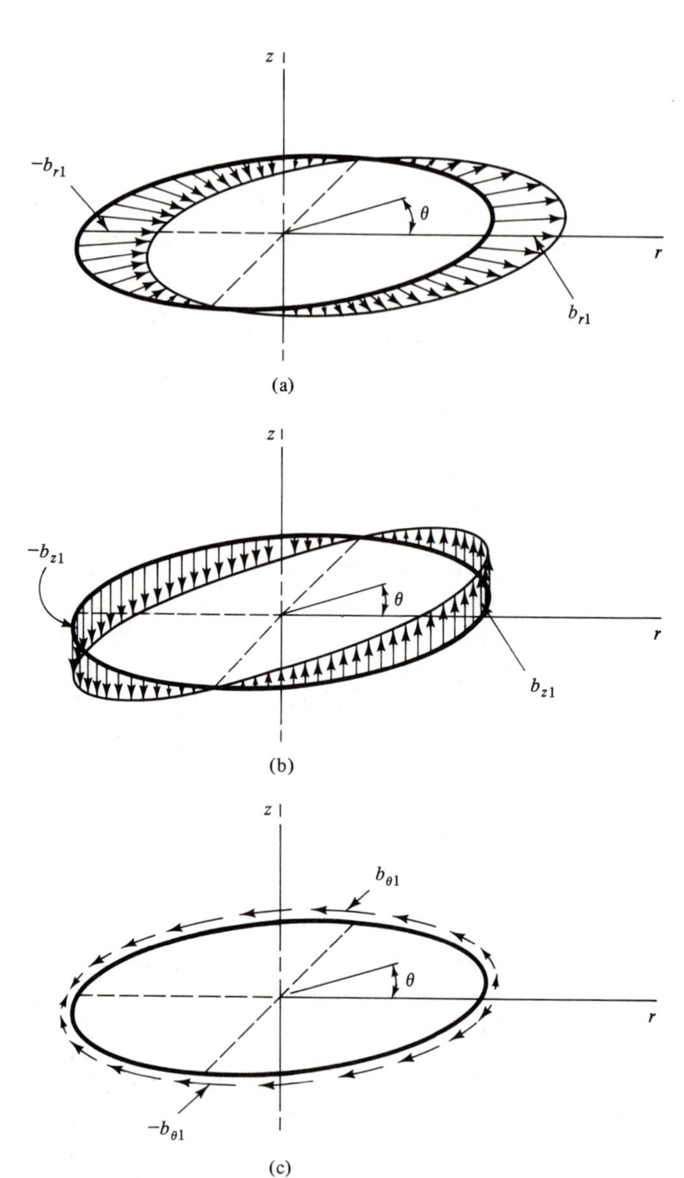

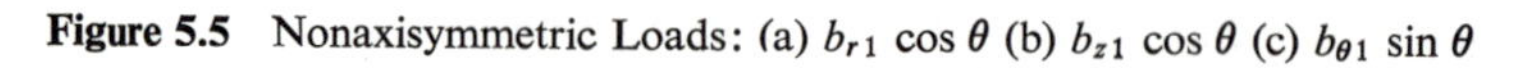

Figure 5.5 Nonaxisymmetric Loads: (a) $b_{r1} \cos\theta$ (b) $b_{z1} \cos\theta$ (c) $b_{\theta 1} \sin\theta$

$$\gamma_{rz} = \frac{\partial u}{\partial z} + \frac{\partial v}{\partial r} \qquad \gamma_{z\theta} = \frac{1}{r}\frac{\partial v}{\partial \theta} + \frac{\partial w}{\partial z} \tag{5.4-2}$$

$$\gamma_{r\theta} = \frac{1}{r}\frac{\partial u}{\partial \theta} + \frac{\partial w}{\partial r} - \frac{w}{r}$$

Here we see that the radius r appears in the denominators of several expressions. From Eqs. (5.4-2) we can form the operator $\mathbf{d}$ as:

$$\mathbf{d} = \begin{bmatrix} \frac{\partial}{\partial r} & 0 & 0 \\ 0 & \frac{\partial}{\partial z} & 0 \\ \frac{1}{r} & 0 & \frac{1}{r}\frac{\partial}{\partial \theta} \\ \frac{\partial}{\partial z} & \frac{\partial}{\partial r} & 0 \\ 0 & \frac{1}{r}\frac{\partial}{\partial \theta} & \frac{\partial}{\partial z} \\ \frac{1}{r}\frac{\partial}{\partial \theta} & 0 & \frac{\partial}{\partial r} - \frac{1}{r} \end{bmatrix} \tag{5.4-3}$$

The stress vector for nonaxisymmetric loads must include $\tau_{z\theta}$ and $\tau_{r\theta}$, as follows:

$$\boldsymbol{\sigma} = \{\sigma_r, \sigma_z, \sigma_\theta, \tau_{rz}, \tau_{z\theta}, \tau_{r\theta}\} \tag{c}$$

Stress-strain relationships from Sec. 5.1 are easily extended to cover six types of stresses and the corresponding strains. For example, if the material is orthotropic [see Eq. (5.1-3)], we add the diagonal terms E_{55} and E_{66} to form a 6×6 matrix **E**. Similarly, for an isotropic material [see Eq. (5.1-4)], we add $E_{55} = E_{66} = e_3$.

The response of an axisymmetric solid to a series of harmonic loads [see Eqs. (5.4-1)] consists of a series of harmonic generic displacements that may be expressed as:

$$u = \sum_{j=0}^{m} u_j \cos j\theta \qquad v = \sum_{j=0}^{m} v_j \cos j\theta$$
$$w = \sum_{j=0}^{m} w_j \sin j\theta \tag{5.4-4}$$

Again, if the loads were antisymmetric with respect to a plane of symmetry, the functions $\sin j\theta$ and $\cos j\theta$ would be interchanged. Applying the operator **d** in Eq. (5.4-3) to Eqs. (5.4-4) expressed in terms of **f**, we find a typical partition of the strain-displacement matrix to be:

$$(\mathbf{B}_i)_j = \begin{bmatrix} f_{i,r}\cos j\theta & 0 & 0 \\ 0 & f_{i,z}\cos j\theta & 0 \\ \frac{f_i}{r}\cos j\theta & 0 & j\frac{f_i}{r}\cos j\theta \\ f_{i,z}\cos j\theta & f_{i,r}\cos j\theta & 0 \\ 0 & -j\frac{f_i}{r}\sin j\theta & f_{i,z}\sin j\theta \\ -j\frac{f_i}{r}\sin j\theta & 0 & \left(f_{i,r} - \frac{f_i}{r}\right)\sin j\theta \end{bmatrix} \tag{5.4-5}$$

where $i = 1, 2, \ldots, n_{en}$ and $j = 0, 1, 2, \ldots, m$.

An element stiffness matrix for each harmonic set of symmetric loads may be written as:

$$\mathbf{K}_j = \int_A \int_0^{2\pi} \mathbf{B}_j^T \mathbf{E}\ \mathbf{B}_j r\, d\theta\, dA$$

$$= k\pi \int_A \mathbf{B}_j^T \mathbf{E}\ \mathbf{B}_j r\, dA \qquad (j = 0, 1, 2, \ldots, m) \qquad (5.4\text{-}6)$$

where $k = 2$ for $j = 0$, and $k = 1$ for $j = 1, 2, \ldots, m$. The latter constant ($k = 1$) appears as a consequence of:

$$\int_0^{2\pi} \cos^2 j\theta\, d\theta = \int_0^{2\pi} \sin^2 j\theta\, d\theta = \pi \qquad \text{(d)}$$

Furthermore, equivalent nodal loads for each harmonic set of symmetric body forces take the form:

$$\mathbf{p}_{bj} = k\pi \int_A \mathbf{f}^T \mathbf{b}_j r\, dA \qquad (j = 0, 1, 2, \ldots, m) \qquad (5.4\text{-}7)$$

And equivalent nodal loads due to initial strains are:

$$\mathbf{p}_{0j} = k\pi \int_A \mathbf{B}_j^T \mathbf{E}\ \boldsymbol{\epsilon}_{0j} r\, dA \qquad (j = 0, 1, 2, \ldots, m) \qquad (5.4\text{-}8)$$

Finally, the stresses for each harmonic response become:

$$\boldsymbol{\sigma}_j = \mathbf{E}(\mathbf{B}_j \mathbf{q}_j - \boldsymbol{\epsilon}_{0j}) \qquad (j = 0, 1, 2, \ldots, m) \qquad (5.4\text{-}9)$$

5.5 PROGRAMS AXSOQ4 AND AXSOQ8 AND APPLICATIONS

Let us now consider two computer programs named AXSOQ4 and AXSOQ8 that analyze axisymmetric, isotropic solids, using the ring elements AXQ4 and AXQ8 described in Sec. 5.3. With very few modifications, Programs PSQ4 and PSQ8 in Sec. 3.6 can be converted to Programs AXSOQ4 and AXSOQ8. The data for PSCST (see Table 2.7 in Sec. 2.5) must be altered further to account for the fact that the continuum to be analyzed is an axisymmetric solid with axisymmetric loads. The structural parameters IPS and t must be deleted, and the nodal coordinates x and y must be replaced by r and z.

The subprogram in AXSOQ4 that generates the element stiffness matrix (named STAXQ4) is practically the same as Subprogram STIFQ4 (see Macro-Flow Chart 3.1) in Sec. 3.6. However, when calculating stiffness terms there is multiplication by $2\pi r$ instead of t. Furthermore, the subprogram that generates the **B** matrix (called BMAXQ4) has the same general logic as Subprogram BMATQ4 (see Macro-Flow Chart 3.2). In the present case BMAXQ4 computes not only **B** and $|\mathbf{J}|$ but also the radius $r(\xi_j, \eta_k)$ of the numerical integration point. The task of extending Program AXSOQ4 to create AXSOQ8 is similar to that mentioned in Sec. 3.6 for extending PSQ4 to PSQ8.

Example 1

Suppose that an aluminum circular cylinder is subjected to an internal pressure b_r (force per unit area), as indicated in Fig. 5.6. Such a condition causes a high value of hoop stress σ_θ in the wall of the cylinder, and a single element AXQ4 may be used to solve the problem. For this example we take the following values of physical parameters:

$$L = 1 \text{ in.} \qquad b_r = 1 \text{ k/in.}^2$$
$$E = 1 \times 10^4 \text{ k/in.}^2 \qquad \nu = 0.3$$

which are expressed in U.S. units.

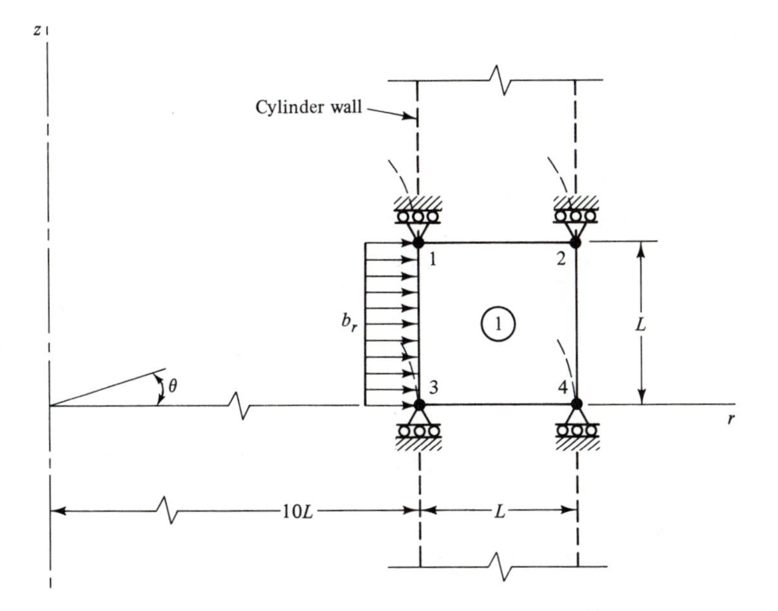

Figure 5.6 Example 1: Element AXQ4 with Hoop Stress

Table 5.1 gives the computer output obtained by applying Program AXSOQ4 to this problem. Due to the loading shown in Fig. 5.6, the hoop stress is approximately equal to 10 k/in.², as it should be.

Example 2

Consider now the spinning axisymmetric solid illustrated in Fig. 5.7(a). The angular velocity of spin is denoted by ω, and the body is symmetric with respect to the r-θ plane. Values of the physical parameters are:

$$L = 2 \text{ in.} \qquad E = 3 \times 10^4 \text{ k/in.}^2 \qquad \nu = 0.3$$
$$\rho = 7.33 \times 10^{-7} \text{ k-sec}^2\text{/in.}^4 \qquad \omega = 1 \times 10^2 \text{ rad/sec}$$

for which the material is steel and U.S. units are used. The *centrifugal body force* (per

TABLE 5.1 Computer Output for Example 1

```
PROGRAM AXSOQ4

***  EXAMPLE 1:  ONE SQUARE ELEMENT WITH HOOP STRESS  ***

STRUCTURAL PARAMETERS
   NN   NE  NRN  NLS            E           PR
    4    1    4    1  1.0000D 04  3.0000D-01

COORDINATES OF NODES
 NODE           R           Z
    1  1.0000D 01  1.0000D 00
    2  1.1000D 01  1.0000D 00
    3  1.0000D 01  0.0000D-01
    4  1.1000D 01  0.0000D-01

ELEMENT INFORMATION
ELEM.    I    J    K    L
    1    3    4    2    1

NODAL RESTRAINTS
 NODE   R1   R2
    1    0    1
    2    0    1
    3    0    1
    4    0    1

NUMBER OF DEGREES OF FREEDOM =    4
NUMBER OF NODAL RESTRAINTS   =    4
NUMBER OF TERMS IN SN        =   20

*********  LOADING NUMBER    1  *********
  NLN  NEA  NEV  IPR
    0    1    0    0

AREA LOADS
    I    J         BA1         BA2         BA3         BA4
    1    3  1.0000D 00  1.0000D 00  0.0000D-01  0.0000D-01

NODAL DISPLACEMENTS
 NODE         DN1         DN2
    1  9.9469D-03 -0.0000D-01
    2  9.5154D-03 -0.0000D-01
    3  9.9469D-03 -0.0000D-01
    4  9.5154D-03  0.0000D-01

ELEMENT STRESSES
ELEM. INT.    R STRESS    Z STRESS    T STRESS   RZ STRESS
    1    1 -2.4082D-01  3.0787D 00  1.0503D 01  0.0000D-01
    1    2 -6.7204D-01  2.6475D 00  9.4969D 00 -7.0890D-15
    1    3 -6.7204D-01  2.6475D 00  9.4969D 00 -7.0890D-15
    1    4 -2.4082D-01  3.0787D 00  1.0503D 01  0.0000D-01

SUPPORT REACTIONS
 NODE         AR1         AR2
    1  0.0000D-01  9.6855D 01
    2  0.0000D-01  9.1641D 01
    3  0.0000D-01 -9.6855D 01
    4  0.0000D-01 -9.1641D 01
```

unit volume) at any point in the solid is:

$$b_r = \rho\omega^2 r$$

as exemplified by Prob. 5.3-3 at the end of the chapter.

To make an approximate analysis of this solid, we discretize the shaded half in Fig. 5.7(a). The network of AXQ4 elements chosen for this problem appears in Fig.

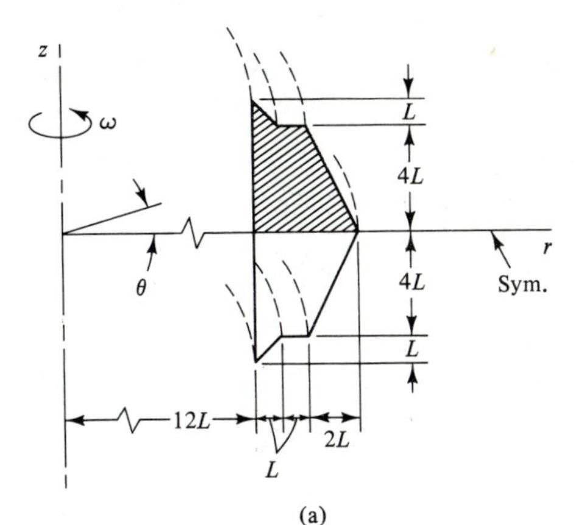

(a)

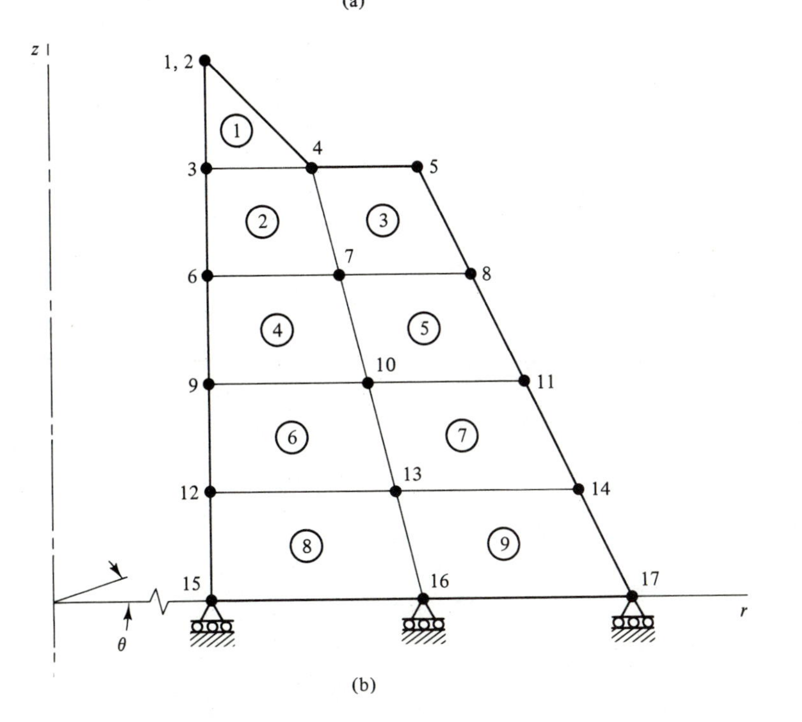

(b)

Figure 5.7 Example 2: (a) Spinning Body (b) AXQ4 Network

5.7(b). Restraints at nodes on the r-θ plane are rollers that prevent translations across the plane of symmetry. Note that element 1 is *specialized to become triangular* by making the coordinates of nodes 1 and 2 equal.

When this example is solved using Program AXSOQ4, the maximum radial displacement of approximately 5×10^{-3} in. occurs at node 15. While the normal stress σ_θ does not vary much over the cross section, its maximum value is found to be about 6 k/in.^{2} at the numerical integration point in element 6 nearest node 12.

Example 3

Figure 5.8 shows an axially loaded concrete pedestal with a network of AXQ8 elements fitted very closely to its curved boundaries. Physical parameters for this problem are:

$$L = 0.4 \text{ m} \qquad b_z = 1 \times 10^4 \text{ kN/m}^2$$
$$E = 2 \times 10^7 \text{ kN/m}^2 \qquad \nu = 0.3$$

where the units are S.I. Nodes 1, 6, 9, . . . , 38 lie on the axis of symmetry and are restrained in the r direction. However, nodes 41 through 45 at the fixed base are restrained in both the r and z directions.

Upon applying Program AXSOQ8 to this example, we find that the maximum value of the normal stress σ_z is slightly over 1×10^4 kN/m² at the integration point of element 4 nearest node 13. Also, a graph of σ_z near the base of the pedestal is given

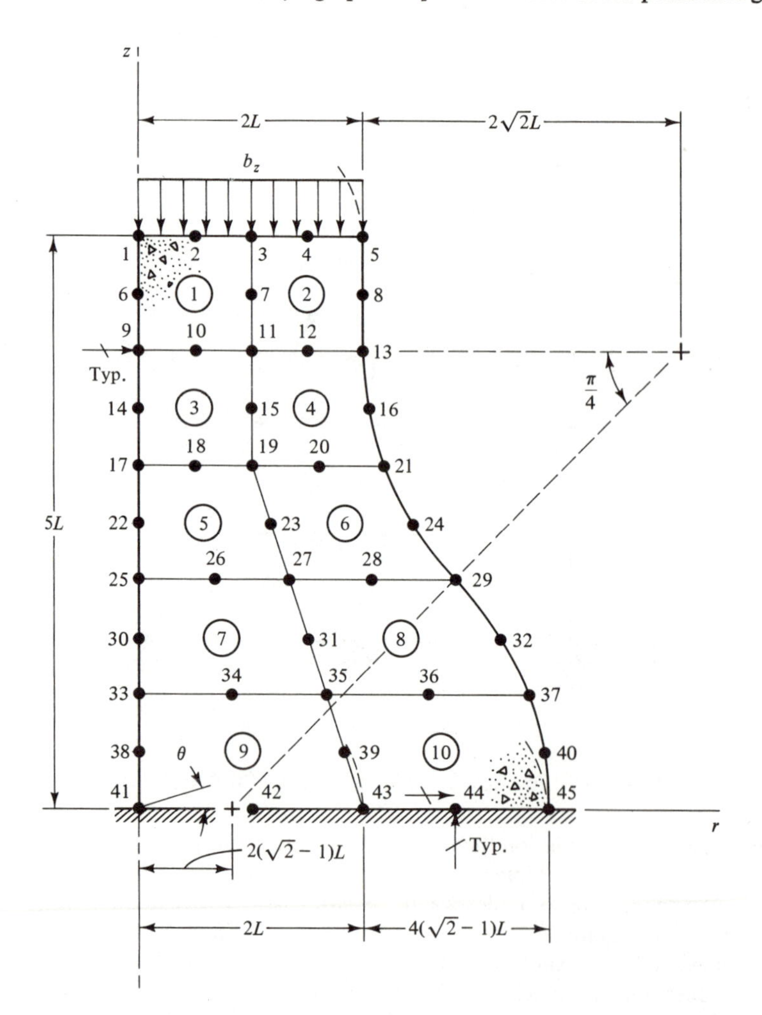

Figure 5.8 Example 3: Axially Loaded Pedestal with AXQ8 Network

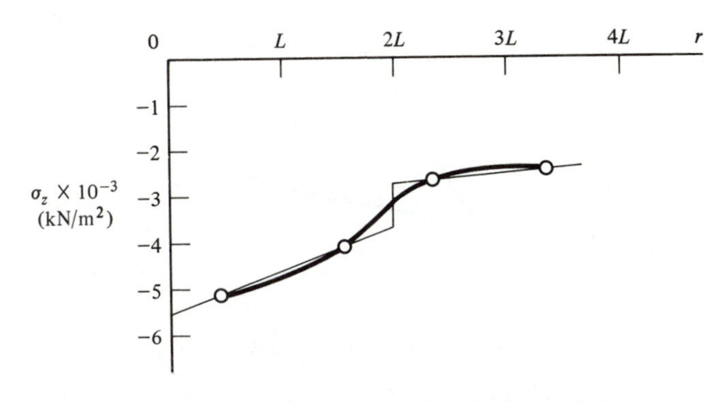

Figure 5.9 Stress σ_z Near r Axis for Example 3

in Fig. 5.9. As before, the straight-line parts of the graph are discontinuous, and more refinement of the network would be required to improve the results.

REFERENCES

1. Timoshenko, S. P., and Goodier, J. N., *Theory of Elasticity*, 3d ed., McGraw-Hill, New York, 1970.
2. Wilson, E. L., "Structural Analysis of Axisymmetric Solids," *AIAA Jour.*, Vol. 3, No. 12, December 1965, pp. 2269–2274.
3. Clough, R. W., "The Finite Element Method in Structural Mechanics," Chap. 7 of *Stress Analysis*, ed. O. C. Zienkiewicz and G. S. Hollister, Wiley, New York, 1965.
4. Zienckiewicz, O. C., *The Finite Element Method*, 3d ed., McGraw-Hill Ltd., London, 1977.
5. Sokolnikoff, I. S., and Redheffer, R. M., *Mathematics of Physics and Modern Engineering*, McGraw-Hill, New York, 1966.
6. Love, A. E. H., *The Mathematical Theory of Elasticity*, 4th ed., Cambridge Univ. Press, 1927.

PROBLEMS

5.2-1. A ring element in an axisymmetric solid has a triangular section, as shown in the figure. Coordinates r_i and z_i of nodes 1, 2, and 3 are (10, 2), (20, 4), and (15, 10), respectively. Find the equivalent nodal loads due to a force P_z (per unit circumferential length) applied in the positive z direction at the midpoint of side 1-3.

5.2-2. The figure shows a linearly varying load applied in the positive r direction along side 2-3 of the triangular section. The intensity of loading is b_{r2} (force per unit area) at point 2 and b_{r3} at point 3. Derive explicit expressions for the equivalent nodal loads.

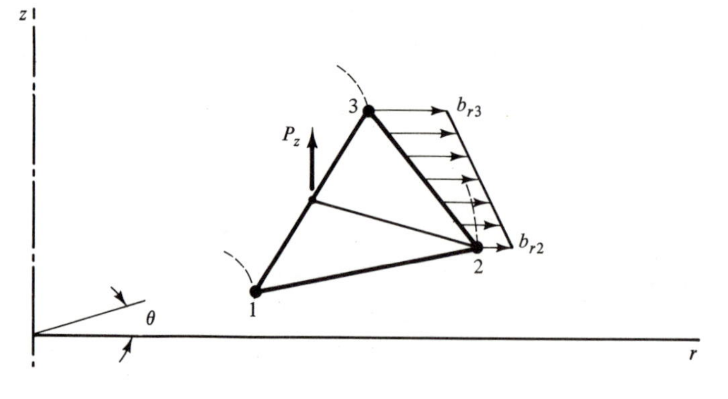

Problems 5.2-1 through 5.2-3

5.2-3. Assume that the axisymmetric element in the figure is subjected to a uniform gravitational loading b_g (force per unit volume) acting in the negative z direction. Using numerical integration with $n = 3$, determine the equivalent nodal loads due to this influence.

5.2-4. A ring element in an axisymmetric solid has a rectangular section, as shown in the figure. Coordinates r_i and z_i of nodes 1, 2, 3, and 4 are (10, 2), (20, 2), (20, 10), and (10, 10), respectively. Find the equivalent nodal loads due to a force of magnitude $P\sqrt{2}$ (per unit circumferential length) applied to the midpoint of side 3-4 at the angle $\pi/4$ (see figure).

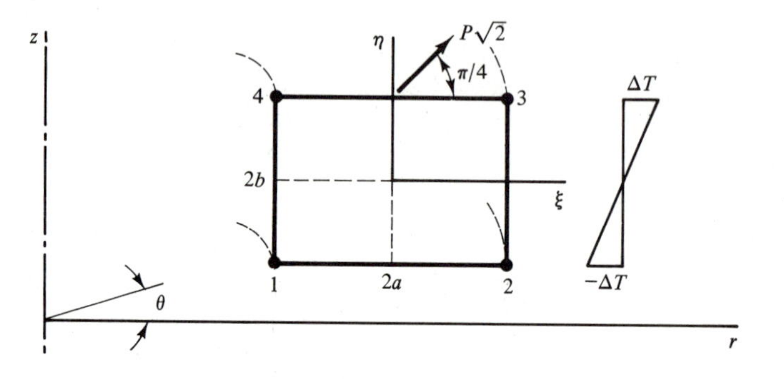

Problems 5.2-4 through 5.2-6

5.2-5. At the right side of the figure a linear temperature variation is indicated, which pertains to the whole element. Assuming that the material is isotropic, determine the equivalent nodal loads due to this influence.

5.2-6. Let the element in the figure be subjected to a uniformly distributed body force b_r (per unit volume) applied in the positive r direction. Find the equivalent nodal loads due to this radial loading.

5.3-1. The figure shows an axisymmetric ring element (AXQ4) with a quadrilateral section. Coordinates r_i and z_i of nodes 1, 2, 3, and 4 are (10, 3), (20, 2), (18, 10), and (11, 9), respectively. Obtain the equivalent nodal loads due to a force $\boldsymbol{P_r}$

(per unit circumferential length) applied in the positive r direction at the mid-point of side 1-4.

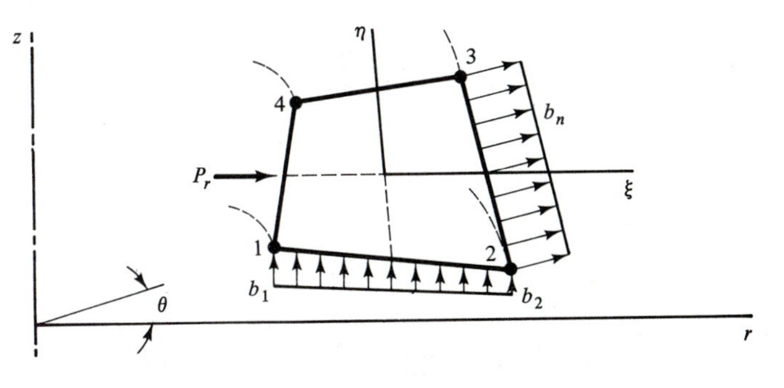

Problems 5.3-1 through 5.3-4

5.3-2. A uniformly distributed loading b_n (force per unit area) is normal to side 2-3 of the AXQ4 element in the figure. Derive explicit expressions for the equivalent nodal loads.

5.3-3. Find equivalent nodal loads at point 1 for the AXQ4 element in the figure due to rotation about the z axis at constant angular velocity ω. Let ρ = mass density, and use numerical integration with $n = 2$ each way.

5.3-4. The figure shows a linearly varying load (force per unit area) applied in the z direction on edge 1-2 of element AXQ4. Find the nonzero equivalent nodal loads due to this influence.

CHAPTER

Flexure in Plates

6.1 FLEXURAL STRESSES AND STRAINS IN PLATES

When a thin plate is subjected to forces applied in a direction normal to its own plane, the plate bends and is said to be in a state of flexure. While this condition of stresses and strains has much in common with flexure in a beam (see Sec. 1.5), the plate is more complicated because it is two-dimensional, whereas a beam is only one-dimensional. In this section we will examine strain-displacement and stress-strain relationships for a plate in flexure. Furthermore, we shall develop generalized stresses (moments) and generalized strains (curvatures) and their mutual relationships. Also, the differential equations of equilibrium will be written for the purpose of calculating internal shearing forces from internal moments.

Figure 6.1 shows an infinitesimal element of a plate in flexure, for which the x-y plane is the *neutral surface*. The height of this element is the finite thickness t of the plate, while its other two dimensions are dx and dy. A typical slice of the infinitesimal element appears at the distance z from the neutral surface. On this slice are shown the types of stresses and strains that constitute the primary contributors to the deformations of a plate in bending. Strains in the plane of the slice are defined as:

$$\epsilon_x = \frac{\partial u}{\partial x} \qquad \epsilon_y = \frac{\partial v}{\partial y} \qquad \gamma_{xy} = \frac{\partial u}{\partial y} + \frac{\partial v}{\partial x} \tag{a}$$

A basic assumption in the flexural theory of thin plates is that normals to the neutral surface remain straight and normal during deformation. Therefore, we

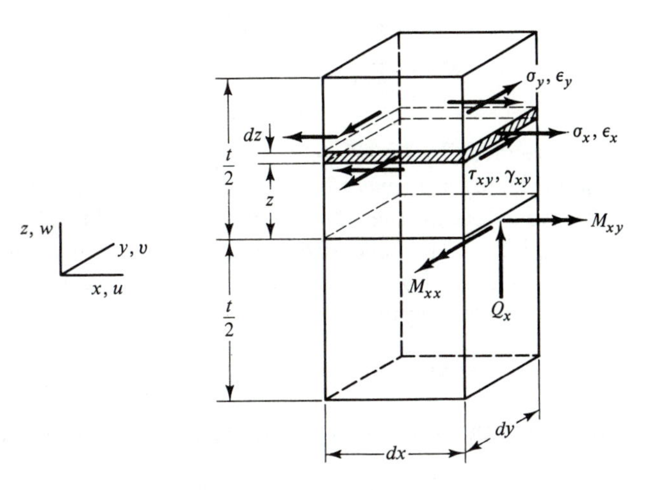

Figure 6.1 Flexure in a Plate

can express the displacements u and v in terms of w, as follows:

$$u = -z\frac{\partial w}{\partial x} \qquad v = -z\frac{\partial w}{\partial y} \tag{b}$$

Substitution of Eqs. (b) into Eqs. (a) yields:

$$\epsilon_x = -z\frac{\partial^2 w}{\partial x^2} \qquad \epsilon_y = -z\frac{\partial^2 w}{\partial y^2} \qquad \gamma_{xy} = -2z\frac{\partial^2 w}{\partial x\,\partial y} \tag{6.1-1}$$

These expressions are the *strain-displacement relationships* for bending in a plate. They involve only one type of translation (w) and three types of strain (ϵ_x, ϵ_y, and γ_{xy}). The other translations (u and v) have linear variations from the neutral surface, as shown by Eqs. (b). Furthermore, the normal strain ϵ_z and the shearing strains γ_{xz} and γ_{yz} are usually omitted in the analysis of thin plates (1).

Stresses shown on the infinitesimal slice in Fig. 6.1 are σ_x, σ_y, and τ_{xy}, which correspond to the strains ϵ_x, ϵ_y, and γ_{xy}. *Stress-strain relationships* for the plate are taken to be the same as those for the case of plane stress (see Sec. 2.1). This usage is justified because the plate is thin and unrestrained in the z direction (except at supports). Thus, for an isotropic material we have:

$$\mathbf{E} = \frac{E}{1-\nu^2}\begin{bmatrix} 1 & \nu & 0 \\ \nu & 1 & 0 \\ 0 & 0 & \lambda \end{bmatrix} \tag{2.1-16 repeated}$$

where

$$\lambda = \frac{1-\nu}{2}$$

If the material is orthotropic with x and y as principal material directions, the stress-strain matrix $\mathbf{E}$ becomes:

$$\mathbf{E} = \begin{bmatrix} E_{11} & E_{12} & 0 \\ E_{21} & E_{22} & 0 \\ 0 & 0 & E_{33} \end{bmatrix} \tag{6.1-2}$$

in which the terms are given by Eq. (2.1-20).

Also shown in Fig. 6.1 are *generalized stresses* (or moments) M_{xx} and M_{xy} acting on the positive x face of the infinitesimal element. (The shearing force Q_x will be considered later.) The generalized stress M_{xx} consists of the moment (per unit width of the plate) of σ_x stresses with respect to the neutral surface. It is obtained from the following integral:

$$M_{xx} = \int_{-t/2}^{t/2} -\sigma_x z\, dz \tag{c}$$

Now we express σ_x in terms of ϵ_x and ϵ_y by using the constants in Eq. (6.1-2). Thus,

$$\sigma_x = E_{11}\epsilon_x + E_{12}\epsilon_y \tag{d}$$

Then substitute Eqs. (6.1-1) for ϵ_x and ϵ_y into Eq. (d), and the resulting expression for σ_x goes into Eq. (c) to yield:

$$M_{xx} = \left(E_{11}\frac{\partial^2 w}{\partial x^2} + E_{12}\frac{\partial^2 w}{\partial y^2}\right)\int_{-t/2}^{t/2} z^2\, dz \tag{e}$$

The integral in this equation produces:

$$\int_{-t/2}^{t/2} z^2\, dz = \frac{t^3}{12} \tag{f}$$

So the final form of M_{xx} becomes:

$$M_{xx} = \left(E_{11}\frac{\partial^2 w}{\partial x^2} + E_{12}\frac{\partial^2 w}{\partial y^2}\right)\frac{t^3}{12} \tag{6.1-3}$$

Similarly, we can develop the generalized stress M_{yy} (moment per unit width of the plate) acting on the positive y face, as follows:

$$M_{yy} = \left(E_{21}\frac{\partial^2 w}{\partial x^2} + E_{22}\frac{\partial^2 w}{\partial y^2}\right)\frac{t^3}{12} \tag{6.1-4}$$

In addition, the twisting moment M_{xy} (per unit width) can be derived as:

$$M_{xy} = \left(2E_{33}\frac{\partial^2 w}{\partial x\, \partial y}\right)\frac{t^3}{12} \tag{6.1-5}$$

If we take the vector $\mathbf{M}$ of generalized stresses to be:

$$\mathbf{M} = \{M_{xx}, M_{yy}, M_{xy}\} \tag{g}$$

and the vector $\boldsymbol{\phi}$ of *generalized strains* (or curvatures) as:

$$\boldsymbol{\phi} = \{\phi_{xx}, \phi_{yy}, \phi_{xy}\} = \{w_{,xx}, w_{,yy}, 2w_{,xy}\} \tag{h}$$

then the *generalized stress-strain* (*or moment-curvature*) *operator* becomes:

$$\bar{\mathbf{E}} = \mathbf{E}\frac{t^3}{12} \tag{6.1-6}$$

Thus, we have $\mathbf{M}$ related to $\boldsymbol{\phi}$ by the expression:

$$\mathbf{M} = \bar{\mathbf{E}}\ \boldsymbol{\phi} \tag{6.1-7}$$

Rotation-of-axes transformations for generalized stresses and strains are the same as in the case of plane stress. This fact may be demonstrated by integration through the thickness of the plate, as follows:

$$\mathbf{M}' = \int_{-t/2}^{t/2} -\boldsymbol{\sigma}' z\, dz \tag{i}$$

where the prime denotes inclined axes as before. Substitution of $\boldsymbol{\sigma}' = \mathbf{T}_\sigma \boldsymbol{\sigma}$ into Eq. (i) produces:

$$\mathbf{M}' = \int_{-t/2}^{t/2} -\mathbf{T}_\sigma \boldsymbol{\sigma} z\, dz \tag{j}$$

in which $\mathbf{T}_\sigma$ is given by Eq. (4.1-21). Then further substitution of $\boldsymbol{\sigma} = \mathbf{E}\ \boldsymbol{\epsilon}$ into Eq. (j) yields:

$$\mathbf{M}' = \int_{-t/2}^{t/2} -\mathbf{T}_\sigma \mathbf{E}\ \boldsymbol{\epsilon} z\, dz \tag{k}$$

The strain-displacement relationships in Eqs. (6.1-1) may be expressed more briefly as:

$$\boldsymbol{\epsilon} = -z\boldsymbol{\phi} \tag{6.1-8}$$

When Eq. (6.1-8) is substituted for $\boldsymbol{\epsilon}$ in Eq. (k), we have:

$$\mathbf{M}' = \mathbf{T}_\sigma \mathbf{E}\ \boldsymbol{\phi} \int_{-t/2}^{t/2} z^2\, dz = \mathbf{T}_\sigma \mathbf{E}\ \boldsymbol{\phi}\frac{t^3}{12} \tag{ℓ}$$

Hence, from Eqs. (6.1-6) and (6.1-7)

$$\mathbf{M}' = \mathbf{T}_\sigma \bar{\mathbf{E}}\ \boldsymbol{\phi} = \mathbf{T}_\sigma \mathbf{M} \tag{6.1-9}$$

Thus, the relationship between $\mathbf{M}'$ and $\mathbf{M}$ is the same as that between $\boldsymbol{\sigma}'$ and $\boldsymbol{\sigma}$.

In order to show also the relationship between $\boldsymbol{\phi}'$ and $\boldsymbol{\phi}$, we equate the complementary virtual strain energy densities (integrated through the thickness) for the primed and unprimed coordinates. Thus,

$$(\delta\mathbf{M}')^{\mathrm{T}}\boldsymbol{\phi}' = \delta\mathbf{M}^{\mathrm{T}}\boldsymbol{\phi} \tag{m}$$

Now substitute the transposed incremental form of Eq. (6.1-9) into Eq. (m) to obtain:

$$\delta\mathbf{M}^{\mathrm{T}}\mathbf{T}_\sigma^{\mathrm{T}}\boldsymbol{\phi}' = \delta\mathbf{M}^{\mathrm{T}}\boldsymbol{\phi} \tag{n}$$

Cancellation of $\delta\mathbf{M}^{\mathrm{T}}$ and premultiplication of Eq. (n) with $\mathbf{T}_\sigma^{-\mathrm{T}}$ gives:

$$\boldsymbol{\phi}' = \mathbf{T}_\epsilon \boldsymbol{\phi} \tag{6.1-10}$$

In this expression we have:

$$\mathbf{T}_\epsilon = \mathbf{T}_\sigma^{-\mathrm{T}} \tag{6.1-11}$$

for which the terms are given by Eq. (4.1-22).

Transformation of $\bar{\mathbf{E}}'$ to $\bar{\mathbf{E}}$ is the same as that for transformation of $\mathbf{E}'$ to $\mathbf{E}$ in Sec. 4.1. Hence, in primed coordinates we can write:

$$\mathbf{M}' = \bar{\mathbf{E}}'\boldsymbol{\phi}' \tag{o}$$

If Eqs. (6.1-9) and (6.1-10) are substituted into Eq. (o), they produce:

$$\mathbf{T}_\sigma\mathbf{M} = \bar{\mathbf{E}}'\mathbf{T}_\epsilon\boldsymbol{\phi} \tag{p}$$

Then we premultiply Eq. (p) by $\mathbf{T}_\sigma^{-1}$ and use Eq. (6.1-11) to find:

$$\mathbf{M} = \mathbf{T}_\epsilon^{\mathrm{T}}\bar{\mathbf{E}}'\mathbf{T}_\epsilon\boldsymbol{\phi} \tag{q}$$

From this expression we conclude that:

$$\bar{\mathbf{E}} = \mathbf{T}_\epsilon^{\mathrm{T}}\bar{\mathbf{E}}'\mathbf{T}_\epsilon \tag{6.1-12}$$

Also,

$$\bar{\mathbf{E}}' = \mathbf{T}_\sigma\bar{\mathbf{E}}\ \mathbf{T}_\sigma^{\mathrm{T}} \tag{6.1-13}$$

We may compare Eqs. (6.1-12) and (6.1-13) with Eqs. (4.1-19) and (4.1-20) to see that they are of the same form.

Figure 6.2 depicts the neutral surface of the infinitesimal plate element, which lies in the x-y plane and is of size dx by dy. Also shown are body forces b_z (in the z direction) and generalized stresses M_{xx}, M_{yy}, Q_x, and so on, acting on the edges. The shearing forces Q_x and Q_y represent integrals of shearing stresses τ_{xz} and τ_{yz} (per unit width) on the x and y faces of the infinitesimal element. Positive senses of moments shown in the figure correspond to positive senses of curvatures, and positive senses of shearing forces correspond to positive slopes. We can write an equilibrium equation that relates the shearing forces to the applied loads. Thus, a summation of forces in the z direction gives:

$$b_z\,dx\,dy - Q_x\,dy + \left(Q_x + \frac{\partial Q_x}{\partial x}\,dx\right)dy - Q_y\,dx + \left(Q_y + \frac{\partial Q_y}{\partial y}\,dy\right)dx = 0$$

From this equation, we find:

$$\frac{\partial Q_x}{\partial x} + \frac{\partial Q_y}{\partial y} + b_z = 0 \tag{6.1-14}$$

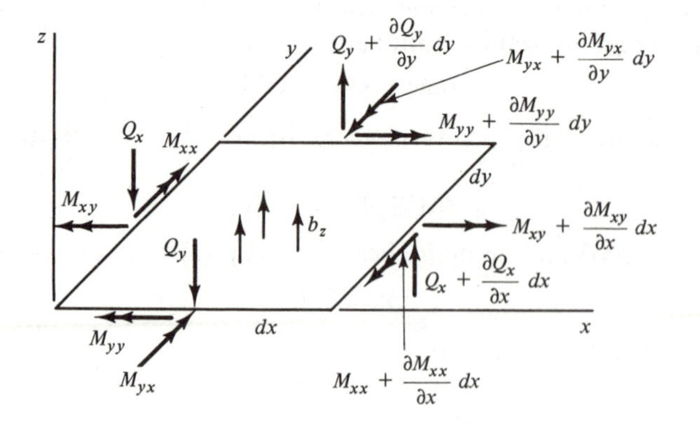

Figure 6.2 Equilibrium in a Plate

We can also write a moment equilibrium equation by summing moments about the y axis, as follows:

$$-b_z\,dx\,dy\,\frac{dx}{2} - \left(Q_x + \frac{\partial Q_x}{\partial x}\,dx\right)dy\,dx + Q_y\,dx\,\frac{dx}{2} - \left(Q_y + \frac{\partial Q_y}{\partial y}\,dy\right)dx\,\frac{dx}{2}$$

$$+ M_{xx}\,dy - \left(M_{xx} + \frac{\partial M_{xx}}{\partial x}\,dx\right)dy + M_{yx}\,dx - \left(M_{yx} + \frac{\partial M_{yx}}{\partial y}\,dy\right)dx = 0$$

If we neglect second-order terms, this expression simplifies to:

$$\frac{\partial M_{xx}}{\partial x} + \frac{\partial M_{yx}}{\partial y} + Q_x \approx 0 \qquad (6.1\text{-}15)$$

Similarly, a summation of moments about the x axis leads to:

$$\frac{\partial M_{yy}}{\partial y} + \frac{\partial M_{xy}}{\partial x} + Q_y \approx 0 \qquad (6.1\text{-}16)$$

Equations (6.1-15) and (6.1-16) provide a method for calculating the shearing forces from derivatives of the moments in a plate subjected to flexure.

6.2 RECTANGULAR ELEMENTS

The simplest type of plate-bending element has rectangular geometry. In this article we discuss two well-known elements that belong to this family. The first element is said to be *nonconforming* because it does not have normal-slope compatibility at the edges. The second is a *conforming* element, however, because it does possess that property. Nevertheless, both elements can model states of constant strain for flexure in a plate, and they also have complete and balanced functions. Therefore, both these rectangles produce convergent results, in accordance with the criteria in Sec. 1.10.

First we shall consider the plate-bending element in Fig. 6.3. It is called the *MZC rectangle* because it was originally developed by Melosh (2) and Zienkiewicz and Cheung (3). As with other elements of this type, it has only one generic displacement, which is w (translation in the z direction). Thus,

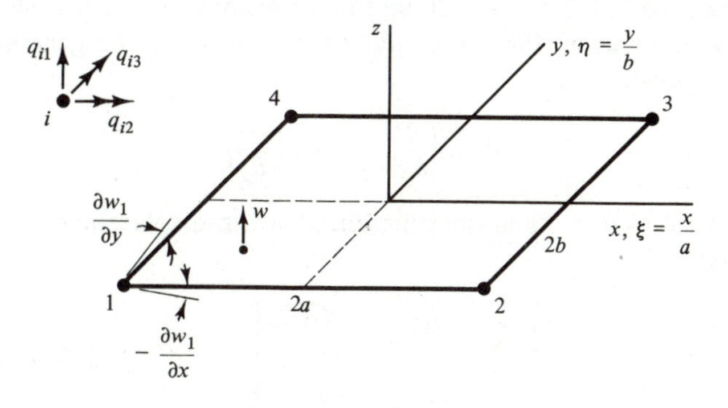

Figure 6.3 MZC Rectangle

$$\mathbf{u} = w \tag{a}$$

Also shown in the figure are the nodal displacements:

$$\mathbf{q}_i = \{q_{i1}, q_{i2}, q_{i3}\} = \left\{w_i, \frac{\partial w_i}{\partial y}, -\frac{\partial w_i}{\partial x}\right\} \qquad (i = 1, 2, 3, 4) \tag{b}$$

The subscripts in vector $\mathbf{q}_i$ are computed as $i1 = 3i - 2$, $i2 = 3i - 1$, and $i3 = 3i$. Note the sign change for $q_{i3} = -\partial w_i/\partial x$ to correlate the slope with a positive nodal rotation. Nodal actions corresponding to the displacements in Eq. (b) are:

$$\mathbf{p}_i = \{p_{i1}, p_{i2}, p_{i3}\} = \{p_{zi}, M_{xi}, M_{yi}\} \qquad (i = 1, 2, 3, 4) \tag{c}$$

The symbol p_{zi} denotes a force in the z direction, but M_{xi} and M_{yi} are moments in the x and y senses. Note that these fictitious moments at the nodes are not the same as the distributed moments in the vector **M** of generalized stresses.

The displacement function chosen for this element is:

$$\begin{aligned} w = c_1 + c_2\xi + c_3\eta + c_4\xi^2 + c_5\xi\eta + c_6\eta^2 \\ + c_7\xi^3 + c_8\xi^2\eta + c_9\xi\eta^2 + c_{10}\eta^3 + c_{11}\xi^3\eta + c_{12}\xi\eta^3 \end{aligned} \tag{6.2-1}$$

which is a complete cubic of ten terms and two quartic terms. From this assumption it is possible to derive the displacement shape functions to be:

$$\mathbf{f}_i = [f_{i1} \quad f_{i2} \quad f_{i3}] \tag{6.2-2a}$$

where

$$\begin{aligned} f_{i1} &= \tfrac{1}{8}(1 + \xi_0)(1 + \eta_0)(2 + \xi_0 + \eta_0 - \xi^2 - \eta^2) \\ f_{i2} &= -\tfrac{1}{8}b\eta_i(1 + \xi_0)(1 - \eta_0)(1 + \eta_0)^2 \\ f_{i3} &= \tfrac{1}{8}a\xi_i(1 - \xi_0)(1 + \eta_0)(1 + \xi_0)^2 \end{aligned} \tag{6.2-2b}$$

and

$$\xi_0 = \xi_i\xi \qquad \eta_0 = \eta_i\eta \qquad (i = 1, 2, 3, 4)$$

For values of ξ_i and η_i pertaining to the corners of a rectangle as well as a quadrilateral, see Table 3.2. Recalling the definitions for generalized strains (or curvatures) discussed in the preceding section, we write the generalized linear differential operator $\bar{\mathbf{d}}$ as:

$$\bar{\mathbf{d}} = \left\{\frac{\partial^2}{\partial x^2}, \frac{\partial^2}{\partial y^2}, \frac{2\partial^2}{\partial x\,\partial y}\right\} \tag{6.2-3}$$

which is devoid of $-z$. Then the generalized strain-displacement matrix $\bar{\mathbf{B}}$ may be stated as:

$$\bar{\mathbf{B}}_i = \bar{\mathbf{d}}\ \mathbf{f}_i = \begin{bmatrix} f_{i1,xx} & f_{i2,xx} & f_{i3,xx} \\ f_{i1,yy} & f_{i2,yy} & f_{i3,yy} \\ 2f_{i1,xy} & 2f_{i2,xy} & 2f_{i3,xy} \end{bmatrix} \qquad (i = 1, 2, 3, 4) \tag{6.2-4a}$$

In particular,

$$\bar{\mathbf{B}}_1 = \frac{1}{4a^2b^2}\begin{bmatrix} 3\xi(1-\eta)b^2 & 0 & (1-3\xi)(1-\eta)ab^2 \\ 3(1-\xi)\eta a^2 & -(1-\xi)(1-3\eta)a^2b & 0 \\ (4-3\xi^2-3\eta^2)ab & (1-\eta)(1+3\eta)ab^2 & -(1-\xi)(1+3\xi)a^2b \end{bmatrix} \tag{6.2-4b}$$

Corresponding to the generalized strains are the generalized stresses, calculated as:

$$\mathbf{M} = \{M_{xx}, M_{yy}, M_{xy}\} = \bar{\mathbf{E}}\ \boldsymbol{\phi} = \bar{\mathbf{E}}\ \bar{\mathbf{B}}\ \mathbf{q} \tag{d}$$

Thus, for an isotropic material, we use Eq. (2.1-16) to find:

$$\bar{\mathbf{E}}\ \bar{\mathbf{B}} = \frac{Et^3}{48a^2b^2(1-\nu^2)}\begin{bmatrix} 3\xi(1-\eta)b^2 + 3\nu(1-\xi)\eta a^2 & \cdots \\ 3\nu\xi(1-\eta)b^2 + 3(1-\xi)\eta a^2 & \cdots \\ \lambda(4-3\xi^2-3\eta^2)ab & \cdots \end{bmatrix} \tag{6.2-5}$$

which is a 3×12 matrix. Then the stiffnesses for the element can be calculated from:

$$\mathbf{K} = \int_A \bar{\mathbf{B}}^{\mathrm{T}}\bar{\mathbf{E}}\ \bar{\mathbf{B}}\, dA = ab\int_{-1}^{1}\int_{-1}^{1} \bar{\mathbf{B}}^{\mathrm{T}}\bar{\mathbf{E}}\ \bar{\mathbf{B}}\, d\xi\, d\eta \tag{6.2-6}$$

for which the results are given in Table 6.1. Similarly, equivalent nodal loads are computed with the formulas:

$$\mathbf{p}_b = \int_A \mathbf{f}^{\mathrm{T}} b_z\, dA = ab\int_{-1}^{1}\int_{-1}^{1} \mathbf{f}^{\mathrm{T}} b_z\, d\xi\, d\eta \tag{6.2-7}$$

$$\mathbf{p}_0 = \int_A \bar{\mathbf{B}}^{\mathrm{T}}\bar{\mathbf{E}}\ \boldsymbol{\phi}_0\, dA = ab\int_{-1}^{1}\int_{-1}^{1} \bar{\mathbf{B}}^{\mathrm{T}}\bar{\mathbf{E}}\ \boldsymbol{\phi}_0\, d\xi\, d\eta \tag{6.2-8}$$

The integrals in Eqs. (6.2-6) through (6.2-8) are over area only, because integration through the thickness has already been accomplished using generalized stresses and strains.

After the elements have been assembled and the structure has been analyzed for nodal displacements, the generalized stresses at selected points in each element may be obtained from:

$$\mathbf{M} = \bar{\mathbf{E}}(\bar{\mathbf{B}}\ \mathbf{q} - \boldsymbol{\phi}_0) \tag{6.2-9}$$

Then the flexural stresses can be found with the following relationship:

$$\boldsymbol{\sigma} = \{\sigma_x, \sigma_y, \tau_{xy}\} = -\frac{12z}{t^3}\mathbf{M} \tag{6.2-10}$$

In addition, the shearing forces derived in Sec. 6.1 (expressed in modified form) are:

$$Q_x = -\frac{\partial M_{xx}}{\partial x} - \frac{\partial M_{yx}}{\partial y} \qquad \text{(6.1-15) modified}$$

$$Q_y = -\frac{\partial M_{yy}}{\partial y} - \frac{\partial M_{xy}}{\partial x} \qquad \text{(6.1-16) modified}$$

TABLE 6.1 Stiffness Matrix for MZC Rectangle

$$\mathbf{K} = \frac{Et^3}{12(1-\nu^2)}(\mathbf{K}_1 + \mathbf{K}_2 + \mathbf{K}_3 + \mathbf{K}_4)$$

$$\mathbf{K}_1 = \frac{b}{6a^3}\begin{bmatrix}
6 & & & & & & & & & & & \\
0 & 0 & & & & & & & & & & \\
-6a & 0 & 8a^2 & & & & & & & & & \\
-6 & 0 & 6a & 6 & & & & & \text{Sym.} & & & \\
0 & 0 & 0 & 0 & 0 & & & & & & & \\
-6a & 0 & 4a^2 & 6a & 0 & 8a^2 & & & & & & \\
-3 & 0 & 3a & 3 & 0 & 3a & 6 & & & & & \\
0 & 0 & 0 & 0 & 0 & 0 & 0 & 0 & & & & \\
-3a & 0 & 2a^2 & 3a & 0 & 4a^2 & 6a & 0 & 8a^2 & & & \\
3 & 0 & -3a & -3 & 0 & -3a & -6 & 0 & -6a & 6 & & \\
0 & 0 & 0 & 0 & 0 & 0 & 0 & 0 & 0 & 0 & 0 & \\
-3a & 0 & 4a^2 & 3a & 0 & 2a^2 & 6a & 0 & 4a^2 & -6a & 0 & 8a^2
\end{bmatrix}$$

$$\mathbf{K}_2 = \frac{a}{6b^3}\begin{bmatrix}
6 & & & & & & & & & & & \\
6b & 8b^2 & & & & & & & & & & \\
0 & 0 & 0 & & & & & & & & & \\
3 & 3b & 0 & 6 & & & & & \text{Sym.} & & & \\
3b & 4b^2 & 0 & 6b & 8b^2 & & & & & & & \\
0 & 0 & 0 & 0 & 0 & 0 & & & & & & \\
-3 & -3b & 0 & -6 & -6b & 0 & 6 & & & & & \\
3b & 2b^2 & 0 & 6b & 4b^2 & 0 & -6b & 8b^2 & & & & \\
0 & 0 & 0 & 0 & 0 & 0 & 0 & 0 & 0 & & & \\
-6 & -6b & 0 & -3 & -3b & 0 & 3 & -3b & 0 & 6 & & \\
6b & 4b^2 & 0 & 3b & 2b^2 & 0 & -3b & 4b^2 & 0 & -6b & 8b^2 & \\
0 & 0 & 0 & 0 & 0 & 0 & 0 & 0 & 0 & 0 & 0 & 0
\end{bmatrix}$$

$$\mathbf{K}_3 = \frac{\nu}{2ab}\begin{bmatrix}
1 & & & & & & & & & & & \\
b & 0 & & & & & & & & & & \\
-a & -2ab & 0 & & & & & & & & & \\
-1 & -b & 0 & 1 & & & & & \text{Sym.} & & & \\
-b & 0 & 0 & b & 0 & & & & & & & \\
0 & 0 & 0 & a & 2ab & 0 & & & & & & \\
1 & 0 & 0 & -1 & 0 & -a & 1 & & & & & \\
0 & 0 & 0 & 0 & 0 & 0 & -b & 0 & & & & \\
0 & 0 & 0 & -a & 0 & 0 & a & -2ab & 0 & & & \\
-1 & 0 & a & 1 & 0 & 0 & -1 & b & 0 & 1 & & \\
0 & 0 & 0 & 0 & 0 & 0 & b & 0 & 0 & -b & 0 & \\
a & 0 & 0 & 0 & 0 & 0 & 0 & 0 & 0 & -a & 2ab & 0
\end{bmatrix}$$

$$\mathbf{K}_4 = \frac{\lambda}{15ab}\begin{bmatrix}
21 & & & & & & & & & & & \\
3b & 8b^2 & & & & & & & & & & \\
-3a & 0 & 8a^2 & & & & & & & & & \\
-21 & -3b & 3a & 21 & & & & & & \text{Sym.} & & \\
-3b & -8b^2 & 0 & 3b & 8b^2 & & & & & & & \\
-3a & 0 & -2a^2 & 3a & 0 & 8a^2 & & & & & & \\
21 & 3b & -3a & -21 & -3b & -3a & 21 & & & & & \\
-3b & 2b^2 & 0 & 3b & -2b^2 & 0 & -3b & 8b^2 & & & & \\
3a & 0 & 2a^2 & -3a & 0 & -8a^2 & 3a & 0 & 8a^2 & & & \\
-21 & -3b & 3a & 21 & 3b & 3a & -21 & 3b & -3a & 21 & & \\
3b & -2b^2 & 0 & -3b & 2b^2 & 0 & 3b & -8b^2 & 0 & -3b & 8b^2 & \\
3a & 0 & -8a^2 & -3a & 0 & 2a^2 & 3a & 0 & -2a^2 & -3a & 0 & 8a^2
\end{bmatrix}$$

Hence, the maximum shearing stresses τ_{xz} and τ_{yz} become:

$$\tau_{xz} = \frac{3Q_x}{2t} \qquad \tau_{yz} = \frac{3Q_y}{2t} \tag{6.2-11}$$

The MZC rectangle is nonconforming because normal slopes are not compatible, and discontinuities occur at adjoining edges. The normal slope along an edge varies cubically and is not uniquely defined by two slopes at the ends of the edge. Slopes at two points determine a linear variation; slopes at three points determine a quadratic variation; and so on.

Example 1

Find the stiffness term K_{11} for the MZC rectangle, assuming isotropic material. If we use only the first column of $\bar{\mathbf{B}}_1$ from Eq. (6.2-4b) and the first column of $\bar{\mathbf{E}}\bar{\mathbf{B}}$ from Eq. (6.2-5), we can evaluate K_{11} from Eq. (6.2-6), as follows:

$$\begin{aligned} K_{11} &= \frac{Et^3}{192a^3b^3(1-\nu^2)} \int_{-1}^{1}\int_{-1}^{1} [9\xi^2(1-\eta)^2b^4 + 18\nu\xi(1-\xi)(1-\eta)\eta a^2b^2 \\ &\qquad + 9(1-\xi)^2\eta^2a^4 + \lambda(4-3\xi^2-3\eta^2)^2a^2b^2]\,d\xi\,d\eta \\ &= \frac{Et^3}{192a^3b^3(1-\nu^2)}\left(16b^4 + 16a^4 + 8\nu a^2b^2 + \frac{112\lambda a^2b^2}{5}\right) \\ &= \frac{Et^3}{12(1-\nu^2)}\left(\frac{b}{a^3} + \frac{a}{b^3} + \frac{\nu}{2ab} + \frac{7\lambda}{5ab}\right) \end{aligned}$$

This formula is the same as that in Table 6.1.

Now let us consider the second type of rectangular plate-bending element, which is shown in Fig. 6.4. Since this element was introduced by Bogner, Fox, and Schmit (4), it will be referred to as the *BFS rectangle*. In this case the nodal displacements are:

$$\begin{aligned} \mathbf{q}_i &= \{q_{i1}, q_{i2}, q_{i3}, q_{i4}\} \\ &= \left\{w_i, \frac{\partial w_i}{\partial y}, -\frac{\partial w_i}{\partial x}, \frac{\partial^2 w_i}{\partial x\,\partial y}\right\} \qquad (i = 1, 2, 3, 4) \end{aligned} \tag{e}$$

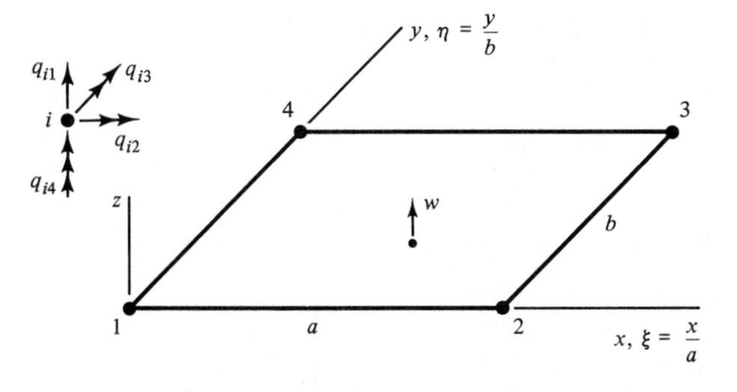

Figure 6.4 BFS Rectangle

The subscripts in vector **q** are calculated as $i1 = 4i - 3$, $i2 = 4i - 2$, $i3 = 4i - 1$, and $i4 = 4i$. We see that the fourth type of nodal displacement is the *warp*, or the cross rate of change of slope. Nodal actions corresponding to the displacements in Eq. (e) are:

$$\mathbf{p}_i = \{p_{i1}, p_{i2}, p_{i3}, p_{i4}\}$$
$$= \{p_{zi}, M_{xi}, M_{yi}, X_{xyi}\} \qquad (i = 1, 2, 3, 4) \qquad \text{(f)}$$

where X_{xyi} is the generalized action (a second moment of force) corresponding to $w_{i,xy}$.

For this element the displacement function is chosen to be the following 16-term polynomial:

$$\begin{aligned} w = c_1 &+ c_2 x &+ c_3 x^2 &+ c_4 x^3 \\ + c_5 y &+ c_6 xy &+ c_7 x^2 y &+ c_8 x^3 y \\ + c_9 y^2 &+ c_{10} xy^2 &+ c_{11} x^2 y^2 &+ c_{12} x^3 y^2 \\ + c_{13} y^3 &+ c_{14} xy^3 &+ c_{15} x^2 y^3 &+ c_{16} x^3 y^3 \end{aligned} \qquad (6.2\text{-}12)$$

This function is a complete cubic of ten terms, plus six other terms below and to the right of the dashed line in Eq. (6.2-12). Displacement shape functions are given in Table 6.2, and those for node 1 are illustrated in Figs. 6.5(a), (b), (c), and (d).

Next, we write the generalized strain-displacement matrix for the BFS rectangle as:

$$\bar{\mathbf{B}}_i = \bar{\mathbf{d}}\ \mathbf{f}_i = \begin{bmatrix} f_{i1,xx} & f_{i2,xx} & f_{i3,xx} & f_{i4,xx} \\ f_{i1,yy} & f_{i2,yy} & f_{i3,yy} & f_{i4,yy} \\ 2f_{i1,xy} & 2f_{i2,xy} & 2f_{i3,xy} & 2f_{i4,xy} \end{bmatrix} \qquad (i = 1, 2, 3, 4) \qquad (6.2\text{-}13a)$$

Evaluation of the first column of $\bar{\mathbf{B}}_1$ gives:

$$\bar{\mathbf{B}}_{1,1} = \frac{1}{a^2 b^2} \begin{bmatrix} -6(1 - 2\xi)(1 - 3\eta^2 + 2\eta^3)b^2 \\ -6(1 - 3\xi^2 + 2\xi^3)(1 - 2\eta)a^2 \\ 72\xi\eta(1 - \xi)(1 - \eta)ab \end{bmatrix} \qquad (6.2\text{-}13b)$$

Of course, there are 15 other columns of matrix $\bar{\mathbf{B}}$ that are not shown here.

TABLE 6.2 Displacement Shape Functions for BFS Rectangle

j	f_j	j	f_j
1	$(1 - 3\xi^2 + 2\xi^3)(1 - 3\eta^2 + 2\eta^3)$	9	$(3\xi^2 - 2\xi^3)(3\eta^2 - 2\eta^3)$
2	$(1 - 3\xi^2 + 2\xi^3)(\eta - 2\eta^2 + \eta^3)b$	10	$-(3\xi^2 - 2\xi^3)(\eta^2 - \eta^3)b$
3	$-(\xi - 2\xi^2 + \xi^3)(1 - 3\eta^2 + 2\eta^3)a$	11	$(\xi^2 - \xi^3)(3\eta^2 - 2\eta^3)a$
4	$(\xi - 2\xi^2 + \xi^3)(\eta - 2\eta^2 + \eta^3)ab$	12	$(\xi^2 - \xi^3)(\eta^2 - \eta^3)ab$
5	$(3\xi^2 - 2\xi^3)(1 - 3\eta^2 + 2\eta^3)$	13	$(1 - 3\xi^2 + 2\xi^3)(3\eta^2 - 2\eta^3)$
6	$(3\xi^2 - 2\xi^3)(\eta - 2\eta^2 + \eta^3)b$	14	$-(1 - 3\xi^2 + 2\xi^3)(\eta^2 - \eta^3)b$
7	$(\xi^2 - \xi^3)(1 - 3\eta^2 + 2\eta^3)a$	15	$-(\xi - 2\xi^2 + \xi^3)(3\eta^2 - 2\eta^3)a$
8	$-(\xi^2 - \xi^3)(\eta - 2\eta^2 + \eta^3)ab$	16	$-(\xi - 2\xi^2 + \xi^3)(\eta^2 - \eta^3)ab$

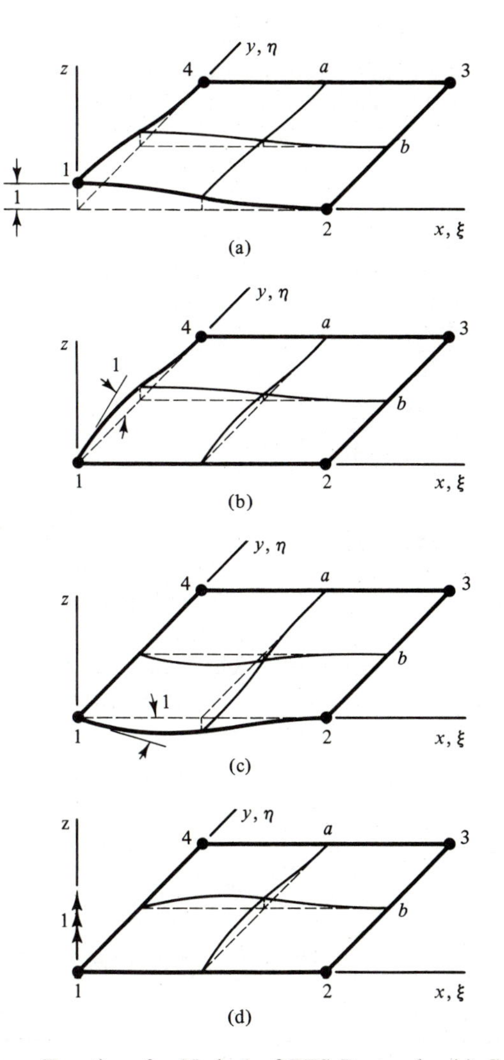

Figure 6.5 Shape Functions for Node 1 of BFS Rectangle: (a) Condition $q_1 = 1$ (b) Condition $q_2 = 1$ (c) Condition $q_3 = 1$ (d) Condition $q_4 = 1$

Formulation of generalized stresses for the BFS rectangle is the same as for the MZC rectangle [see Eq. (d)]. However, stiffnesses and equivalent nodal loads [see Eqs. (6.2-6) through (6.2-8)] must be calculated using zero as the lower limit on the integrals. Results for the stiffness matrix were obtained in this manner and tabulated by Von Rieseman (5). Final stresses may be found using Eqs. (6.2-9) through (6.2-11).

The BFS rectangle is said to be conforming because it has normal-slope compatibility at all edges. That is, the normal slope along a given edge has a cubic variation that is controlled by four parameters, which are the normal slope and warp at each end. This element gives greater accuracy than the MZC rectangle because of a higher-order displacement function and a larger number of nodal displacements. The improvement is not due to the fact that the BFS rectangle has normal-slope compatibility. In fact, the enforcement of this criterion makes an element somewhat stiffer than it should be (9).

Example 2

For the BFS rectangle, determine the equivalent nodal loads at point 1 due to a uniformly distributed load b_z (force per unit area) applied in the positive z direction. Using Eq. (6.2-7) (slightly modified) and the first four displacement shape functions from Table 6.2, we have:

$$\mathbf{p}_{b1} = ab \int_0^1 \int_0^1 \{f_1, f_2, f_3, f_4\} b_z \, d\xi \, d\eta$$

$$= \frac{ab}{144} \{36, 6b, -6a, ab\} b_z$$

6.3 TRIANGULAR ELEMENTS

Due to their geometric versatility, triangular elements constitute a useful subset of plate-bending elements. As in the preceding section, we discuss a nonconforming and a conforming element. The nonconforming triangle appears in Fig. 6.6. It is called the *CKZ triangle* because it was formulated by Cheung, King, and Zienkiewicz (6, 7). Nodal displacements for this element are:

$$\mathbf{q}_i = \{q_{i1}, q_{i2}, q_{i3}\} = \{w_i, w_{i,y}, -w_{i,x}\} \qquad (i = 1, 2, 3) \qquad \text{(a)}$$

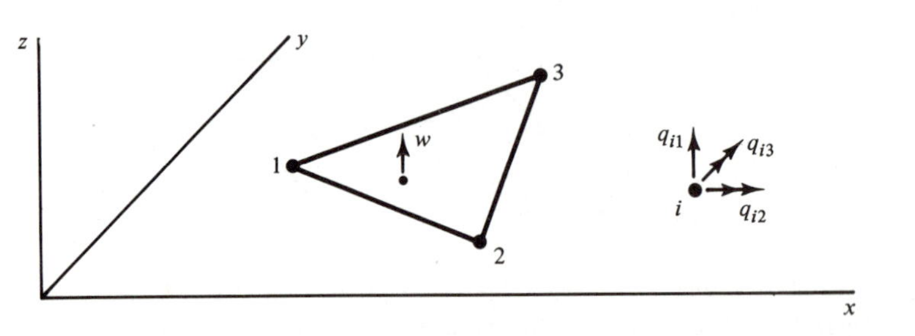

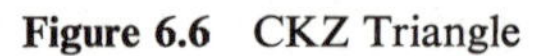
Figure 6.6 CKZ Triangle

Table 6.3 contains the assumed displacement shape functions in area coordinates. To obtain the generalized strain-displacement matrix $\bar{\mathbf{B}}$, we must differentiate the functions in Table 6.3 in accordance with Eq. (6.2-3). For this purpose the chain rules for differentiation of a function in area coordinates with respect to x and y are given by Eqs. (3.2-12). Explicit area integrals have been evaluated for element stiffnesses and equivalent nodal loads, and the results appear in (7). However, because the exact formulas are rather lengthy, it is easier to use numerical integration.

TABLE 6.3 Displacement Shape Functions for CKZ Triangle

j	f_j
1	$\xi_1 + \xi_1^2\xi_2 + \xi_1^2\xi_3 - \xi_1\xi_2^2 - \xi_1\xi_3^2$
2	$b_3(\xi_1^2\xi_2 + \alpha) - b_2(\xi_3\xi_1^2 + \alpha)$
3	$a_3(\xi_1^2\xi_2 + \alpha) - a_2(\xi_3\xi_1^2 + \alpha)$
4	$\xi_2 + \xi_2^2\xi_3 + \xi_2^2\xi_1 - \xi_2\xi_3^2 - \xi_2\xi_1^2$
5	$b_1(\xi_2^2\xi_3 + \alpha) - b_3(\xi_1\xi_2^2 + \alpha)$
6	$a_1(\xi_2^2\xi_3 + \alpha) - a_3(\xi_1\xi_2^2 + \alpha)$
7	$\xi_3 + \xi_3^2\xi_1 + \xi_3^2\xi_2 - \xi_3\xi_1^2 - \xi_3\xi_2^2$
8	$b_2(\xi_3^2\xi_1 + \alpha) - b_1(\xi_2\xi_3^2 + \alpha)$
9	$a_2(\xi_3^2\xi_1 + \alpha) - a_1(\xi_2\xi_3^2 + \alpha)$

$$\alpha = \frac{\xi_1\xi_2\xi_3}{2}$$

$$a_1 = x_{23} \qquad a_2 = x_{31} \qquad a_3 = x_{12}$$
$$b_1 = -y_{23} \qquad b_2 = -y_{31} \qquad b_3 = -y_{12}$$

A conforming triangle worthy of note is the *HCT triangle* shown in Fig. 6.7. The development of this element is attributed to Hsieh, Clough, and Tocher (8). As shown in the figure, the triangle is divided into three subtriangles, which are labeled a, b, and c. Point 4 is the centroid of triangle 1-2-3, and normal-slope compatibility is enforced at midedge nodes between the subtriangles. The displacement function w_a chosen for subtriangle a is:

$$w_a = c_{a1} + c_{a2}\bar{x} + c_{a3}\bar{y} + c_{a4}\bar{x}^2 + c_{a5}\bar{x}\bar{y} + c_{a6}\bar{y}^2 + c_{a7}\bar{x}^3 + c_{a8}\bar{x}\bar{y}^2 + c_{a9}\bar{y}^3 \qquad \text{(b)}$$

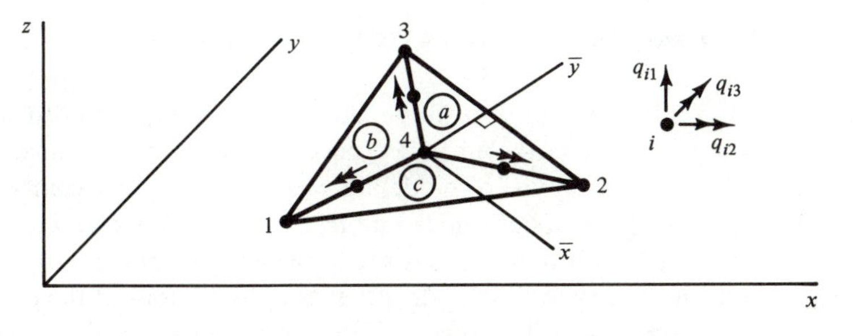

Figure 6.7 HCT Triangle

In this function the variables $\bar{x}$ and $\bar{y}$ are local coordinates with their origin at point 4 and with the $\bar{y}$ axis perpendicular to edge 2-3 (see the figure). Two similar functions w_b and w_c are also defined for subtriangles b and c. Note that omission of the $\bar{x}^2\bar{y}$ term in Eq. (b) guarantees that the normal slope varies linearly along the exterior boundary. Subelement stiffnesses are derived for triangles a, b, and c, which are then transformed from local to global axes. During this transformation, the stiffness matrix for the whole triangle is assembled, and interior nodal displacements are eliminated. The resulting matrix is of size 9×9, which is the same as for the CKZ triangle. The HCT triangle is less accurate than the CKZ element, however, because normal-slope compatibility is gained at the expense of unduly high stiffnesses (9).

Four subtriangles may be used to compose a plate-bending quadrilateral with normal-slope compatibility. Figure 6.8 indicates how subtriangles a, b, c, and d were employed to form the *CF quadrilateral*, which was devised by Clough and Felippa (10). In this configuration the interior nodal displacements were eliminated to produce a stiffness matrix of size 12×12. This element is more accurate than the HCT triangle, but only because the number of nodal displacements is larger. When compared with other quadrilaterals having only twelve displacement coordinates at exterior nodes, the CF quadrilateral proves to be the most accurate (10); but its formulation appears to be the most tedious.

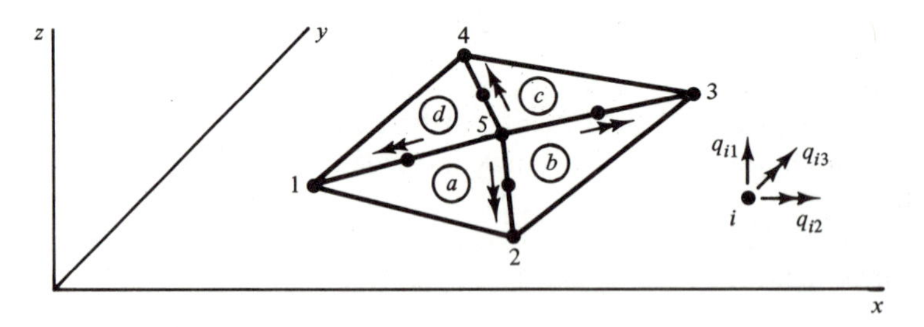

Figure 6.8 CF Quadrilateral

6.4 QUADRILATERAL AS SPECIALIZATION OF HEXAHEDRON

We can specialize a hexahedral solid finite element of the type studied in Sec. 4.4 to serve as a plate or a shell element by making one dimension small compared to the other two. This type of modeling was introduced by Ahmad et al. (11) and applies to analyses of both thick and thin shells. In Chapter 7 a general shell element will be developed on this basis. For the analysis of plates in flexure, however, it is also necessary to restrict the other two dimensions of the modified element to lie in a single plane. This section is devoted to the specialization of the isoparametric hexahedron H20 to become a plate-bending quadrilateral element called PBQ8.

Figure 6.9(a) shows the original H20 element, which has quadratic interpolation formulas defining its geometry. In order to understand the constraints

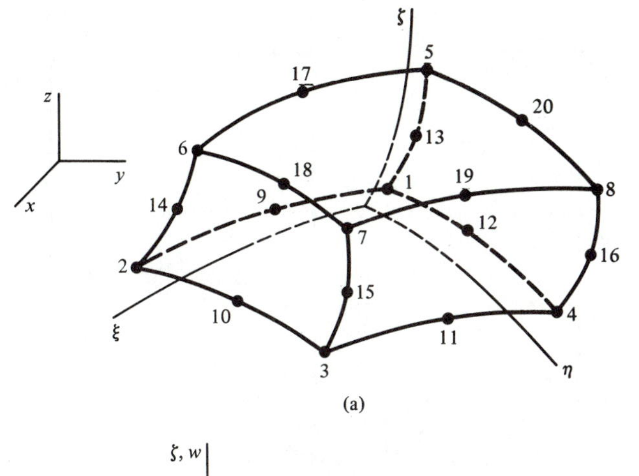

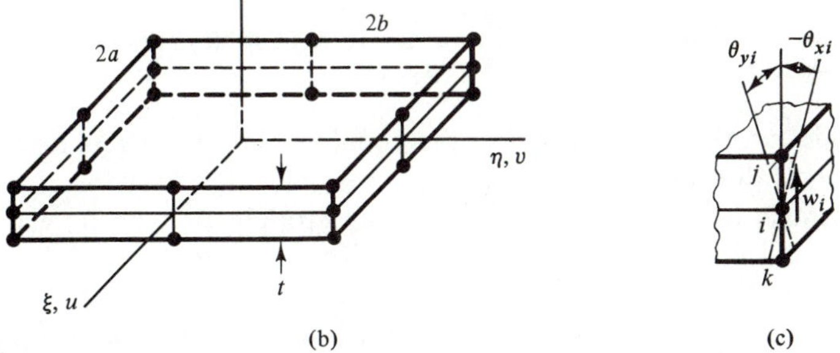

Figure 6.9 Specialization of Hexahedron: (a) Element H20 (b) Rectangular Parent PQR8 of Element PBQ8 Before Constraints (c) Constrained Nodal Displacements

needed to convert it to a plate-bending element, we first form a flat rectangular solid by making the natural coordinates ξ, η, and ζ orthogonal and the ζ dimension small. The resulting element appears in Fig. 6.9(b) as the rectangular parent PQR8 of element PBQ8 before constraints. Note that groups of three nodes occur at the corners, while pairs of nodes are at midedge locations of element PQR8. By invoking certain constraints, we can convert each group and pair of nodes to a single node on the middle surface, as shown in Fig. 6.9(c). The nodal displacements indicated at point i in that figure are:

$$\mathbf{q}_i = \{q_{i1}, q_{i2}, q_{i3}\} = \{w_i, \theta_{xi}, \theta_{yi}\} \qquad (i = 1, 2, \ldots, 8) \tag{a}$$

where θ_{xi} and θ_{yi} are small positive rotations about the x and y axes. Relationships between nodal displacements at a corner of element PQR8, a midedge of PQR8, and a node of element PBQ8 can be seen more clearly in Figs. 6.10(a), (b), and (c). The two types of constraints to be introduced are:

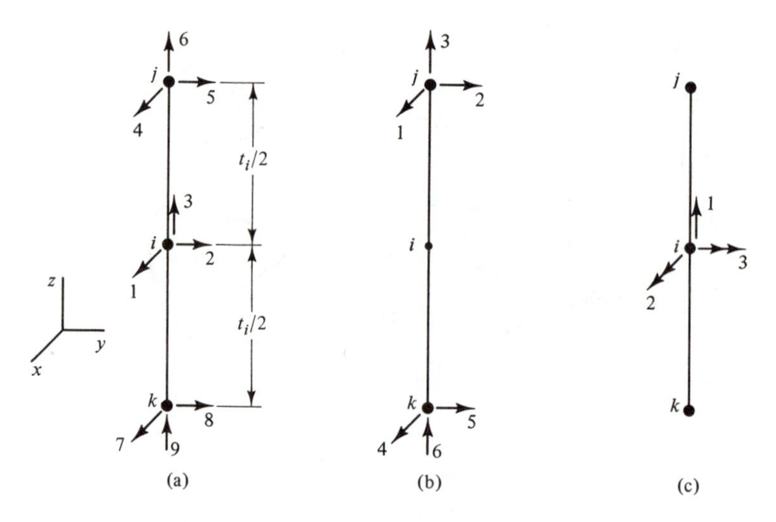

Figure 6.10 Nodal Displacements: (a) Corner of PQR8 (b) Midedge of PQR8 (c) Node of PBQ8

1. Nodes on the same normal to the middle surface have equal translations in the ζ direction.
2. Normals to the middle surface remain straight (but no longer normal) during deformation.

Using these criteria, we can relate the nine nodal translations in Fig. 6.10(a) to the three nodal displacements in Fig. 6.10(c) by the following 9×3 *constraint matrix*:

$$\mathbf{G}_{ai} = \begin{bmatrix} 0 & 0 & 0 \\ 0 & 0 & 0 \\ 1 & 0 & 0 \\ 0 & 0 & \frac{t_i}{2} \\ 0 & -\frac{t_i}{2} & 0 \\ 1 & 0 & 0 \\ 0 & 0 & -\frac{t_i}{2} \\ 0 & \frac{t_i}{2} & 0 \\ 1 & 0 & 0 \end{bmatrix} \tag{b}$$

Similarly, the six nodal translations in Fig. 6.10(b) are related to the three nodal displacements in Fig. 6.10(c) by the constraint matrix:

$$\mathbf{G}_{bi} = \begin{bmatrix} 0 & 0 & \frac{t_i}{2} \\ 0 & -\frac{t_i}{2} & 0 \\ 1 & 0 & 0 \\ 0 & 0 & -\frac{t_i}{2} \\ 0 & \frac{t_i}{2} & 0 \\ 1 & 0 & 0 \end{bmatrix} \tag{c}$$

which is of size 6×3. If we were to apply each of these constraint matrices in four locations, we would be able to reduce the number of nodal displacements from $(4)(9) + (4)(6) = 60$ to $(8)(3) = 24$. Instead of following this path, however, we will pursue a more direct formulation of element PBQ8 in the manner described by Cook (12).

Figure 6.11 shows element PBQ8 with its neutral surface lying in the x-y plane. Its geometry is defined to be the same as that for element Q8 in Sec. 3.4. Thus,

$$x = \sum_{i=1}^{8} f_i x_i \qquad y = \sum_{i=1}^{8} f_i y_i \qquad (z = 0) \tag{3.4-13 repeated}$$

where

$$\begin{aligned} f_i &= \tfrac{1}{4}(1 + \xi_0)(1 + \eta_0)(-1 + \xi_0 + \eta_0) && (i = 1, 2, 3, 4) \\ f_i &= \tfrac{1}{2}(1 - \xi^2)(1 + \eta_0) && (i = 5, 7) \\ f_i &= \tfrac{1}{2}(1 + \xi_0)(1 - \eta^2) && (i = 6, 8) \end{aligned} \tag{3.4-12 repeated}$$

Generic displacements at any point off the neutral surface are:

$$\mathbf{u} = \{u, v, w\} \tag{d}$$

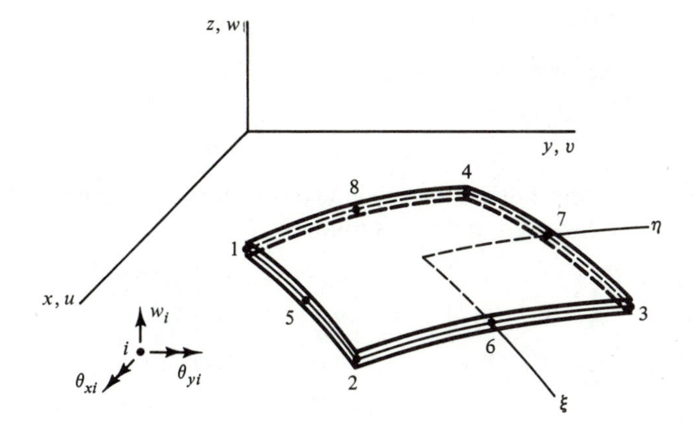

Figure 6.11 Element PBQ8

We assume that w, θ_x, and θ_y vary quadratically over the element, so that

$$u = z\theta_y = z\sum_{i=1}^{8} f_i\theta_{yi}$$
$$v = -z\theta_x = -z\sum_{i=1}^{8} f_i\theta_{xi} \tag{6.4-1a}$$
$$w = \sum_{i=1}^{8} f_i w_i$$

in which the displacement shape functions f_i are the same as those in Eqs. (3.4-12). Note that the rotations θ_x and θ_y are chosen independently of w and are not related to it by differentiation. In this case the displacement shape functions may be displayed in matrix form as:

$$\mathbf{f}_i = \begin{bmatrix} 0 & 0 & z \\ 0 & -z & 0 \\ 1 & 0 & 0 \end{bmatrix} f_i \qquad (i = 1, 2, \ldots, 8) \tag{6.4-1b}$$

The 3×3 Jacobian matrix required in this formulation is:

$$\mathbf{J} = \begin{bmatrix} x_{,\xi} & y_{,\xi} & 0 \\ x_{,\eta} & y_{,\eta} & 0 \\ 0 & 0 & 1 \end{bmatrix} \tag{6.4-2a}$$

where

$$x_{,\xi} = \sum_{i=1}^{8} f_{i,\xi} x_i \quad \text{etc.}$$

The inverse of $\mathbf{J}$ becomes:

$$\mathbf{J}^{-1} = \mathbf{J}^* = \begin{bmatrix} \xi_{,x} & \eta_{,x} & 0 \\ \xi_{,y} & \eta_{,y} & 0 \\ 0 & 0 & 1 \end{bmatrix} \tag{6.4-2b}$$

We need certain derivatives with respect to local coordinates, which are collected into the following 3×3 matrix:

$$\begin{bmatrix} u_{,\xi} & v_{,\xi} & w_{,\xi} \\ u_{,\eta} & v_{,\eta} & w_{,\eta} \\ u_{,z} & v_{,z} & w_{,z} \end{bmatrix} = \sum_{i=1}^{8} \begin{bmatrix} zf_{i,\xi}\theta_{yi} & -zf_{i,\xi}\theta_{xi} & f_{i,\xi}w_i \\ zf_{i,\eta}\theta_{yi} & -zf_{i,\eta}\theta_{xi} & f_{i,\eta}w_i \\ f_i\theta_{yi} & -f_i\theta_{xi} & 0 \end{bmatrix} \tag{6.4-3}$$

Transformation of these derivatives to global coordinates is accomplished using the inverse of the Jacobian matrix, as follows:

$$\begin{bmatrix} u_{,x} & v_{,x} & w_{,x} \\ u_{,y} & v_{,y} & w_{,y} \\ u_{,z} & v_{,z} & w_{,z} \end{bmatrix} = \mathbf{J}^{-1} \begin{bmatrix} u_{,\xi} & v_{,\xi} & w_{,\xi} \\ u_{,\eta} & v_{,\eta} & w_{,\eta} \\ u_{,z} & v_{,z} & w_{,z} \end{bmatrix}$$
$$= \sum_{i=1}^{8} \begin{bmatrix} za_i\theta_{yi} & -za_i\theta_{xi} & a_i w_i \\ zb_i\theta_{yi} & -zb_i\theta_{xi} & b_i w_i \\ f_i\theta_{yi} & -f_i\theta_{xi} & 0 \end{bmatrix} \tag{6.4-4}$$

in which

$$a_i = J^*_{11} f_{i,\xi} + J^*_{12} f_{i,\eta} \qquad b_i = J^*_{21} f_{i,\xi} + J^*_{22} f_{i,\eta} \tag{e}$$

The five types of nonzero strains to be considered for element PBQ8 are:

$$\boldsymbol{\epsilon} = \begin{bmatrix} \epsilon_x \\ \epsilon_y \\ \gamma_{xy} \\ \gamma_{xz} \\ \gamma_{yz} \end{bmatrix} = \begin{bmatrix} u_{,x} \\ v_{,y} \\ u_{,y} + v_{,x} \\ u_{,z} + w_{,x} \\ v_{,z} + w_{,y} \end{bmatrix} \tag{f}$$

By inspection of the second version of this strain vector, we can assemble the ith part of matrix **B** from terms in Eq. (6.4-4) as:

$$\mathbf{B}_i = \begin{bmatrix} 0 & 0 & za_i \\ 0 & -zb_i & 0 \\ 0 & -za_i & zb_i \\ a_i & 0 & f_i \\ b_i & -f_i & 0 \end{bmatrix} \qquad (i = 1, 2, \ldots, 8) \tag{6.4-5}$$

Stresses corresponding to the strains in Eq. (f) are:

$$\boldsymbol{\sigma} = \{\sigma_x, \sigma_y, \tau_{xy}, \tau_{xz}, \tau_{yz}\} \tag{g}$$

Then the stress-strain relationships for either an orthotropic or an isotropic material may be expressed by:

$$\mathbf{E} = \begin{bmatrix} E_{11} & E_{12} & 0 & 0 & 0 \\ E_{21} & E_{22} & 0 & 0 & 0 \\ 0 & 0 & E_{33} & 0 & 0 \\ 0 & 0 & 0 & E_{44} & 0 \\ 0 & 0 & 0 & 0 & E_{55} \end{bmatrix} \tag{6.4-6}$$

Assuming that x and y are principal material directions in the orthotropic case, we may write the terms in matrix **E** as:

$$\begin{aligned} E_{11} &= \frac{E_x}{1 - \nu_{xy}\nu_{yx}} \qquad & E_{12} &= \frac{\nu_{xy}E_x}{1 - \nu_{xy}\nu_{yx}} \qquad & E_{22} &= \frac{E_y}{1 - \nu_{xy}\nu_{yx}} \\ E_{33} &= G_{xy} & E_{44} &= \frac{G_{xz}}{1.2} & E_{55} &= \frac{G_{yz}}{1.2} \end{aligned} \tag{h}$$

The first four terms are drawn from Eq. (2.1-20), and the last two are divided by the *form factor* 1.2 to account for the fact that the transverse shearing strains produce too little strain energy (1). On the other hand, if the material is isotropic, the constants in Eqs. (h) become:

$$E_x = E_y = E \qquad \nu_{xy} = \nu_{yx} = \nu \qquad G_{xy} = G_{xz} = G_{yz} = \frac{E}{2(1 + \nu)} \tag{i}$$

As a preliminary matter before formulating element stiffnesses, we partition matrix $\mathbf{B}$ and factor z from the upper part, as follows:

$$\mathbf{B} = \begin{bmatrix} \mathbf{B}_A \\ \mathbf{B}_B \end{bmatrix} = \begin{bmatrix} z\bar{\mathbf{B}}_A \\ \mathbf{B}_B \end{bmatrix} \tag{j}$$

where $\mathbf{B}_A$ has three rows and $\mathbf{B}_B$ has two rows. Similarly, matrix $\mathbf{E}$ may be partitioned into:

$$\mathbf{E} = \begin{bmatrix} \mathbf{E}_A & \mathbf{0} \\ \mathbf{0} & \mathbf{E}_B \end{bmatrix} \tag{k}$$

in which $\mathbf{E}_A$ is of size 3×3 and $\mathbf{E}_B$ is of size 2×2. Then the stiffness matrix for element PBQ8 is written as:

$$\mathbf{K} = \int_V \mathbf{B}^{\mathrm{T}}\mathbf{E}\ \mathbf{B}\, dV = \int_V [z\bar{\mathbf{B}}_A^{\mathrm{T}} \quad \mathbf{B}_B^{\mathrm{T}}] \begin{bmatrix} \mathbf{E}_A & \mathbf{0} \\ \mathbf{0} & \mathbf{E}_B \end{bmatrix} \begin{bmatrix} z\bar{\mathbf{B}}_A \\ \mathbf{B}_B \end{bmatrix} dV$$

$$= \int_V (z^2\bar{\mathbf{B}}_A^{\mathrm{T}}\mathbf{E}_A\bar{\mathbf{B}}_A + \mathbf{B}_B^{\mathrm{T}}\mathbf{E}_B\mathbf{B}_B)\, dV \tag{ℓ}$$

Integration through the thickness yields:

$$\mathbf{K} = \int_A (\bar{\mathbf{B}}_A^{\mathrm{T}}\bar{\mathbf{E}}_A\bar{\mathbf{B}}_A + \mathbf{B}_B^{\mathrm{T}}\bar{\mathbf{E}}_B\mathbf{B}_B)\, dA \tag{m}$$

where

$$\bar{\mathbf{E}}_A = \mathbf{E}_A \frac{t^3}{12} \qquad \bar{\mathbf{E}}_B = \mathbf{E}_B t \tag{n}$$

Thus,

$$\mathbf{K} = \int_A \bar{\mathbf{B}}^{\mathrm{T}}\bar{\mathbf{E}}\ \bar{\mathbf{B}}\, dA = \int_{-1}^{1}\int_{-1}^{1} \bar{\mathbf{B}}^{\mathrm{T}}\bar{\mathbf{E}}\ \bar{\mathbf{B}}\,|\mathbf{J}|\, d\xi\, d\eta \tag{6.4-7}$$

which must be evaluated numerically. Matrix $\bar{\mathbf{E}}$ in Eq. (6.4-7) has the form:

$$\bar{\mathbf{E}} = \begin{bmatrix} \bar{\mathbf{E}}_A & \mathbf{0} \\ \mathbf{0} & \bar{\mathbf{E}}_B \end{bmatrix} \tag{6.4-8}$$

And the ith part of matrix $\bar{\mathbf{B}}$ is:

$$\bar{\mathbf{B}}_i = \begin{bmatrix} 0 & 0 & a_i \\ 0 & -b_i & 0 \\ 0 & -a_i & b_i \\ \hdashline a_i & 0 & f_i \\ b_i & -f_i & 0 \end{bmatrix} \qquad (i = 1, 2, \ldots, 8) \tag{6.4-9}$$

Nonzero terms in the upper partition of $\bar{\mathbf{B}}_i$ are curvatures per unit rotation. Note that these rates of change of θ_x and θ_y are not the same as second derivatives of w with respect to x and y. Furthermore, the lower partition of $\bar{\mathbf{B}}_i$ contains angles per unit displacement. In the first column are rates of change of w, while columns 2 and 3 consist of generic rotations of the normal due to unit values of nodal rotations.

Equivalent nodal loads due to body forces on element PBQ8 are calculated as:

$$\mathbf{p}_b = \int_V \mathbf{f}^{\mathrm{T}}\mathbf{b}\, dV = \int_{-1}^{1}\int_{-1}^{1} \mathbf{f}^{\mathrm{T}}\mathbf{b}\,|\mathbf{J}|\, d\xi\, d\eta \tag{6.4-10}$$

in which

$$\mathbf{b} = \{0, 0, b_z\} \tag{o}$$

and b_z is force per unit area in the z direction. Also, equivalent nodal loads caused by initial strains are:

$$\mathbf{p}_0 = \int_V \mathbf{B}^{\mathrm{T}}\mathbf{E}\, \boldsymbol{\epsilon}_0\, dV = \int_{-1}^{1}\int_{-1}^{1} \bar{\mathbf{B}}^{\mathrm{T}}\bar{\mathbf{E}}\, \boldsymbol{\phi}_0\,|\mathbf{J}|\, d\xi\, d\eta \tag{6.4-11}$$

where

$$\boldsymbol{\phi}_0 = \{\phi_{xx0}, \phi_{yy0}, \phi_{xy0}, 0, 0\} \tag{p}$$

Of course, the integrals in Eqs. (6.4-10) and (6.4-11) must be evaluated numerically. Finally, the generalized stresses in vector $\mathbf{M}$ may be computed from:

$$\mathbf{M} = \{M_{xx}, M_{yy}, M_{xy}, Q_x, Q_y\} = \bar{\mathbf{E}}(\bar{\mathbf{B}}\, \mathbf{q} - \boldsymbol{\phi}_0) \tag{6.4-12}$$

And the actual stresses are obtained using Eqs. (6.2-10) and (6.2-11).

Example 1

For the rectangular parent of element PBQ8 (after constraints are imposed), find the stiffness term K_{11}, assuming that the material is isotropic. For this term Eq. (6.4-7) specializes to:

$$K_{11} = ab\int_{-1}^{1}\int_{-1}^{1} \bar{\mathbf{B}}_{1,1}^{\mathrm{T}}\bar{\mathbf{E}}\, \bar{\mathbf{B}}_{1,1}\, d\xi\, d\eta \tag{q}$$

in which the symbol $\bar{\mathbf{B}}_{1,1}$ represents the first column of submatrix $\bar{\mathbf{B}}_1$. From Eqs. (6.4-5), (h), and (i) we have:

$$\bar{\mathbf{E}}\, \bar{\mathbf{B}}_{1,1} = \frac{Et}{2.4(1+\nu)}\{0, 0, 0, a_1, b_1\}$$

Then Eqs. (e) give:

$$a_1 = J_{11}^* f_{1,\xi} + 0 = \frac{1}{4a}(1-\eta)(2\xi+\eta)$$

$$b_1 = 0 + J_{22}^* f_{1,\eta} = \frac{1}{4b}(1-\xi)(2\eta+\xi)$$

Substitution of these expressions into Eq. (q) yields:

$$K_{11} = \frac{abEt}{38.4(1+\nu)}\int_{-1}^{1}\int_{-1}^{1}\left[\frac{1}{a^2}(1-\eta)^2(2\xi+\eta)^2 + \frac{1}{b^2}(1-\xi)^2(2\eta+\xi)^2\right] d\xi\, d\eta$$

By integrating and simplifying the results, we obtain:

$$K_{11} = \frac{13Et}{54ab(1+\nu)}(a^2+b^2)$$

Example 2

Suppose that the rectangular parent of element PBQ8 (with constraints) is subjected to a uniformly distributed loading b_z (force per unit area) applied in the positive z

direction. Determine the equivalent nodal forces in the z direction at points 1 and 5 due to this influence.

For obtaining the force at node 1, we specialize Eq. (6.4-10) to become:

$$p_{b1} = ab \int_{-1}^{1} \int_{-1}^{1} f_1 b_z \, d\xi \, d\eta \tag{r}$$

Substituting the function f_1 from Eqs. (3.4-12) into Eq. (r) and integrating, we find:

$$p_{b1} = -\frac{ab}{3} b_z$$

Similarly, the equivalent nodal force at point 5 is:

$$p_{b13} = ab \int_{-1}^{1} \int_{-1}^{1} f_5 b_z \, d\xi \, d\eta \tag{s}$$

Upon substituting f_5 from Eqs. (3.4-12) into Eq. (s) and integrating, we obtain:

$$p_{b13} = \frac{4ab}{3} b_z$$

Note that the equivalent nodal forces have opposite signs in this case.

In addition to the serendipity element discussed herein, other quadrilaterals for plate bending are of interest. For example, the straight-sided bilinear displacement quadrilateral of Hughes et al. (13) is simpler to use but cannot model curved boundaries or be adapted to general shells. The Lagrange (12) and heterosis (14) elements both have interior nodes and require more numerical integration points for good accuracy.

6.5 ANNULAR ELEMENT

The special geometry of a *circular plate* can be modeled exactly with annular and circular finite elements. Figure 6.12(a) shows the annular type of element that can be used for such a plate, except at its center, where a circular element would evidently be required. However, it has been shown (15) that a hole of small diameter can be used in place of a circular element without significant loss of accuracy in the results. We can also achieve this effect with an annular element having $r_1 = 0$ by using numerical integration, for which the radius of the first integration point is nonzero. In addition, *annular segments* (with $\theta < 2\pi$) prove to be useful for modeling exactly a plate having that form (16). However, in this section only annular elements (with $\theta = 2\pi$) will be discussed for both axisymmetric and nonaxisymmetric loads.

First, we assume that the annular element in Fig. 6.12(a) is part of a circular or annular plate that is subjected to axisymmetric loads. The element has only one generic displacement. Thus,

$$\mathbf{u} = v \tag{a}$$

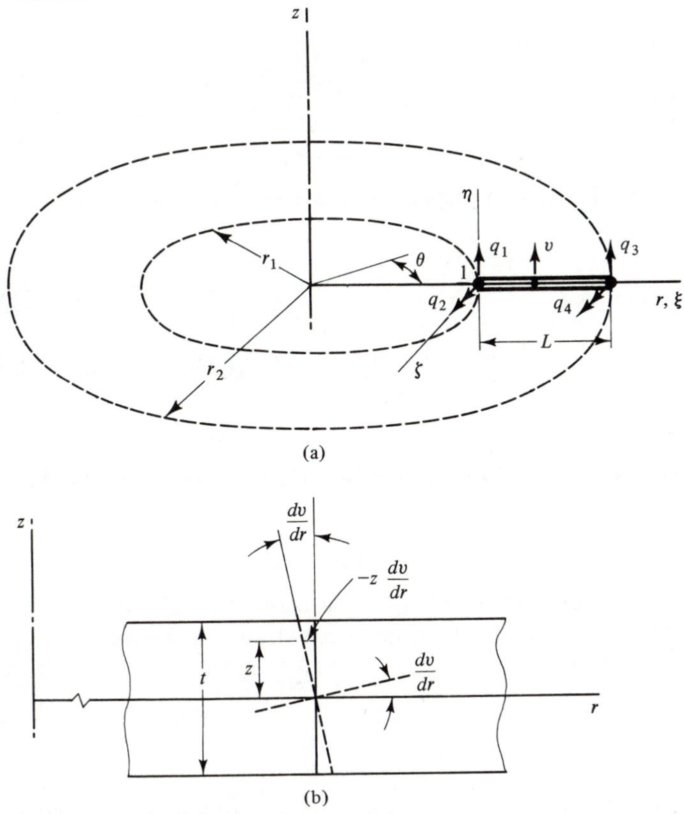

Figure 6.12 Circular Plate: (a) Annular Element (b) Radial Slope

At two nodal circles the following nodal displacements are also indicated:

$$\mathbf{q} = \{q_1, q_2, q_3, q_4\} = \{v_1, v_{1,r}, v_2, v_{2,r}\} \tag{b}$$

Displacement shape functions chosen for this element are the same as those used for the flexural element in Sec. 1.5. That is,

$$\begin{aligned}\mathbf{f} &= [f_1 \quad f_2 \quad f_3 \quad f_4] \\ &= [1 - 3\xi^2 + 2\xi^3 \quad (\xi - 2\xi^2 + \xi^3)L \quad 3\xi^2 - 2\xi^3 \quad -(\xi^2 - \xi^3)L]\end{aligned} \tag{6.5-1}$$

where $\xi = (r - r_1)/L$ and $L = r_2 - r_1$.

From Fig. 6.12(b) we can deduce the strain-displacement relationships to be:

$$\epsilon_r = -z\frac{d^2v}{dr^2} \qquad \epsilon_\theta = -\frac{z}{r}\frac{dv}{dr} \tag{c}$$

Because the term $-z$ can be factored from both of these expressions, a generalized linear differential operator $\bar{\mathbf{d}}$ can be written as:

$$\bar{\mathbf{d}} = \begin{bmatrix} \dfrac{d^2}{dr^2} \\ \dfrac{1}{r}\dfrac{d}{dr} \end{bmatrix} \tag{6.5-2}$$

Then the generalized strain-displacement matrix $\bar{\mathbf{B}}$ is found to be:

$$\bar{\mathbf{B}} = \bar{\mathbf{d}}\ \mathbf{f}$$

$$= \begin{bmatrix} -\dfrac{6}{L^2}(1-2\xi) & -\dfrac{2}{L}(2-3\xi) & \dfrac{6}{L^2}(1-2\xi) & -\dfrac{2}{L}(1-3\xi) \\ -\dfrac{6}{Lr}(\xi-\xi^2) & \dfrac{1}{r}(1-4\xi+3\xi^2) & \dfrac{6}{Lr}(\xi-\xi^2) & -\dfrac{1}{r}(2\xi-3\xi^2) \end{bmatrix} \tag{6.5-3}$$

which is also devoid of $-z$.

Generalized stress-strain relationships for this element can be expressed as:

$$\bar{\mathbf{E}} = \mathbf{E}\frac{t^3}{12} = \begin{bmatrix} E_{11} & E_{12} \\ E_{21} & E_{22} \end{bmatrix}\frac{t^3}{12} \tag{6.5-4}$$

for either an orthotropic or an isotropic material. This operator relates the generalized stresses in $\mathbf{M}$ to the generalized strains in $\boldsymbol{\phi}$, where

$$\mathbf{M} = \{M_{rr}, M_{\theta\theta}\} \tag{6.5-5}$$

and

$$\boldsymbol{\phi} = \left\{\frac{d^2v}{dr^2}, \frac{1}{r}\frac{dv}{dr}\right\} \tag{6.5-6}$$

In this case the first term in $\boldsymbol{\phi}$ is a type of curvature, but the second term is not. Nevertheless, both terms may be characterized as generalized strains.

Now we can write the stiffness matrix for the element as follows:

$$\mathbf{K} = L\int_0^1\int_0^{2\pi} \bar{\mathbf{B}}^{\mathrm{T}}\bar{\mathbf{E}}\ \bar{\mathbf{B}}\ r\,d\theta\,d\xi = 2\pi L\int_0^1 \bar{\mathbf{B}}^{\mathrm{T}}\bar{\mathbf{E}}\ \bar{\mathbf{B}}\ r\,d\xi \tag{6.5-7}$$

Similarly, equivalent nodal loads due to body forces become:

$$\mathbf{p}_b = 2\pi L\int_0^1 \mathbf{f}^{\mathrm{T}} b_z r\,d\xi \tag{6.5-8}$$

And those caused by initial strains are:

$$\mathbf{p}_0 = 2\pi L\int_0^1 \bar{\mathbf{B}}^{\mathrm{T}}\bar{\mathbf{E}}\ \boldsymbol{\phi}_0 r\,d\xi \tag{6.5-9}$$

Note that the second term in $\boldsymbol{\phi}_0$ must be obtained from the first by integration with respect to r [see Eq. (6.5-6)].

Next, we shall consider a circular plate that is subjected to nonaxisymmetric loads. As noted in Sec. 5.4, any set of loads on a solid of revolution (in this case a plate) can be decomposed into two sets, which are symmetric and antisymmetric with respect to a plane of symmetry containing the axis of revolution. For symmetric loads on a circular plate, Fourier decomposition into m harmonic terms produces:

$$b_z = \sum_{j=0}^{m} b_{zj} \cos j\theta \tag{6.5-10}$$

where b_{zj} is a function of r only, and $j = 0$ is the axisymmetric case. In addition, the responses to these harmonic loads become:

$$v = \sum_{j=0}^{m} v_j \cos j\theta \tag{6.5-11}$$

If the loads were antisymmetric with respect to the plane of symmetry, the function $\cos j\theta$ in Eqs. (6.5-10) and (6.5-11) would be replaced with $\sin j\theta$.

Strain-displacement relationships for nonaxisymmetric loads are (1):

$$\begin{aligned} \epsilon_r &= -z\frac{\partial^2 v}{\partial r^2} \qquad \epsilon_\theta = -z\left(\frac{1}{r^2}\frac{\partial^2 v}{\partial \theta^2} + \frac{1}{r}\frac{\partial v}{\partial r}\right) \\ \gamma_{r\theta} &= -2z\left(\frac{1}{r}\frac{\partial^2 v}{\partial r\,\partial \theta} - \frac{1}{r^2}\frac{\partial v}{\partial \theta}\right) \end{aligned} \tag{6.5-12}$$

When v_j in Eq. (6.5-11) is expressed in terms of $\mathbf{f}$ [see Eq. (6.5-1)], the operations in Eqs. (6.5-12) yield the generalized strain-displacement matrix $\bar{\mathbf{B}}_j$, which is typified by:

$$(\bar{\mathbf{B}}_i)_j = \begin{bmatrix} f_{i,rr} \cos j\theta \\ -j^2 \frac{f_i}{r^2} \cos j\theta + \frac{f_{i,r}}{r} \cos j\theta \\ -j \frac{f_{i,r}}{r} \sin j\theta + j \frac{f_i}{r^2} \sin j\theta \end{bmatrix} \tag{6.5-13}$$

where $i = 1, 2, 3, 4$ and $j = 0, 1, 2, \ldots, m$.

The generalized stress-strain matrix in Eq. (6.5-4) can be expanded for the present case by adding element E_{33}. Then the 3×3 operator $\bar{\mathbf{E}}$ will relate the generalized stresses

$$\mathbf{M} = \{M_{rr}, M_{\theta\theta}, M_{r\theta}\} \tag{6.5-14}$$

to the generalized strains:

$$\boldsymbol{\phi} = \left\{\frac{\partial^2 v}{\partial r^2}, \frac{1}{r^2}\frac{\partial^2 v}{\partial \theta^2} + \frac{1}{r}\frac{\partial v}{\partial r}, 2\left(\frac{1}{r}\frac{\partial^2 v}{\partial r\,\partial \theta} - \frac{1}{r^2}\frac{\partial v}{\partial \theta}\right)\right\} \tag{6.5-15}$$

An element stiffness matrix for each harmonic set of symmetric loads may now be written as:

$$\mathbf{K}_j = k\pi L \int_0^1 \bar{\mathbf{B}}_j^{\mathrm{T}} \bar{\mathbf{E}}\ \bar{\mathbf{B}}_j r\, d\xi \qquad (j = 0, 1, 2, \ldots, m) \tag{6.5-16}$$

where $k = 2$ for $j = 0$ and $k = 1$ for $j = 1, 2, \ldots, m$. In addition, equivalent nodal loads for each harmonic set of symmetric body forces have the form:

$$\mathbf{p}_{bj} = k\pi L \int_0^1 \mathbf{f}^{\mathrm{T}} b_{zj} r\, d\xi \qquad (j = 0, 1, 2, \ldots, m) \tag{6.5-17}$$

And equivalent nodal loads due to initial strains are:

$$\mathbf{p}_{0j} = k\pi L \int_0^1 \bar{\mathbf{B}}_j^{\mathrm{T}} \bar{\mathbf{E}}\ \boldsymbol{\phi}_{0j} r\, d\xi \qquad (j = 0, 1, 2, \ldots, m) \tag{6.5-18}$$

Finally, the generalized stresses for each harmonic response become:

$$\mathbf{M}_j = \bar{\mathbf{E}}(\bar{\mathbf{B}}_j \mathbf{q}_j - \boldsymbol{\phi}_{0j}) \qquad (j = 0, 1, 2, \ldots, m) \tag{6.5-19}$$

6.6 PROGRAM PBQ8 AND APPLICATIONS

In this section we deal with a computer program named PBQ8 that analyzes isotropic plates in bending using the element described in Sec. 6.4. This program may be constructed by modifying Program PSQ8 from Sec. 3.6, because the geometric interpolation functions and the displacement shape functions are the same for both. However, in Program PBQ8 the matrices **f**, **B**, **E**, and so on, must be handled according to the expressions given in Sec. 6.4.

Example 1

In Fig. 6.13 a square cantilever plate is modeled by a single PBQ8 element. To simplify the results, rotational restraints in the x direction are shown at points 1, 3, 4, and 5 (by double-headed arrows with small cross lines on their shafts). The figure also implies that restraints against translations in the z direction and against rotations in the x and y senses exist at the fixed points 6, 7, and 8. In this problem we assume the following values of structural parameters:

$$a = 5 \text{ in.} \qquad t = 0.2 \text{ in.}$$

$$E = 1 \times 10^4 \text{ k/in.}^2 \qquad \nu = 0.3$$

where the units are U.S. and the material is aluminum.

Forces equal to P, $4P$, and P (equivalent to a uniform edge load) are applied in the z direction at nodes 1, 2, and 3 in Fig. 6.13. We let the value of the first and third forces be $P = 0.01$ k, and this set of three forces becomes our first loading case. A second load set, consisting of the moments M_y, $4M_y$, and M_y, will also be applied at points 1, 2, and 3. These moments act in the y sense, and we take the value of $M_y =$ 0.1 k-in.

Table 6.4 shows the computer output from Program PBQ8 for this example. Structural data on the first page of the table are followed by the results for each of the two loading cases mentioned. Note that the transverse shearing stresses τ_{xz} and τ_{yz} in

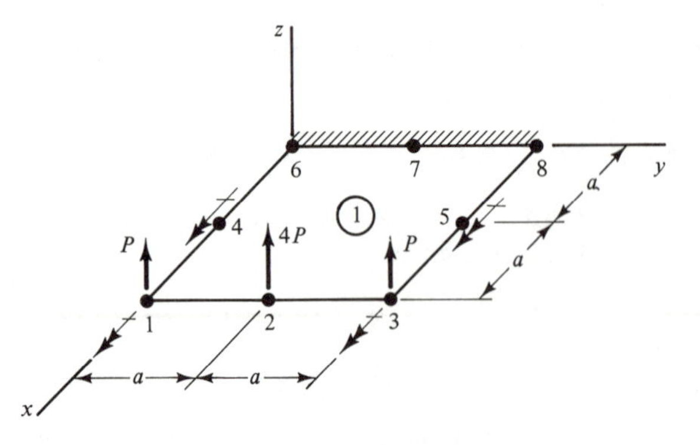

Figure 6.13 Example 1: Cantilever Element PBQ8 with Tip Loads

TABLE 6.4 Computer Output for Example 1

```
PROGRAM PBQ8

***  EXAMPLE 1:  CANTILEVER ELEMENT WITH TIP LOADS  ***

STRUCTURAL PARAMETERS
  NN   NE  NRN  NLS           E           PR            T
   8    1    7    2  1.0000D 04  3.0000D-01  2.0000D-01

COORDINATES OF NODES
 NODE          X           Y
    1  1.0000D 01  0.0000D-01
    2  1.0000D 01  5.0000D 00
    3  1.0000D 01  1.0000D 01
    4  5.0000D 00  0.0000D-01
    5  5.0000D 00  1.0000D 01
    6  0.0000D-01  0.0000D-01
    7  0.0000D-01  5.0000D 00
    8  0.0000D-01  1.0000D 01

ELEMENT INFORMATION
ELEM.   N1   N2   N3   N4   N5   N6   N7   N8
    1    1    3    8    6    2    5    7    4

NODAL RESTRAINTS
 NODE   R1   R2   R3
    1    0    1    0
    3    0    1    0
    4    0    1    0
    5    0    1    0
    6    1    1    1
    7    1    1    1
    8    1    1    1

NUMBER OF DEGREES OF FREEDOM =   11
NUMBER OF NODAL RESTRAINTS   =   13
NUMBER OF TERMS IN SN        =   98

**********  LOADING NUMBER    1  **********
  NLN  NEL  NEA  IPR
    3    0    0    0

ACTIONS AT NODES
 NODE         AN1         AN2         AN3
    1  1.0000D-02  0.0000D-01  0.0000D-01
    2  4.0000D-02  0.0000D-01  0.0000D-01
    3  1.0000D-02  0.0000D-01  0.0000D-01

NODAL DISPLACEMENTS
 NODE         DN1         DN2         DN3
    1  2.7309D-01 -0.0000D-01 -4.0950D-02
    2  2.7309D-01 -2.4989D-15 -4.0950D-02
    3  2.7309D-01 -0.0000D-01 -4.0950D-02
    4  8.5359D-02 -0.0000D-01 -3.0712D-02
    5  8.5359D-02 -0.0000D-01 -3.0712D-02
    6  0.0000D-01  0.0000D-01  0.0000D-01
    7  0.0000D-01  0.0000D-01  0.0000D-01
    8  0.0000D-01  0.0000D-01  0.0000D-01

ELEMENT STRESSES
ELEM. INT.    X STRESS    Y STRESS   XY STRESS   XZ STRESS   YZ STRESS
```

TABLE 6.4 (cont.)

1	1	-1.9019D 00	-5.7058D-01	2.4161D-13	4.5000D-02	-8.3400D-15
1	2	-1.9019D 00	-5.7058D-01	2.0746D-14	4.5000D-02	-1.2510D-14
1	3	-7.0981D 00	-2.1294D 00	9.7953D-14	4.5000D-02	1.2510D-14
1	4	-7.0981D 00	-2.1294D 00	1.5775D-13	4.5000D-02	1.2510D-14

SUPPORT REACTIONS

NODE	AR1	AR2	AR3
1	0.0000D-01	1.5437D-15	0.0000D-01
3	0.0000D-01	4.7310D-14	0.0000D-01
4	0.0000D-01	-6.0000D-02	0.0000D-01
5	0.0000D-01	6.0000D-02	0.0000D-01
6	-1.0000D-02	-3.0000D-02	1.0000D-01
7	-4.0000D-02	-4.2076D-14	4.0000D-01
8	-1.0000D-02	3.0000D-02	1.0000D-01

********** LOADING NUMBER 2 **********

NLN	NEL	NEA	IPR
3	0	0	0

ACTIONS AT NODES

NODE	AN1	AN2	AN3
1	0.0000D-01	0.0000D-01	1.0000D-01
2	0.0000D-01	0.0000D-01	4.0000D-01
3	0.0000D-01	0.0000D-01	1.0000D-01

NODAL DISPLACEMENTS

NODE	DN1	DN2	DN3
1	-4.0950D-01	-0.0000D-01	8.1900D-02
2	-4.0950D-01	3.3333D-15	8.1900D-02
3	-4.0950D-01	-0.0000D-01	8.1900D-02
4	-1.0237D-01	-0.0000D-01	4.0950D-02
5	-1.0237D-01	-0.0000D-01	4.0950D-02
6	0.0000D-01	0.0000D-01	0.0000D-01
7	0.0000D-01	0.0000D-01	0.0000D-01
8	0.0000D-01	0.0000D-01	0.0000D-01

ELEMENT STRESSES

ELEM.	INT.	X STRESS	Y STRESS	XY STRESS	XZ STRESS	YZ STRESS
1	1	9.0000D 00	2.7000D 00	-6.4085D-13	3.5445D-13	4.5870D-14
1	2	9.0000D 00	2.7000D 00	2.9203D-13	2.4603D-13	8.3400D-15
1	3	9.0000D 00	2.7000D 00	-4.5682D-14	2.5020D-13	-8.3400D-15
1	4	9.0000D 00	2.7000D 00	-2.9609D-13	2.8773D-13	-2.9190D-14

SUPPORT REACTIONS

NODE	AR1	AR2	AR3
1	0.0000D-01	3.0000D-02	0.0000D-01
3	0.0000D-01	-3.0000D-02	0.0000D-01
4	0.0000D-01	1.2000D-01	0.0000D-01
5	0.0000D-01	-1.2000D-01	0.0000D-01
6	-4.6197D-14	3.0000D-02	-1.0000D-01
7	-2.5953D-13	5.1048D-14	-4.0000D-01
8	-4.9664D-14	-3.0000D-02	-1.0000D-01

the last two columns of the stress output are equal to the average values multiplied by the factor 1.5. Upon checking the displacements, stresses, and reactions in the output, we find that the program gives correct results.

Example 2

Next, we shall analyze a fixed, square plate with a concentrated load P applied in the z direction at its center. Figure 6.14 shows a quarter of the plate discretized into a 4×4 network of PBQ8 elements. Because of symmetry about the x and y axes,

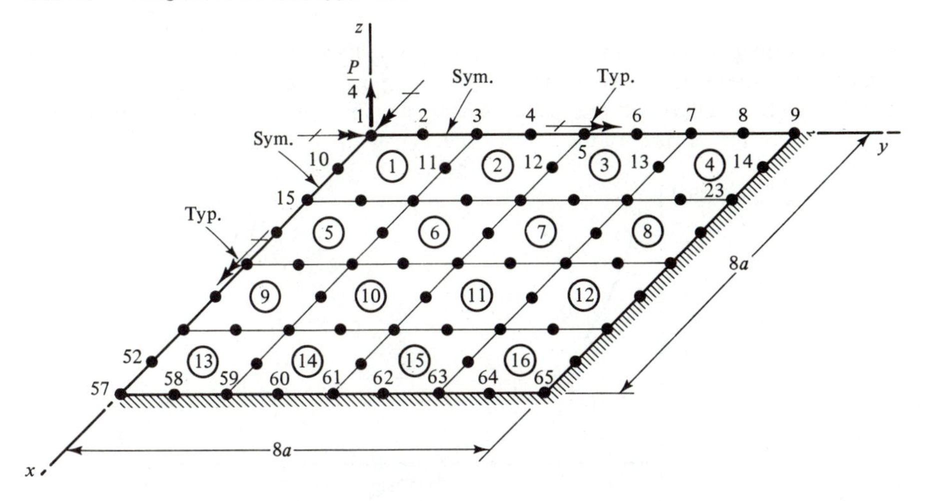

Figure 6.14 Example 2: Fixed, Square Plate with Central Load and PBQ8 Network

restraints against rotations in the x sense are shown at nodes 1, 10, 15, . . . , 52; and restraints against rotations in the y sense appear at nodes 1, 2, 3, . . . , 8. It is implied that nodes on the fixed boundary are restrained against displacements of all types. Physical parameters in this example are:

$$a = 1 \text{ in.} \qquad t = 0.25 \text{ in.} \qquad P = 1 \text{ k}$$

$$E = 3 \times 10^4 \text{ k/in.}^2 \qquad \nu = 0.3$$

for which the material is steel and the units are U.S.

Let us compute the deflection at the center of the plate for a series of PBQ8 networks. Table 6.5 summarizes the results from four runs of Program PBQ8, using networks of 1×1, 2×2, 3×3, and 4×4 elements. Errors indicated in the table are found by comparing the calculated values with the classical result (17) of $w_1 = 0.033397$ inch. Because this value of w_1 does not include shearing deformations, the numerical results have a positive error for the 4×4 network. The errors listed in the table correlate well with those given by Abel and Desai (18) for other elements.

TABLE 6.5 Center Deflection for Example 2

Network	Center Deflection (in.)	% Error
1 × 1	0.037457	+12.1
2 × 2	0.026520	−20.6
3 × 3	0.032972	− 1.27
4 × 4	0.033538	+ 0.422

Another item of interest is the variation of flexural stress in the plate. Figure 6.15 gives a graph of the stress σ_y (for a 4×4 mesh) at the upper surface of the plate along the row of integration points nearest to the y axis. This stress becomes very large at the center of the plate, where it theoretically approaches infinity. On the other hand, at the fixed boundary of the plate (where $y = 8a$), the stress is approximately -12 k/in.^{2}, which is close to the correct value (17).

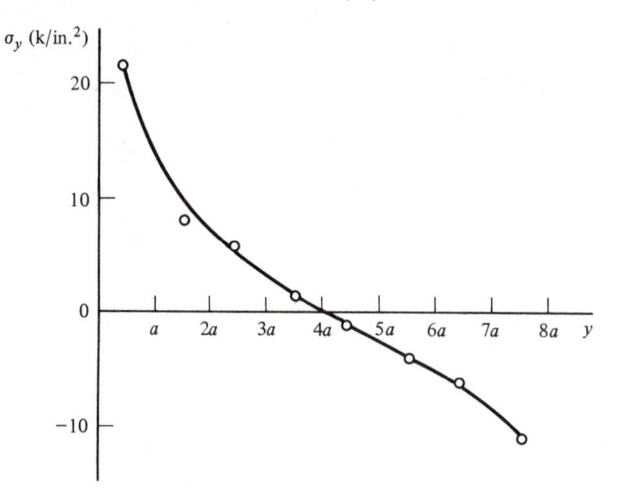

Figure 6.15 Stress σ_y Near y Axis for Example 2

Example 3

Suppose that a brass ship rudder is subjected to a uniformly distributed pressure b_z (force per unit area), as depicted in Fig. 6.16. To fit the curved boundaries, a network of PBQ8 elements is used. Node points 57 through 65 of the network are considered to be fixed, whereas all other nodes are unrestrained. The following physical parameters will be used:

$$L = 0.2 \text{ m} \qquad t = 0.025 \text{ m} \qquad b_z = 10 \text{ kN/m}^2$$
$$E = 1 \times 10^8 \text{ kN/m}^2 \qquad \nu = 0.3$$

where the units are S.I. Also, the coordinates of nodes on the curved boundaries have the values:

Point	1	2	8	9	10	14	56
x Coord.	$0.3L$	$0.08L$	$0.09L$	$0.28L$	$0.62L$	$0.60L$	$3.72L$
y Coord.	$0.3L$	$0.62L$	$3.48L$	$3.90L$	$0.08L$	$4.28L$	$6.10L$

When the data for this problem are run using Program PBQ8, the maximum deflection on the free edge of the rudder is found to be $w_3 = 3.0955 \times 10^{-3}$ m (at node 3). Figure 6.17 shows a graph of the flexural stress σ_x on the positive z surface along the row of integration points just below line 3-59. The maximum value of this stress approaches -30×10^3 kN/m^2 at the fixed edge of the rudder.

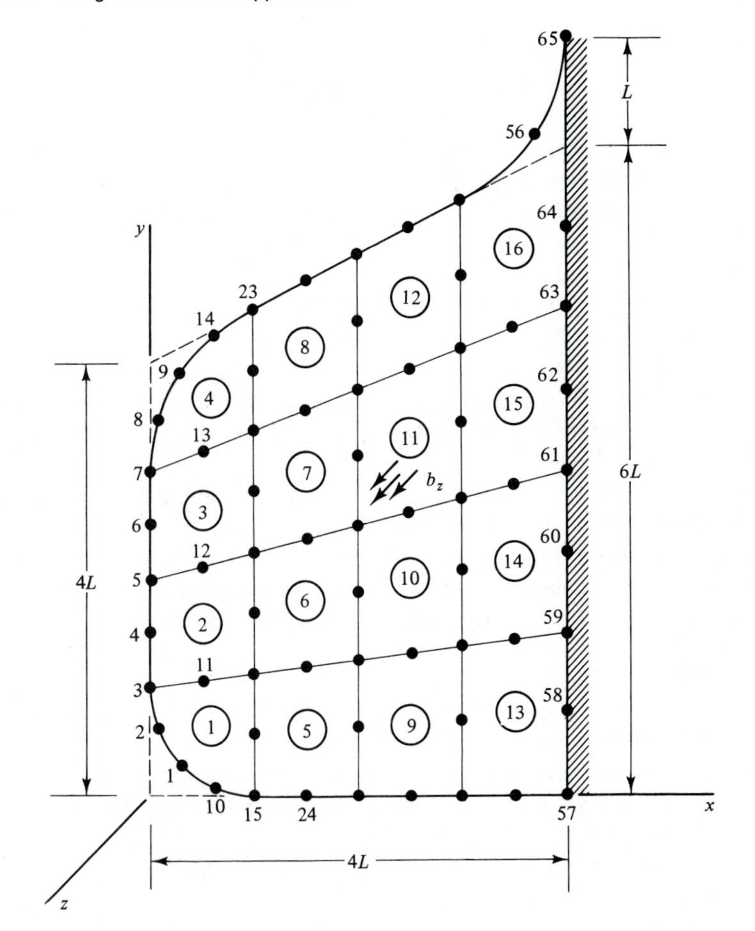

Figure 6.16 Example 3: Ship Rudder with Uniform Pressure and PBQ8 Network

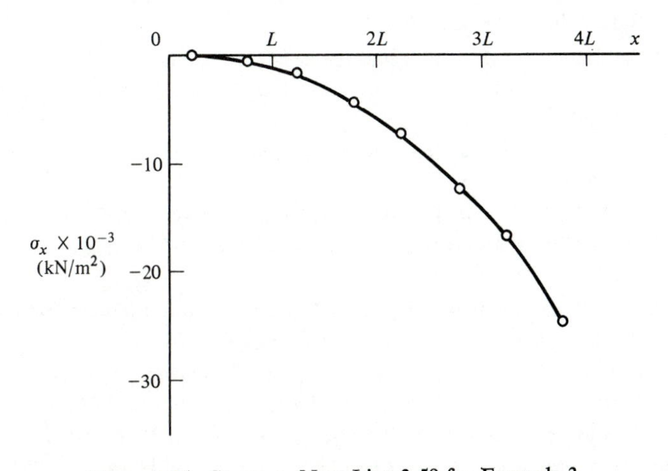

Figure 6.17 Stress σ_x Near Line 3-59 for Example 3

REFERENCES

1. Timoshenko, S. P., and Woinowsky-Krieger, S., *Theory of Plates and Shells*, 2d ed., McGraw-Hill, New York, 1959.
2. Melosh, R. J., "Basis of Derivation of Matrices for the Direct Stiffness Method," *AIAA Jour.*, Vol. 1, No. 7, July 1963, pp. 1631–1637.
3. Zienkiewicz, O. C., and Cheung, Y. K., "The Finite Element Method for Analysis of Elastic Isotropic and Orthotropic Slabs," *Proc. Inst. Civ. Eng.*, Vol. 28, 1964, pp. 471–488.
4. Bogner, F. K., Fox, R. L., and Schmit, L. A., Jr., "The Generation of Interelement-Compatible Stiffness and Mass Matrices by the Use of Interpolation Formulas," *Proc. Conf. Mat. Meth. Struc. Mech.*, AFIT, Wright-Patterson AF Base, Ohio, 1965, pp. 397–443.
5. Von Rieseman, W. A., "Large Deflections of Elastic Beams and Plates Using the Finite Element Method," Ph.D. Thesis, Dept. Civil Eng., Stanford Univ., 1968.
6. Bazeley, G. P., Cheung, Y. K., Irons, B. M., and Zienkiewicz, O. C., "Triangular Elements in Plate Bending: Conforming and Nonconforming Solutions," *Proc. Conf. Mat. Meth. Struc. Mech.*, AFIT, Wright-Patterson AF Base, Ohio, 1965, pp. 547–576.
7. Cheung, Y. K., King, I. P., and Zienkiewicz, O. C., "Slab Bridges With Arbitrary Shape and Support Condition—A General Method of Analysis Based on Finite Elements," *Proc. Inst. Civ. Eng.*, Vol. 40, 1968, pp. 9–36.
8. Clough, R. W., and Tocher, J. L., "Finite Element Stiffness Matrices for Analysis of Plates in Bending," *Proc. Conf. Mat. Meth. Struc. Mech.*, AFIT, Wright-Patterson AF Base, Ohio, 1965, pp. 515–545.
9. Zienkiewicz, O. C., *The Finite Element Method*, 3d ed., McGraw-Hill Ltd., London, 1977.
10. Clough, R. W., and Felippa, C. A., "A Refined Quadrilateral Element for Analysis of Plate Bending," *Proc. 2d Conf. Mat. Meth. Struc. Mech.*, AFIT, Wright-Patterson AF Base, Ohio, 1968, pp. 399–440.
11. Ahmad, S., Irons, B. M., and Zienkiewicz, O. C., "Analysis of Thick and Thin Shell Structures by Curved Finite Elements," *Int. Jour. Num. Meth. Eng.*, Vol. 3, No. 4, 1971, pp. 575–586.
12. Cook, R. D., *Concepts and Applications of Finite Element Analysis*, 2d ed., Wiley, New York, 1981.
13. Hughes, T. J. R., Taylor, R. L., and Kanoknukulchai, W., "A Simple and Efficient Finite Element for Plate Bending," *Int. Jour. Num. Meth. Eng.*, Vol. 11, 1977, pp. 1529–1543.
14. Hughes, T. J. R., and Cohen, M., "The 'Heterosis' Finite Element for Plate Bending," *Jour. Comp. and Strucs.*, Vol. 9, 1978, pp. 445–450.
15. Kirkhope, J., and Wilson, G. J., "Vibration of Circular and Annular Plates Using Finite Elements," *Int. Jour. Num. Meth. Eng.*, Vol. 4, No. 2, 1972, pp. 181–193.
16. Sawko, F., and Merriman, P. A., "An Annular Segment Finite Element for Plate Bending," *Int. Jour. Num. Meth. Eng.*, Vol. 3, No. 1, 1971, pp. 119–129.

17. Szilard, R., *Theory and Analysis of Plates*, Prentice-Hall, Englewood Cliffs, N.J., 1974.

18. Abel, J. F., and Desai, C. S., "Comparison of Finite Elements for Plate Bending," *Jour. Struc. Div.*, ASCE, Vol. 98, No. ST9, September 1972, pp. 2143–2148.

PROBLEMS

6.2-1. Derive a general expression for the restraint stiffness matrix $\mathbf{K}_r$ of a flexural plate element on an elastic foundation, as indicated in the figure. For this purpose, assume that the displacement shape functions in $\mathbf{f}$ are available. Let the foundation restraint action be r_z, which is defined as a force in the z direction (per unit area of the plate) due to a unit amount of displacement w.

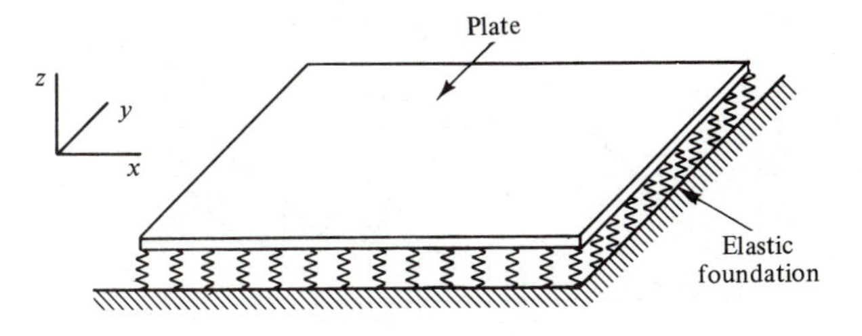

Problem 6.2-1

6.2-2. For the MZC rectangular element shown in the figure, derive the equivalent nodal loads at point 1 due to a uniformly distributed load b_z (force per unit area) in the positive z direction.

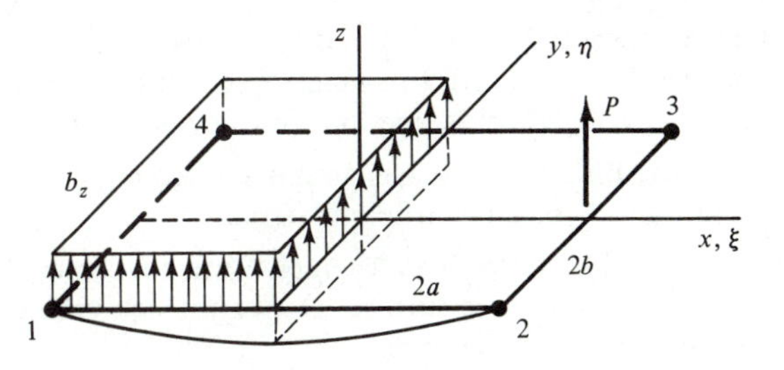

Problems 6.2-2 through 6.2-8

6.2-3. Find the equivalent nodal loads at point 2 for the MZC rectangle in the figure, due to a concentrated force P applied in the positive z direction at the midpoint of edge 2-3.

6.2-4. Assume that an MZC rectangular element is subjected to a uniform loading b_z (force per unit area) in the positive z direction, distributed over half the area (see figure). Determine the equivalent nodal loads at point 4 due to this condition.

6.2-5. Derive the stiffness term K_{12} for the MZC rectangle, assuming isotropic material.

6.2-6. Derive the stiffness term K_{13} for the MZC rectangle, assuming isotropic material.

6.2-7. Given that the MZC rectangle has an initial curvature ϕ_{xx0} (see figure), find the equivalent nodal loads at point 1 for an isotropic material.

6.2-8. Let the MZC rectangle be subjected to a linear temperature gradient through its thickness. If the temperature change at the upper surface is ΔT and that at the lower surface is $-\Delta T$, derive the equivalent nodal loads at point 1, assuming isotropic material.

6.2-9. For the BFS rectangular element shown in the figure, determine the equivalent nodal loads at point 3 due to a uniformly distributed load b_z (force per unit area) applied in the positive z direction.

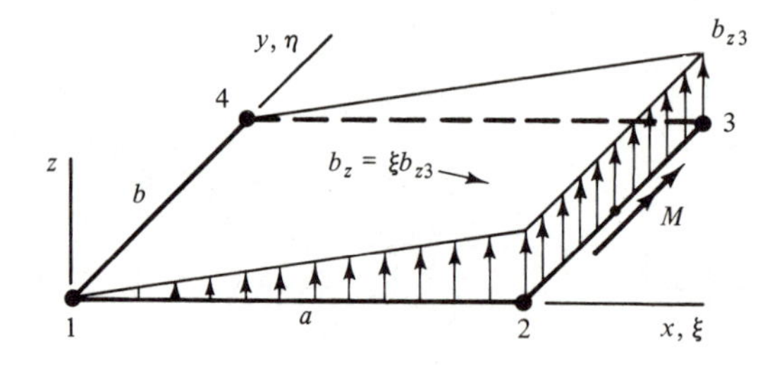

Problems 6.2-9 through 6.2-13

6.2-10. The figure shows a BFS rectangle subjected to a loading b_z (force per unit area) that varies linearly in the ξ direction. Find the equivalent nodal loads at point 1 due to this condition.

6.2-11. Assume that a concentrated moment M is applied in the y direction at the midpoint of edge 2-3 on the BFS rectangle (see figure). Derive the equivalent nodal loads at point 2. (*Hint:* Shape functions for $\partial w/\partial x$ are required.)

6.2-12. Assuming that the BFS rectangle has an initial warp ϕ_{xy0}, find the equivalent nodal loads at point 4 for an isotropic material.

6.2-13. Suppose that a BFS rectangle has a temperature change of ΔT at its upper surface and $-\Delta T$ at its lower surface. Derive the equivalent nodal loads at point 3, assuming isotropic material.

6.4-1. The figure shows the rectangular parent of element PBQ8 (with constraints). Derive the stiffness term K_{12} for this element, assuming that the material is isotropic.

6.4-2. For the rectangular parent of element PBQ8 (see figure), derive the stiffness term K_{13}, assuming isotropic material.

6.4-3. A concentrated force P is applied in the positive z direction at point $(\xi, \eta) = (-\frac{1}{2}, -\frac{1}{3})$ on the rectangular parent of element PBQ8 (see figure). Find the

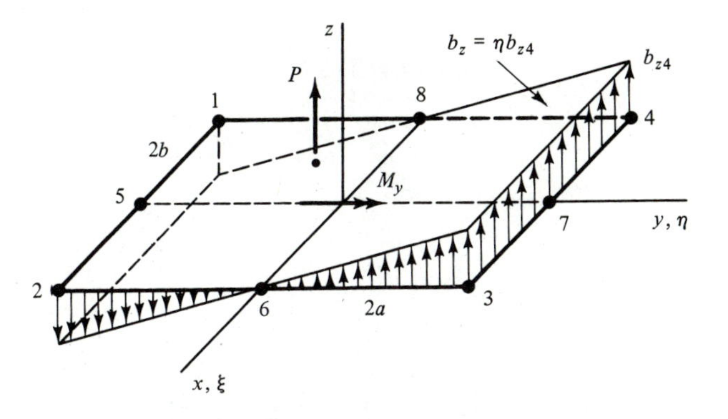

Problems 6.4-1 through 6.4-5

equivalent nodal forces in the z direction at points 1, 2, 3, and 4 due to this load.

6.4-4. Repeat Prob. 6.4.3 for points 5, 6, 7, and 8.

6.4-5. Let the rectangular parent of element PBQ8 be subjected to a loading b_z (force per unit area) that varies linearly with respect to η (see figure). Determine the equivalent nodal forces in the z direction at points 3 and 7 due to this influence.

6.4-6. A concentrated moment M_y is applied in the positive y direction at the center of the rectangular parent of element PBQ8 (see figure). Find all the equivalent nodal loads due to this influence. (*Hint:* The generic rotations of the normal to the neutral surface are:

$$\theta_x = \sum_{i=1}^{8} f_i \theta_{xi} \quad \text{and} \quad \theta_y = \sum_{i=1}^{8} f_i \theta_{yi}.)$$

CHAPTER

7

General Shells

7.1 INTRODUCTION

Finite elements for the analysis of shells with general geometry take a variety of forms, depending upon the theory used. Some elements are based upon the thin-shell theories of classical mechanics, consisting of the analysis of deep or shallow shells. A commonly used theory of *deep shells* is based upon the strain-displacement relationships of Novozhilov (1, 2). On the other hand, a specialized theory of *shallow shells* follows the simplified strain-displacement relationships of Vlasov (3). The latter method is more approximate than the former, but accurate results have been obtained, even when shallow-shell concepts were applied to deep shells (4).

Instead of using the classical theories, we can analyze general shells with finite elements based on previous formulations. The simplest approach consists of using flat facets (5, 6), such as the triangle shown in Figs. 7.1(a) and (b). The first figure depicts the *membrane components* of generic and nodal displacements, and the second figure indicates the *flexural components*. These two sets of components are uncoupled within a particular element, but interactions of membrane and flexural components occur between adjacent elements that do not lie in the same plane. Other aspects of flat-facet shell elements are discussed in the next section.

Curved elements for shell analysis can be devised by specializing solid elements to be thin in one direction, while introducing constraint conditions on nodal displacements (7, 8). As examples, the hexahedron and the pentahedron in Figs. 7.2(a) and (b) can be specialized to become quadrilateral and triangular

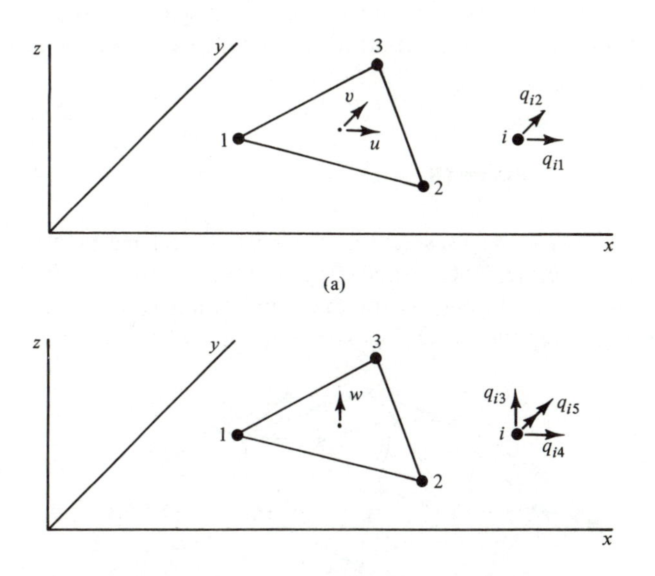

Figure 7.1 Flat-Facet Element: (a) Membrane Components (b) Flexural Components

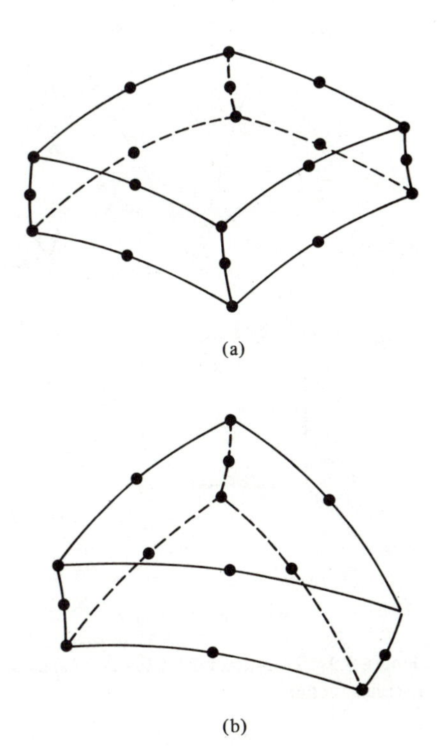

Figure 7.2 Specialization of Solids: (a) Hexahedron (b) Pentahedron

shell elements that are curved in three-dimensional space. Section 7.3 covers the particular task of specializing a hexahedral solid into a curved quadrilateral shell element.

7.2 FLAT-FACET ELEMENTS

Figure 7.3(a) shows a general shell that has been discretized by using a network of triangular finite elements. Appearing in Fig. 7.3(b) is a combination of generic and nodal displacements for the membrane (plane-stress) and flexural (plate-bending) components. One possible mixture of formulations consists of

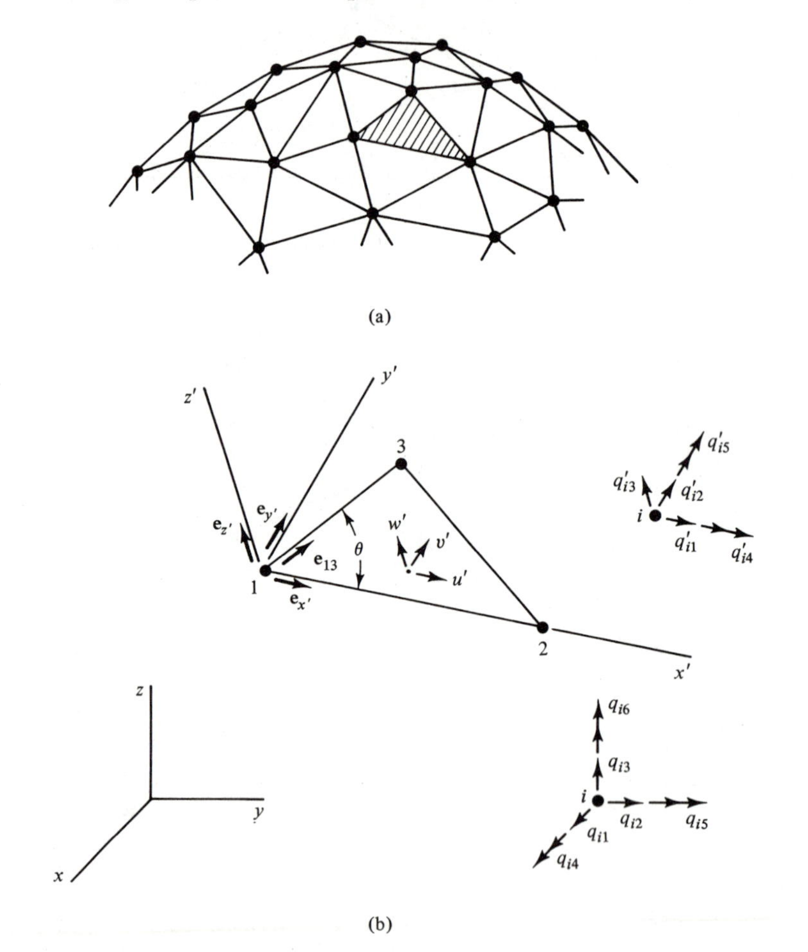

Figure 7.3 Triangular Flat Facets: (a) Discretized Shell (b) Local and Global Displacements

using the CST element for plane stress (see Sec. 2.2) and the CKZ triangle for plate bending (see Sec. 6.3). In such a combination there would be five nodal displacements in the directions of local axes [see the primed displacements in Fig. 7.3(b)].

Before assembling the triangular elements shown in Fig. 7.3(a), it is necessary to convert actions and displacements from local directions to global directions, as indicated in Fig. 7.3(b). For this purpose we can use the following rotation-of-axes transformation of nodal displacements:

$$\mathbf{q}'_i = \hat{\mathbf{R}}_i \mathbf{q}_i \qquad (i = 1, 2, 3) \tag{7.2-1}$$

In this expression the vector of local displacements $\mathbf{q}'_i$ has five terms, which are:

$$\mathbf{q}'_i = \{q'_{i1}, q'_{i2}, \ldots, q'_{i5}\} \tag{a}$$

However, the vector of global displacements $\mathbf{q}_i$ has the six terms [see Fig. 7.3(b)]:

$$\mathbf{q}_i = \{q_{i1}, q_{i2}, \ldots, q_{i6}\} \tag{b}$$

Thus, we see that the rotation transformation matrix $\hat{\mathbf{R}}_i$ is of size 5×6, as follows:

$$\hat{\mathbf{R}}_i = \begin{bmatrix} \lambda_{11} & \lambda_{12} & \lambda_{13} & 0 & 0 & 0 \\ \lambda_{21} & \lambda_{22} & \lambda_{23} & 0 & 0 & 0 \\ \lambda_{31} & \lambda_{32} & \lambda_{33} & 0 & 0 & 0 \\ 0 & 0 & 0 & \lambda_{11} & \lambda_{12} & \lambda_{13} \\ 0 & 0 & 0 & \lambda_{21} & \lambda_{22} & \lambda_{23} \end{bmatrix} \tag{c}$$

This matrix contains rows of direction cosines of the local (primed) axes with respect to the global (unprimed) axes, as explained in Sec. 1.7.

Knowing the coordinates of nodes 1 and 2 in Fig. 7.3(b), we can find the direction cosines (or components) of a unit vector $\mathbf{e}_{x'}$ in the direction of the x' axis. Thus,

$$\lambda_{11} = \frac{x_{12}}{L_{12}} \qquad \lambda_{12} = \frac{y_{12}}{L_{12}} \qquad \lambda_{13} = \frac{z_{12}}{L_{12}} \tag{d}$$

Similarly, the coordinates of points 1 and 3 yield the components of a unit vector $\mathbf{e}_{13}$ in the direction of side 1-3 of the triangle.

$$c_{x13} = \frac{x_{13}}{L_{13}} \qquad c_{y13} = \frac{y_{13}}{L_{13}} \qquad c_{z13} = \frac{z_{13}}{L_{13}} \tag{e}$$

Then the components of a unit vector $\mathbf{e}_{z'}$ in the direction of axis z' [see Fig. 7.3(b)] may be found as the result of the normalized vector product:

$$\mathbf{e}_{z'} = \frac{\mathbf{e}_{x'} \times \mathbf{e}_{13}}{\sin\theta} \tag{f}$$

which gives λ_{31}, λ_{32}, and λ_{33}. Finally, the components of a unit vector $\mathbf{e}_{y'}$ in the direction of axis y' are obtained from:

$$\mathbf{e}_{y'} = \mathbf{e}_{z'} \times \mathbf{e}_{x'} \tag{g}$$

which yields λ_{21}, λ_{22}, and λ_{23}.

The transpose of the rotation-of-axes tranformation matrix $\hat{\mathbf{R}}_i$ in Eq. (c) serves to convert nodal actions from local to global directions. Thus,

$$\mathbf{p}_i = \hat{\mathbf{R}}_i^{\mathrm{T}} \mathbf{p}'_i \qquad (i = 1, 2, 3) \tag{7.2-2}$$

In addition, submatrix $\mathbf{K}'_{ij}$ of the element stiffness matrix may be transformed as follows:

$$\mathbf{K}_{ij} = \hat{\mathbf{R}}_i^{\mathrm{T}} \mathbf{K}'_{ij} \hat{\mathbf{R}}_j \qquad (i = 1, 2, 3; j = 1, 2, 3) \tag{7.2-3}$$

which generates a 6×6 array from a 5×5 matrix.

While flat-facet elements are easy to formulate and use, their accuracy is not very good because of modeling defects. Slope discontinuities normal to adjacent edges result in poor values for internal stresses. Therefore, the curved type of element described in the next section is preferred.

7.3 SPECIALIZATION OF HEXAHEDRON

In this section we specialize the isoparametric hexahedron H20 to become a thin, curved quadrilateral element for the analysis of general shells. Development of the shell element SHQ8 is similar to the technique used in obtaining element PBQ8 (see Sec. 6.4) for plate bending. However, the constraint conditions are modified because two additional translations, u_i and v_i, occur at each node of the shell element. Thus, the constraint matrix $\mathbf{G}_{ai}$ for a corner node of the rectangular parent element has two more columns than before, as follows:

$$\mathbf{G}_{ai} = \begin{bmatrix} 1 & 0 & 0 & 0 & 0 \\ 0 & 1 & 0 & 0 & 0 \\ 0 & 0 & 1 & 0 & 0 \\ 1 & 0 & 0 & 0 & \frac{t_i}{2} \\ 0 & 1 & 0 & -\frac{t_i}{2} & 0 \\ 0 & 0 & 1 & 0 & 0 \\ 1 & 0 & 0 & 0 & -\frac{t_i}{2} \\ 0 & 1 & 0 & \frac{t_i}{2} & 0 \\ 0 & 0 & 1 & 0 & 0 \end{bmatrix} \tag{a}$$

When we compare this 9×5 matrix with Eq. (b) in Sec. 6.4, we see that columns 1 and 2 have been added. Similarly, the constraint matrix $\mathbf{G}_{bi}$ for a midedge node of the rectangular parent becomes:

$$\mathbf{G}_{bi} = \begin{bmatrix} 1 & 0 & 0 & 0 & \frac{t_i}{2} \\ 0 & 1 & 0 & -\frac{t_i}{2} & 0 \\ 0 & 0 & 1 & 0 & 0 \\ 1 & 0 & 0 & 0 & -\frac{t_i}{2} \\ 0 & 1 & 0 & \frac{t_i}{2} & 0 \\ 0 & 0 & 1 & 0 & 0 \end{bmatrix} \qquad \text{(b)}$$

which is a 6 × 5 array that can be compared with Eq. (c) in Sec. 6.4. With five displacements at each of eight nodes, element SHQ8 has (8)(5) = 40 nodal displacements.

As with the plate element PBQ8, the general shell element SHQ8 will be formulated directly (7, 8). Figure 7.4(a) shows the geometric layout of element SHQ8, in which the coordinates of any point are:

$$\begin{bmatrix} x \\ y \\ z \end{bmatrix} = \sum_{i=1}^{8} f_i \begin{bmatrix} x_i \\ y_i \\ z_i \end{bmatrix} + \sum_{i=1}^{8} f_i \zeta \frac{t_i}{2} \begin{bmatrix} \ell_{3i} \\ m_{3i} \\ n_{3i} \end{bmatrix} \qquad \text{(7.3-1)}$$

The interpolation functions f_i appearing in this equation are given by Eqs. (3.4-12). In addition, the terms ℓ_{3i}, m_{3i}, and n_{3i} are the direction cosines of a vector $\mathbf{V}_{3i}$ that is normal to the middle surface and spans the thickness t_i of the shell at node i. Figure 7.4(b) shows this vector, which is obtained as:

$$\mathbf{V}_{3i} = \begin{bmatrix} x_j - x_k \\ y_j - y_k \\ z_j - z_k \end{bmatrix} = \begin{bmatrix} \ell_{3i} \\ m_{3i} \\ n_{3i} \end{bmatrix} t_i \qquad \text{(7.3-2)}$$

Points j and k in the figure are at the surfaces of the shell. In a computer program either the coordinates of points j and k or the direction cosines for $\mathbf{V}_{3i}$ must be given as data.

Generic displacements at any point in the shell element are taken to be in the directions of global axes. Thus,

$$\mathbf{u} = \{u, v, w\} \qquad \text{(c)}$$

On the other hand, nodal displacements consist of these same translations (in global directions) as well as two small rotations α_i and β_i about two local

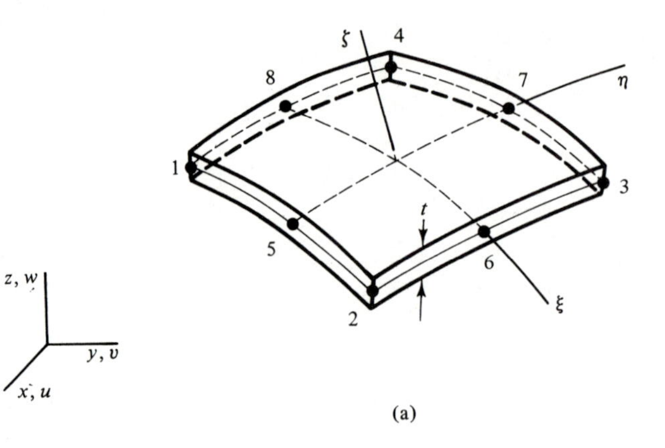

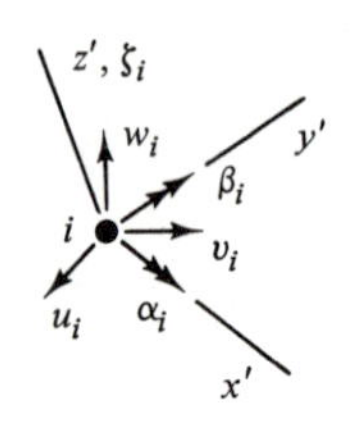

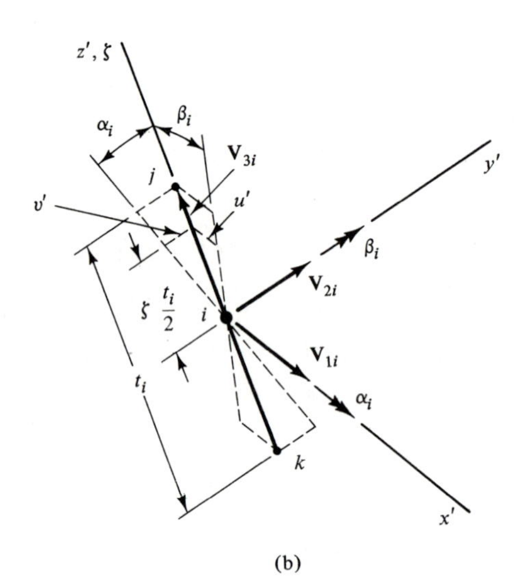

Figure 7.4 (a) Element SHQ8 (b) Nodal Vectors

tangential axes x' and y', as indicated in Fig. 7.4(a). Hence,

$$\mathbf{q}_i = \{u_i, v_i, w_i, \alpha_i, \beta_i\} \qquad (i = 1, 2, \ldots, 8) \tag{d}$$

Generic displacements in terms of nodal displacements are:

$$\begin{bmatrix} u \\ v \\ w \end{bmatrix} = \sum_{i=1}^{8} f_i \begin{bmatrix} u_i \\ v_i \\ w_i \end{bmatrix} + \sum_{i=1}^{8} f_i \zeta \frac{t_i}{2} \boldsymbol{\mu}_i \begin{bmatrix} \alpha_i \\ \beta_i \end{bmatrix} \tag{7.3-3}$$

In this formula the symbol $\boldsymbol{\mu}_i$ denotes the following matrix:

$$\boldsymbol{\mu}_i = \begin{bmatrix} -\ell_{2i} & \ell_{1i} \\ -m_{2i} & m_{1i} \\ -n_{2i} & n_{1i} \end{bmatrix} \tag{7.3-4}$$

Column 1 in this array contains negative values of the direction cosines of the second tangential vector $\mathbf{V}_{2i}$, and column 2 has the direction cosines for the first tangential vector $\mathbf{V}_{1i}$ [see Fig. 7.4(b)]. These vectors are orthogonal to the vector $\mathbf{V}_{3i}$ and to each other, but the choice for the direction of one of them is arbitrary. To settle the choice, we let:

$$\mathbf{V}_{1i} = \mathbf{e}_y \times \mathbf{V}_{3i} \tag{e}$$

Then

$$\mathbf{V}_{2i} = \mathbf{V}_{3i} \times \mathbf{V}_{1i} \tag{f}$$

[If $\mathbf{V}_{3i}$ is parallel to $\mathbf{e}_y$ in Eq. (e), the latter is replaced by $\mathbf{e}_z$.] Figure 7.4(b) shows local generic translations u' and v' (in the directions of $\mathbf{V}_{1i}$ and $\mathbf{V}_{2i}$) due to the nodal rotations β_i and α_i, respectively. Their values are:

$$u' = \zeta \frac{t_i}{2} \beta_i \qquad v' = -\zeta \frac{t_i}{2} \alpha_i \tag{g}$$

Contributions of these terms to the generic displacements at any point are given by the second summation in Eq. (7.3-3).

The 3×3 Jacobian matrix required for this element is:

$$\mathbf{J} = \begin{bmatrix} x_{,\xi} & y_{,\xi} & z_{,\xi} \\ x_{,\eta} & y_{,\eta} & z_{,\eta} \\ x_{,\zeta} & y_{,\zeta} & z_{,\zeta} \end{bmatrix} \tag{7.3-5a}$$

We find the derivatives in matrix $\mathbf{J}$ from Eq. (7.3-1), as follows:

$$x_{,\xi} = \sum_{i=1}^{8} f_{i,\xi} x_i + \sum_{i=1}^{8} f_{i,\xi} \zeta \frac{t_i}{2} \ell_{3i}$$

$$x_{,\eta} = \sum_{i=1}^{8} f_{i,\eta} x_i + \sum_{i=1}^{8} f_{i,\eta} \zeta \frac{t_i}{2} \ell_{3i}$$

$$x_{,\zeta} = \sum_{i=1}^{8} f_i \frac{t_i}{2} \ell_{3i} \qquad \text{and so on}$$

The inverse of $\mathbf{J}$ becomes:

$$\mathbf{J}^{-1} = \mathbf{J}^* = \begin{bmatrix} \xi_{,x} & \eta_{,x} & \zeta_{,x} \\ \xi_{,y} & \eta_{,y} & \zeta_{,y} \\ \xi_{,z} & \eta_{,z} & \zeta_{,z} \end{bmatrix} \tag{7.3-5b}$$

We need certain derivatives of the generic displacements [see Eq. (7.3-3)] with respect to local coordinates. These derivatives are listed in a column vector of nine terms, as follows:

$$\begin{bmatrix} u_{,\xi} \\ u_{,\eta} \\ u_{,\zeta} \\ v_{,\xi} \\ v_{,\eta} \\ v_{,\zeta} \\ w_{,\xi} \\ w_{,\eta} \\ w_{,\zeta} \end{bmatrix} = \sum_{i=1}^{8} \begin{bmatrix} f_{i,\xi} & 0 & 0 & -\zeta f_{i,\xi}\ell_{2i} & \zeta f_{i,\xi}\ell_{1i} \\ f_{i,\eta} & 0 & 0 & -\zeta f_{i,\eta}\ell_{2i} & \zeta f_{i,\eta}\ell_{1i} \\ 0 & 0 & 0 & -f_i\ell_{2i} & f_i\ell_{1i} \\ 0 & f_{i,\xi} & 0 & -\zeta f_{i,\xi}m_{2i} & \zeta f_{i,\xi}m_{1i} \\ 0 & f_{i,\eta} & 0 & -\zeta f_{i,\eta}m_{2i} & \zeta f_{i,\eta}m_{1i} \\ 0 & 0 & 0 & -f_i m_{2i} & f_i m_{1i} \\ 0 & 0 & f_{i,\xi} & -\zeta f_{i,\xi}n_{2i} & \zeta f_{i,\xi}n_{1i} \\ 0 & 0 & f_{i,\eta} & -\zeta f_{i,\eta}n_{2i} & \zeta f_{i,\eta}n_{1i} \\ 0 & 0 & 0 & -f_i n_{2i} & f_i n_{1i} \end{bmatrix} \begin{bmatrix} u_i \\ v_i \\ w_i \\ \frac{t_i}{2}\alpha_i \\ \frac{t_i}{2}\beta_i \end{bmatrix} \tag{7.3-6}$$

Transformation of these derivatives to global coordinates requires that the inverse of the Jacobian matrix be applied. Therefore,

$$\begin{bmatrix} u_{,x} \\ u_{,y} \\ \cdots \\ w_{,z} \end{bmatrix} = \begin{bmatrix} \mathbf{J}^* & \mathbf{O} & \mathbf{O} \\ \mathbf{O} & \mathbf{J}^* & \mathbf{O} \\ \mathbf{O} & \mathbf{O} & \mathbf{J}^* \end{bmatrix} \begin{bmatrix} u_{,\xi} \\ u_{,\eta} \\ \cdots \\ w_{,\zeta} \end{bmatrix} \tag{7.3-7a}$$

Multiplying the terms in this equation, we obtain:

$$\begin{bmatrix} u_{,x} \\ u_{,y} \\ u_{,z} \\ v_{,x} \\ v_{,y} \\ v_{,z} \\ w_{,x} \\ w_{,y} \\ w_{,z} \end{bmatrix} = \sum_{i=1}^{8} \begin{bmatrix} a_i & 0 & 0 & -d_i\ell_{2i} & d_i\ell_{1i} \\ b_i & 0 & 0 & -e_i\ell_{2i} & e_i\ell_{1i} \\ c_i & 0 & 0 & -g_i\ell_{2i} & g_i\ell_{1i} \\ 0 & a_i & 0 & -d_i m_{2i} & d_i m_{1i} \\ 0 & b_i & 0 & -e_i m_{2i} & e_i m_{1i} \\ 0 & c_i & 0 & -g_i m_{2i} & g_i m_{1i} \\ 0 & 0 & a_i & -d_i n_{2i} & d_i n_{1i} \\ 0 & 0 & b_i & -e_i n_{2i} & e_i n_{1i} \\ 0 & 0 & c_i & -g_i n_{2i} & g_i n_{1i} \end{bmatrix} \begin{bmatrix} u_i \\ v_i \\ w_i \\ \alpha_i \\ \beta_i \end{bmatrix} \tag{7.3-7b}$$

in which

$$\begin{aligned} a_i &= J^*_{11} f_{i,\xi} + J^*_{12} f_{i,\eta} & d_i &= \frac{t_i}{2}(a_i\zeta + J^*_{13} f_i) \\ b_i &= J^*_{21} f_{i,\xi} + J^*_{22} f_{i,\eta} & e_i &= \frac{t_i}{2}(b_i\zeta + J^*_{23} f_i) \\ c_i &= J^*_{31} f_{i,\xi} + J^*_{32} f_{i,\eta} & g_i &= \frac{t_i}{2}(c_i\zeta + J^*_{33} f_i) \end{aligned} \tag{h}$$

For element SHQ8 we consider six types of nonzero strains, as follows:

$$\boldsymbol{\epsilon} = \begin{bmatrix} \epsilon_x \\ \epsilon_y \\ \epsilon_z \\ \gamma_{xy} \\ \gamma_{yz} \\ \gamma_{zx} \end{bmatrix} = \begin{bmatrix} u_{,x} \\ v_{,y} \\ w_{,z} \\ u_{,y} + v_{,x} \\ v_{,z} + w_{,y} \\ w_{,x} + u_{,z} \end{bmatrix} \tag{i}$$

Noting the second version of this strain vector, we may construct the ith part of matrix $\mathbf{B}$ from terms in Eq. (7.3-7b) as:

$$\mathbf{B}_i = \begin{bmatrix} a_i & 0 & 0 & -d_i\ell_{2i} & d_i\ell_{1i} \\ 0 & b_i & 0 & -e_im_{2i} & e_im_{1i} \\ 0 & 0 & c_i & -g_in_{2i} & g_in_{1i} \\ b_i & a_i & 0 & -e_i\ell_{2i} - d_im_{2i} & e_i\ell_{1i} + d_im_{1i} \\ 0 & c_i & b_i & -g_im_{2i} - e_in_{2i} & g_im_{1i} + e_in_{1i} \\ c_i & 0 & a_i & -d_in_{2i} - g_i\ell_{2i} & d_in_{1i} + g_i\ell_{1i} \end{bmatrix} \tag{7.3-8}$$

$$(i = 1, 2, \ldots, 8)$$

Stress-strain relationships in local directions for either orthotropic or isotropic materials take the form:

$$\boldsymbol{\sigma}' = \mathbf{E}'\boldsymbol{\epsilon}' \tag{j}$$

Or,

$$\begin{bmatrix} \sigma_{x'} \\ \sigma_{y'} \\ \sigma_{z'} \\ \tau_{x'y'} \\ \tau_{y'z'} \\ \tau_{z'x'} \end{bmatrix} = \begin{bmatrix} E_{x'x'} & E_{x'y'} & 0 & 0 & 0 & 0 \\ E_{y'x'} & E_{y'y'} & 0 & 0 & 0 & 0 \\ 0 & 0 & 0 & 0 & 0 & 0 \\ 0 & 0 & 0 & G_{x'y'} & 0 & 0 \\ 0 & 0 & 0 & 0 & \dfrac{G_{y'z'}}{1.2} & 0 \\ 0 & 0 & 0 & 0 & 0 & \dfrac{G_{z'x'}}{1.2} \end{bmatrix} \begin{bmatrix} \epsilon_{x'} \\ \epsilon_{y'} \\ \epsilon_{z'} \\ \gamma_{x'y'} \\ \gamma_{y'z'} \\ \gamma_{z'x'} \end{bmatrix} \tag{7.3-9}$$

The local (primed) axes appear in Figs. 7.4(a) and (b). We can transform matrix $\mathbf{E}'$ from local to global directions using the 6×6 operator $\mathbf{T}_\epsilon$ from Eq. (4.1-15a). Thus,

$$\mathbf{E} = \mathbf{T}_\epsilon^{\mathrm{T}}\mathbf{E}'\mathbf{T}_\epsilon \tag{7.3-10}$$

To evaluate matrix $\mathbf{T}_\epsilon$ at an integration point, we must find the direction cosines for vectors $\mathbf{V}_1$, $\mathbf{V}_2$, and $\mathbf{V}_3$ at the point. This may be done with the following sequence of calculations:

$$\mathbf{e}_1 = (\mathbf{J}_1)_{\text{norm.}} \qquad \mathbf{e}_3 = (\mathbf{J}_1 \times \mathbf{J}_2)_{\text{norm.}} \qquad \mathbf{e}_2 = \mathbf{e}_3 \times \mathbf{e}_1 \tag{k}$$

In these expressions the vector $(\mathbf{J}_1)_{\text{norm.}}$ denotes the first row of the Jacobian matrix normalized to unit length, and so on.

Equation (7.3-10) would be more efficient if the third row and column of matrix $\mathbf{E}'$ (corresponding to $\sigma_{z'}$ and $\epsilon_{z'}$) were deleted, along with the third row of matrix $\mathbf{T}_\epsilon$. Also, as an alternative to that equation, we could use:

$$\mathbf{B}' = \mathbf{T}_\epsilon \mathbf{B} \tag{ℓ}$$

but this approach usually requires more arithmetic operations. Nevertheless, forming matrix $\mathbf{B}'$ in a computer program can make the calculation of stresses more convenient [see Eq. (7.3-14)].

Now we are ready to formulate the element stiffness matrix, as follows:

$$\mathbf{K} = \int_{-1}^{1} \int_{-1}^{1} \int_{-1}^{1} \mathbf{B}^{\mathrm{T}}\mathbf{E}\ \mathbf{B}\,|\mathbf{J}|\,d\xi\,d\eta\,d\zeta \tag{m}$$

In this expression matrices $\mathbf{B}$ and $\mathbf{J}$ are functions of ξ, η, and ζ. Integration of Eq. (m) through the thickness* of the element gives:

$$\begin{aligned}\mathbf{K} &= \int_{-1}^{1} \int_{-1}^{1} \int_{-1}^{1} (\mathbf{B}_a + \zeta\mathbf{B}_b)^{\mathrm{T}}\mathbf{E}(\mathbf{B}_a + \zeta\mathbf{B}_b)\,|\mathbf{J}|\,d\xi\,d\eta\,d\zeta \\ &= \int_{-1}^{1} \int_{-1}^{1} (2\mathbf{B}_a^{\mathrm{T}}\mathbf{E}\ \mathbf{B}_a + \tfrac{2}{3}\mathbf{B}_b^{\mathrm{T}}\mathbf{E}\ \mathbf{B}_b)\,|\mathbf{J}|\,d\xi\,d\eta\end{aligned} \tag{7.3-11}$$

Here the matrices $\mathbf{B}_a$ and $\mathbf{B}_b$ are both of size 6×40, but the latter array contains only terms that are to be multiplied by ζ. The integrals in Eq. (7.3-11) must be evaluated numerically, using two integration points in each of the ξ and η directions (8). In this process the factor 2 multiplies $t_i/2$, which is a common term in the third row of $|\mathbf{J}|$.

Equivalent nodal loads due to body forces on element SHQ8 may be found as:

$$\mathbf{p}_b = \int_V \mathbf{f}^{\mathrm{T}}\mathbf{b}\,dV = 2\int_{-1}^{1} \int_{-1}^{1} \mathbf{f}^{\mathrm{T}}\mathbf{b}\,|\mathbf{J}|\,d\xi\,d\eta \tag{7.3-12}$$

In this expression the load vector $\mathbf{b}$ is assumed to contain components of force (per unit volume) that are uniform through the thickness of the shell. Thus,

$$\mathbf{b} = \{b_x, b_y, b_z\} \tag{n}$$

Also, the shape functions in Eq. (7.3-12) are drawn from the first part of Eq. (7.3-3), as follows:

$$\mathbf{f}_i = \begin{bmatrix} 1 & 0 & 0 & 0 & 0 \\ 0 & 1 & 0 & 0 & 0 \\ 0 & 0 & 1 & 0 & 0 \end{bmatrix} f_i \qquad (i = 1, 2, \ldots, 8) \tag{o}$$

Note that such body forces do not cause any equivalent nodal moments. In

*Terms in matrix $\mathbf{J}$ containing ζ are neglected for the purpose of simplifying integration through the thickness.

addition, the equivalent nodal loads caused by initial strains are:

$$\mathbf{p}_0 = \int_V \mathbf{B}^{\mathrm{T}}\mathbf{E}\ \boldsymbol{\epsilon}_0\, dV = \int_V \mathbf{B}^{\mathrm{T}}\mathbf{T}_\epsilon^{\mathrm{T}}\mathbf{E}'\boldsymbol{\epsilon}_0'\, dV$$
$$= \int_{-1}^{1}\int_{-1}^{1}\int_{-1}^{1} (\mathbf{B}_a + \zeta\mathbf{B}_b)^{\mathrm{T}}\mathbf{T}_\epsilon^{\mathrm{T}}\mathbf{E}'\boldsymbol{\epsilon}_0'\,|\mathbf{J}|\, d\xi\, d\eta\, d\zeta$$
$$= 2\int_{-1}^{1}\int_{-1}^{1} \mathbf{B}_a^{\mathrm{T}}\mathbf{T}_\epsilon^{\mathrm{T}}\mathbf{E}'\boldsymbol{\epsilon}_0'\,|\mathbf{J}|\, d\xi\, d\eta \tag{7.3-13}$$

in which

$$\boldsymbol{\epsilon}_0' = \{\epsilon_{x'0}, \epsilon_{y'0}, \ldots, \gamma_{z'x'0}\} \tag{p}$$

After the nodal displacements in the vector $\mathbf{q}$ have been obtained, stresses in the element may be calculated from:

$$\boldsymbol{\sigma}' = \mathbf{E}'(\mathbf{T}_\epsilon\mathbf{B}\ \mathbf{q} - \boldsymbol{\epsilon}_0') \tag{7.3-14}$$

Such stresses should be determined at the sampling points for numerical integration (8).

Element SHQ8 for shell analysis can be further specialized to become a *membrane element* if we omit certain terms in the formulation. Figure 7.5 shows such an element, in which only nodal translations (u_i, v_i, and w_i) and membrane strains ($\epsilon_{x'}$, $\epsilon_{y'}$, and $\gamma_{x'y'}$) are of interest. The thickness of a membrane is usually constant. Because nodal rotations are not considered, we can omit columns 4 and 5 from submatrix $\mathbf{B}_i$, which leaves:

$$\mathbf{B}_i = \begin{bmatrix} a_i & 0 & 0 \\ 0 & b_i & 0 \\ 0 & 0 & c_i \\ b_i & a_i & 0 \\ 0 & c_i & b_i \\ c_i & 0 & a_i \end{bmatrix} \qquad (i = 1, 2, \ldots, 8) \tag{7.3-15}$$

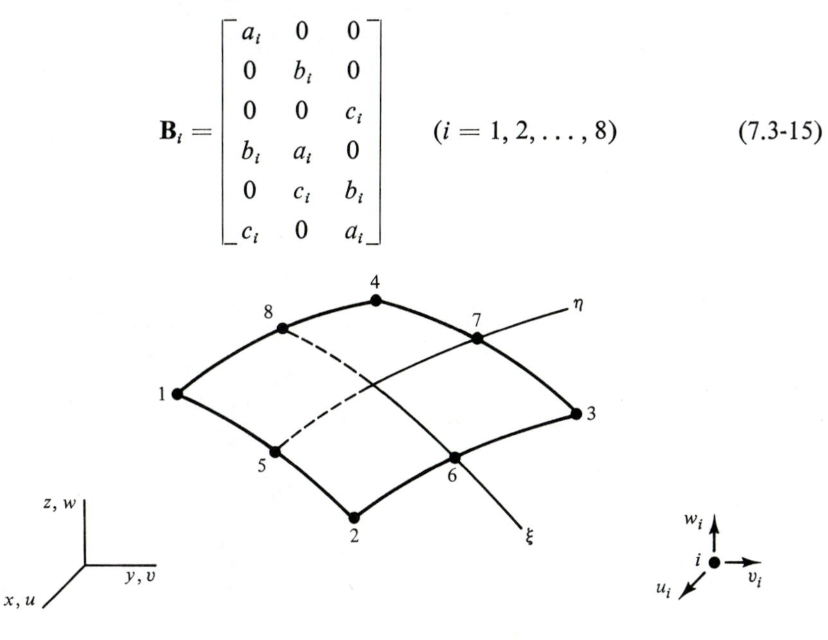

Figure 7.5 Membrane Element

where a_i, b_i, and c_i are given in Eqs. (h). Also, matrix $\mathbf{E}'$ can be reduced to:

$$\mathbf{E}' = \begin{bmatrix} E_{x'x'} & E_{x'y'} & 0 \\ E_{y'x'} & E_{y'y'} & 0 \\ 0 & 0 & G_{x'y'} \end{bmatrix} \tag{7.3-16}$$

because only $\sigma_{x'}$, $\sigma_{y'}$, and $\tau_{x'y'}$ need be calculated. In this case we would delete the third, fifth, and sixth rows of matrix $\mathbf{T}_\epsilon$ as well.

7.4 PROGRAM SHQ8 AND APPLICATIONS

Element SHQ8, developed in the preceding section, constitutes the basis of a computer program for the analysis of general shells. This program, named SHQ8, uses the formulations given by Eqs. (7.3-8) through (7.3-11) to construct the stiffness matrix of an isotropic element. For this purpose, additional structural data needed consist of the direction cosines ℓ_3, m_3, and n_3 of the normal vector $\mathbf{V}_3$ [see Eq. (7.3-2)] at every node of a discretized shell. Also, if the thickness of the shell varies, the parameter t at every node must be given as well. Stresses at numerical integration points are computed in local directions, in accordance with Eq. (7.3-14).

Example 1

To check the program, let us first examine the cantilever element SHQ8 in Fig. 7.6. This element is fixed at nodes 6, 7, and 8 but is only partially restrained at nodes 1, 3, 4, and 5 in order to simplify the results. Forces P_z, $4P_z$, and P_z are applied in the

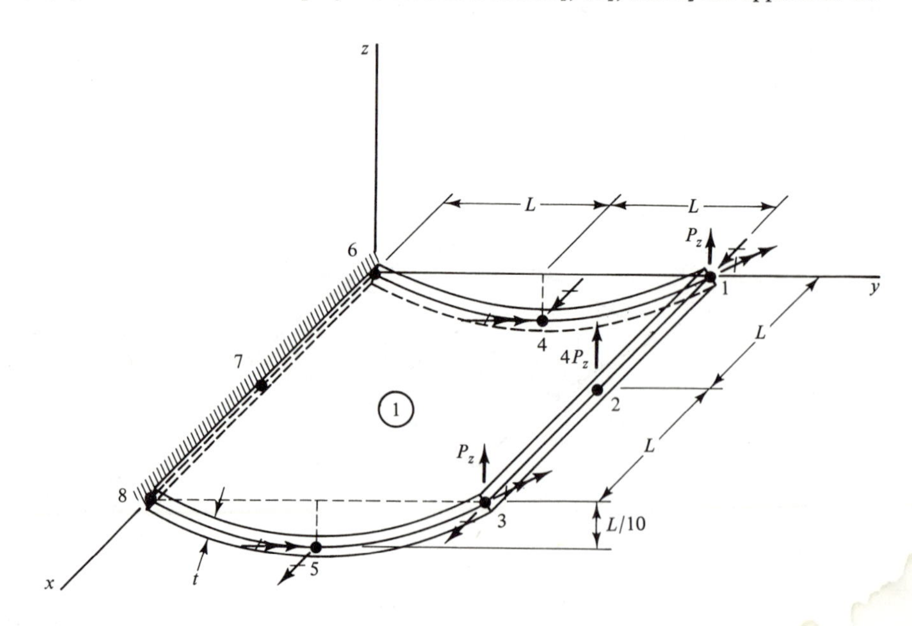

Figure 7.6 Example 1: Cantilever Element SHQ8 with Tip Loads

z direction at nodes 1, 2, and 3, as shown in the figure. These nodal loads are equivalent to a uniformly distributed line load along edge 1-3. Physical parameters in this problem are:

$$L = 0.2 \text{ m} \qquad t = 0.05 \text{ m} \qquad E = 4 \times 10^7 \text{ kN/m}^2$$

$$\nu = 0.3 \qquad P_z = 5 \text{ kN}$$

for which the material is magnesium and the units are S.I.

Table 7.1 gives the computer output from Program SHQ8 for this example. The structural information on the first page of the table includes direction cosines for the vectors $\mathbf{V}_1$, $\mathbf{V}_2$, and $\mathbf{V}_3$ at each of the nodes. Those for $\mathbf{V}_3$ are input as data, whereas those for $\mathbf{V}_1$ and $\mathbf{V}_2$ [see Eq. (7.3-4)] are computed internally. In this problem a uniform value of thickness, given in the structural data, is assigned to each of the nodes. Table 7.1 also shows the data and results for the load set under consideration. Displacements, stresses, and reactions appearing in the results are approximately correct. Note that for each integration point the membrane stresses (MS) are printed on the first line, whereas the flexural stresses (FS) are printed on the second line. As in the plate-bending program PBQ8, the transverse shearing stresses $\tau_{x'z'}$ and $\tau_{y'z'}$ in the second line are equal to the average values multiplied by the factor 1.5.

Example 2

The second example is a doubly symmetric cylindrical roof shell subjected to a uniform dead load. Figure 7.7 shows a quarter of the roof with a 2 × 2 network of SHQ8 ele-

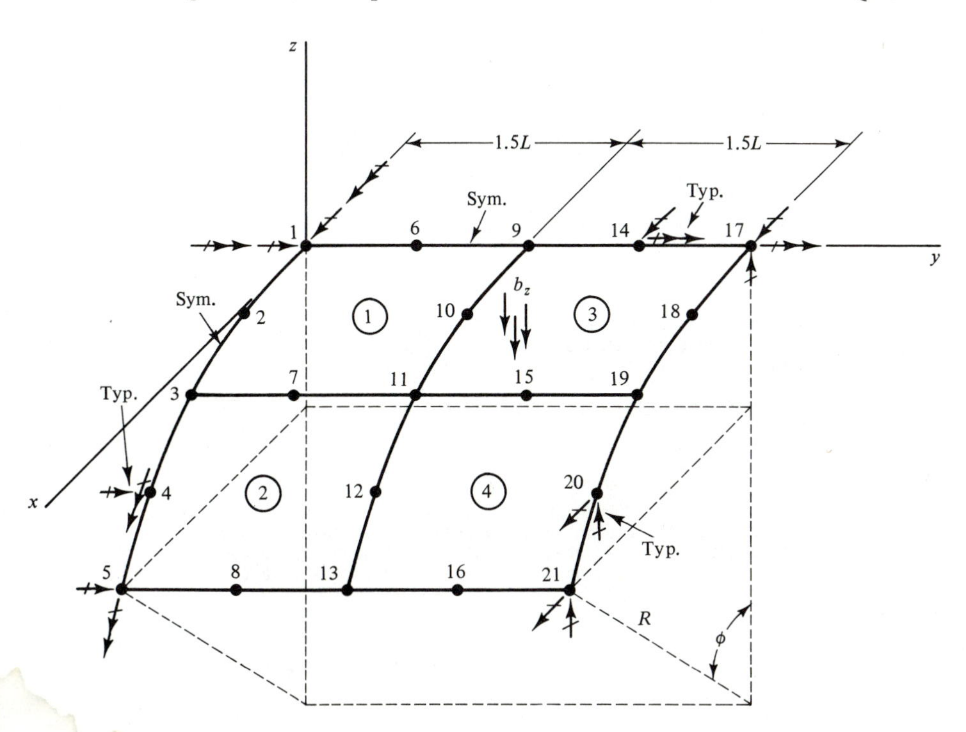

Figure 7.7 Example 2: Cylindrical Roof with Dead Load and SHQ8 Network

TABLE 7.1 Computer Output for Example 1

```
PROGRAM SHQ8

***  EXAMPLE 1:  CANTILEVER ELEMENT WITH TIP LOADS  ***

STRUCTURAL PARAMETERS
   NN   NE  NRN  NLS           E            PR            T
    8    1    7    1  4.0000D 07  3.0000D-01  5.0000D-02

COORDINATES OF NODES
 NODE          X           Y           Z           T
    1  0.0000D-01  4.0000D-01  0.0000D-01  5.0000D-02
    2  2.0000D-01  4.0000D-01  0.0000D-01  5.0000D-02
    3  4.0000D-01  4.0000D-01  0.0000D-01  5.0000D-02
    4  0.0000D-01  2.0000D-01 -2.0000D-02  5.0000D-02
    5  4.0000D-01  2.0000D-01 -2.0000D-02  5.0000D-02
    6  0.0000D-01  0.0000D-01  0.0000D-01  5.0000D-02
    7  2.0000D-01  0.0000D-01  0.0000D-01  5.0000D-02
    8  4.0000D-01  0.0000D-01  0.0000D-01  5.0000D-02

TANGENTIAL AND NORMAL VECTORS AT THE NODES
 NODE   V1 X   V1 Y   V1 Z   V2 X   V2 Y   V2 Z   V3 X   V3 Y   V3 Z
    1  1.000  0.000  0.000  0.000  0.981  0.194  0.000 -0.194  0.981
    2  1.000  0.000  0.000  0.000  0.981  0.194  0.000 -0.194  0.981
    3  1.000  0.000  0.000  0.000  0.981  0.194  0.000 -0.194  0.981
    4  1.000  0.000  0.000  0.000  1.000  0.000  0.000  0.000  1.000
    5  1.000  0.000  0.000  0.000  1.000  0.000  0.000  0.000  1.000
    6  1.000  0.000  0.000  0.000  0.981 -0.194  0.000  0.194  0.981
    7  1.000  0.000  0.000  0.000  0.981 -0.194  0.000  0.194  0.981
    8  1.000  0.000  0.000  0.000  0.981 -0.194  0.000  0.194  0.981

ELEMENT INFORMATION
ELEM.   N1   N2   N3   N4   N5   N6   N7   N8
    1    6    8    3    1    7    5    2    4

NODAL RESTRAINTS
 NODE   R1   R2   R3   R4   R5
    1    1    0    0    0    1
    3    1    0    0    0    1
    4    1    0    0    0    1
    5    1    0    0    0    1
    6    1    1    1    1    1
    7    1    1    1    1    1
    8    1    1    1    1    1

NUMBER OF DEGREES OF FREEDOM =   17
NUMBER OF NODAL RESTRAINTS   =   23
NUMBER OF TERMS IN SN        =  217

**********  LOADING NUMBER    1  **********
  NLN  NEL  NEA  NEV  NEH  IPR
    3    0    0    0    0    0

ACTIONS AT NODES
 NODE         AN1         AN2         AN3         AN4         AN5
    1  0.0000D-01  0.0000D-01  5.0000D 00  0.0000D-01  0.0000D-01
    2  0.0000D-01  0.0000D-01  2.0000D 01  0.0000D-01  0.0000D-01
    3  0.0000D-01  0.0000D-01  5.0000D 00  0.0000D-01  0.0000D-01

NODAL DISPLACEMENTS
 NODE         DN1         DN2         DN3         DN4         DN5
```

TABLE 7.1 (cont.)

1	0.0000D-01	-1.7074D-04	3.5061D-03	1.2667D-02	-0.0000D-01
2	-1.1968D-19	-1.7074D-04	3.5061D-03	1.3109D-02	-7.5515D-17
3	-0.0000D-01	-1.7074D-04	3.5061D-03	1.2667D-02	-0.0000D-01
4	-0.0000D-01	8.9211D-05	1.1128D-03	9.6075D-03	-0.0000D-01
5	-0.0000D-01	8.9211D-05	1.1128D-03	9.6075D-03	0.0000D-01
6	0.0000D-01	0.0000D-01	0.0000D-01	0.0000D-01	0.0000D-01
7	0.0000D-01	0.0000D-01	0.0000D-01	0.0000D-01	0.0000D-01
8	0.0000D-01	0.0000D-01	0.0000D-01	0.0000D-01	0.0000D-01

ELEMENT STRESSES

ELEM.	INT.	X' STRESS	Y' STRESS	X'Y' STRESS	X'Z' STRESS	Y'Z' STRESS	
1	1	-5.0471D 01	-1.6824D 02	-1.2688D-11			MS
		-1.6752D 04	-5.5840D 04	-1.9744D 02	2.2359D 03	-2.7974D-11	FS
1	2	-5.0471D 01	-1.6824D 02	-4.2909D-12			MS
		-1.6752D 04	-5.5840D 04	1.9744D 02	2.2359D 03	-5.5299D-11	FS
1	3	5.0471D 01	1.6824D 02	4.7198D-11			MS
		-4.4588D 03	-1.4863D 04	7.7055D 02	2.2359D 03	2.2865D-11	FS
1	4	5.0471D 01	1.6824D 02	3.7145D-11			MS
		-4.4588D 03	-1.4863D 04	-7.7055D 02	2.2359D 03	5.1343D-11	FS

SUPPORT REACTIONS

NODE	AR1	AR2	AR3	AR4	AR5
1	-2.9332D-01	0.0000D-01	0.0000D-01	0.0000D-01	-4.0345D-02
3	2.9332D-01	0.0000D-01	0.0000D-01	0.0000D-01	4.0345D-02
4	-5.3668D-13	0.0000D-01	0.0000D-01	0.0000D-01	1.2180D 00
5	-6.3324D-13	0.0000D-01	0.0000D-01	0.0000D-01	-1.2180D 00
6	2.9332D-01	2.0031D-13	-5.0000D 00	-2.0326D 00	6.0158D-01
7	-1.2114D-13	-1.4926D-15	-2.0000D 01	-7.7676D 00	-5.7759D-14
8	-2.9332D-01	6.4058D-13	-5.0000D 00	-2.0326D 00	-6.0158D-01

ments. The thickness of the shell is constant, and x-z and y-z are planes of symmetry. Restraints at a node on a plane of symmetry prevent translation across the plane and rotation in the plane, as indicated at nodes 4 and 14. On the other hand, restraints at the ends of the shell prevent translations in the x and z directions, as at node 20. The dead load b_z (force per unit area) is shown acting in the negative z direction in Fig. 7.7. The following physical parameters are given:

$$L = 100 \text{ in.} \qquad R = 3L \qquad t = 3 \text{ in.} \qquad \phi = 40^\circ$$

$$E = 3 \times 10^3 \text{ k/in.}^2 \qquad \nu = 0.3 \qquad b_z = 6.25 \times 10^{-4} \text{ k/in.}^2$$

where the units are U.S. and the material is reinforced concrete.

When the data for this example are run on Program SHQ8, the results show that the maximum x and z translations of $u = -1.97$ in. and $w = -3.69$ in. occur at node 5. Also, the sum of membrane and flexural stresses in the x' direction at the integration point nearest node 1 is $\sigma_{x'} = -1.530$ k/in.^{2} (at the lower surface).

Example 3

Figure 7.8 depicts a quarter of a doubly symmetric subsurface storage tank with a network of SHQ8 elements. The tank is fixed at its base and has thicknesses t_1 at nodes 1 through 65 and t_2 at nodes 66 through 105. Restraints at a node on a plane of symmetry prevent translation across the plane and rotation in the plane, as at nodes 5 and 85. The surrounding fluid (of unit weight w_f) is water, and its surface lies at $z_f = 8L$. We know the following physical parameters:

$$L = 1 \text{ m} \qquad t_1 = 0.05 \text{ m} \qquad t_2 = 0.03 \text{ m}$$

$$E = 2 \times 10^8 \text{ kN/m}^2 \qquad \nu = 0.3 \qquad w_f = 10 \text{ kN/m}^3$$

in which the material of the shell is steel and the units are S.I.

This shell consists of a circular cylindrical portion (elements 1 through 16) connected to a spherical portion (elements 17 through 28). In the latter part, node 89 lies at the intersection of three great circles, which are the curves 61-89, 85-89, and 89-93. Locations of other nodes are found by equalizing central angles.

To calculate *equivalent nodal loads due to hydrostatic pressure*, we must first find the components of the pressure normal to the neutral surface of the shell at a numerical

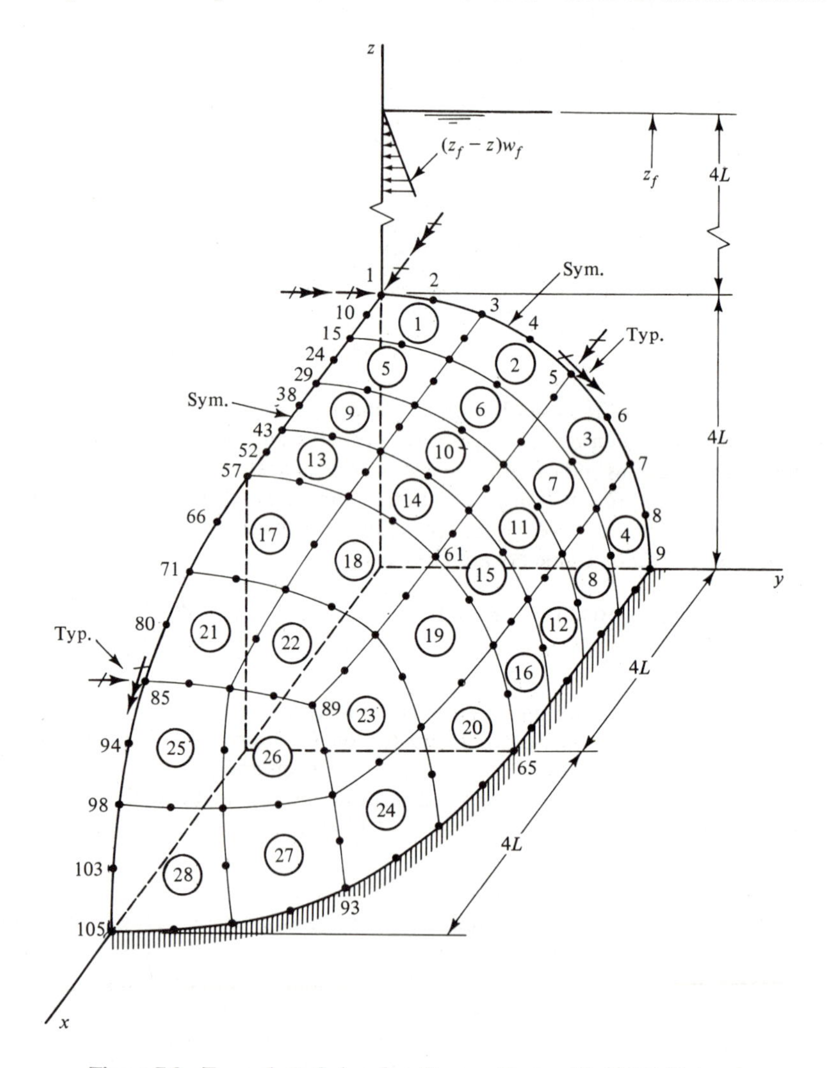

Figure 7.8 Example 3: Subsurface Storage Tank with SHQ8 Network

integration point. Thus,

$$\mathbf{b}_f = \pm(z_f - z_{j,k})w_f\mathbf{e}_\zeta^{\mathrm{T}} \tag{a}$$

In this expression the coordinate $z_{j,k}$ of an integration point is subtracted from that of the fluid surface, z_f. Then the result is multiplied by the unit weight of the fluid w_f and the column vector $\mathbf{e}_\zeta^{\mathrm{T}}$. This vector is the transpose of the row vector $\mathbf{e}_\zeta$, containing the three direction cosines of the normal to the neutral surface. That is,

$$\mathbf{e}_\zeta = \frac{1}{c}[x_{,\zeta} \quad y_{,\zeta} \quad z_{,\zeta}] \tag{b}$$

in which

$$c = \sqrt{(x_{,\zeta})^2 + (y_{,\zeta})^2 + (z_{,\zeta})^2} \tag{c}$$

The sign of vector $\mathbf{b}_f$ is positive if the positive senses of the hydrostatic pressure and the vector $\mathbf{e}_\zeta$ coincide. In the case of opposite senses, the sign will be negative. Now we may calculate the equivalent nodal loads due to hydrostatic pressure as:

$$\mathbf{p}_f = \int_A \mathbf{f}^{\mathrm{T}}\mathbf{b}_f \, dA = \int_{-1}^{1}\int_{-1}^{1} \mathbf{f}^{\mathrm{T}}\mathbf{b}_f |\mathbf{J}'| \, d\xi \, d\eta \tag{d}$$

In this equation the expression for dA is:

$$dA = |\mathbf{J}'| \, d\xi \, d\eta \tag{e}$$

where

$$|\mathbf{J}'| = (\mathbf{V}_\xi \times \mathbf{V}_\eta) \cdot \mathbf{e}_\zeta = \begin{vmatrix} x_{,\xi} & y_{,\xi} & z_{,\xi} \\ x_{,\eta} & y_{,\eta} & z_{,\eta} \\ \dfrac{x_{,\zeta}}{c} & \dfrac{y_{,\zeta}}{c} & \dfrac{z_{,\zeta}}{c} \end{vmatrix} \tag{f}$$

Note that the modified Jacobian matrix $\mathbf{J}'$ is similar to the 3×3 matrix $\mathbf{J}$ in Eq. (7.3-5a), but the third row of $\mathbf{J}'$ is normalized to have unit length (to transform area instead of volume).

By running the data for this submerged shell on Program SHQ8, we find the maximum translational deflection of $w_1 = 5.63 \times 10^{-4}$ m at node 1. In the cylindrical portion of the shell, the maximum normal stress is $\sigma_{y'} = -10.6 \times 10^3$ kN/m^2. This value occurs at the inner surface of element 12 for the integration point nearest node 37. A similar normal stress intensity of $\sigma_{y'} = -8.92 \times 10^3$ kN/m^2 is also found in the spherical portion. This stress exists at the outer surface of element 27 for the integration point nearest node 91. Of course, the principal normal stresses will be somewhat higher than these values.

REFERENCES

1. Novozhilov, V. V., *Thin Shell Theory*, 2d ed., P. Noordhoff Ltd., Gröningen, Netherlands, 1964.
2. Dupuis, G., and Goël, J., "A Curved Finite Element for Thin Elastic Shells," *Int. Jour. Sol. Struc.*, Vol. 6, No. 11, 1970, pp. 1413–1428.
3. Vlasov, V. Z., "General Theory of Shells and Its Applications in Engineering," *NASA TTF*-99, April 1964.

4. Cowper, G. R., Lindberg, G. M., and Olson, M. D., "A Shallow Shell Finite Element of Triangular Shape," *Int. Jour. Sol. Struc.*, Vol. 6, No. 8, 1970, pp. 1133–1156.
5. Greene, B. E., Strome, D. R., and Weikel, R. C., "Application of the Stiffness Method to the Analysis of Shell Structures," *Proc. Av. Conf.*, ASME, Los Angeles, March 1961.
6. Zienkiewicz, O. C., *The Finite Element Method*, 3d ed., McGraw-Hill Ltd., London, 1977.
7. Ahmad, S., Irons, B. M., and Zienkiewicz, O. C., "Analysis of Thick and Thin Shell Structures by Curved Finite Elements," *Int. Jour. Num. Meth. Eng.*, Vol. 2, No. 3, 1970, pp. 419–451.
8. Cook, R. D., *Concepts and Applications of Finite Element Analysis*, 2d ed., Wiley, New York, 1981.

CHAPTER

8

Axisymmetric Shells

8.1 INTRODUCTION

In 1963 Grafton and Strome (1) pioneered the development of finite elements for axisymmetric shells with axisymmetric loads. Their element consists of a ring in the shape of a conical frustum, as shown in Fig. 8.1(a). Thus, the element has a straight section that can be used to model approximately any axisymmetric shell, and Sec. 8.2 is devoted to that topic.

As with flat facets for general shells, the ring element with a straight section does not model axisymmetric shells very accurately, unless the shell itself happens to be conical. For this reason other works appeared in the technical literature using elements with curved sections, as illustrated in Fig. 8.1(b). In the paper by Jones and Strome (2), for example, the meridional axis of the element has the formula:

$$\cos \gamma = c_1 + c_2 \xi + c_3 \xi^2 \tag{a}$$

The symbol γ in this expression denotes the angle between the axis of symmetry and a tangent to the axis of the section [see Fig. 8.1(a)]. Alternatively, this angle can be measured between the radial axis r and the normal η to the axis of the section, as indicated in Figs. 8.1(a) and (b). Another formulation was presented by Stricklin et al. (3), who used the following geometric definition:

$$\gamma = c_1 + c_2 \xi + c_3 \xi^2 \tag{b}$$

However, both these approaches lead to rather complicated expressions that are difficult to implement.

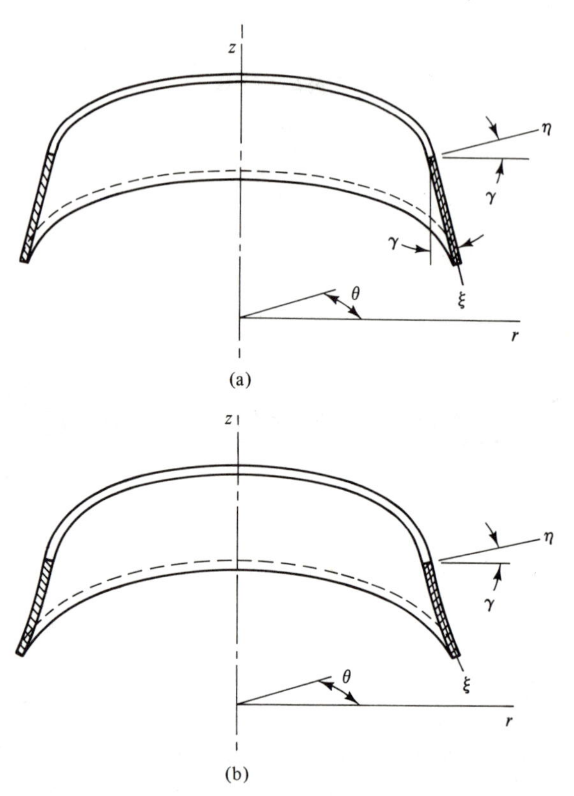

Figure 8.1 Axisymmetric Shell Elements: (a) Straight Section (b) Curved Section

A method of greater generality and more simplicity became available with the advent of isoparametric elements. Ahmad et al. (4) recognized that isoparametric elements for axisymmetric solids can be specialized to become elements for shells as well. This may be accomplished by making one cross-sectional dimension of the ring element small and by inducing constraints to reduce the number of independent nodal displacements. This topic is developed for a particular type of isoparametric cross section in Sec. 8.3. As with solids, nonaxisymmetric loads may also be of interest for axisymmetric shells. These types of loads are covered in Sec. 8.4.

8.2 STRAIGHT SECTION

Because a conical frustum with a straight section can model a conical shell exactly, this element is deemed worthy of consideration (1). Figure 8.2(a) shows local inclined axes x', y', and z', as well as two of their dimensionless counterparts, ξ and η. Generic displacements in the directions of local (primed) axes

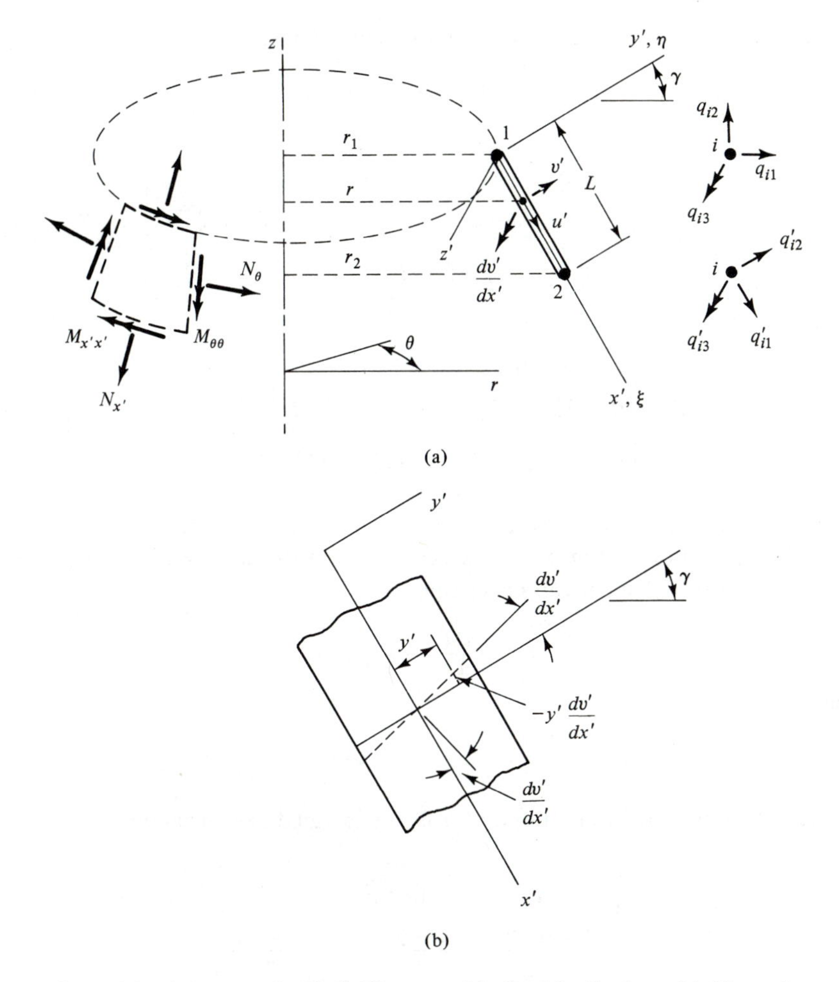

Figure 8.2 Axisymmetric Shell Element with Straight Section: (a) Element (b) Slope Change

consist of two translations and a slope. Hence,

$$\mathbf{u}' = \left\{u', v', \frac{dv'}{dx'}\right\} \tag{a}$$

as shown in the figure. Nodal displacements (at points 1 and 2) in local directions are of the same types, as follows:

$$\mathbf{q}_i = \{q'_{i1}, q'_{i2}, q'_{i3}\} = \left\{u'_i, v'_i, \frac{dv'_i}{dx'}\right\} \qquad (i = 1, 2) \tag{b}$$

Displacement shape functions are chosen to be linear for the membrane components and cubic for the flexural components. Thus, if we let:

$$\mathbf{f}' = [\mathbf{f}'_1 \quad \mathbf{f}'_2] \tag{c}$$

then

$$\mathbf{f}'_1 = \begin{bmatrix} 1-\xi & 0 & 0 \\ 0 & 1 - 3\xi^2 + 2\xi^3 & (\xi - 2\xi^2 + \xi^3)L \\ 0 & -\dfrac{6}{L}(\xi - \xi^2) & 1 - 4\xi + 3\xi^2 \end{bmatrix} \tag{8.2-1a}$$

and

$$\mathbf{f}'_2 = \begin{bmatrix} \xi & 0 & 0 \\ 0 & 3\xi^2 - 2\xi^3 & -(\xi^2 - \xi^3)L \\ 0 & \dfrac{6}{L}(\xi - \xi^2) & -2\xi + 3\xi^2 \end{bmatrix} \tag{8.2-1b}$$

which separates the functions for node 1 from those for node 2.

By inspection of Fig. 8.2(b), we can see that the strain-displacement relationships for this element can be written as:

$$\epsilon_{x'} = \frac{du'}{dx'} - y' \frac{d^2v'}{(dx')^2} \tag{d}$$

and

$$\epsilon_\theta = \epsilon_{z'} = \frac{1}{r}\left(u' \sin\gamma + v' \cos\gamma - y' \frac{dv'}{dx'} \sin\gamma\right) \tag{e}$$

Hence, the linear differential operator $\mathbf{d}'$ for inclined axes becomes:

$$\mathbf{d}' = \begin{bmatrix} \dfrac{d}{dx'} & -y' \dfrac{d^2}{(dx')^2} & 0 \\ \dfrac{\sin\gamma}{r} & \dfrac{\cos\gamma}{r} & -\dfrac{y' \sin\gamma}{r} \end{bmatrix} \tag{8.2-2}$$

If the strain-displacement matrix $\mathbf{B}'$ is also separated into two parts, we have:

$$\mathbf{B}' = [\mathbf{B}'_1 \quad \mathbf{B}'_2] \tag{f}$$

where

$$\mathbf{B}'_1 = \mathbf{d}'\mathbf{f}'_1$$

$$= \begin{bmatrix} -\dfrac{1}{L} & \dfrac{6}{L^2}(1-2\xi)y' & \dfrac{2}{L}(2-3\xi)y' \\ \dfrac{1}{r}\{(1-\xi)\sin\gamma\} & \begin{array}{l}\dfrac{1}{Lr}\{(1-3\xi^2+2\xi^3)L\cos\gamma \\ \quad + 6(\xi-\xi^2)y'\sin\gamma\}\end{array} & \begin{array}{l}\dfrac{1}{r}\{(\xi-2\xi^2+\xi^3)L\cos\gamma \\ \quad - (1-4\xi+3\xi^2)y'\sin\gamma\}\end{array} \end{bmatrix} \tag{8.2-3a}$$

and

$$\mathbf{B}_2' = \mathbf{d}'\mathbf{f}_2'$$

$$= \begin{bmatrix} \dfrac{1}{L} & -\dfrac{6}{L^2}(1-2\xi)y' & \dfrac{2}{L}(1-3\xi)y' \\ \dfrac{1}{r}(\xi \sin\gamma) & \dfrac{1}{Lr}\{(3\xi^2 - 2\xi^3)L\cos\gamma & \dfrac{1}{r}\{-(\xi^2-\xi^3)L\cos\gamma \\ & -6(\xi-\xi^2)y'\sin\gamma\} & +(2\xi-3\xi^2)y'\sin\gamma\} \end{bmatrix} \tag{8.2-3b}$$

Each of these matrices is of size 2 × 3.

Stress-strain relationships for either an orthotropic or an isotropic material are:

$$\boldsymbol{\sigma}' = \begin{bmatrix} \sigma_{x'} \\ \sigma_{z'} \end{bmatrix} = \begin{bmatrix} E_{x'x'} & E_{x'z'} \\ E_{z'x'} & E_{z'z'} \end{bmatrix} \begin{bmatrix} \epsilon_{x'} \\ \epsilon_{z'} \end{bmatrix} = \mathbf{E}'\boldsymbol{\epsilon}' \tag{8.2-4}$$

In the orthotropic case, the principal material directions are implied to be x' and z'.

Having evolved the necessary matrices, we can now form the stiffness matrix (in local directions) for the element as:

$$\mathbf{K}' = \int_V (\mathbf{B}')^{\mathrm{T}}\mathbf{E}'\mathbf{B}'\, dV \tag{8.2-5}$$

In addition, the equivalent nodal loads due to body forces (in local directions) are:

$$\mathbf{p}_b' = \int_V (\mathbf{f}')^{\mathrm{T}}\mathbf{b}'\, dV \tag{8.2-6}$$

where

$$\mathbf{b}' = \{b_{x'}, b_{y'}, 0\} \tag{g}$$

And those due to initial strains become:

$$\mathbf{p}_0' = \int_V (\mathbf{B}')^{\mathrm{T}}\mathbf{E}'\boldsymbol{\epsilon}_0'\, dV \tag{8.2-7}$$

in which

$$\boldsymbol{\epsilon}_0' = \{\epsilon_{x'0}, \epsilon_{z'0}\} \tag{h}$$

By defining appropriate *generalized stresses and strains*, we can integrate Eqs. (8.2-5), (8.2-6), and (8.2-7) through the thickness and around the circumference of the element. Figure 8.2(a) shows the generalized stresses, which are listed as follows:

$$\mathbf{N} = \{N_{x'}, N_\theta\} \qquad \mathbf{M} = \{M_{x'x'}, M_{\theta\theta}\} \tag{i}$$

Vector $\mathbf{N}$ contains forces (per unit width), and vector $\mathbf{M}$ consists of moments (per unit width). Altogether, we have the generalized stresses:

$$\bar{\boldsymbol{\sigma}}' = \{\mathbf{N}, \mathbf{M}\} \tag{j}$$

Corresponding generalized strains are obtained from Eqs. (d) and (e) as:

$$\boldsymbol{\epsilon}' = \left\{\frac{du'}{dx'}, \frac{1}{r}(u'\sin\gamma + v'\cos\gamma)\right\} \tag{k}$$

and

$$\boldsymbol{\phi}' = \left\{\frac{d^2v'}{(dx')^2}, \frac{1}{r}\frac{dv'}{dx'}\sin\gamma\right\} \tag{ℓ}$$

Thus, the *membrane strains* are given in Eq. (k), and the *flexural strains* (devoid of $-y'$) appear in Eq. (ℓ). Altogether, the generalized strains are:

$$\bar{\boldsymbol{\epsilon}}' = \{\boldsymbol{\epsilon}', \boldsymbol{\phi}'\} \tag{m}$$

Furthermore, the generalized linear differential operator $\bar{\mathbf{d}}'$ becomes:

$$\bar{\mathbf{d}}' = \begin{bmatrix} \dfrac{d}{dx'} & 0 & 0 \\ \dfrac{\sin\gamma}{r} & \dfrac{\cos\gamma}{r} & 0 \\ 0 & \dfrac{d^2}{(dx')^2} & 0 \\ 0 & 0 & \dfrac{\sin\gamma}{r} \end{bmatrix} \tag{8.2-8}$$

Applying this operator to the two parts of matrix $\mathbf{f}$ given in Eqs. (8.2-1a) and (8.2-1b), we obtain the following generalized strain-displacement relationships:

$$\bar{\mathbf{B}}_1' = \bar{\mathbf{d}}'\mathbf{f}_1$$

$$= \begin{bmatrix} -\dfrac{1}{L} & 0 & 0 \\ \dfrac{1}{r}(1-\xi)\sin\gamma & \dfrac{1}{r}(1-3\xi^2+2\xi^3)\cos\gamma & \dfrac{1}{r}(\xi-2\xi^2+\xi^3)L\cos\gamma \\ 0 & -\dfrac{6}{L^2}(1-2\xi) & -\dfrac{2}{L}(2-3\xi) \\ 0 & -\dfrac{6}{Lr}(\xi-\xi^2)\sin\gamma & \dfrac{1}{r}(1-4\xi+3\xi^2)\sin\gamma \end{bmatrix} \tag{8.2-9a}$$

and

$$\bar{\mathbf{B}}_2' = \bar{\mathbf{d}}'\mathbf{f}_2$$

$$= \begin{bmatrix} \dfrac{1}{L} & 0 & 0 \\ \dfrac{\xi}{r}\sin\gamma & \dfrac{1}{r}(3\xi^2-2\xi^3)\cos\gamma & -\dfrac{1}{r}(\xi^2-\xi^3)L\cos\gamma \\ 0 & \dfrac{6}{L^2}(1-2\xi) & -\dfrac{2}{L}(1-3\xi) \\ 0 & \dfrac{6}{Lr}(\xi-\xi^2)\sin\gamma & -\dfrac{1}{r}(2\xi-3\xi^2)\sin\gamma \end{bmatrix} \tag{8.2-9b}$$

Each of these matrices is of size 4×3.

In accordance with the preceding definitions, the *generalized stress-strain relationships* may be written as:

$$\bar{\boldsymbol{\sigma}}' = \bar{\mathbf{E}}'\bar{\boldsymbol{\epsilon}}' \tag{n}$$

where

$$\bar{\mathbf{E}}' = \begin{bmatrix} \mathbf{E}'t & \mathbf{0} \\ \mathbf{0} & \dfrac{\mathbf{E}'t^3}{12} \end{bmatrix} \tag{8.2-10}$$

and matrix $\mathbf{E}'$ was given in Eq. (8.2-4). Then the element stiffness matrix [Eq. (8.2-5)] becomes:

$$\begin{aligned} \mathbf{K}' &= L \int_0^1 \int_0^{2\pi} (\bar{\mathbf{B}}')^{\mathrm{T}} \bar{\mathbf{E}}' \bar{\mathbf{B}}' r \, d\theta \, d\xi \\ &= 2\pi L \int_0^1 (\bar{\mathbf{B}}')^{\mathrm{T}} \bar{\mathbf{E}}' \bar{\mathbf{B}}' r \, d\xi \end{aligned} \tag{8.2-11}$$

Also, the equivalent nodal loads due to body forces [Eq. (8.2-6)] are:

$$\mathbf{p}'_b = 2\pi L \int_0^1 (\mathbf{f}')^{\mathrm{T}} \mathbf{b}' t \, r \, d\xi \tag{8.2-12}$$

And those due to initial strains (Eq. 8.2-7) are:

$$\mathbf{p}'_0 = 2\pi L \int_0^1 (\mathbf{B}')^{\mathrm{T}} \bar{\mathbf{E}}' \bar{\boldsymbol{\epsilon}}'_0 r \, d\xi \tag{8.2-13}$$

For this element a rotation-of-axes transformation matrix is required for converting nodal actions and displacements from global directions to local directions. This matrix has the form:

$$\hat{\mathbf{R}} = \begin{bmatrix} \mathbf{R} & \mathbf{O} \\ \mathbf{O} & \mathbf{R} \end{bmatrix} \tag{8.2-14}$$

in which

$$\mathbf{R} = \begin{bmatrix} \lambda_{11} & \lambda_{12} & 0 \\ \lambda_{21} & \lambda_{22} & 0 \\ 0 & 0 & 1 \end{bmatrix} \tag{8.2-15}$$

where λ_{11}, λ_{12}, and so on, are the direction cosines of the local axes with respect to the global axes. Using matrix $\hat{\mathbf{R}}$, we can transform nodal actions and displacements as follows:

$$\mathbf{p}' = \hat{\mathbf{R}}\,\mathbf{p} \qquad \mathbf{q}' = \hat{\mathbf{R}}\,\mathbf{q} \tag{8.2-16}$$

Furthermore, the element stiffness matrix converts from local to global directions by the formula:

$$\mathbf{K} = \hat{\mathbf{R}}^{\mathrm{T}} \mathbf{K}' \hat{\mathbf{R}} \tag{8.2-17}$$

These kinds of transformations must precede the assembly process.

8.3 CURVED SECTION

As shown by Ahmad et al. (4), it is possible to specialize a ring element with an isoparametric cross section to become an axisymmetric shell element by making one dimension small compared to the other. In this section we demonstrate the procedure by specializing element AXQ8 from Sec. 5.3 to form a shell element called AXSH3.

Figure 8.3(a) shows the axisymmetric solid element AXQ8, for which the cross section is an isoparametric quadrilateral with eight nodes. As the first step in the process, we make the axes ξ and η orthogonal and reduce the η dimension to the thickness t. Thus, we form the rectangular parent AXSR3 of element AXSH3 (before constraints), as shown in Fig. 8.3(b). Next, we may introduce constraints to refer the displacements at each group and pair of nodes to those of a single node on the middle surface, as depicted in Fig. 8.3(c). The nodal displacements indicated at point i in that figure are:

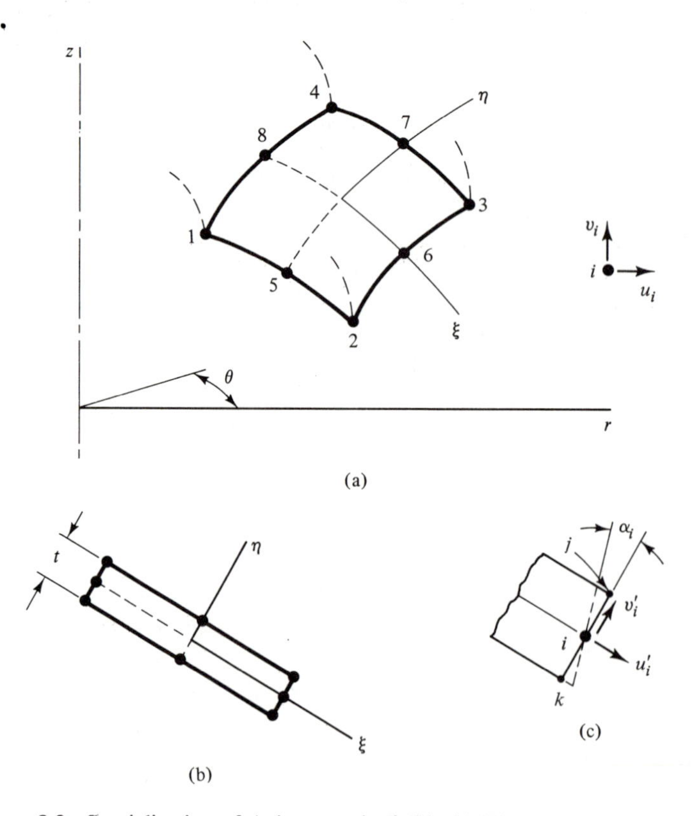

Figure 8.3 Specialization of Axisymmetric Solid: (a) Element AXQ8 (b) Rectangular Parent AXSR3 of Element AXSH3 Before Constraints (c) Constrained Nodal Displacements

$$\mathbf{q}_i = \{q_{i1}, q_{i2}, q_{i3}\} = \{u_i, v_i, \alpha_i\} \qquad (i = 1, 2, 3) \tag{a}$$

where α_i is a small positive rotation about an axis normal to the ξ-η plane. Figures 8.4(a), (b), and (c) show relationships between nodal displacements at an end of element AXSR3, the middle of AXSR3, and a node of element AXSH3. The two types of constraints to be invoked are:

1. Nodes on the same normal to the middle surface have equal translations in the η direction.
2. Normals to the middle surface remain straight (but no longer normal) during deformation.

With these criteria we can relate the six nodal translations in Fig. 8.4(a) to the three nodal displacements in Fig. 8.4(c) by the following 6×3 constraint matrix:

$$\mathbf{G}_{ai} = \begin{bmatrix} 1 & 0 & 0 \\ 0 & 1 & 0 \\ 1 & 0 & -\frac{t_i}{2} \\ 0 & 1 & 0 \\ 1 & 0 & \frac{t_i}{2} \\ 0 & 1 & 0 \end{bmatrix} \tag{b}$$

Similarly, the four nodal translations in Fig. 8.4(b) are related to the three nodal displacements in Fig. 8.4(c) by the constraint matrix:

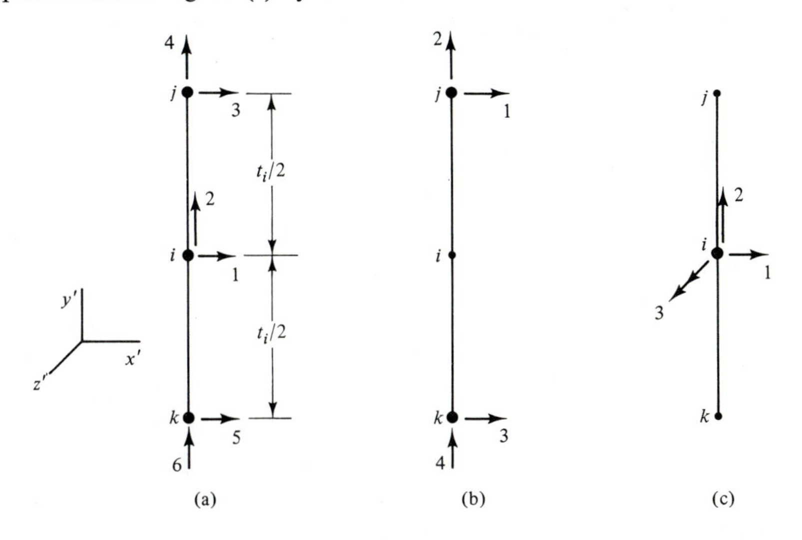

Figure 8.4 Nodal Displacements: (a) End of AXSR3 (b) Middle of AXSR3 (c) Node of AXSH3

$$\mathbf{G}_{bi} = \begin{bmatrix} 1 & 0 & -\frac{t_i}{2} \\ 0 & 1 & 0 \\ 1 & 0 & \frac{t_i}{2} \\ 0 & 1 & 0 \end{bmatrix} \tag{c}$$

which is of size 4×3. If we were to apply Eq. (b) at the ends and Eq. (c) at the middle, we could reduce the number of nodal displacements from 16 to 9. However, we will take a more direct approach, which is similar to those in Secs. 6.4 and 7.3 for plates and general shell elements.

Figures 8.5(a) and (b) show element AXSH3, for which the coordinates of any point may be stated as:

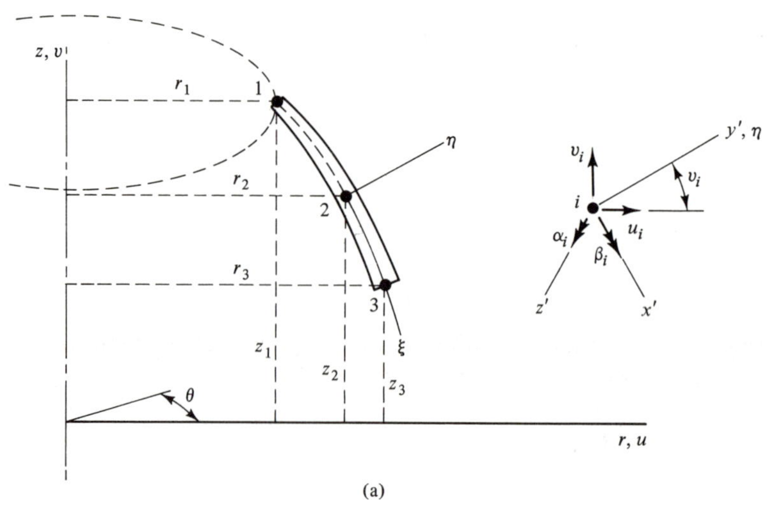

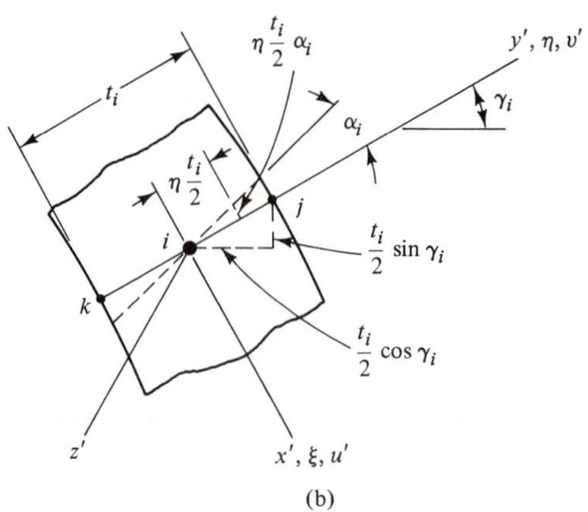

Figure 8.5 (a) Element AXSH3 (b) Nodal Rotation

$$\begin{bmatrix} r \\ z \end{bmatrix} = \sum_{i=1}^{3} f_i \begin{bmatrix} r_i \\ z_i \end{bmatrix} + \sum_{i=1}^{3} f_i \eta \frac{t_i}{2} \begin{bmatrix} \cos \gamma_i \\ \sin \gamma_i \end{bmatrix} \tag{8.3-1}$$

where

$$f_1 = -\frac{\xi(1-\xi)}{2} \qquad f_2 = 1 - \xi^2 \qquad f_3 = \frac{\xi(1+\xi)}{2} \tag{8.3-2}$$

Generic displacements at any point in the element are:

$$\mathbf{u} = \{u, v\} \tag{d}$$

These displacements can be expressed in terms of the nodal displacements u_i, v_i, and α_i, as follows:

$$\begin{bmatrix} u \\ v \end{bmatrix} = \sum_{i=1}^{3} f_i \begin{bmatrix} u_i \\ v_i \end{bmatrix} + \sum_{i=1}^{3} f_i \eta \frac{t_i}{2} \begin{bmatrix} -\sin \gamma_i \\ \cos \gamma_i \end{bmatrix} \alpha_i \tag{8.3-3}$$

Geometric details justifying these expressions appear in Fig. 8.5(b). The Jacobian matrix required for this element is:

$$\mathbf{J} = \begin{bmatrix} r_{,\xi} & z_{,\xi} \\ r_{,\eta} & z_{,\eta} \end{bmatrix} \tag{8.3-4a}$$

in which

$$r_{,\xi} = \sum_{i=1}^{3} f_{i,\xi} r_i + \sum_{i=1}^{3} f_{i,\xi} \eta \frac{t_i}{2} \cos \gamma_i \qquad \text{and so on.}$$

The inverse of $\mathbf{J}$ becomes:

$$\mathbf{J}^{-1} = \mathbf{J}^* = \begin{bmatrix} \xi_{,r} & \eta_{,r} \\ \xi_{,z} & \eta_{,z} \end{bmatrix} \tag{8.3-4b}$$

We will need derivatives of the generic displacements with respect to the local coordinates, as follows:

$$\begin{bmatrix} u_{,\xi} \\ u_{,\eta} \\ v_{,\xi} \\ v_{,\eta} \end{bmatrix} = \sum_{i=1}^{3} \begin{bmatrix} f_{i,\xi} & 0 & -f_{i,\xi}\eta \sin \gamma_i \\ 0 & 0 & -f_i \sin \gamma_i \\ 0 & f_{i,\xi} & f_{i,\xi}\eta \cos \gamma_i \\ 0 & 0 & f_i \cos \gamma_i \end{bmatrix} \begin{bmatrix} u_i \\ v_i \\ \frac{t_i}{2}\alpha_i \end{bmatrix} \tag{8.3-5}$$

These derivatives are transformed to global coordinates by the operation:

$$\begin{bmatrix} u_{,r} \\ u_{,z} \\ v_{,r} \\ v_{,z} \end{bmatrix} = \begin{bmatrix} \mathbf{J}^* & \mathbf{O} \\ \mathbf{O} & \mathbf{J}^* \end{bmatrix} \begin{bmatrix} u_{,\xi} \\ u_{,\eta} \\ v_{,\xi} \\ v_{,\eta} \end{bmatrix}$$

$$= \sum_{i=1}^{3} \begin{bmatrix} a_i & 0 & -d_i \sin \gamma_i \\ b_i & 0 & -e_i \sin \gamma_i \\ 0 & a_i & d_i \cos \gamma_i \\ 0 & b_i & e_i \cos \gamma_i \end{bmatrix} \begin{bmatrix} u_i \\ v_i \\ \alpha_i \end{bmatrix} \tag{8.3-6}$$

where

$$
\begin{aligned}
a_i &= J^*_{11} f_{i,\xi} \qquad & d_i &= \frac{(a_i\eta + J^*_{12} f_i)t_i}{2} \\
b_i &= J^*_{21} f_{i,\xi} \qquad & e_i &= \frac{(b_i\eta + J^*_{22} f_i)t_i}{2}
\end{aligned}
\tag{e}
$$

We consider four types of nonzero strains for element AXSH3. They are:

$$
\boldsymbol{\epsilon} = \begin{bmatrix} \epsilon_r \\ \epsilon_z \\ \epsilon_\theta \\ \gamma_{rz} \end{bmatrix} = \begin{bmatrix} u_{,r} \\ v_{,z} \\ \dfrac{u}{r} \\ u_{,z} + v_{,r} \end{bmatrix}
\tag{f}
$$

Using the second version of this strain vector, we form the ith part of matrix $\mathbf{B}$ from terms in Eq. (8.3-6) as:

$$
\mathbf{B}_i = \begin{bmatrix} a_i & 0 & -d_i \sin \gamma_i \\ 0 & b_i & e_i \cos \gamma_i \\ \dfrac{f_i}{r} & 0 & -\dfrac{1}{2r}(f_i \eta t_i \sin \gamma_i) \\ b_i & a_i & d_i \cos \gamma_i - e_i \sin \gamma_i \end{bmatrix}
\tag{8.3-7}
$$

$$(i = 1, 2, 3)$$

For either orthotropic or isotropic materials, the stress-strain relationships in local coordinates become:

$$
\boldsymbol{\sigma}' = \mathbf{E}' \boldsymbol{\epsilon}'
\tag{g}
$$

Or,

$$
\begin{bmatrix} \sigma_{x'} \\ \sigma_{y'} \\ \sigma_{z'} \\ \tau_{x'y'} \end{bmatrix} = \begin{bmatrix} E_{x'x'} & 0 & E_{x'z'} & 0 \\ 0 & 0 & 0 & 0 \\ E_{z'x'} & 0 & E_{z'z'} & 0 \\ 0 & 0 & 0 & \dfrac{G_{x'y'}}{1.2} \end{bmatrix} \begin{bmatrix} \epsilon_{x'} \\ \epsilon_{y'} \\ \epsilon_{z'} \\ \gamma_{x'y'} \end{bmatrix}
\tag{8.3-8}
$$

for which the primed axes are shown in Figs. 8.5(a) and (b). Matrix $\mathbf{E}'$ can be transformed from local to global directions using the operation:

$$
\mathbf{E} = \mathbf{T}_\epsilon^{\mathrm{T}} \mathbf{E}' \mathbf{T}_\epsilon
\tag{7.3-10}
$$

repeated

The matrix $\mathbf{T}_\epsilon$ in this expression is of size 4×4. It is taken from the upper left part of the matrix in Eq. (4.1-15a) and specialized as follows:

$$
\mathbf{T}_\epsilon = \begin{bmatrix} \ell_1^2 & m_1^2 & 0 & \ell_1 m_1 \\ \ell_2^2 & m_2^2 & 0 & \ell_2 m_2 \\ 0 & 0 & 1 & 0 \\ 2\ell_1 \ell_2 & 2m_1 m_2 & 0 & \ell_1 m_2 + \ell_2 m_1 \end{bmatrix}
\tag{8.3-9}
$$

Equation (7.3-10) would be more efficient if the second row and column of matrix $\mathbf{E}'$ (corresponding to $\sigma_{y'}$ and $\epsilon_{y'}$) were deleted, along with the second row of the operator $\mathbf{T}_\epsilon$ in Eq. (8.3-9). To evaluate matrix $\mathbf{T}_\epsilon$ at an integration point, we must find the direction cosines for unit vectors $\mathbf{e}_1$ and $\mathbf{e}_2$ in the directions of x' and y' at the point. This may be done as follows:

$$\mathbf{e}_1 = (\mathbf{J}_1)_{\text{norm.}} = \{\ell_1, m_1\} \qquad \mathbf{e}_2 = \{\ell_2, m_2\} = \{-m_1, \ell_1\} \tag{h}$$

The element stiffness matrix may now be formulated as:

$$\mathbf{K} = \int_{-1}^{1}\int_{-1}^{1}\int_{0}^{2\pi} \mathbf{B}^{\mathrm{T}}\mathbf{E}\ \mathbf{B}\,|\,\mathbf{J}\,|\,d\xi\,d\eta\,r\,d\theta \tag{i}$$

in which matrices $\mathbf{B}$ and $\mathbf{J}$ are functions of ξ and η only. Integration of Eq. (i) through the thickness* and around the circumference of the element produces:

$$\begin{aligned}\mathbf{K} &= 2\pi \int_{-1}^{1}\int_{-1}^{1} (\mathbf{B}_a + \eta\mathbf{B}_b)^{\mathrm{T}}\mathbf{E}(\mathbf{B}_a + \eta\mathbf{B}_b)\,|\,\mathbf{J}\,|\,r\,d\xi\,d\eta \\ &= 2\pi \int_{-1}^{1} (2\mathbf{B}_a^{\mathrm{T}}\mathbf{E}\ \mathbf{B}_a + \tfrac{2}{3}\mathbf{B}_b^{\mathrm{T}}\mathbf{E}\ \mathbf{B}_b)\,|\,\mathbf{J}\,|\,r\,d\xi\end{aligned} \tag{8.3-10}$$

In this equation the matrices $\mathbf{B}_a$ and $\mathbf{B}_b$ are both of size 4×9, but the latter array contains only terms that are to be multiplied by η. The remaining integral in Eq. (8.3-10) must be evaluated numerically, using two integration points in the ξ direction (5). In this process the factor 2 multiplies $t_i/2$, which is a common term in the second row of $|\,\mathbf{J}\,|$.

Equivalent nodal loads due to body forces on element AXSH3 may be found as:

$$\mathbf{p}_b = \int_V \mathbf{f}^{\mathrm{T}}\mathbf{b}\,dV = 4\pi \int_{-1}^{1} \mathbf{f}^{\mathrm{T}}\mathbf{b}\,|\,\mathbf{J}\,|\,r\,d\xi \tag{8.3-11}$$

From Eq. (8.3-3) we have:

$$\mathbf{f}_i = \begin{bmatrix} 1 & 0 & 0 \\ 0 & 1 & 0 \end{bmatrix} f_i \qquad (i = 1, 2, 3) \tag{j}$$

And the components of load (per unit volume, constant through the thickness) are:

$$\mathbf{b} = \{b_r, b_z\} \tag{k}$$

Also, the equivalent nodal loads caused by initial strains become:

$$\begin{aligned}\mathbf{p}_0 &= \int_V \mathbf{B}^{\mathrm{T}}\mathbf{E}\ \boldsymbol{\epsilon}_0\,dV = \int_V \mathbf{B}^{\mathrm{T}}\mathbf{T}_\epsilon^{\mathrm{T}}\mathbf{E}'\boldsymbol{\epsilon}_0'\,dV \\ &= \int_{-1}^{1}\int_{-1}^{1}\int_{0}^{2\pi} (\mathbf{B}_a + \eta\mathbf{B}_b)^{\mathrm{T}}\mathbf{T}_\epsilon^{\mathrm{T}}\mathbf{E}'\boldsymbol{\epsilon}_0'\,|\,\mathbf{J}\,|\,d\xi\,d\eta\,r\,d\theta \\ &= 4\pi \int_{-1}^{1} \mathbf{B}_a^{\mathrm{T}}\mathbf{T}_\epsilon^{\mathrm{T}}\mathbf{E}'\boldsymbol{\epsilon}_0'\,|\,\mathbf{J}\,|\,r\,d\xi\end{aligned} \tag{8.3-12}$$

*Terms in r and matrix $\mathbf{J}$ containing η are neglected for the purpose of simplifying integration through the thickness.

in which

$$\boldsymbol{\epsilon}_0' = \{\epsilon_{x'0}, \epsilon_{y'0}, \epsilon_{z'0}, \gamma_{x'y'0}\} \tag{ℓ}$$

After solving for the nodal displacements in the vector $\mathbf{q}$, we can find stresses in the element using:

$$\boldsymbol{\sigma}' = \mathbf{E}'(\mathbf{T}_\epsilon \mathbf{B}\ \mathbf{q} - \boldsymbol{\epsilon}_0') \tag{7.3-14}$$

repeated

These stresses should be calculated at the numerical integration points (5).

As with element SHQ8 for general shell analysis, we can further specialize element AXSH3 for axisymmetric shells to become a membrane element by omitting certain terms. Figure 8.6 shows this type of *axisymmetric membrane element*, in which only nodal translations (u_i and v_i) and positive normal strains ($\epsilon_{x'}$ and $\epsilon_{z'}$) are of interest. Ordinarily, the thickness of such an element is constant. When nodal rotations are not considered, we must omit column 3 from submatrix $\mathbf{B}_i$, leaving:

$$\mathbf{B}_i = \begin{bmatrix} a_i & 0 \\ 0 & b_i \\ \dfrac{f_i}{r} & 0 \\ b_i & a_i \end{bmatrix} \qquad (i = 1, 2, 3) \tag{8.3-13}$$

where a_i and b_i are given by Eqs. (e). Also, because only $\sigma_{x'}$ and $\sigma_{z'}$ need be calculated, matrix $\mathbf{E}'$ can be reduced to:

$$\mathbf{E}' = \begin{bmatrix} E_{x'x'} & E_{x'z'} \\ E_{z'x'} & E_{z'z'} \end{bmatrix} \tag{8.3-14}$$

In this instance we must delete rows 2 and 4 of matrix $\mathbf{T}_\epsilon$ as well.

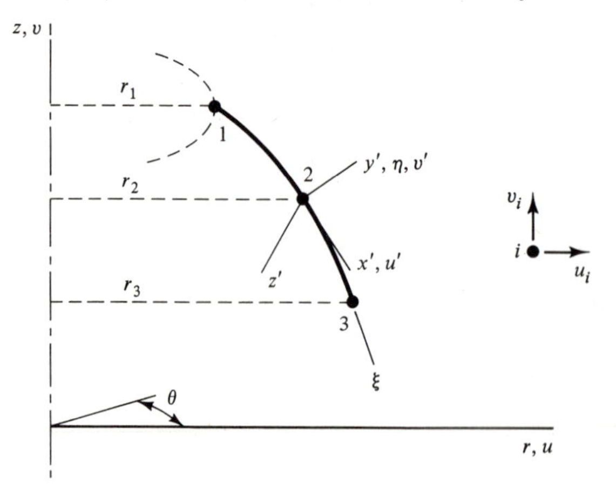

Figure 8.6 Axisymmetric Membrane Element

It is also possible to derive a *curved beam element* from the axisymmetric shell element AXSH3. Figure 8.7 shows such a member lying in the x-y plane. Nodal displacements for the curved beam are the same as those for AXSH3. In this case we must omit the third row from submatrix $\mathbf{B}_i$, which leaves:

$$\mathbf{B}_i = \begin{bmatrix} a_i & 0 & -d_i \sin \gamma_i \\ 0 & b_i & e_i \cos \gamma_i \\ b_i & a_i & d_i \cos \gamma_i - e_i \sin \gamma_i \end{bmatrix} \qquad (8.3\text{-}15)$$

$$(i = 1, 2, 3)$$

where a_i, b_i, d_i, and e_i are defined by Eqs. (e). In addition, matrix $\mathbf{E}'$ can be reduced to:

$$\mathbf{E}' = \begin{bmatrix} E_{x'x'} & 0 \\ 0 & \dfrac{G_{x'y'}}{1.2} \end{bmatrix} \qquad (8.3\text{-}16)$$

because only $\sigma_{x'}$ and $\tau_{x'y'}$ need be calculated. The matrix $\mathbf{T}_\epsilon$ must also be reduced by deleting rows 2 and 3 and column 3.

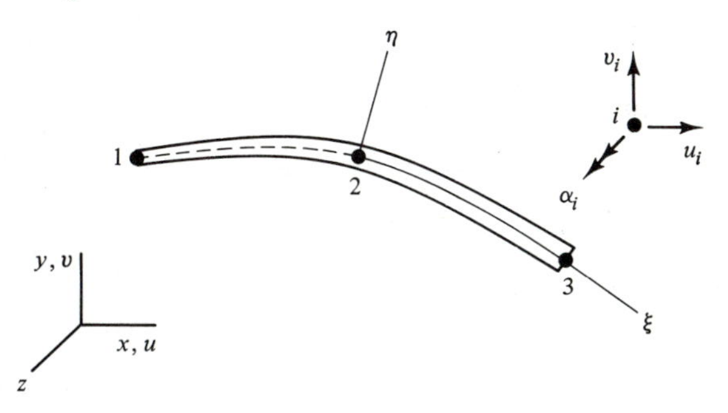

Figure 8.7 Curved Beam Element

8.4 NONAXISYMMETRIC LOADS

For many axisymmetric shells the loads are not distributed axisymmetrically (4, 5, 6). In Sec. 5.4 we expressed the loads that are symmetric with respect to a plane containing the axis of revolution as the Fourier components:

$$b_r = \sum_{j=0}^{m} b_{rj} \cos j\theta \qquad b_z = \sum_{j=0}^{m} b_{zj} \cos j\theta \qquad \begin{matrix}(5.4\text{-}1)\\ \text{repeated}\end{matrix}$$

$$b_\theta = \sum_{j=0}^{m} b_{\theta j} \sin j\theta$$

If the loads were antisymmetric with respect to the plane, the functions $\sin j\theta$ and $\cos j\theta$ would be interchanged. The response of an axisymmetric shell to this series of harmonic loads consists of a series of harmonic generic displacements.

For element AXSH3 these displacements are expressed as follows:

$$\begin{bmatrix} u \\ v \\ w \end{bmatrix} = \sum_{j=1}^{m} \begin{bmatrix} \cos j\theta & 0 & 0 \\ 0 & \cos j\theta & 0 \\ 0 & 0 & \sin j\theta \end{bmatrix}$$

$$\times \left\{ \sum_{i=1}^{3} f_i \begin{bmatrix} u_i \\ v_i \\ w_i \end{bmatrix}_j + \sum_{i=1}^{3} f_i \eta \frac{t_i}{2} \begin{bmatrix} -\sin \gamma_i & 0 \\ \cos \gamma_i & 0 \\ 0 & 1 \end{bmatrix} \begin{bmatrix} \alpha_i \\ \beta_i \end{bmatrix}_j \right\} \tag{8.4-1}$$

where the angle β_i is a small rotation about the x' axis [see Fig. 8.5(a)]. As before, if the loads were antisymmetric with respect to a plane of symmetry, the functions $\sin j\theta$ and $\cos j\theta$ would be interchanged.

For this analysis we must determine the derivatives of the generic displacements in Eq. (8.4-1) with respect to local coordinates. Thus,

$$\begin{bmatrix} u_{,\xi} \\ u_{,\eta} \\ u_{,\theta} \\ v_{,\xi} \\ v_{,\eta} \\ v_{,\theta} \\ w_{,\xi} \\ w_{,\eta} \\ w_{,\theta} \end{bmatrix}_j = \sum_{i=1}^{3} \begin{bmatrix} f_{i,\xi}c_j & 0 & 0 & -f_{i,\xi}\eta s_i c_j & 0 \\ 0 & 0 & 0 & -f_i s_i c_j & 0 \\ -jf_i s_j & 0 & 0 & jf_i \eta s_i s_j & 0 \\ 0 & f_{i,\xi}c_j & 0 & f_{i,\xi}\eta c_i c_j & 0 \\ 0 & 0 & 0 & f_i c_i c_j & 0 \\ 0 & -jf_i s_j & 0 & -jf_i \eta c_i s_j & 0 \\ 0 & 0 & f_{i,\xi}s_j & 0 & f_{i,\xi}\eta s_j \\ 0 & 0 & 0 & 0 & f_i s_j \\ 0 & 0 & jf_i c_j & 0 & jf_i \eta c_j \end{bmatrix} \begin{bmatrix} u_i \\ v_i \\ w_i \\ \frac{t_i}{2}\alpha_i \\ \frac{t_i}{2}\beta_i \end{bmatrix}_j \tag{8.4-2}$$

In the coefficient matrix of this expression, the following abbreviations are used:

$$s_i = \sin \gamma_i \qquad s_j = \sin j\theta$$
$$c_i = \cos \gamma_i \qquad c_j = \cos j\theta \tag{a}$$

The required Jacobian matrix is:

$$\mathbf{J} = \begin{bmatrix} r_{,\xi} & z_{,\xi} & 0 \\ r_{,\eta} & z_{,\eta} & 0 \\ 0 & 0 & 1 \end{bmatrix} \tag{8.4-3a}$$

In this case the inverse of $\mathbf{J}$ is seen to be:

$$\mathbf{J}^{-1} = \mathbf{J}^* = \begin{bmatrix} \xi_{,r} & \eta_{,r} & 0 \\ \xi_{,z} & \eta_{,z} & 0 \\ 0 & 0 & 1 \end{bmatrix} \tag{8.4-3b}$$

Using this inverse matrix, we can transform the derivatives in Eq. (8.4-2) to global coordinates, as follows:

$$\begin{bmatrix} u_{,r} \\ u_{,z} \\ u_{,\theta} \\ \cdots \\ w_{,\theta} \end{bmatrix}_j = \begin{bmatrix} \mathbf{J}^* & \mathbf{O} & \mathbf{O} \\ \mathbf{O} & \mathbf{J}^* & \mathbf{O} \\ \mathbf{O} & \mathbf{O} & \mathbf{J}^* \end{bmatrix} \begin{bmatrix} u_{,\xi} \\ u_{,\eta} \\ u_{,\theta} \\ \cdots \\ w_{,\theta} \end{bmatrix}_j \tag{8.4-4a}$$

Multiplying the terms in this equation produces:

$$\begin{bmatrix} u_{,r} \\ u_{,z} \\ u_{,\theta} \\ v_{,r} \\ v_{,z} \\ v_{,\theta} \\ w_{,r} \\ w_{,z} \\ w_{,\theta} \end{bmatrix}_j = \sum_{i=1}^{3} \begin{bmatrix} a_i c_j & 0 & 0 & -d_i s_i c_j & 0 \\ b_i c_j & 0 & 0 & -e_i s_i c_j & 0 \\ -j f_i s_j & 0 & 0 & \frac{1}{2}(j f_i \eta s_i s_j t_i) & 0 \\ 0 & a_i c_j & 0 & d_i c_i c_j & 0 \\ 0 & b_i c_j & 0 & e_i c_i c_j & 0 \\ 0 & -j f_i s_j & 0 & -\frac{1}{2}(j f_i \eta c_i s_j t_i) & 0 \\ 0 & 0 & a_i s_j & 0 & \frac{d_i s_j t_i}{2} \\ 0 & 0 & b_i s_j & 0 & \frac{e_i s_j t_i}{2} \\ 0 & 0 & j f_i c_j & 0 & \frac{j f_i \eta c_j t_i}{2} \end{bmatrix} \begin{bmatrix} u_i \\ v_i \\ w_i \\ \alpha_i \\ \beta_i \end{bmatrix}_j \tag{8.4-4b}$$

where the constants a_i, b_i, d_i, and e_i are given by Eqs. (e) in Sec. 8.3.

For nonaxisymmetric loads on element AXSH3, we consider six types of nonzero strains. Thus,

$$\boldsymbol{\epsilon}_j = \begin{bmatrix} \epsilon_r \\ \epsilon_z \\ \epsilon_\theta \\ \gamma_{rz} \\ \gamma_{z\theta} \\ \gamma_{r\theta} \end{bmatrix}_j = \begin{bmatrix} u_{,r} \\ v_{,z} \\ \frac{1}{r}(u + w_{,\theta}) \\ u_{,z} + v_{,r} \\ w_{,z} + \frac{v_{,\theta}}{r} \\ \frac{u_{,\theta}}{r} + w_{,r} - \frac{w}{r} \end{bmatrix}_j \tag{8.4-5}$$

The strain-displacement relationships shown in the second form of this vector are the same as those in Eqs. (5.4-2). Using these relationships and Eq. (8.4-4b), we may construct the ith part of matrix $\mathbf{B}$ for the jth harmonic response, as follows:

$$
(\mathbf{B}_i)_j = \begin{bmatrix}
a_i c_j & 0 & 0 & -d_i s_i c_j & 0 \\
0 & b_i c_j & 0 & e_i c_i c_j & 0 \\
\dfrac{f_i c_j}{r} & 0 & \dfrac{j f_i c_j}{r} & -\dfrac{1}{2r}(f_i \eta s_i c_j t_i) & \dfrac{1}{2r}(j f_i \eta c_j t_i) \\
b_i c_j & a_i c_j & 0 & (d_i c_i - e_i s_i) c_j & 0 \\
0 & -\dfrac{j f_i s_j}{r} & b_i s_j & -\dfrac{1}{2r}(j f_i \eta c_i s_j t_i) & \dfrac{1}{2}(e_i s_j t_i) \\
-\dfrac{j f_i s_j}{r} & 0 & \left(a_i - \dfrac{f_i}{r}\right) s_j & \dfrac{1}{2r}(j f_i \eta s_i s_j t_i) & \left(d_i - \dfrac{f_i \eta}{r}\right) \dfrac{s_j t_i}{2}
\end{bmatrix} \tag{8.4-6}
$$

where $i = 1, 2, 3$ and $j = 0, 1, 2, \ldots, m$.

Stress-strain relationships for the present analysis are the same as those in Eqs. (7.3-9) and (7.3-10). Therefore, we can formulate the element stiffness matrix $\mathbf{K}_j$ as:

$$
\mathbf{K}_j = k\pi \int_{-1}^{1} \int_{-1}^{1} \mathbf{B}_j^{\mathrm{T}} \mathbf{E} \ \mathbf{B}_j |\mathbf{J}| r \, d\xi \, d\eta \tag{8.4-7}
$$

where $k = 2$ for $j = 0$ and $k = 1$ for $j = 1, 2, \ldots, m$. Furthermore, equivalent nodal loads for each harmonic set of symmetric body forces have the form:

$$
\mathbf{p}_j = k\pi \int_{-1}^{1} \int_{-1}^{1} \mathbf{f}^{\mathrm{T}} \mathbf{b}_j |\mathbf{J}| r \, d\xi \, d\eta \qquad (j = 0, 1, 2, \ldots, m) \tag{8.4-8}
$$

where

$$
\mathbf{f}_i = \begin{bmatrix} 1 & 0 & 0 & 0 & 0 \\ 0 & 1 & 0 & 0 & 0 \\ 0 & 0 & 1 & 0 & 0 \end{bmatrix} f_i \qquad (i = 1, 2, 3) \tag{b}
$$

and the body forces (per unit volume, constant through the thickness) are:

$$
\mathbf{b}_j = \{b_{rj}, b_{zj}, b_{\theta j}\} \tag{c}
$$

In addition, the equivalent nodal loads due to initial strains may be expressed as:

$$
\mathbf{p}_{0j} = k\pi \int_{-1}^{1} \int_{-1}^{1} \mathbf{B}_j^{\mathrm{T}} \mathbf{T}_\epsilon^{\mathrm{T}} \mathbf{E}' \boldsymbol{\epsilon}'_{0j} |\mathbf{J}| r \, d\xi \, d\eta \qquad (j = 0, 1, 2, \ldots, m) \tag{8.4-9}
$$

in which the initial local strains consist of:

$$
\boldsymbol{\epsilon}'_{0j} = \{\epsilon_{x'0}, \epsilon_{y'0}, \epsilon_{z'0}, \gamma_{x'y'0}, \gamma_{y'z'0}, \gamma_{x'z'0}\} \tag{d}
$$

Finally, the stresses for each harmonic response become:

$$
\boldsymbol{\sigma}'_j = \mathbf{E}'(\mathbf{T}_\epsilon \mathbf{B}_j \mathbf{q}_j - \boldsymbol{\epsilon}'_{0j}) \qquad (j = 0, 1, 2, \ldots, m) \tag{8.4-10}
$$

These stresses are also in the directions of local axes.

8.5 PROGRAM AXSH3 AND APPLICATIONS

In Sec. 8.3 we developed the element named AXSH3. Let us now consider a computer program that uses this element for the analysis of axisymmetric shells. Program AXSH3 applies the expressions given as Eqs. (8.3-7) through (8.3-10) to generate the stiffness matrix of an isotropic element. As in the general shell program SHQ8 (see Sec. 7.4), the thickness of the axisymmetric shell may be different at each node. Also, the cosine and sine of the angle γ [see Figs. 8.5(a) and (b)] must be given for each node. Stresses are computed in local directions at numerical integration points, as indicated by Eq. (7.3-14). The following problems are all axisymmetric shells with axisymmetric loads.

Example 1

For the first example, we take a single element AXSH3 with hoop stress, as illustrated in Fig. 8.8. This problem consists of a circular cylinder that has an internal pressure b_r (force per unit area). The primary objective is to compute the hoop stress σ_θ in the wall of the cylinder. For this purpose, the following values of physical parameters are assumed:

$$L = 0.3 \text{ m} \qquad b_r = 7 \times 10^2 \text{ kN/m}^2$$
$$E = 1 \times 10^8 \text{ kN/m}^2 \qquad \nu = 0.3$$

where the units are S.I. and the material is copper.

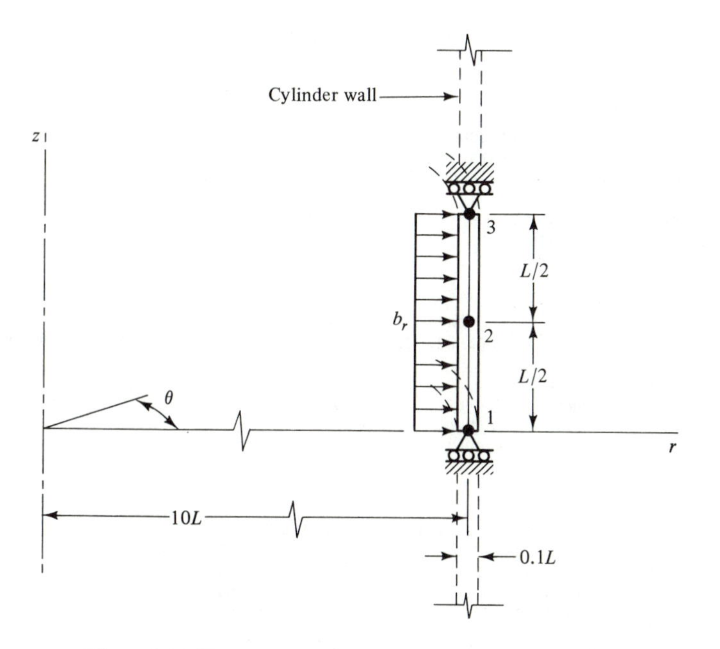

Figure 8.8 Example 1: Element AXSH3 with Hoop Stress

The computer output from Program AXSH3 for this example appears in Table 8.1. For area loads the first three decimal numbers on the line are force intensities in the r direction at the first, second, and third nodes of the element (allowing quadratic distribution). Similarly, the last three numbers on the line give force intensities in the z direction. Due to the internal pressure, the stress σ_θ in the element is 7×10^4 kN/m^2, as it should be. For each integration point the membrane stresses (MS) are printed on the first line, and flexural stresses (FS) are printed on the second line. As before, the transverse shearing stress $\tau_{x'y'}$ in the second line is equal to the average value multiplied by the factor 1.5.

TABLE 8.1 Computer Output for Example 1

```
PROGRAM AXSH3

***  EXAMPLE 1:  ONE ELEMENT WITH HOOP STRESS  ***

STRUCTURAL PARAMETERS
   NN   NE  NRN  NLS           E            PR            T
    3    1    2    1  1.0000D 08  3.0000D-01  3.0000D-03

COORDINATES OF NODES
 NODE           R           Z  COS(GAMMA)  SIN(GAMMA)           T
    1  3.0000D-01  0.0000D-01  1.0000D 00  0.0000D-01  3.0000D-03
    2  3.0000D-01  1.5000D-02  1.0000D 00  0.0000D-01  3.0000D-03
    3  3.0000D-01  3.0000D-02  1.0000D 00  0.0000D-01  3.0000D-03

ELEMENT INFORMATION
ELEM.   N1   N2   N3
    1    3    2    1

NODAL RESTRAINTS
 NODE   R1   R2   R3
    1    0    1    0
    3    0    1    0

NUMBER OF DEGREES OF FREEDOM =    7
NUMBER OF NODAL RESTRAINTS   =    2
NUMBER OF TERMS IN SN        =   37

**********  LOADING NUMBER    1  **********
  NLN  NEA  NEV  NEH  IPR
    0    1    0    0    0

AREA LOADS
ELEM.         BA1         BA2         BA3         BA4         BA5         BA6
    1  7.0000D 02  7.0000D 02  7.0000D 02  0.0000D-01  0.0000D-01  0.0000D-01

NODAL DISPLACEMENTS
 NODE         DN1         DN2         DN3
    1  1.9110D-04 -0.0000D-01 -1.2438D-15
    2  1.9110D-04  1.2870D-19 -1.1443D-15
    3  1.9110D-04 -0.0000D-01 -1.0743D-15

ELEMENT STRESSES
ELEM. INT.  X' STRESS   Z' STRESS X'Y' STRESS
    1    1  2.1000D 04  7.0000D 04              MS
            7.4445D-10  2.2333D-10  6.6881D-11  FS
    1    2  2.1000D 04  7.0000D 04              MS
            1.1181D-09  3.3543D-10 -1.8958D-12  FS

SUPPORT REACTIONS
 NODE         AR1         AR2         AR3
    1  0.0000D-01 -1.1875D 02  0.0000D-01
    3  0.0000D-01  1.1875D 02  0.0000D-01
```

Example 2

Figure 8.9(a) shows a hyperboloidal cooling tower constructed of reinforced concrete. Prestress cables apply a ring load of intensity P_r (force per unit circumferential length) in the radial direction at the top edge of the tower. The wall has a constant thickness

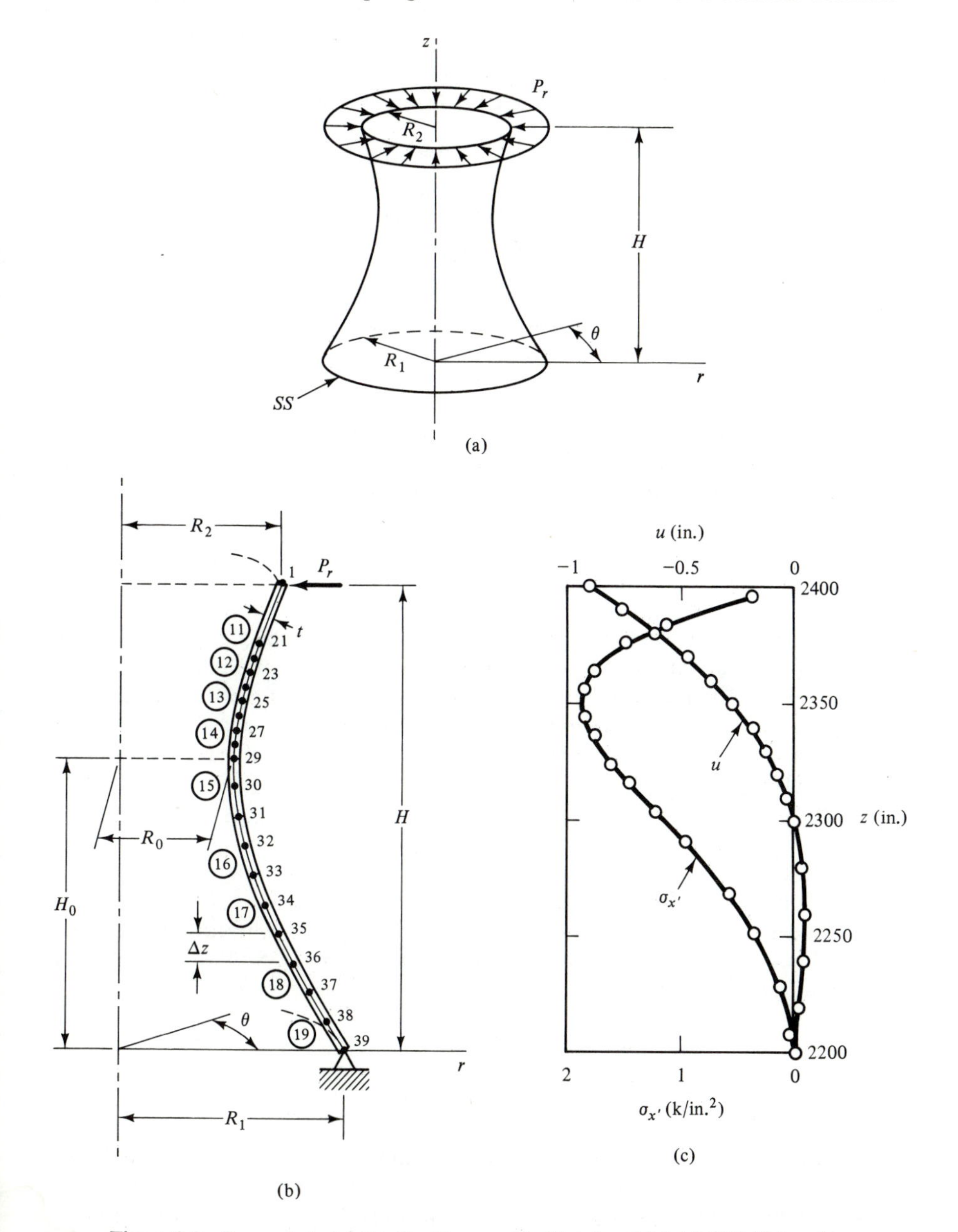

Figure 8.9 Example 2: (a) Cooling Tower with Ring Load (b) AXSH3 Network (c) Deflection and Stress Graphs

t and is simply supported at the base. Geometric details of this problem appear more clearly in Fig. 8.9(b), which shows the dimensions as well as a network of AXSH3 elements. Essential physical parameters are known to be:

$$R_0 = 600 \text{ in.} \qquad R_1 = 2R_0 \qquad H = 4R_0 \qquad H_0 = 2.5R_0$$
$$t = 8 \text{ in.} \qquad E = 3 \times 10^3 \text{ k/in.}^2 \qquad \nu = 0.3 \qquad P = 1 \text{ k/in.}$$

in which the units are U.S.

From the hyperbolic geometry, we locate nodes with the formula:

$$r^2 = 0.48(z - H_0)^2 + R_0^2$$

Nodes at the top of the tower are spaced very closely because of the concentrated ring load. However, those near the bottom have more space between, as indicated in Fig. 8.9(b). The increments Δz between nodes are:

Nodes	1–11	11–21	21–29	29–39
Δz (in.)	10	20	75	150

By running the given information on Program AXSH3, we find (as expected) that the maximum translational deflection is $u_1 = -0.904$ in. at node 1. Also, the maximum flexural stress is $\sigma_{x'} = 1.876$ k/in.^{2} in element 3 at the integration point near node 5. Graphs of the translation u and the stress $\sigma_{x'}$ near the top of the tower are given in Fig. 8.9(c). Both these quantities die out to negligible values within 200 inches of the top.

Example 3

Suppose that a spherical water tank is supported at its midheight, as indicated in Fig. 8.10(a). If the tank is full, the magnitude of the hydrostatic pressure varies as shown in Fig. 8.10(b). The latter figure also depicts a network of AXSH3 elements and gives other geometric details. Note that the tank is restrained against translations at midheight, and the thickness t is assumed to be constant. Physical parameters for this shell are given as:

$$R = 10 \text{ m} \qquad t = 0.04 \text{ m} \qquad E = 2 \times 10^8 \text{ kN/m}^2$$
$$\nu = 0.3 \qquad w_f = 10 \text{ kN/m}^3$$

for which the material is steel and the units are S.I.

Nodes near the support at midheight are close together, whereas those near the top and bottom are farther apart. Because the section in Fig. 8.10(b) is circular, we can locate nodes using an incremental central angle $\Delta\phi$, as follows:

Nodes	1–11	11–21	21–41	41–51	51–61
$\Delta\phi$ (°)	4.5	3.0	1.5	3.0	4.5

Axisymmetric shells make excellent fluid storage tanks. *Equivalent nodal loads due to hydrostatic pressure* may be calculated in a manner very similar to the method described in Example 3 of Sec. 7.4. For element AXSH3, however, the components of the pressure normal to the neutral surface of the shell at the jth integration point are

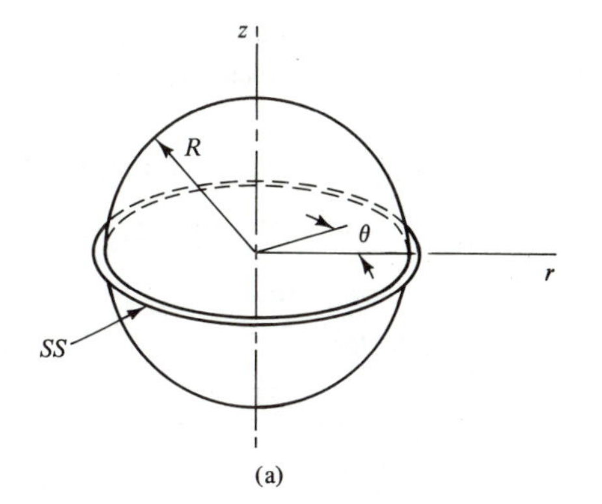

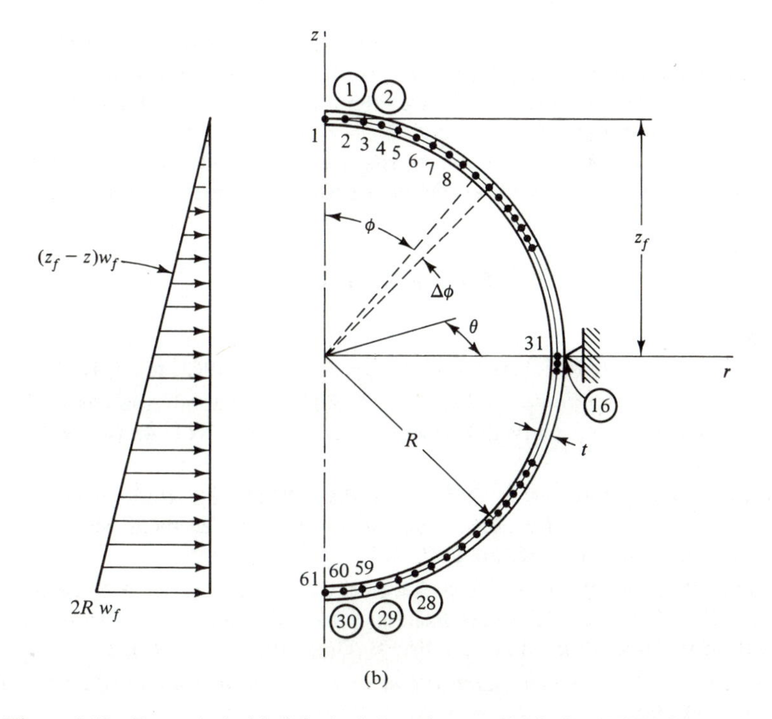

Figure 8.10 Example 3: (a) Spherical Containment Tank (b) AXSH3 Network

computed from:

$$\mathbf{b}_f = \pm(z_f - z_j)w_f\mathbf{e}_\eta^{\mathrm{T}} \tag{a}$$

In this formula the vector $\mathbf{e}_\eta$ contains two direction cosines of the normal to the neutral surface. Thus,

$$\mathbf{e}_\eta = \frac{1}{c}[r_{,\eta} \quad z_{,\eta}] \tag{b}$$

where

$$c = \sqrt{(r_{,\eta})^2 + (z_{,\eta})^2} \tag{c}$$

As before, the sign of vector $\mathbf{b}_f$ will be positive when the positive sense of the hydrostatic pressure is the same as that of the vector $\mathbf{e}_\eta$. Then the equivalent nodal loads become:

$$\mathbf{p}_f = \int_A \mathbf{f}^{\mathrm{T}}\mathbf{b}_f \, dA = 2\pi \int_{-1}^{1} \mathbf{f}^{\mathrm{T}}\mathbf{b}_f r \,|\mathbf{J}'|\, d\xi \tag{d}$$

Now the expression for dA is:

$$dA = 2\pi r \,|\mathbf{J}'|\, d\xi \tag{e}$$

The determinant of $\mathbf{J}'$ may be calculated as:

$$|\mathbf{J}'| = (\mathbf{e}_\zeta \times \mathbf{V}_\xi) \cdot \mathbf{e}_\eta = \begin{vmatrix} 1 & 0 & 0 \\ 0 & r_{,\xi} & z_{,\xi} \\ 0 & \dfrac{r_{,\eta}}{c} & \dfrac{z_{,\eta}}{c} \end{vmatrix} = \begin{vmatrix} r_{,\xi} & z_{,\xi} \\ \dfrac{r_{,\eta}}{c} & \dfrac{z_{,\eta}}{c} \end{vmatrix} \tag{f}$$

In this case the 2×2 matrix $\mathbf{J}'$ is the same as $\mathbf{J}$ in Eq. (8.3-4a), except that the second row is normalized to have unit length (to transform arc length instead of area).

When the data for this example are submitted to Program AXSH3, they produce a maximum translational displacement of $v_{61} = -1.53 \times 10^{-3}$ m at node 61, which is the bottom of the tank. In addition, the maximum value of normal stress $\sigma_{x'}$ is found to be 3.376×10^4 kN/m^2 at the inner surface of element 16 near node 31.

REFERENCES

1. Grafton, P. E., and Strome, D. R., "Analysis of Axisymmetrical Shells by the Direct Stiffness Method," *AIAA Jour.*, Vol. 1, No. 10, 1963, pp. 2342–2347.
2. Jones, R. E., and Strome, D. R., "Direct Stiffness Method Analysis of Shells of Revolution Utilizing Curved Elements," *AIAA Jour.*, Vol. 4, No. 9, 1966, pp. 1519–1525.
3. Stricklin, J. A., Navaratna, D. R., and Pian, T. H. H., "Improvements on the Analysis of Shells of Revolution by the Matrix Displacement Method," *AIAA Jour.*, Vol. 4, No. 11, 1966, pp. 2069–2072.
4. Ahmad, S., Irons, B. M., and Zienckiewicz, O. C., "Curved Thick Shell and Membrane Elements with Particular Reference to Axisymmetric Problems," *Proc. 2d Conf. Mat. Meth. Struc. Mech.*, WPAFB, Ohio, 1968, pp. 539–572.
5. Cook, R. D., *Concepts and Applications of Finite Element Analysis*, 2d ed., Wiley, New York, 1981.
6. Percy, J. H., Pian, T. H. H., Klein, S., and Navaratna, D. R., "Application of Matrix Displacement Method to Linear Elastic Analysis of Shells of Revolution," *AIAA Jour.*, Vol. 3, No. 11, 1965, pp. 2138–2145.

CHAPTER

9

Vibrational Analysis

9.1 INTRODUCTION

Any linearly elastic continuum will have natural frequencies and modes of vibration that can be investigated by considering the mass of the body as well as its stiffness. Figure 9.1 shows an infinitesimal element in Cartesian coordinates with applied body forces that vary with time t. These forces are $b_x(t)\,dV$, $b_y(t)\,dV$, and $b_z(t)\,dV$. The figure also indicates inertial body forces $\rho\ddot{u}\,dV$, $\rho\ddot{v}\,dV$, and $\rho\ddot{w}\,dV$, where the double dots mean second derivatives of the generic translations u, v, and w with respect to time. The symbol ρ in these expressions represents the *mass density*, which is defined as the inertial force per unit acceleration per unit volume. Note that the inertial forces act in directions that are opposite to the positive senses of the accelerations. We can write equations of small, undamped motion for a generic point within a finite element by using the principle of virtual work, as follows:

$$\int_V \delta\boldsymbol{\epsilon}^{\mathrm{T}}\boldsymbol{\sigma}\,dV = \delta\mathbf{q}^{\mathrm{T}}\mathbf{p}(t) + \int_V \delta\mathbf{u}^{\mathrm{T}}\mathbf{b}(t)\,dV - \int_V \delta\mathbf{u}^{\mathrm{T}}\rho\ddot{\mathbf{u}}\,dV \qquad (9.1\text{-}1)$$

The inertial term appears last in this equation and carries a negative sign. Now assume that

$$\mathbf{u} = \mathbf{f}\,\mathbf{q} \qquad \ddot{\mathbf{u}} = \mathbf{f}\,\ddot{\mathbf{q}} \qquad (\mathrm{a})$$

which implies that acceleration shape functions are the same as displacement shape functions $\mathbf{f}$. Substitution of Eqs. (a) and other previously derived relationships into Eq. (9.1-1) gives:

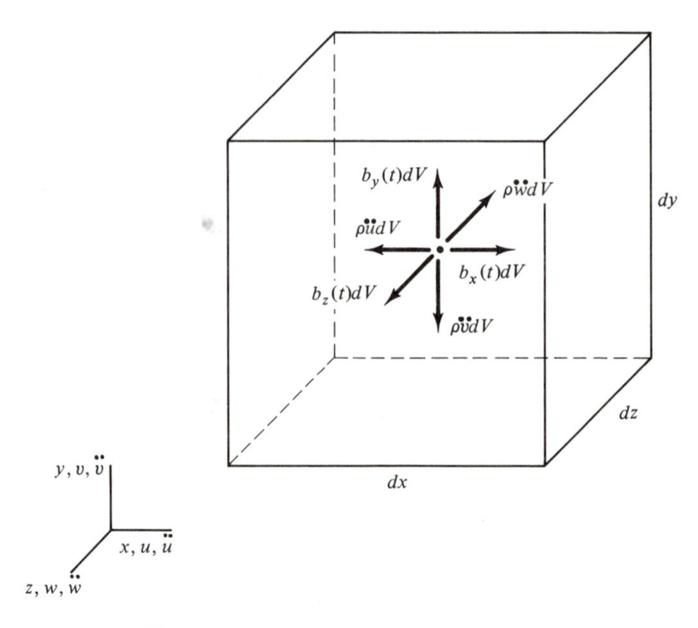

Figure 9.1 Applied and Inertial Body Forces

$$\delta\mathbf{q}^{\mathrm{T}} \int_V \mathbf{B}^{\mathrm{T}}\mathbf{E}\ \mathbf{B}\,dV\,\mathbf{q} = \delta\mathbf{q}^{\mathrm{T}}\mathbf{p} + \delta\mathbf{q}^{\mathrm{T}} \int_V \mathbf{f}^{\mathrm{T}}\mathbf{b}\,dV - \delta\mathbf{q}^{\mathrm{T}} \int_V \rho\mathbf{f}^{\mathrm{T}}\mathbf{f}\,dV\,\ddot{\mathbf{q}} \tag{b}$$

Cancellation of $\delta\mathbf{q}^{\mathrm{T}}$ and rearrangement of this equation produces:

$$\mathbf{M}\ \ddot{\mathbf{q}} + \mathbf{K}\ \mathbf{q} = \mathbf{p} + \mathbf{p}_b \tag{9.1-2}$$

in which the *equivalent* (or *consistent*) *mass matrix* (1, 2) for the element is:

$$\mathbf{M} = \int_V \rho\mathbf{f}^{\mathrm{T}}\mathbf{f}\,dV \tag{9.1-3}$$

This matrix is symmetric because it is formed using the product of $\mathbf{f}$ and its transpose. Alternatively, the consistent mass matrix may be written as:

$$\mathbf{M} = \mathbf{h}^{-\mathrm{T}} \int_V \rho\mathbf{g}^{\mathrm{T}}\mathbf{g}\,dV\,\mathbf{h}^{-1} \tag{9.1-4}$$

For a concentrated mass M, no integration is required; and Eq. (9.1-3) simplifies to:

$$\mathbf{M} = M\mathbf{f}^{\mathrm{T}}\mathbf{f} \tag{9.1-5}$$

wherever matrix $\mathbf{f}$ pertains to generic translations.

A consistent mass matrix, as given by Eqs. (9.1-3), (9.1-4), or (9.1-5), contains inertial actions at the nodes of an element due to unit accelerations of the nodes. Thus, by using the virtual work principle, we replace distributed inertial actions with energy-equivalent inertial actions concentrated at the nodes. As for stiffnesses and equivalent nodal loads, these inertial actions are entirely fictitious and are used only for analytical purposes.

If it is necessary to transform a consistent mass matrix from local to global directions, we do the following:

$$\mathbf{M} = \hat{\mathbf{R}}^{\mathrm{T}}\mathbf{M}'\hat{\mathbf{R}} \tag{9.1-6}$$

which is the same rotation-of-axes transformation as that derived for stiffnesses in Sec. 1.7 [see Eq. (1.7-18)]. Then the assembly of masses for the whole structure takes the form:

$$\mathbf{M}_N = \sum_{i=1}^{n_e} \mathbf{M}_i \tag{c}$$

as described in Sec. 1.8 for stiffnesses and equivalent nodal loads. The symbol $\mathbf{M}_N$ in Eq. (c) represents the consistent mass matrix for all the nodes in the assemblage. Adding inertial actions to the left-hand side of Eq. (1.8-2), we obtain:

$$\mathbf{M}_N\ddot{\mathbf{D}}_N + \mathbf{S}_N\mathbf{D}_N = \mathbf{A}_N + \mathbf{A}_{Nb} \tag{9.1-7}$$

in which $\ddot{\mathbf{D}}_N$ is a vector of accelerations for all the nodes in the structure. The vector $\mathbf{A}_{N0}$ of equivalent nodal loads due to initial strains is omitted from Eq. (9.1-7) because that influence is considered to be a static problem.

Next, we can set the terms on the right-hand side of Eq. (9.1-7) equal to zero, which yields:

$$\mathbf{M}_N\ddot{\mathbf{D}}_N + \mathbf{S}_N\mathbf{D}_N = \mathbf{O} \tag{9.1-8}$$

This expression is the undamped *equation of motion for free vibrations* of a discretized continuum. The determination of natural frequencies and mode shapes from this equation will be covered in Sec. 9.3.

In many problems it is sufficiently accurate to merely lump tributary masses at the nodes of a discretized continuum (3). In such a case the *lumped-mass matrix* $\mathbf{M}_L$ for the whole structure is:

$$\mathbf{M}_L = \begin{bmatrix} \mathbf{M}_1 & \mathbf{O} & \cdots & \mathbf{O} & \cdots & \mathbf{O} \\ \mathbf{O} & \mathbf{M}_2 & \cdots & \mathbf{O} & \cdots & \mathbf{O} \\ \cdots & \cdots & \cdots & \cdots & \cdots & \cdots \\ \mathbf{O} & \mathbf{O} & \cdots & \mathbf{M}_j & \cdots & \mathbf{O} \\ \cdots & \cdots & \cdots & \cdots & \cdots & \cdots \\ \mathbf{O} & \mathbf{O} & \cdots & \mathbf{O} & \cdots & \mathbf{M}_{n_n} \end{bmatrix} \tag{9.1-9}$$

where n_n is the number of nodes. The submatrix $\mathbf{M}_j$ in Eq. (9.1-9) denotes a small diagonal array defined as:

$$\mathbf{M}_j = M_j\mathbf{I}_0 \tag{d}$$

In this expression M_j is the tributary mass lumped at node j, and $\mathbf{I}_0$ is an identity matrix with 1 replaced by 0 wherever a nontranslational displacement occurs. Thus, the lumped-mass approach has the advantage that the mass matrix $\mathbf{M}_L$ is always diagonal.

9.2 CONSISTENT MASS MATRICES

Using the formulas developed in the preceding article, we can now obtain consistent mass matrices for various finite elements. We begin with one-dimensional elements (see Sec. 1.5), of which the axial element is the simplest. Figure 9.2(a) shows the cross-sectional translation u for an axial element, in which the corresponding acceleration $\ddot{u}$ (multiplied by the distributed mass) produces a distributed axial inertia force. Instead of handling the distributed mass directly, we will generate the fictitious nodal masses contained in the consistent mass matrix $\mathbf{M}$. Toward that end, we write the geometric matrix $\mathbf{g}$ for the axial

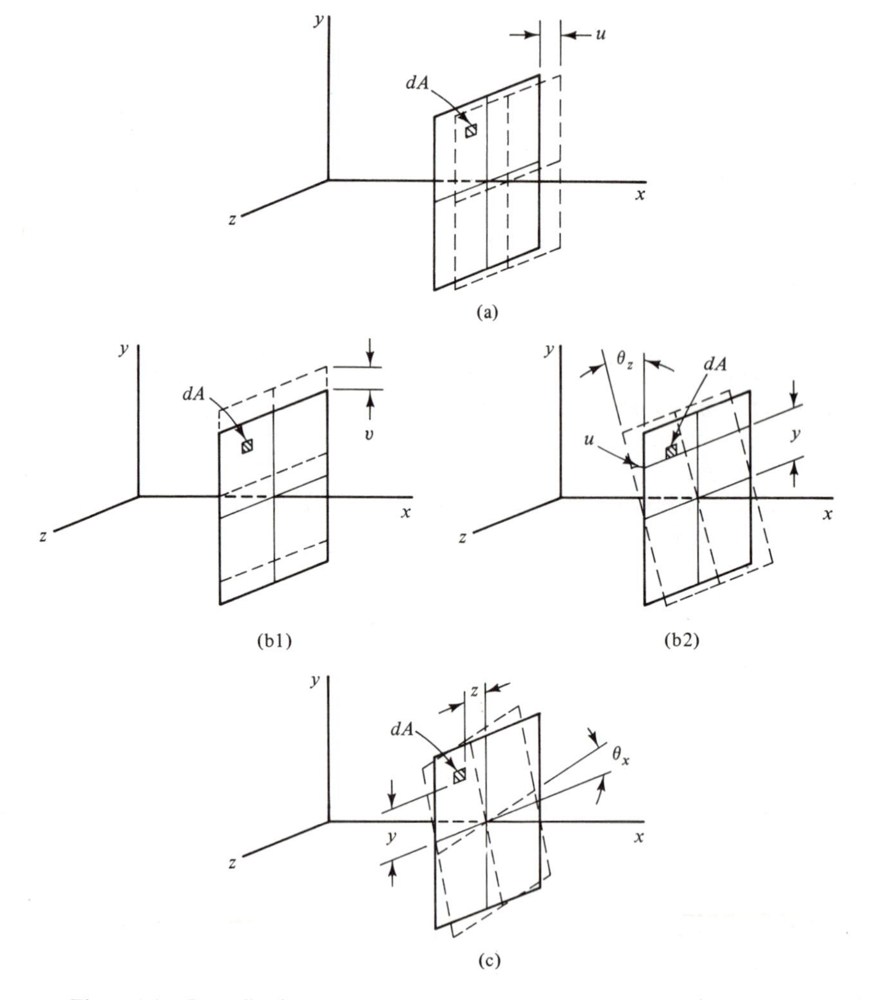

Figure 9.2 Cross-Sectional Displacements: (a) Axial Element (b) Flexural Element (c) Torsional Element

element as:

$$\mathbf{g} = [1 \quad x] \tag{a}$$

Then premultiply matrix $\mathbf{g}$ by its own transpose to obtain:

$$\mathbf{g}^{\mathrm{T}}\mathbf{g} = \begin{bmatrix} 1 & x \\ x & x^2 \end{bmatrix} \tag{b}$$

Integration of this product over the length L of the element produces:

$$\int_0^L \mathbf{g}^{\mathrm{T}}\mathbf{g}\, dx = \begin{bmatrix} L & \frac{1}{2}L^2 \\ \frac{1}{2}L^2 & \frac{1}{3}L^3 \end{bmatrix} \tag{c}$$

From Sec. 1.4 we have:

$$\mathbf{h}^{-1} = \begin{bmatrix} 1 & 0 \\ -\dfrac{1}{L} & \dfrac{1}{L} \end{bmatrix} \tag{d}$$

Using Eqs. (c) and (d) in Eq. (9.1-4) yields:

$$\mathbf{M} = \rho A \mathbf{h}^{-\mathrm{T}} \int_0^L \mathbf{g}^{\mathrm{T}}\mathbf{g}\, dx\, \mathbf{h}^{-1} = \frac{\rho A L}{6}\begin{bmatrix} 2 & 1 \\ 1 & 2 \end{bmatrix} \tag{9.2-1}$$

in which the constant cross-sectional area A is:

$$A = \int_A dA \tag{e}$$

We see from Eq. (9.2-1) that the sum of all of the terms in matrix $\mathbf{M}$ is ρAL, as it should be.

The second type of one-dimensional element is the beam, or flexural element. A typical cross section of this type of member translates in the y direction, as shown in Fig. 9.2(b1). However, the section also rotates about its neutral axis, as indicated in Fig. 9.2(b2). The *translational inertia* terms are much more important than the rotational terms, so they will be considered first. From Sec. 1.5 we have the geometric matrix:

$$\mathbf{g} = [1 \quad x \quad x^2 \quad x^3] \tag{f}$$

Then the product $\mathbf{g}^{\mathrm{T}}\mathbf{g}$ becomes:

$$\mathbf{g}^{\mathrm{T}}\mathbf{g} = \begin{bmatrix} 1 & x & x^2 & x^3 \\ x & x^2 & x^3 & x^4 \\ x^2 & x^3 & x^4 & x^5 \\ x^3 & x^4 & x^5 & x^6 \end{bmatrix} \tag{g}$$

And the integral of this product over the length L is:

$$\int_0^L \mathbf{g}^{\mathrm{T}}\mathbf{g}\, dx = \begin{bmatrix} L & \frac{1}{2}L^2 & \frac{1}{3}L^3 & \frac{1}{4}L^4 \\ \frac{1}{2}L^2 & \frac{1}{3}L^3 & \frac{1}{4}L^4 & \frac{1}{5}L^5 \\ \frac{1}{3}L^3 & \frac{1}{4}L^4 & \frac{1}{5}L^5 & \frac{1}{6}L^6 \\ \frac{1}{4}L^4 & \frac{1}{5}L^5 & \frac{1}{6}L^6 & \frac{1}{7}L^7 \end{bmatrix} \tag{h}$$

From Sec. 1.5 we also use:

$$\mathbf{h}^{-1} = \frac{1}{L^3}\begin{bmatrix} L^3 & 0 & 0 & 0 \\ 0 & L^3 & 0 & 0 \\ -3L & -2L^2 & 3L & -L^2 \\ 2 & L & -2 & L \end{bmatrix} \tag{i}$$

Substitution of Eqs. (h) and (i) into Eq. (9.1-4) produces:

$$\mathbf{M}_t = \frac{\rho AL}{420}\begin{bmatrix} 156 & 22L & 54 & -13L \\ 22L & 4L^2 & 13L & -3L^2 \\ 54 & 13L & 156 & -22L \\ -13L & -3L^2 & -22L & 4L^2 \end{bmatrix} \tag{9.2-2}$$

which is the *consistent mass matrix for translational inertia* in a prismatic beam.

Rotational inertia (or *rotary inertia*) terms for a beam can be deduced from Fig. 9.2(b2), where the translation u in the x direction of a point on the cross section is seen to be:

$$u = -y\theta_z \tag{j}$$

In this expression,

$$\theta_z = v_{,x} = \mathbf{f}_{,x}\mathbf{q} = \mathbf{g}_{,x}\mathbf{h}^{-1}\mathbf{q} \tag{k}$$

By using the principle of virtual work as before and integrating over the cross section, we find:

$$\mathbf{M}_r = \int_0^L \rho I \mathbf{f}_{,x}^{\mathrm{T}}\mathbf{f}_{,x}\, dx \tag{9.2-3}$$

in which the moment of inertia I of the cross section about the neutral axis is:

$$I = \int_A y^2\, dA \tag{ℓ}$$

Alternatively, we can write Eq. (9.2-3) as:

$$\mathbf{M}_r = \mathbf{h}^{-\mathrm{T}} \int_0^L \rho I \mathbf{g}_{,x}^{\mathrm{T}}\mathbf{g}_{,x}\, dx\, \mathbf{h}^{-1} \tag{9.2-4}$$

To evaluate this formula, we need the matrix $\mathbf{g}_{,x}$, which is obtained from Eq. (f), as follows:

$$\mathbf{g}_{,x} = [0 \quad 1 \quad 2x \quad 3x^2] \tag{m}$$

Premultiplication of matrix $\mathbf{g}_{,x}$ by its own transpose gives:

$$\mathbf{g}_{,x}^{\mathrm{T}}\mathbf{g}_{,x} = \begin{bmatrix} 0 & 0 & 0 & 0 \\ 0 & 1 & 2x & 3x^2 \\ 0 & 2x & 4x^2 & 6x^3 \\ 0 & 3x^2 & 6x^3 & 9x^4 \end{bmatrix} \tag{n}$$

Then integration of this product over the length L yields:

$$\int_0^L \mathbf{g}_{,x}^{\mathrm{T}}\mathbf{g}_{,x}\,dx = \begin{bmatrix} 0 & 0 & 0 & 0 \\ 0 & L & L^2 & L^3 \\ 0 & L^2 & \frac{4}{3}L^3 & \frac{3}{2}L^4 \\ 0 & L^3 & \frac{3}{2}L^4 & \frac{9}{5}L^5 \end{bmatrix} \tag{o}$$

Finally, substitution of Eqs. (i) and (o) into Eq. (9.2-4) results in:

$$\mathbf{M}_r = \frac{\rho I}{30L}\begin{bmatrix} 36 & 3L & -36 & 3L \\ 3L & 4L^2 & -3L & -L^2 \\ -36 & -3L & 36 & -3L \\ 3L & -L^2 & -3L & 4L^2 \end{bmatrix} \tag{9.2-5}$$

which is the *consistent mass matrix for rotational inertia* in a prismatic beam.

Alternatively, we could derive the formula for $\mathbf{M}_r$, as given by Eq. (9.2-3), using the generalized inertial moment $-\rho I\ddot{\theta}_z\,dx$ and the corresponding virtual rotation $\delta\theta_z$ of the cross section. Additional contributions to matrix $\mathbf{M}$ due to shearing deformations have also been developed and are given in (4).

The third type of one-dimensional element is the torsional element, for which the rotated cross section appears in Fig. 9.2(c). Due to the small rotation θ_x, there are two components of translation at any point on the cross section, which are:

$$v = -z\theta_x \qquad w = y\theta_x \tag{p}$$

The value of θ_x is:

$$\theta_x = \mathbf{f}\;\mathbf{q} = \mathbf{g}\;\mathbf{h}^{-1}\mathbf{q} \tag{q}$$

where matrices $\mathbf{g}$ and $\mathbf{h}^{-1}$ are given by Eqs. (a) and (d). Application of Eq. (9.1-3) and integration over the cross section yields:

$$\mathbf{M} = \int_0^L \rho J\mathbf{f}^{\mathrm{T}}\mathbf{f}\,dx \tag{9.2-6}$$

In this expression the symbol J represents the polar moment of inertia, as follows:

$$J = \int_A (y^2 + z^2)\,dA \tag{r}$$

The alternative form of Eq. (9.2-6) is:

$$\mathbf{M} = \mathbf{h}^{-\mathrm{T}}\int_0^L \rho J\mathbf{g}^{\mathrm{T}}\mathbf{g}\,dx\,\mathbf{h}^{-1} \tag{9.2-7}$$

Evaluation of this expression leads to:

$$\mathbf{M} = \frac{\rho JL}{6}\begin{bmatrix} 2 & 1 \\ 1 & 2 \end{bmatrix} \tag{9.2-8}$$

which is the consistent mass matrix for a torsional element of constant cross section. This matrix could also be derived using the generalized inertial moment $-\rho J\ddot{\theta}_x\,dx$ and the corresponding virtual rotation $\delta\theta_x$.

We now consider two-dimensional finite elements of the types shown in Figs. 9.3(a), (b), and (c). The first of these figures illustrates the constant strain triangle (CST), which was thoroughly discussed in Sec. 2.2. For this element we have:

$$\mathbf{g} = \begin{bmatrix} 1 & x & y & 0 & 0 & 0 \\ 0 & 0 & 0 & 1 & x & y \end{bmatrix} \tag{s}$$

Premultiplication of matrix **g** with its own transpose yields:

$$\mathbf{g}^{\mathrm{T}}\mathbf{g} = \begin{bmatrix} \mathbf{W} & \mathbf{O} \\ \mathbf{O} & \mathbf{W} \end{bmatrix} \tag{t}$$

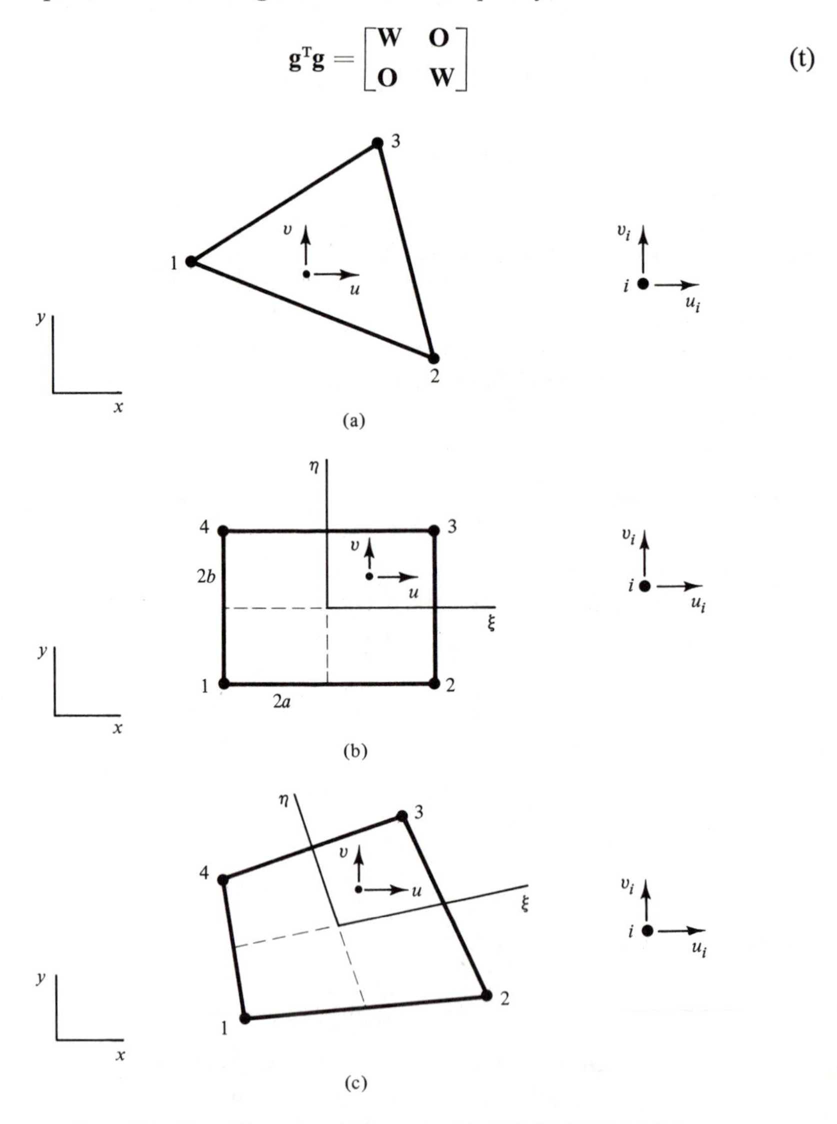

Figure 9.3 Two-Dimensional Elements: (a) CST (b) BDR (c) Q4

in which the submatrix $\mathbf{W}$ is:

$$\mathbf{W} = \begin{bmatrix} 1 & x & y \\ x & x^2 & xy \\ y & xy & y^2 \end{bmatrix} \tag{u}$$

As before, let

$$\mathbf{h}^{-1} = \mathbf{h}_T^{-1}\mathbf{T} = \begin{bmatrix} \mathbf{h}_1^{-1} & \mathbf{O} \\ \mathbf{O} & \mathbf{h}_1^{-1} \end{bmatrix} \mathbf{T} \tag{v}$$

where the matrices $\mathbf{T}$ and $\mathbf{h}_1^{-1}$ are given by Eqs. (g) and (j) in Sec. 2.2. Substitution of Eq. (v) into Eq. (9.1-4) produces:

$$\mathbf{M} = \mathbf{T}^{\mathrm{T}}\mathbf{M}_T\mathbf{T} \tag{9.2-9}$$

in which

$$\mathbf{M}_T = \mathbf{h}_T^{-\mathrm{T}} \int_V \rho \mathbf{g}^{\mathrm{T}}\mathbf{g}\, dV\, \mathbf{h}_T^{-1} \tag{9.2-10}$$

After the integration and multiplication shown, the resulting form of matrix $\mathbf{M}_T$ is:

$$\mathbf{M}_T = \begin{bmatrix} \mathbf{M}_1 & \mathbf{O} \\ \mathbf{O} & \mathbf{M}_1 \end{bmatrix} \tag{9.2-11}$$

where

$$\mathbf{M}_1 = \frac{\rho V}{12} \begin{bmatrix} 2 & 1 & 1 \\ 1 & 2 & 1 \\ 1 & 1 & 2 \end{bmatrix} \tag{9.2-12}$$

and $V = At$. Then the final arrangement of terms in matrix $\mathbf{M}$ is given by Eq. (9.2-9). As with the axial element, we observe that adding all the terms in matrix $\mathbf{M}_1$ gives the total mass of the CST element. However, in this case $\mathbf{M}_1$ appears twice—once for the x direction and once for the y direction.

The geometric difficulties involved when applying Eq. (9.2-10) to a triangular element can be avoided by using area coordinates, as described in Sec. 3.2. For this purpose we rewrite Eq. (9.2-10) as:

$$\mathbf{M}_T = \int_V \rho \mathbf{f}_T^{\mathrm{T}}\mathbf{f}_T\, dV \tag{9.2-13}$$

in which

$$\mathbf{f}_T = \mathbf{g}\ \mathbf{h}_T^{-1} \tag{w}$$

Then for the constant strain triangle we have:

$$\mathbf{f}_T = \begin{bmatrix} \xi_1 & \xi_2 & \xi_3 & 0 & 0 & 0 \\ 0 & 0 & 0 & \xi_1 & \xi_2 & \xi_3 \end{bmatrix} \tag{x}$$

If this array is premultiplied by its own transpose, the result is:

$$\mathbf{f}_T^{\mathrm{T}}\mathbf{f}_T = \begin{bmatrix} \mathbf{X} & \mathbf{O} \\ \mathbf{O} & \mathbf{X} \end{bmatrix} \tag{y}$$

where the submatrix **X** is seen to be

$$\mathbf{X} = \begin{bmatrix} \xi_1^2 & \xi_1\xi_2 & \xi_1\xi_3 \\ \xi_1\xi_2 & \xi_2^2 & \xi_2\xi_3 \\ \xi_1\xi_3 & \xi_2\xi_3 & \xi_3^2 \end{bmatrix} \tag{z}$$

Integrating $\rho\mathbf{X}$ over the volume using Eq. (3.2-13), we find the terms given previously in Eq. (9.2-12).

The bilinear displacement rectangle (BDR) appears in Fig. 9.3(b). For this element the matrix **g** in dimensionless coordinates is:

$$\mathbf{g} = \begin{bmatrix} 1 & \xi & \eta & \xi\eta & 0 & 0 & 0 & 0 \\ 0 & 0 & 0 & 0 & 1 & \xi & \eta & \xi\eta \end{bmatrix} \tag{a$'$}$$

Then the product $\mathbf{g}^{\mathrm{T}}\mathbf{g}$ becomes:

$$\mathbf{g}^{\mathrm{T}}\mathbf{g} = \begin{bmatrix} \mathbf{W} & \mathbf{O} \\ \mathbf{O} & \mathbf{W} \end{bmatrix} \tag{b$'$}$$

In this case the submatrix **W** is as follows:

$$\mathbf{W} = \begin{bmatrix} 1 & \xi & \eta & \xi\eta \\ \xi & \xi^2 & \xi\eta & \xi^2\eta \\ \eta & \xi\eta & \eta^2 & \xi\eta^2 \\ \xi\eta & \xi^2\eta & \xi\eta^2 & \xi^2\eta^2 \end{bmatrix} \tag{c$'$}$$

Integration of $\rho\mathbf{W}$ with respect to ξ and η produces nonzero terms only in diagonal positions. By using $\mathbf{h}_T^{-1}$ from Sec. 2.3 as shown in Eq. (9.2-10), we find that the result is of the same form as Eq. (9.2-11). However, the submatrix $\mathbf{M}_1$ is now:

$$\mathbf{M}_1 = \frac{\rho V}{36}\begin{bmatrix} 4 & 2 & 1 & 2 \\ 2 & 4 & 2 & 1 \\ 1 & 2 & 4 & 2 \\ 2 & 1 & 2 & 4 \end{bmatrix} \tag{9.2-14}$$

in which $V = 4abt$. Finally, application of the 8×8 rearrangement operator **T** [see Eq. (g) of Sec. 2.3] in Eq. (9.2-9) produces the desired form of matrix **M**.

Figure 9.3(c) shows the isoparametric quadrilateral (Q4) that was discussed in Sec. 3.4. To obtain consistent mass terms for this element, it is necessary to use numerical integration. Thus, when Gaussian quadrature is applied twice to Eq. (9.1-3), it becomes:

$$\mathbf{M} = \sum_{k=1}^{n}\sum_{j=1}^{n} R_j R_k \rho t \mathbf{f}^{\mathrm{T}}(\xi_j, \eta_k)\mathbf{f}(\xi_j, \eta_k)\,|\mathbf{J}(\xi_j, \eta_k)| \tag{9.2-15}$$

For this case the use of $n = 2$ and $R_j = R_k = 1$ provides exact results if ρ and t are constants.

Some commonly used three-dimensional elements are depicted in Figs. 9.4(a), (b), and (c). In the first of these figures we see the constant strain

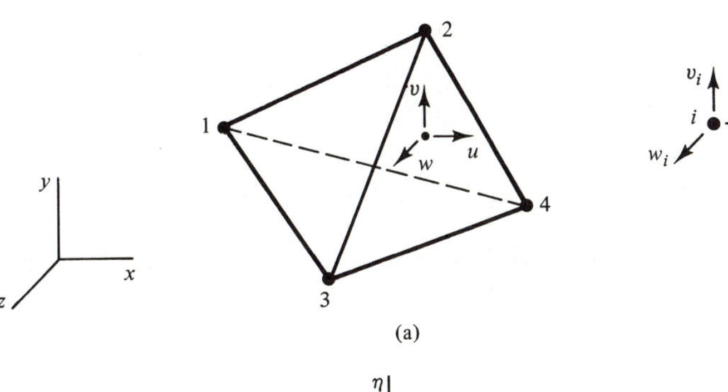

(a)

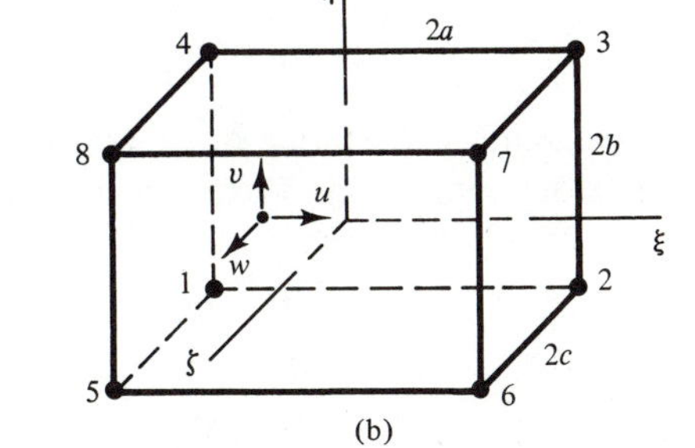

(b)

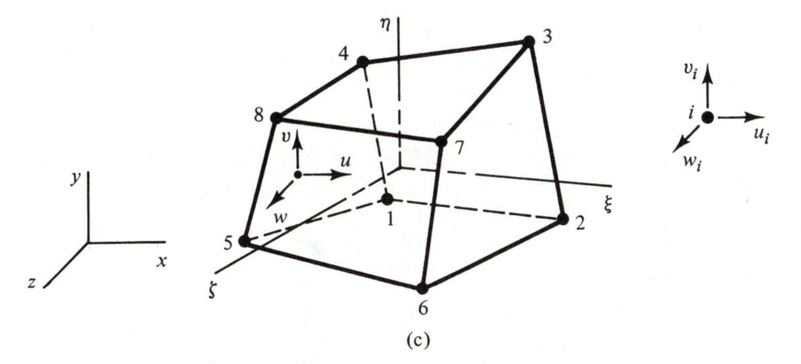

(c)

Figure 9.4 Three-Dimensional Elements: (a) Tet-4 (b) RS8 (c) H8

tetrahedron (Tet-4), having a node at each vertex. From Sec. 4.5 the geometric matrix for this element is:

$$\mathbf{g} = \begin{bmatrix} 1 & x & y & z & 0 & 0 & 0 & 0 & 0 & 0 & 0 & 0 \\ 0 & 0 & 0 & 0 & 1 & x & y & z & 0 & 0 & 0 & 0 \\ 0 & 0 & 0 & 0 & 0 & 0 & 0 & 0 & 1 & x & y & z \end{bmatrix} \tag{d$'$}$$

When matrix **g** is premultiplied by its own transpose, we obtain:

$$\mathbf{g}^{\mathrm{T}}\mathbf{g} = \begin{bmatrix} \mathbf{W} & \mathbf{O} & \mathbf{O} \\ \mathbf{O} & \mathbf{W} & \mathbf{O} \\ \mathbf{O} & \mathbf{O} & \mathbf{W} \end{bmatrix} \tag{e$'$}$$

where the submatrix $\mathbf{W}$ is the following 4×4 array:

$$\mathbf{W} = \begin{bmatrix} 1 & x & y & z \\ x & x^2 & xy & xz \\ y & xy & y^2 & yz \\ z & xz & yz & z^2 \end{bmatrix} \tag{f'}$$

Then the application of Eq. (9.2-10) produces submatrices of the type:

$$\mathbf{M}_1 = \frac{\rho V}{20}\begin{bmatrix} 2 & 1 & 1 & 1 \\ 1 & 2 & 1 & 1 \\ 1 & 1 & 2 & 1 \\ 1 & 1 & 1 & 2 \end{bmatrix} \tag{9.2-16}$$

As before, we see that the sum of the terms in submatrix $\mathbf{M}_1$ equals the total mass of element Tet-4. However, in this three-dimensional case $\mathbf{M}_1$ appears three times in matrix $\mathbf{M}_T$, as follows:

$$\mathbf{M}_T = \begin{bmatrix} \mathbf{M}_1 & \mathbf{O} & \mathbf{O} \\ \mathbf{O} & \mathbf{M}_1 & \mathbf{O} \\ \mathbf{O} & \mathbf{O} & \mathbf{M}_1 \end{bmatrix} \tag{9.2-17}$$

For element Tet-4 we can form a 12×12 rearrangement operator $\mathbf{T}$ that is similar to the 6×6 operator used for the triangular element discussed earlier. Finally, the application of matrix $\mathbf{T}$ in Eq. (9.2-9) gives the consistent mass matrix $\mathbf{M}$ in its preferred arrangement.

An easier approach for the tetrahedron consists of using volume coordinates, as discussed in Sec. 4.2. Then for the Tet-4 element we have:

$$\mathbf{f}_T = \begin{bmatrix} \xi_1 & \xi_2 & \xi_3 & \xi_4 & 0 & 0 & 0 & 0 & 0 & 0 & 0 & 0 \\ 0 & 0 & 0 & 0 & \xi_1 & \xi_2 & \xi_3 & \xi_4 & 0 & 0 & 0 & 0 \\ 0 & 0 & 0 & 0 & 0 & 0 & 0 & 0 & \xi_1 & \xi_2 & \xi_3 & \xi_4 \end{bmatrix} \tag{g'}$$

Premultiplication of matrix $\mathbf{f}_T$ with its own transpose yields:

$$\mathbf{f}_T^{\mathrm{T}}\mathbf{f}_T = \begin{bmatrix} \mathbf{X} & \mathbf{O} & \mathbf{O} \\ \mathbf{O} & \mathbf{X} & \mathbf{O} \\ \mathbf{O} & \mathbf{O} & \mathbf{X} \end{bmatrix} \tag{h'}$$

in which the submatrix $\mathbf{X}$ is:

$$\mathbf{X} = \begin{bmatrix} \xi_1^2 & \xi_1\xi_2 & \xi_1\xi_3 & \xi_1\xi_4 \\ \xi_1\xi_2 & \xi_2^2 & \xi_2\xi_3 & \xi_2\xi_4 \\ \xi_1\xi_3 & \xi_2\xi_3 & \xi_3^2 & \xi_3\xi_4 \\ \xi_1\xi_4 & \xi_2\xi_4 & \xi_3\xi_4 & \xi_4^2 \end{bmatrix} \tag{i'}$$

By integrating the terms in $\rho\mathbf{X}$ over the volume using Eq. (4.2-16), we again obtain the matrix $\mathbf{M}_1$ given in Eq. (9.2-16).

Shown in Fig. 9.4(b) is the rectangular solid (RS8) having eight corner nodes. For this element the discussion in Sec. 4.4 implies that the geometric matrix in dimensionless coordinates is the following 3×24 array:

$$\mathbf{g} = \begin{bmatrix} 1 & \xi & \eta & \zeta & \xi\eta & \eta\zeta & \xi\zeta & \xi\eta\zeta & \cdots & 0 \\ 0 & 0 & 0 & 0 & 0 & 0 & 0 & 0 & \cdots & 0 \\ 0 & 0 & 0 & 0 & 0 & 0 & 0 & 0 & \cdots & \xi\eta\zeta \end{bmatrix} \tag{j'}$$

When this matrix is premultiplied by its own transpose, the result is a 24×24 matrix $\mathbf{g}^{\mathrm{T}}\mathbf{g}$, consisting of three identical submatrices in diagonal positions [see Eq. (e′)]. Each of these 8×8 submatrices contains the terms:

$$\mathbf{W} = \begin{bmatrix} 1 & & & & & & & \\ \xi & \xi^2 & & & & \text{Sym.} & & \\ \eta & \xi\eta & \eta^2 & & & & & \\ \zeta & \xi\zeta & \eta\zeta & \zeta^2 & & & & \\ \xi\eta & \xi^2\eta & \xi\eta^2 & \xi\eta\zeta & \xi^2\eta^2 & & & \\ \eta\zeta & \xi\eta\zeta & \eta^2\zeta & \eta\zeta^2 & \xi\eta^2\zeta & \eta^2\zeta^2 & & \\ \xi\zeta & \xi^2\zeta & \xi\eta\zeta & \xi\zeta^2 & \xi^2\eta\zeta & \xi\eta\zeta^2 & \xi^2\zeta^2 & \\ \xi\eta\zeta & \xi^2\eta\zeta & \xi\eta^2\zeta & \xi\eta\zeta^2 & \xi^2\eta^2\zeta & \xi\eta^2\zeta^2 & \xi^2\eta\zeta^2 & \xi^2\eta^2\zeta^2 \end{bmatrix} \tag{k'}$$

Now application of Eq. (9.2-10) produces submatrices of the following type:

$$\mathbf{M}_1 = \frac{\rho V}{216} \begin{bmatrix} 8 & & & & & & & \\ 4 & 8 & & & & & & \\ 2 & 4 & 8 & & \text{Sym.} & & & \\ 4 & 2 & 4 & 8 & & & & \\ 4 & 2 & 1 & 2 & 8 & & & \\ 2 & 4 & 2 & 1 & 4 & 8 & & \\ 1 & 2 & 4 & 2 & 2 & 4 & 8 & \\ 2 & 1 & 2 & 4 & 4 & 2 & 4 & 8 \end{bmatrix} \tag{9.2-18}$$

where $V = 8abc$. For this case it is easy to construct a 24×24 rearrangement operator $\mathbf{T}$ that is similar to the 8×8 operator used with the rectangular element discussed before. Then Eqs. (9.2-17) and (9.2-9) can be applied to determine the final arrangement of matrix $\mathbf{M}$.

Figure 9.4(c) shows the isoparametric hexahedron (H8) that was first considered in Sec. 4.4. In order to find consistent mass terms for this element, we apply Gaussian quadrature three times to Eq. (9.1-3), as follows:

$$\mathbf{M} = \sum_{\ell=1}^{n} \sum_{k=1}^{n} \sum_{j=1}^{n} R_j R_k R_\ell \rho \mathbf{f}_{jk\ell}^{\mathrm{T}} \mathbf{f}_{jk\ell} \, |\mathbf{J}_{jk\ell}| \tag{9.2-19}$$

In this expression for numerical integration, the matrix $\mathbf{f}_{jk\ell}$ and the determinant $|\mathbf{J}_{jk\ell}|$ are evaluated at each integration point $(\xi_j, \eta_k, \zeta_\ell)$. If the parameters

$n = 2$ and $R_j = R_k = R_\ell = 1$ are used and ρ is constant, Eq. (9.2-19) yields exact results for matrix **M**.

Elements for axisymmetric solids discussed in Chapter 5 have consistent mass matrices for axisymmetric and nonaxisymmetric vibrations, as follows:

$$\mathbf{M}_j = \int_0^{2\pi} \int_A \rho \mathbf{f}^{\mathrm{T}}\mathbf{f} \cos^2 j\theta \; r \, d\theta \, dA \tag{ℓ'}$$

for $j = 0, 1, 2, \ldots, m$. Integration of Eq. (ℓ') with respect to θ yields:

$$\mathbf{M}_j = k\pi \int_A \rho \mathbf{f}^{\mathrm{T}}\mathbf{f} r \, dA \tag{9.2-20}$$

where $k = 2$ for $j = 0$ and $k = 1$ for $j = 1, 2, \ldots, m$. Note that replacement of $\cos j\theta$ with $\sin j\theta$ in Eq. (ℓ') does not change the result.

Now we consider plates in flexure, for which both translational and rotational inertia terms can be developed. As with a beam, the translational terms are much more important and will be discussed first. For this purpose, we integrate Eq. (9.1-3) through the thickness t of the plate to obtain:

$$\mathbf{M}_t = \int_A \rho t \mathbf{f}^{\mathrm{T}}\mathbf{f} \, dA \tag{9.2-21}$$

The alternative form of this equation is:

$$\mathbf{M}_t = \mathbf{h}^{-\mathrm{T}} \int_A \rho t \mathbf{g}^{\mathrm{T}}\mathbf{g} \, dA \, \mathbf{h}^{-1} \tag{9.2-22}$$

Equations (9.2-21) and (9.2-22) represent the consistent mass matrix for *translational inertia in a plate* subjected to bending.

Rotational inertia terms for a plate in flexure are analogous to those for a beam, which were discussed previously. That is, the translation u in the x direction of a point in a plate is:

$$u = -zw_{,x} = -z\mathbf{f}_{,x}\mathbf{q} = -z\mathbf{g}_{,x}\mathbf{h}^{-1}\mathbf{q} \tag{m$'$}$$

where z is the distance of the point from the neutral surface. Similarly, the translation v in the y direction of the same point may be written as:

$$v = -zw_{,y} = -z\mathbf{f}_{,y}\mathbf{q} = -z\mathbf{g}_{,y}\mathbf{h}^{-1}\mathbf{q} \tag{n$'$}$$

Using the principle of virtual work as before and integrating with Eqs. (m$'$) and (n$'$) through the thickness, we find:

$$\mathbf{M}_r = \mathbf{M}_{r1} + \mathbf{M}_{r2} \tag{9.2-23}$$

where

$$\mathbf{M}_{r1} = \int_A \rho \frac{t^3}{12} \mathbf{f}_{,x}^{\mathrm{T}}\mathbf{f}_{,x} \, dA \tag{9.2-24a}$$

and

$$\mathbf{M}_{r2} = \int_A \rho \frac{t^3}{12} \mathbf{f}_{,y}^{\mathrm{T}}\mathbf{f}_{,y} \, dA \tag{9.2-24b}$$

These formulas for $\mathbf{M}_{r1}$ and $\mathbf{M}_{r2}$ represent consistent mass matrices for rotational inertias about two orthogonal axes. Their alternative expressions are:

$$\mathbf{M}_{r1} = \mathbf{h}^{-\mathrm{T}} \int_A \rho \frac{t^3}{12} \mathbf{g}_{,x}^{\mathrm{T}} \mathbf{g}_{,x} \, dA \, \mathbf{h}^{-1} \tag{9.2-25a}$$

and

$$\mathbf{M}_{r2} = \mathbf{h}^{-\mathrm{T}} \int_A \rho \frac{t^3}{12} \mathbf{g}_{,y}^{\mathrm{T}} \mathbf{g}_{,y} \, dA \, \mathbf{h}^{-1} \tag{9.2-25b}$$

Because the rotational inertia terms in plates are usually very small compared to the translational inertia terms, Eqs. (9.2-23) through (9.2-25b) are seldom needed. However, they can be developed and added to $\mathbf{M}_t$ whenever desired.*

The MZC plate-bending rectangle discussed in Sec. 6.2 appears again in Fig. 9.5(a). For this element the (1×12) geometric matrix is:

$$\mathbf{g} = [1 \quad \xi \quad \eta \quad \xi^2 \quad \xi\eta \quad \eta^2 \quad \xi^3 \quad \xi^2\eta \quad \xi\eta^2 \quad \eta^3 \quad \xi^3\eta \quad \xi\eta^3] \tag{o$'$}$$

Terms in the consistent mass matrix $\mathbf{M}_t$ for translational inertia in the plate (5) can be obtained from either Eq. (9.2-21) or Eq. (9.2-22) and are listed in Table 9.1.

Similarly, the BFS rectangle from Sec. 6.2 is shown in Fig. 9.5(b). The geometric matrix for this element has 16 terms, as follows:

$$\mathbf{g} = [1 \quad \xi \quad \eta \quad \xi^2 \quad \xi\eta \quad \eta^2 \quad \xi^3 \quad \xi^2\eta \quad \xi\eta^2 \quad \eta^3 \quad \xi^3\eta \quad \xi^2\eta^2 \quad \xi\eta^3 \quad \xi^3\eta^2 \quad \xi^2\eta^3 \quad \xi^3\eta^3] \tag{p$'$}$$

Results for the matrix $\mathbf{M}_t$ are given in (6).

Pictured in Fig. 9.5(c) is the CKZ triangle, which was discussed in Sec. 6.3. Terms in the matrix $\mathbf{M}_t$ for this element may be found by numerical integration (using the constants in Table 3.1). Thus,

$$\mathbf{M}_t = A \sum_{j=1}^{n} W_j \rho_j t_j \mathbf{f}_j^{\mathrm{T}} \mathbf{f}_j \tag{9.2-26}$$

in which ρ_j and t_j can vary at each integration point.

Another plate-bending element requiring numerical integration is the quadratic quadrilateral PBQ8 of Sec. 6.4. In this case the formula for $\mathbf{M}$ is:

$$\mathbf{M} = \sum_{k=1}^{n} \sum_{j=1}^{n} R_j R_k \rho_{jk} t_{jk} \mathbf{f}_{jk}^{\mathrm{T}} \mathbf{f}_{jk} \, |\mathbf{J}_{jk}| \tag{9.2-27}$$

in which the three rows of matrix $\mathbf{f}$ [see Eq. (6.4-1b)] produce $\mathbf{M}_{r1}$, $\mathbf{M}_{r2}$, and $\mathbf{M}_t$, respectively. If the parameters ρ and t are constant, a value of $n = 2$ each way will usually suffice.

*For element PBQ8 the generic rotations of the normal to the neutral surface are:

$$\theta_x = \sum_{i=1}^{8} f_i \theta_{xi} \qquad \theta_y = \sum_{i=1}^{8} f_i \theta_{yi}$$

Therefore, we must use the functions f_i $(i = 1, 2, \ldots, 8)$ in Eqs. (9.2-24) instead of $\mathbf{f}_{,x}$ and $\mathbf{f}_{,y}$.

TABLE 9.1 Consistent Mass Matrix for MZC Rectangle

$$
\mathbf{M}_t = \frac{\rho tab}{3150}\begin{bmatrix}
1727 & & & & & & & & & & & \\
461b & 160b^2 & & & & & & & & & & \\
-461a & -126ab & 160a^2 & & & & & & & & \text{Sym.} & \\
613 & 199b & -274a & 1727 & & & & & & & & \\
199b & 80b^2 & -84ab & 461b & 160b^2 & & & & & & & \\
274a & 84ab & -120a^2 & 461a & 126ab & 160a^2 & & & & & & \\
197 & 116b & -116a & 613 & 274b & 199a & 1727 & & & & & \\
-116b & -60b^2 & 56ab & -274b & -120b^2 & -84ab & -461b & 160b^2 & & & & \\
116a & 56ab & -60a^2 & 199a & 84ab & 80a^2 & 461a & -126ab & 160a^2 & & & \\
613 & 274b & -199a & 197 & 116b & 116a & 613 & -199b & 274a & 1727 & & \\
-274b & -120b^2 & 84ab & -116b & -60b^2 & -56ab & -199b & 80b^2 & -84ab & -461b & 160b^2 & \\
-199a & -84ab & 80a^2 & -116a & -56ab & -60a^2 & -274a & 84ab & -120a^2 & -461a & 126ab & 160a^2
\end{bmatrix}
$$

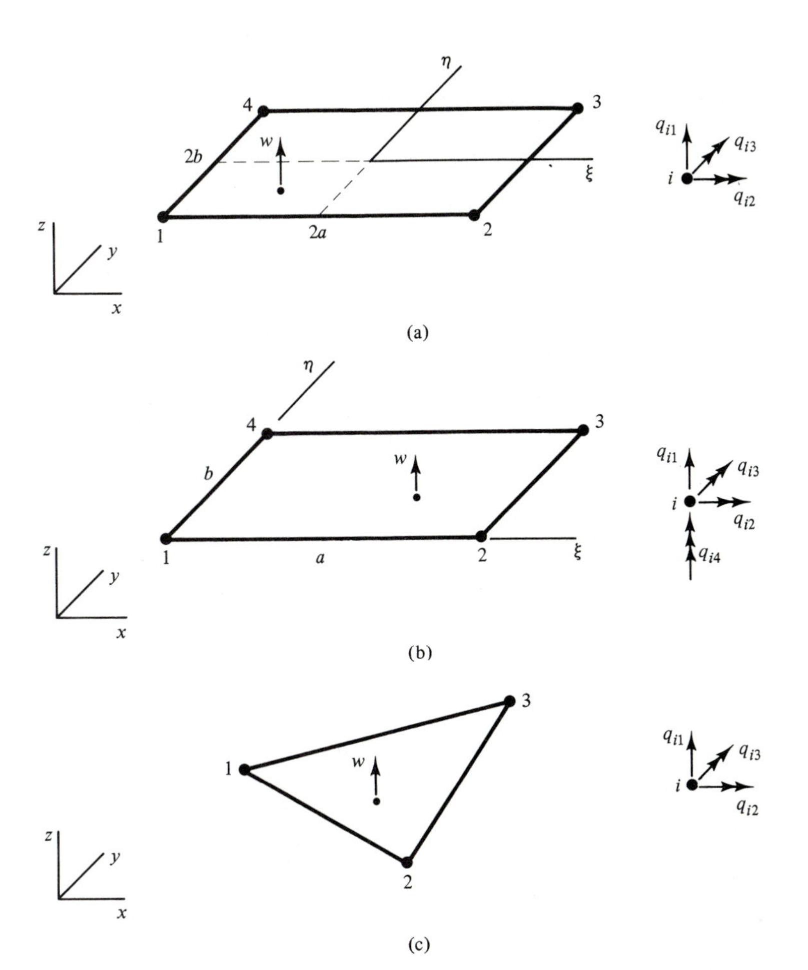

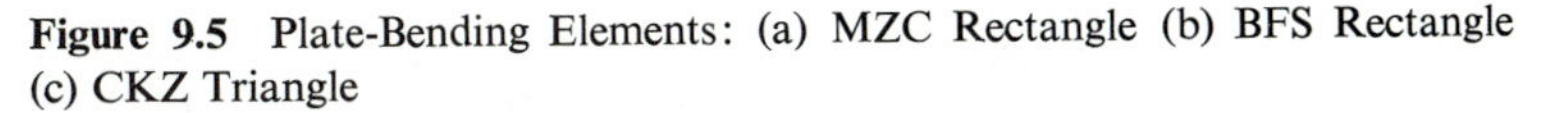

Figure 9.5 Plate-Bending Elements: (a) MZC Rectangle (b) BFS Rectangle (c) CKZ Triangle

The annular element discussed in Sec. 6.5 has a consistent mass matrix $\mathbf{M}_{tj}$ for axisymmetric and nonaxisymmetric vibrations, as follows:

$$\mathbf{M}_{tj} = L \int_0^{2\pi} \int_0^1 \rho t \mathbf{f}^{\mathrm{T}} \mathbf{f} r \cos^2 j\theta \, d\xi \, d\theta \tag{q'}$$

for $j = 0, 1, 2, \ldots, m$. Integration of Eq. (q′) with respect to θ yields:

$$\mathbf{M}_{tj} = k\pi L \int_0^1 \rho t \mathbf{f}^{\mathrm{T}} \mathbf{f} r \, d\xi \tag{9.2-28}$$

where $k = 2$ for $j = 0$ and $k = 1$ for $j = 1, 2, \ldots, m$. Actual use of Eq. (9.2-28) is seldom required because vibrations of circular and annular membranes and plates are well documented in the classical literature (7).

Example 1

Find the consistent mass term $(M_t)_{2,2}$ for the MZC rectangle shown in Fig. 9.5(a). For this purpose we use the displacement shape function:

$$f_2 = \frac{b}{8}(1 - \xi)(1 + \eta)(1 - \eta)^2$$

which is drawn from Eqs. (6.2-2). Then,

$$(M_t)_{2,2} = \rho tab \int_{-1}^{1} \int_{-1}^{1} f_2^2 \, d\xi \, d\eta$$

$$= \frac{\rho tab^3}{64} \int_{-1}^{1} \int_{-1}^{1} (1 - \xi)^2(1 + \eta)^2(1 - \eta)^4 \, d\xi \, d\eta$$

$$= \frac{16}{315} \rho tab^3$$

This term is the same as that in Table 9.1.

Example 2

For the BFS rectangle shown in Fig. 9.5(b), determine the consistent mass term $(M_t)_{9,10}$. From Table 6.2 we have the displacement shape functions:

$$f_9 = (3\xi^2 - 2\xi^3)(3\eta^2 - 2\eta^3) \qquad f_{10} = -(3\xi^2 - 2\xi^3)(\eta^2 - \eta^3)b$$

Thus,

$$(M_t)_{9,10} = \rho tab \int_0^1 \int_0^1 f_9 f_{10} \, d\xi \, d\eta$$

$$= -\rho tab^2 \int_0^1 \int_0^1 (3\xi^2 - 2\xi^3)^2(3\eta^2 - 2\eta^3)(\eta^2 - \eta^3) \, d\xi \, d\eta$$

$$= -\frac{143}{7350} \rho tab^2$$

9.3 EIGENVALUE PROBLEM FOR VIBRATIONS

In Sec. 9.1 the equation of motion for free, undamped vibrations of a discretized continuum was developed as Eq. (9.1-8). When we were deriving that equation, nothing was mentioned about support restraints. If there are none, the structure can have rigid-body motions as well as vibrational motions. Because of the assumption that Eq. (9.1-8) is linear, any rigid-body rotations must be small, while there need not be such a limitation on rigid-body translations. Let us now assume that restraints may exist at some of the nodes, so that some or all of the rigid-body motions are prevented. For such circumstances, we must revert to the notation of Sec. 1.8 to distinguish free nodal displacements (with the subscript F) from restrained nodal displacements (with the subscript R). Then Eq. (9.1-8) may be rewritten so that it pertains only to free displacements, as follows:

$$\mathbf{M}_{FF}\ddot{\mathbf{D}}_F + \mathbf{S}_{FF}\mathbf{D}_F = \mathbf{O} \qquad (9.3\text{-}1)$$

Having clarified this point, we now drop the subscript F for the purpose of simplifying later operations. Thus,

$$\mathbf{M}\,\ddot{\mathbf{D}} + \mathbf{S}\,\mathbf{D} = \mathbf{O} \tag{9.3-2}$$

In this differential equation it is understood that the vector $\mathbf{D}$ contains only free nodal displacements.

Equation (9.3-2) has a known solution (7) that may be stated as:

$$\mathbf{D}_i = \mathbf{\Phi}_i \sin(\omega_i t + \alpha_i) \qquad (i = 1, 2, \ldots, n) \tag{9.3-3}$$

where n is the *number of degrees of freedom*. In this harmonic expression, $\mathbf{\Phi}_i$ is a vector of nodal amplitudes (or *mode shape*) for the ith mode of vibration. The symbol ω_i represents the *angular frequency* of mode i, and α_i denotes the *phase angle*. By differentiating Eq. (9.3-3) twice with respect to the time t, we also find:

$$\ddot{\mathbf{D}}_i = -\omega_i^2 \mathbf{\Phi}_i \sin(\omega_i t + \alpha_i) \tag{9.3-4}$$

Substitution of Eqs. (9.3-3) and (9.3-4) into Eq. (9.3-2) allows cancellation of the term $\sin(\omega_i t + \alpha_i)$, which leaves:

$$(\mathbf{S} - \omega_i^2 \mathbf{M})\mathbf{\Phi}_i = \mathbf{O} \tag{9.3-5}$$

This manipulation has the effect of separating the variable time from those of space, and we are left with a set of n homogeneous algebraic equations in the form of Eq. (9.3-5).

The *algebraic eigenvalue problem* represented by Eq. (9.3-5) may be changed to the standard, symmetric form:

$$(\mathbf{A} - \lambda_i \mathbf{I})\mathbf{X}_i = \mathbf{O} \tag{9.3-6}$$

in which $\mathbf{A}$ is a symmetric matrix and $\mathbf{I}$ is an identity matrix. The symbol λ_i denotes the ith *eigenvalue*, and $\mathbf{X}_i$ is the corresponding *eigenvector* for a system of n homogeneous equations. We can put Eq. (9.3-5) into the form of Eq. (9.3-6) by factoring either matrix $\mathbf{S}$ or matrix $\mathbf{M}$, using the Cholesky square-root method (see Appendix C.1). We choose to factor $\mathbf{S}$ for an important reason that will soon be apparent. Thus,

$$\mathbf{S} = \mathbf{U}^{\mathrm{T}}\mathbf{U} \tag{a}$$

where the factor $\mathbf{U}$ is an upper triangular matrix. Substitute Eq. (a) into Eq. (9.3-5) to obtain:

$$(\mathbf{U}^{\mathrm{T}}\mathbf{U} - \omega_i^2 \mathbf{M})\mathbf{\Phi}_i = \mathbf{O} \tag{b}$$

Then premultiply Eq. (b) by $\mathbf{U}^{-\mathrm{T}}$ and insert $\mathbf{I} = \mathbf{U}^{-1}\mathbf{U}$ after matrix $\mathbf{M}$:

$$\mathbf{U}^{-\mathrm{T}}(\mathbf{U}^{\mathrm{T}}\mathbf{U} - \omega_i^2 \mathbf{M}\;\mathbf{U}^{-1}\mathbf{U})\mathbf{\Phi}_i = \mathbf{O} \tag{c}$$

Rewriting terms in reverse order, we find:

$$(\mathbf{M}_U - \lambda_i \mathbf{I})\mathbf{\Phi}_{Ui} = \mathbf{O} \tag{9.3-7}$$

where

$$\mathbf{M}_U = \mathbf{U}^{-\mathrm{T}}\mathbf{M}\mathbf{U}^{-1} \qquad \lambda_i = \frac{1}{\omega_i^2} \qquad \mathbf{\Phi}_{Ui} = \mathbf{U}\;\mathbf{\Phi}_i \tag{9.3-8}$$

Equation (9.3-7) is now in the standard, symmetric form of the eigenvalue problem given by Eq. (9.3-6). The matrix $\mathbf{A}$ is represented by $\mathbf{M}_U$ [see Eqs. (9.3-8)], which is guaranteed to be symmetric. In addition, we see that the eigenvalue λ_i is equal to the reciprocal of the square of the angular frequency.

This is the consequence of choosing to factor **S** and is numerically advantageous because the highest eigenvalue (corresponding to the lowest frequency) is usually extracted first. The eigenvector $\mathbf{\Phi}_{Ui}$ in Eq. (9.3-7) is related to $\mathbf{\Phi}_i$ by the last expression in Eqs. (9.3-8). This constitutes a change in coordinates to a new set where the stiffness matrix is equal to **I**. After the eigenvalues and eigenvectors have been found from Eq. (9.3-7), the angular frequencies and mode shapes can be determined as:

$$\omega_i = \frac{1}{\sqrt{\lambda_i}} \qquad \mathbf{\Phi}_i = \mathbf{U}^{-1}\mathbf{\Phi}_{Ui} \tag{9.3-9}$$

If the stiffness matrix is semidefinite, it cannot be factored as described here, because at least one rigid-body mode (with $\omega_i = 0$) is present. In such a case it would be possible to factor the mass matrix instead, using the reduction method in Sec. 9.4 if necessary to achieve a positive-definite matrix. Due to symmetry of the coefficient matrices in Eq. (9.3-5), the eigenvectors $\mathbf{\Phi}_i$ are orthogonal with respect to both **S** and **M** (7). Solution of the algebraic eigenvalue problem will be discussed next.

From the theory of homogeneous algebraic equations (8), nontrivial solutions for Eq. (9.3-6) exist only if the following condition is fulfilled:

$$|\mathbf{A} - \lambda_i \mathbf{I}| = 0 \tag{9.3-10}$$

Expansion of the determinant in this expression yields a polynomial of order n called the *characteristic equation*. The n roots λ_i of this equation are the *characteristic values*, or eigenvalues. Substitution of these values (one at a time) into the homogeneous equations [Eq. (9.3-6)] produces the *characteristic vectors* (or eigenvectors) $\mathbf{X}_i$, within arbitrary constants. Alternatively (7), each eigenvector may be found as a column of the adjoint matrix $\mathbf{H}_i^a$ of the *characteristic matrix* $\mathbf{H}_i$, obtained from Eq. (9.3-6) as follows:

$$\mathbf{H}_i \mathbf{X}_i = \mathbf{O} \tag{9.3-11}$$

where

$$\mathbf{H}_i = \mathbf{A} - \lambda_i \mathbf{I} \tag{9.3-12}$$

In general, the process of directly extracting the roots of the characteristic equation must be done iteratively and is not efficient for large problems. If the eigenvalues and eigenvectors of all the modes are to be found, it is best to use Householder transformations (9) and convert matrix **A** to tridiagonal form. Then the final values of λ_i and the vectors $\mathbf{X}_i$ can be determined by iteration with the QR algorithm. On the other hand, if only a few of the modes are desired, the method of inverse iteration with spectral shifting is more efficient (10).

9.4 REDUCTION OF STIFFNESS AND MASS MATRICES

The concept of matrix condensation was introduced in Sec. 2.4 for the purpose of making internal nodal displacements dependent upon those at the external nodes of a finite element. The basic idea of matrix condensation is simply

Gaussian elimination of chosen displacements to reduce the size of a problem. In static analysis no loss of accuracy results from such a reduction because the dependent displacements are recovered (exactly) in the back-substitution phase. In dynamic (or vibrational) analysis a similar type of condensation can be used to reduce the number of degrees of freedom (11), but a new type of approximation is involved.

We begin the discussion with *stiffness* (or *static*) *reduction* and rewrite Eq. (9.3-2) in expanded form, as follows:

$$\begin{bmatrix} \mathbf{M}_{AA} & \mathbf{M}_{AB} \\ \mathbf{M}_{BA} & \mathbf{M}_{BB} \end{bmatrix} \begin{bmatrix} \ddot{\mathbf{D}}_A \\ \ddot{\mathbf{D}}_B \end{bmatrix} + \begin{bmatrix} \mathbf{S}_{AA} & \mathbf{S}_{AB} \\ \mathbf{S}_{BA} & \mathbf{S}_{BB} \end{bmatrix} \begin{bmatrix} \mathbf{D}_A \\ \mathbf{D}_B \end{bmatrix} = \begin{bmatrix} \mathbf{O} \\ \mathbf{O} \end{bmatrix} \tag{9.4-1}$$

In this equation the subscript A denotes the displacements that are to be eliminated, while the subscript B refers to those that will be retained. Let the accelerations $\ddot{\mathbf{D}}_A$ and $\ddot{\mathbf{D}}_B$ be null, and write the remaining static equations as two sets:

$$\mathbf{S}_{AA}\mathbf{D}_A + \mathbf{S}_{AB}\mathbf{D}_B = \mathbf{O} \tag{9.4-2a}$$

$$\mathbf{S}_{BA}\mathbf{D}_A + \mathbf{S}_{BB}\mathbf{D}_B = \mathbf{O} \tag{9.4-2b}$$

Solve for the vector $\mathbf{D}_A$ in Eq. (9.4-2a), as follows:

$$\mathbf{D}_A = -\mathbf{S}_{AA}^{-1}\mathbf{S}_{AB}\mathbf{D}_B \tag{9.4-3}$$

Substitute Eq. (9.4-3) into Eq. (9.4-2b) to obtain:

$$\mathbf{S}_{BB}^{*}\mathbf{D}_B = \mathbf{O} \tag{9.4-4}$$

in which

$$\mathbf{S}_{BB}^{*} = \mathbf{S}_{BB} - \mathbf{S}_{BA}\mathbf{S}_{AA}^{-1}\mathbf{S}_{AB} \tag{9.4-5}$$

From Eq. (9.4-4) we see that Eqs. (9.4-2) have been reduced to a smaller set, having the same order as $\mathbf{S}_{BB}$. Moreover, Eq. (9.4-3) may now be viewed as the back-substitution process that is necessary to find vector $\mathbf{D}_A$ from $\mathbf{D}_B$.

Turning next to the vibrational problem in Eq. (9.4-1), we shall now consider *mass* (or *dynamic*) *reduction*. As a new approximation, assume that the acceleration vector $\ddot{\mathbf{D}}_A$ is dependent upon $\ddot{\mathbf{D}}_B$ in the same manner that the vector $\mathbf{D}_A$ is related to $\mathbf{D}_B$ in Eq. (9.4-3). Thus,

$$\ddot{\mathbf{D}}_A = -\mathbf{S}_{AA}^{-1}\mathbf{S}_{AB}\ddot{\mathbf{D}}_B \tag{9.4-6}$$

Then equate the virtual work done by the inertial actions in the reduced system to that of the inertial actions in the original system. Thus,

$$\begin{aligned} \delta\mathbf{D}_B^{\mathrm{T}}\mathbf{M}_{BB}^{*}\ddot{\mathbf{D}}_B &= \delta\mathbf{D}^{\mathrm{T}}\mathbf{M}\;\ddot{\mathbf{D}} \\ &= [\delta\mathbf{D}_A^{\mathrm{T}} \quad \delta\mathbf{D}_B^{\mathrm{T}}] \begin{bmatrix} \mathbf{M}_{AA} & \mathbf{M}_{AB} \\ \mathbf{M}_{BA} & \mathbf{M}_{BB} \end{bmatrix} \begin{bmatrix} \ddot{\mathbf{D}}_A \\ \ddot{\mathbf{D}}_B \end{bmatrix} \end{aligned} \tag{9.4-7}$$

But from Eq. (9.4-3) we have:

$$\delta\mathbf{D}_A = -\mathbf{S}_{AA}^{-1}\mathbf{S}_{AB}\,\delta\mathbf{D}_B \tag{a}$$

Substitute Eqs. (a) and (9.4-6) into Eq. (9.4-7) to find:

$$\delta\mathbf{D}_B^{\mathrm{T}}\mathbf{M}_{BB}^{*}\ddot{\mathbf{D}}_B = \delta\mathbf{D}_B^{\mathrm{T}}\mathbf{T}_B^{\mathrm{T}}\mathbf{M}\ \mathbf{T}_B\ddot{\mathbf{D}}_B \tag{b}$$

Cancellation of $\delta\mathbf{D}_B^{\mathrm{T}}$ in this expression yields:

$$\mathbf{M}_{BB}^{*}\ddot{\mathbf{D}}_B = \mathbf{T}_B^{\mathrm{T}}\mathbf{M}\ \mathbf{T}_B\ddot{\mathbf{D}}_B \tag{9.4-8}$$

In this equation the transformation matrix $\mathbf{T}_B$ is:

$$\mathbf{T}_B = \begin{bmatrix} -\mathbf{S}_{AA}^{-1}\mathbf{S}_{AB} \\ \mathbf{I}_B \end{bmatrix} \tag{9.4-9}$$

where $\mathbf{I}_B$ is an identity matrix of the same order as $\mathbf{M}_{BB}$. From Eq. (9.4-8) we see that:

$$\mathbf{M}_{BB}^{*} = \mathbf{T}_B^{\mathrm{T}}\mathbf{M}\ \mathbf{T}_B \tag{9.4-10}$$

Due to the virtual work equality in Eq. (9.4-7), the mass terms in matrix $\mathbf{M}_{BB}^{*}$ are energy-equivalent to those in the original mass matrix $\mathbf{M}$. However, the reduction in size caused by Eq. (9.4-10) represents an additional approximation inherent to the method.

Example

The cantilever beam in Fig. 9.6 is composed of two prismatic flexural elements, both of which have the same flexural rigidity EI. For this arrangement of elements, the nodal stiffness matrix for the unrestrained beam is:

$$\mathbf{S}_N = \frac{EI}{\ell^3}\begin{bmatrix} 12 & 6\ell & -12 & 6\ell & 0 & 0 \\ 6\ell & 4\ell^2 & -6\ell & 2\ell^2 & 0 & 0 \\ -12 & -6\ell & 24 & 0 & -12 & 6\ell \\ 6\ell & 2\ell^2 & 0 & 8\ell^2 & -6\ell & 2\ell^2 \\ 0 & 0 & -12 & -6\ell & 12 & -6\ell \\ 0 & 0 & 6\ell & 2\ell^2 & -6\ell & 4\ell^2 \end{bmatrix}\begin{matrix} 1 \\ 2 \\ 3 \\ 4 \\ 5 \\ 6 \end{matrix} \tag{c}$$

$$\quad\quad\quad\quad 1 \quad 2 \quad 3 \quad 4 \quad 5 \quad 6$$

Similarly, the nodal mass matrix takes the form:

$$\mathbf{M}_N = \frac{\rho A\ell}{420}\begin{bmatrix} 156 & 22\ell & 54 & -13\ell & 0 & 0 \\ 22\ell & 4\ell^2 & 13\ell & -3\ell^2 & 0 & 0 \\ 54 & 13\ell & 312 & 0 & 54 & -13\ell \\ -13\ell & -3\ell^2 & 0 & 8\ell^2 & 13\ell & -3\ell^2 \\ 0 & 0 & 54 & 13\ell & 156 & -22\ell \\ 0 & 0 & -13\ell & -3\ell^2 & -22\ell & 4\ell^2 \end{bmatrix}\begin{matrix} 1 \\ 2 \\ 3 \\ 4 \\ 5 \\ 6 \end{matrix} \tag{d}$$

$$\quad\quad\quad\quad 1 \quad 2 \quad 3 \quad 4 \quad 5 \quad 6$$

In Eqs. (c) and (d) the contributions of elements 1 and 2 to matrices $\mathbf{S}_N$ and $\mathbf{M}_N$ are enclosed in dashed boxes. Furthermore, the displacement coordinates for the problem (see Fig. 9.6) are indicated at the right and below the matrices.

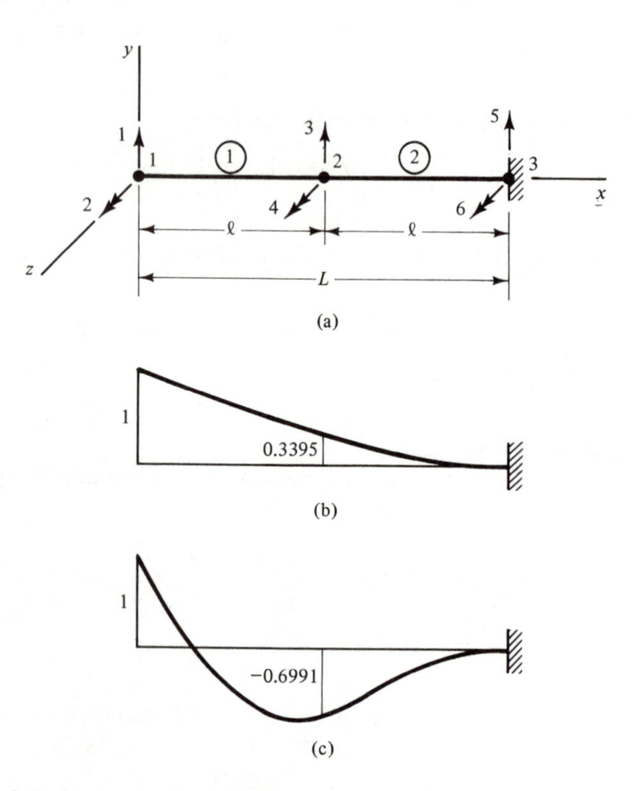

Figure 9.6 (a) Two-Element Cantilever Beam (b) First Mode (c) Second Mode

The objective of this example is to show how we can eliminate the rotational degrees of freedom while retaining the translational degrees of freedom in a beam. As the first step, we remove the fifth and sixth rows and columns from matrices $\mathbf{S}_N$ and $\mathbf{M}_N$ because displacements five and six in the figure are restrained by supports. Then the remaining 4×4 arrays are rearranged to put the rotational terms before the translational terms, as follows:

$$\mathbf{S} = \begin{bmatrix} \mathbf{S}_{AA} & \mathbf{S}_{AB} \\ \mathbf{S}_{BA} & \mathbf{S}_{BB} \end{bmatrix} = \frac{EI}{\ell^3} \left[\begin{array}{cc|cc} 4\ell^2 & 2\ell^2 & 6\ell & -6\ell \\ 2\ell^2 & 8\ell^2 & 6\ell & 0 \\ \hline 6\ell & 6\ell & 12 & -12 \\ -6\ell & 0 & -12 & 24 \end{array} \right] \begin{array}{c} 2 \\ 4 \\ 1 \\ 3 \end{array} \qquad \text{(e)}$$

$$\begin{array}{cccc} 2 & 4 & 1 & 3 \end{array}$$

$$\mathbf{M} = \begin{bmatrix} \mathbf{M}_{AA} & \mathbf{M}_{AB} \\ \mathbf{M}_{BA} & \mathbf{M}_{BB} \end{bmatrix} = \frac{\rho A\ell}{420}\left[\begin{array}{cc|cc} 4\ell^2 & -3\ell^2 & 22\ell & 13\ell \\ -3\ell^2 & 8\ell^2 & -13\ell & 0 \\ \hline 22\ell & -13\ell & 156 & 54 \\ 13\ell & 0 & 54 & 312 \end{array}\right]\begin{matrix} 2 \\ 4 \\ 1 \\ 3 \end{matrix} \qquad \text{(f)}$$
$$\begin{matrix} 2 & 4 & 1 & 3 \end{matrix}$$

The inverse of submatrix $\mathbf{S}_{AA}$, taken from Eq. (e), is:

$$\mathbf{S}_{AA}^{-1} = \frac{\ell}{14EI}\begin{bmatrix} 4 & -1 \\ -1 & 2 \end{bmatrix} \qquad \text{(g)}$$

Substituting this array and the other submatrices of $\mathbf{S}$ from Eq. (e) into Eq. (9.4-5), we obtain:

$$\mathbf{S}_{BB}^{*} = \frac{12EI}{\ell^3}\begin{bmatrix} 1 & -1 \\ -1 & 2 \end{bmatrix} - \frac{18EI}{7\ell^3}\begin{bmatrix} 4 & -3 \\ -3 & 4 \end{bmatrix} = \frac{6EI}{7\ell^3}\begin{bmatrix} 2 & -5 \\ -5 & 16 \end{bmatrix}\begin{matrix} 1 \\ 3 \end{matrix} \qquad \text{(h)}$$
$$\begin{matrix} 1 & 3 \end{matrix}$$

which is the reduced stiffness matrix.

To reduce the mass matrix, we form the transformation matrix $\mathbf{T}_B$ by evaluating $-\mathbf{S}_{AA}^{-1}\mathbf{S}_{AB}$ and substituting it into Eq. (9.4-9). Thus,

$$-\mathbf{S}_{AA}^{-1}\mathbf{S}_{AB} = \frac{3}{7\ell}\begin{bmatrix} -3 & 4 \\ -1 & -1 \end{bmatrix} \qquad \text{(i)}$$

And

$$\mathbf{T}_B = \frac{1}{7\ell}\left[\begin{array}{cc} -9 & 12 \\ -3 & -3 \\ \hline 7\ell & 0 \\ 0 & 7\ell \end{array}\right] \qquad \text{(j)}$$

In this case the submatrix $\mathbf{I}_B$ in the lower partition of matrix $\mathbf{T}_B$ is of order 2 because there are two translational displacements remaining (numbers 1 and 3). Substitution of matrix $\mathbf{M}$ from Eq. (f) and $\mathbf{T}_B$ from Eq. (j) into Eq. (9.4-10) yields:

$$\mathbf{M}_{BB}^{*} = \frac{\rho A\ell}{20580}\begin{bmatrix} 5652 & 3615 \\ 3615 & 18336 \end{bmatrix}\begin{matrix} 1 \\ 3 \end{matrix} \qquad \text{(k)}$$
$$\begin{matrix} 1 & 3 \end{matrix}$$

which is the reduced mass matrix.

For convenience in subsequent calculations, the reduced stiffness matrix in Eq. (h) is expressed numerically as:

$$\mathbf{S}_{BB}^{*} = \frac{EI}{\ell^3}\begin{bmatrix} 1.7143 & -4.2857 \\ -4.2857 & 13.714 \end{bmatrix} \qquad (\ell)$$

Also, the reduced mass matrix in Eq. (k) is rewritten as:

$$\mathbf{M}^*_{BB} = \rho A\ell \begin{bmatrix} 0.27464 & 0.17566 \\ 0.17566 & 0.89096 \end{bmatrix} \tag{m}$$

Using $\mathbf{S}^*_{BB}$ and $\mathbf{M}^*_{BB}$ to set up the eigenvalue problem in the form of Eq. (9.3-5), we have:

$$\begin{bmatrix} S_{11} - \omega_i^2 M_{11} & S_{12} - \omega_i^2 M_{12} \\ S_{21} - \omega_i^2 M_{21} & S_{22} - \omega_i^2 M_{22} \end{bmatrix} \begin{bmatrix} \Phi_{1,i} \\ \Phi_{2,i} \end{bmatrix} = \begin{bmatrix} 0 \\ 0 \end{bmatrix} \qquad (i = 1, 2) \tag{n}$$

When the determinant of the coefficient matrix in Eq. (n) is set equal to zero, the characteristic equation becomes:

$$a(\omega_i^2)^2 + b\omega_i^2 + c = 0 \tag{o}$$

where

$$\begin{aligned} a &= M_{11}M_{22} - M_{12}^2 \\ b &= -(M_{11}S_{22} + M_{22}S_{11} - 2M_{12}S_{12}) \\ c &= S_{11}S_{22} - S_{12}^2 \end{aligned} \tag{p}$$

The solution of Eq. (o) is:

$$\omega_i^2 = \frac{-b \mp \sqrt{b^2 - 4ac}}{2a} \qquad (i = 1, 2) \tag{q}$$

In this problem,

$$a = 0.21384(\rho A\ell)^2 \qquad b = -6.7994 \frac{\rho AEI}{\ell^2} \qquad c = 5.1427 \left(\frac{EI}{\ell^3}\right)^2$$

Substitution of these values and $\ell = L/2$ into Eq. (q) gives:

$$\omega_{1,2} = 3.522, 22.28 \frac{1}{L^2}\sqrt{\frac{EI}{\rho A}} \tag{r}$$

These angular frequencies are in error by +0.17 percent and +1.1 percent, respectively, and are upper bounds of the exact values (7). The corresponding mode shapes may be found by substituting ω_1^2 and ω_2^2 into the homogeneous equations [see Eq. (n)]. Thus,

$$\mathbf{\Phi} = [\mathbf{\Phi}_1 \quad \mathbf{\Phi}_2] = \begin{bmatrix} 1.0000 & 1.0000 \\ 0.3395 & -0.6991 \end{bmatrix} \tag{s}$$

In this *modal matrix* each modal vector is listed columnwise and normalized with respect to its larger component. Sketches of the mode shapes appear in Figs. 9.6(b) and (c).

Frequency coefficients μ_i for prismatic beams with various end conditions are summarized in Table 9.2. In each case the beam is modeled by four flexural elements, and the results for the consistent mass approach (with and without elimination of rotations) are compared with those for the lumped-mass method (with elimination of rotations). The table shows that the consistent mass model produces much better accuracy than the lumped-mass model in beam analysis.

TABLE 9.2 Frequency Coefficients μ_i for Prismatic Beams Modeled by Four Elements

	Support Conditions	Mode	Exact (7)	CM-TR	% Error	CM-TO	% Error	LM-TO	% Error
Simple	L	1	9.870	9.872	+0.020	9.873	+0.030	9.867	−0.030
		2	39.48	39.63	+0.38	39.76	+0.71	39.19	−0.73
		3	88.83	90.45	+1.8	94.03	+5.8	83.21	−6.3
Free		1	22.37	22.41	+0.18	22.46	+0.40	18.91	−15
		2	61.67	62.06	+0.63	63.12	+2.4	48.00	−22
		3	120.9	121.9	+0.83	122.4	+1.2	86.84	−28
Fixed		1	22.37	22.40	+0.13	22.41	+0.18	22.30	−0.31
		2	61.67	62.24	+0.92	62.77	+1.8	59.25	−3.9
		3	120.9	123.5	+2.2	124.8	+3.2	97.40	−19
Cantilever		1	3.516	3.516	+0.00	3.516	+0.00	3.418	−2.8
		2	22.03	22.06	+0.14	22.09	+0.27	20.09	−8.8
		3	61.70	62.18	+0.83	62.97	+2.1	53.20	−14
Propped		1	15.42	15.43	+0.065	15.43	+0.065	15.40	−0.13
		2	49.97	50.28	+0.62	50.56	+1.2	49.05	−1.8
		3	104.2	106.6	+2.3	110.5	+6.0	91.53	−12

$$\omega_i = \frac{\mu_i}{L^2}\sqrt{\frac{EI}{\rho A}}$$

CM: Consistent Masses
LM: Lumped Masses
TR: Translations and Rotations
TO: Translations Only

9.5 PROGRAM VIBPQ8 AND APPLICATIONS

The program discussed in this section computes vibrational frequencies and mode shapes for isotropic plates in bending, using element PBQ8. This program is named VIBPQ8. It incorporates the stiffness matrix from Eq. (6.4-7) and the consistent mass matrix from Eq. (9.2-27). Structural data for VIBPQ8 remain almost the same as those for the static analysis program named PBQ8. However, in the list of structural parameters the number of loading systems (NLS) is replaced by the number of modes (NMOD), and the mass density (RHO) is included at the end of the list.

Logic in a program for vibrational analysis is different from that in a program for static analysis. We must generate the consistent mass matrix as well as the stiffness matrix for all the nodal degrees of freedom in the structure. These arrays are formed as square matrices and then reduced by matrix condensation, as described in Sec. 9.4. For plates in bending, the rotational displacements at nodes are eliminated; and the translational displacements are retained. Subsequently, the eigenvalues and eigenvectors are calculated for the standard, symmetric form, as explained in Sec. 9.3. The computer output includes the mode number, the angular frequency, and the mode shape for each of the vibrational modes desired. The logic for Program VIBPQ8 is shown in the following outline:

1. Read and print structural data
   (a) Problem identification
   (b) Structural parameters
   (c) Nodal coordinates
   (d) Element information
   (e) Nodal restraints
   (f) Calculate displacement indexes
2. Generate stiffness and mass matrices
   (a) Element stiffness matrix
   (b) Element mass matrix ($\mathbf{M}_t + \mathbf{M}_{r1} + \mathbf{M}_{r2}$)
   (c) Transfer to structural stiffness and mass matrices
3. Reduce stiffness and mass matrices
   (a) Calculate reduced stiffness matrix
   (b) Determine transformation matrix
   (c) Calculate reduced mass matrix
4. Determine eigenvalues and eigenvectors
   (a) Factor stiffness matrix
   (b) Invert stiffness factor
   (c) Transform eigenvalue problem to standard, symmetric form
   (d) Solve for eigenvalues and eigenvectors

   (e) Back-transform eigenvectors
   (f) Normalize eigenvectors (with respect to largest terms)
5. Print results (for each mode)
   (a) Mode number
   (b) Angular frequency
   (c) Vibrational mode shape

Example 1

Figure 9.7 shows a square cantilever plate that is modeled by a single PBQ8 element. The nodal restraints are the same as those in Example 1 of Sec. 6.6. No loads are shown, and we assume the following values of structural parameters:

$$a = 6 \text{ in.} \qquad t = 0.2 \text{ in.} \qquad E = 1.4 \times 10^4 \text{ k/in.}^2$$
$$\nu = 0.3 \qquad \rho = 7.77 \times 10^{-7} \text{ k-sec}^2/\text{in.}^4$$

for which the material is bronze and the units are U.S.

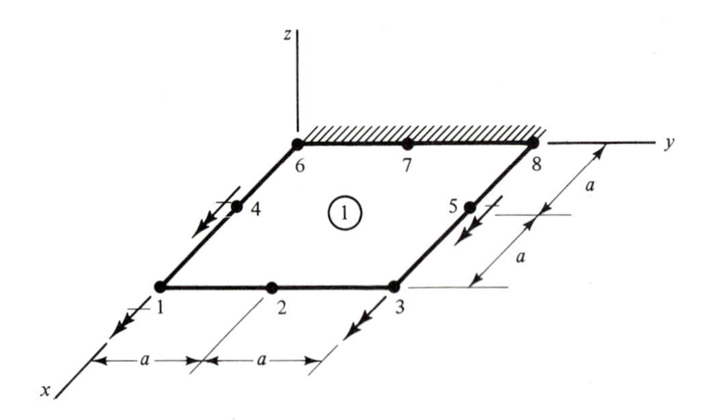

Figure 9.7 Example 1: Cantilever Element PBQ8

The computer output from Program VIBPQ8 for the first three modes of vibration appears in Table 9.3. We see from the table that the angular frequency of the fundamental (or first) mode is calculated to be $\omega_1 = 204.93$ rad/sec. The theoretical value (7) for no shearing deformations or rotational inertias is given by:

$$\omega_1 = \frac{3.516}{(2a)^2}\sqrt{\frac{Et^2}{12(1 - \nu^2)\rho}} = 198.36 \text{ rad/sec}$$

Thus, the finite-element solution differs from this value by +3.31 percent. We also note that the mode shapes are either symmetric or antisymmetric with respect to the plane of symmetry passing through nodes 2 and 7.

Example 2

Now let us consider the free vibrations of a simply supported rectangular plate with a 3×3 network of PBQ8 elements, as illustrated in Fig. 9.8. Because the plate is symmetric with respect to both the x-z and y-z planes, only a quarter of the plate is shown

TABLE 9.3 Computer Output for Example 1

```
PROGRAM VIBPQ8

***  EXAMPLE 1:  CANTILEVER ELEMENT  ***

STRUCTURAL PARAMETERS
   NN   NE  NRN NMOD           E             PR             T           RHO
    8    1    7    3  1.40000D 04  3.00000D-01  2.00000D-01  7.77000D-07

COORDINATES OF NODES
 NODE           X            Y
    1  1.20000D 01  0.00000D-01
    2  1.20000D 01  6.00000D 00
    3  1.20000D 01  1.20000D 01
    4  6.00000D 00  0.00000D-01
    5  6.00000D 00  1.20000D 01
    6  0.00000D-01  0.00000D-01
    7  0.00000D-01  6.00000D 00
    8  0.00000D-01  1.20000D 01

ELEMENT INFORMATION
ELEM.   N1   N2   N3   N4   N5   N6   N7   N8
    1    1    3    8    6    2    5    7    4

NODAL RESTRAINTS
 NODE   R1   R2   R3
    1    0    1    0
    3    0    1    0
    4    0    1    0
    5    0    1    0
    6    1    1    1
    7    1    1    1
    8    1    1    1

NUMBER OF TRANSLATIONAL D.O.F             =    5
NUMBER OF ROTATIONAL D.O.F.               =    6
NUMBER OF TRANSLATIONAL RESTRAINTS        =    3
NUMBER OF ROTATIONAL RESTRAINTS           =   10

MODE     1
ANGULAR FREQUENCY   2.04930 02
 NODE      DN1      NODE      DN1      NODE      DN1      NODE      DN1
    1  1.0000          2  1.0000          3  1.0000          4  0.3258
    5  0.3258          6  0.0000          7  0.0000          8  0.0000

MODE     2
ANGULAR FREQUENCY   8.93480 02
 NODE      DN1      NODE      DN1      NODE      DN1      NODE      DN1
    1  1.0000          2  0.0000          3 -1.0000          4  0.4969
    5 -0.4969          6  0.0000          7  0.0000          8  0.0000

MODE     3
ANGULAR FREQUENCY   2.22970 03
 NODE      DN1      NODE      DN1      NODE      DN1      NODE      DN1
    1  1.0000          2  1.0000          3  1.0000          4 -0.5758
    5 -0.5758          6  0.0000          7  0.0000          8  0.0000
```

in the figure. Such a plate has vibrational modes that are symmetric and antisymmetric with respect to the planes of symmetry. Therefore, it is necessary to analyze a quarter of the plate four times, with nodes on the x and y axes restrained as indicated in Table 9.4. The rotational restraints in Fig. 9.8 at nodes on axes x and y illustrate the symmetric-symmetric case.

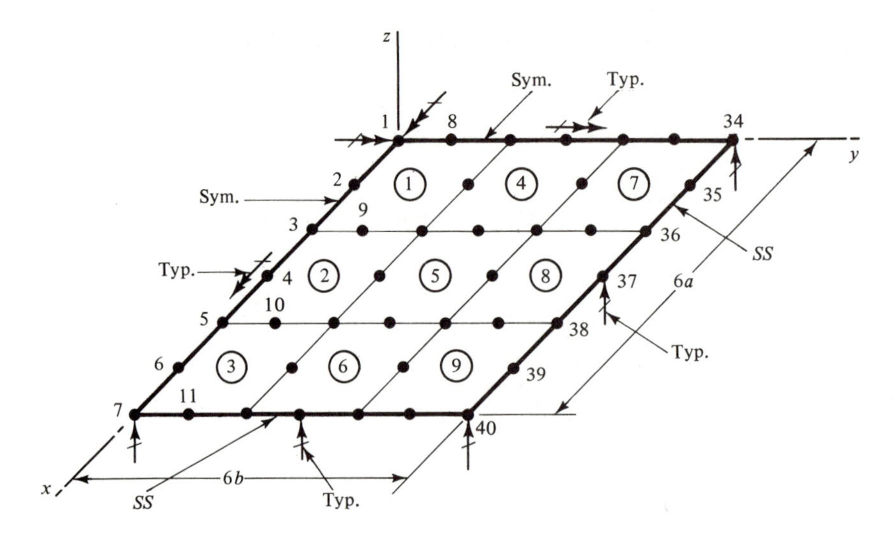

Figure 9.8 Example 2: Simply Supported Rectangular Plate with PBQ8 Network

TABLE 9.4 Restraint Conditions for Doubly Symmetric Plate

Deformation	Symmet.-Symmet.		Symmet.-Antisym.	
Node Location	x Axis	y Axis	x Axis	y Axis
z Translation	0	0	0	1
x Rotation	1	0	1	1
y Rotation	0	1	0	0
Deformation	**Antisym.-Symmet.**		**Antisym.-Antisym.**	
Node Location	x Axis	y Axis	x Axis	y Axis
z Translation	1	0	1	1
x Rotation	0	0	0	1
y Rotation	1	1	1	0

1 = Restrained; 0 = Unrestrained

In this problem the following values are assumed for structural parameters:

$a = 1.5$ in. $\qquad b = 1$ in. $\qquad t = 0.15$ in.

$E = 3 \times 10^4$ k/in.^{2} $\qquad \nu = 0.3$ $\qquad \rho = 7.33 \times 10^{-7}$ k-sec^2/in.^{4}

where the units are U.S. and the material is steel.

Table 9.5 summarizes the output obtained from Program VIBPQ8 for the four analyses mentioned. The first number in each box of the table is the result of applying the known formula (12):

TABLE 9.5 Angular Frequencies for Example 2

n_x \ n_y		1	2	3	4
1	Theory*	909.13	2797.3	5944.3	10350
	FE Result†	903.30	2785.9	6011.7	10755
	% Error	−0.641	−0.408	1.13	3.91
2	Theory	1748.3	3636.5	6783.5	11189
	FE Result	1726.7	3584.2	6885.9	11642
	% Error	−1.24	−1.44	1.51	4.05
3	Theory	3147.0	5035.2	8182.2	12588
	FE Result	3191.5	5217.4	8490.8	14095
	% Error	1.41	3.62	3.77	12.0
4	Theory	5105.1	6993.3	10140	14546
	FE Result	5343.0	7518.2	12056	17623
	% Error	4.66	7.51	18.9	21.2

*From Ref. 12, with no shearing deformations or rotational inertias.
†Using $\mathbf{M}_t$, $\mathbf{M}_{r1}$, and $\mathbf{M}_{r2}$, with elimination of rotations.

$$\omega_{n_x n_y} = \pi^2 \left(\frac{n_x^2}{L_x^2} + \frac{n_y^2}{L_y^2} \right) \sqrt{\frac{Et^2}{12(1 - \nu^2)\rho}}$$

The symbol n_x in this expression denotes the number of half sine waves of z translations in the x direction, and n_y is the number in the y direction. Also, L_x and L_y are the overall dimensions of the plate in the x and y directions.

The second number in each box of Table 9.5 is the result from Program VIBPQ8, and the third number is the percentage difference of the finite-element solution from the theoretical result. The minimum value of this discrepancy (−0.408 percent) occurs in the antisymmetric-symmetric case for which $n_x = 1$ and $n_y = 2$.

Example 3

As a third example we take the quarter of a circular plate with a square opening given in Fig. 9.9. This plate is discretized into a network of PBQ8 elements that model the circular boundary very closely. Edges of the plate along the x and y axes are completely fixed; other edges are free (unrestrained); and the line 17-21 lies in a plane of symmetry. Values assumed for the structural parameters are:

$$a_1 = 0.2 \text{ m} \qquad a_2 = 0.3 \text{ m} \qquad t = 1 \times 10^{-2} \text{ m}$$

$$E = 7 \times 10^7 \text{ kN/m}^2 \qquad \nu = 0.3 \qquad \rho = 2.7 \text{ Mg/m}^3$$

in which the material is aluminum and the units are S.I.

When the data for this example are run with Program VIBPQ8, the first two angular frequencies are found to be:

$$\omega_1 = 52.06 \text{ rad/sec} \qquad \omega_2 = 139.8 \text{ rad/sec}$$

The shape of the first mode is symmetric with respect to line 17-21, but that of the second mode is antisymmetric.

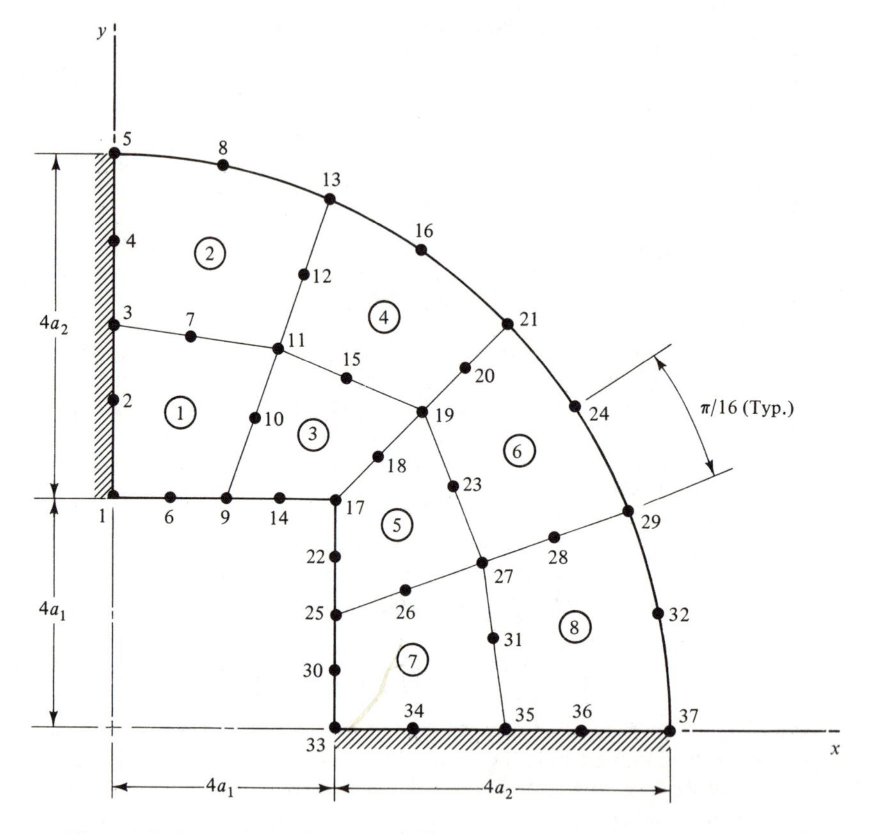

Figure 9.9 Example 3: Quarter of Circular Plate with Square Opening and PBQ8 Network

REFERENCES

1. Archer, J. S., "Consistent Mass Matrix for Distributed Systems," *Proc. ASCE,* Vol. 89, No. ST4, 1963, pp. 161–178.
2. Zienkiewicz, O. C., *The Finite Element Method*, 3d ed., McGraw-Hill Ltd., London, 1977.
3. Clough, R. W., "Analysis of Structural Vibrations and Dynamic Response," *Rec. Adv. Mat. Meth. Struc. Anal. Des.*, ed. by R. H. Gallagher, Y. Yamada, and J. T. Oden, Univ. Ala. Press, Huntsville, Ala., 1971, pp. 25–45.
4. Archer, J. S., "Consistent Matrix Formulations for Structural Analysis Using Finite-Element Techniques," *AIAA Jour.*, Vol. 3, No. 10, 1965, pp. 1910–1918.
5. Przemieniecki, J. S., "Equivalent Mass Matrices for Rectangular Plates in Bending," *AIAA Jour.*, Vol. 4, No. 5, 1966, pp. 949–950.
6. Bogner, F. K., Fox, R. L., and Schmit, L. A., Jr., "The Generation of Interelement-Compatible Stiffness and Mass Matrices by the Use of Interpolation For-

mulas," *Proc. Conf. Mat. Meth. Struc. Mech.*, AFIT, Wright-Patterson AF Base, 1965, pp. 397–443.

7. Timoshenko, S. P., Young, D. H., and Weaver, W., Jr., *Vibration Problems in Engineering*, 4th ed., Wiley, New York, 1974.
8. Gere, J. M., and Weaver, W., Jr., *Matrix Algebra for Engineers*, 2d ed., Brooks-Cole, Monterey, Ca., 1983.
9. Wilkinson, J. H., *The Algebraic Eigenvalue Problem*, Clarendon Press, Oxford, 1965.
10. Bathe, K. J., *Finite Element Procedures in Engineering Analysis*, Prentice-Hall, Englewood Cliffs, N.J., 1982.
11. Guyan, R. J., "Reduction of Stiffness and Mass Matrices," *AIAA Jour.*, Vol. 3, No. 2, 1965, p. 380.
12. Szilard, R., *Theory and Analysis of Plates*, Prentice-Hall, Englewood Cliffs, N.J., 1974.

PROBLEMS

9.2-1. Derive the consistent mass matrix for an axial element of uniform mass density, assuming that its cross-sectional area varies linearly as $A = A_0(1 + x/L)$.

9.2-2. For an axial element, determine the equivalent masses at the nodes due to a concentrated mass M at point $x = L/4$.

9.2-3. For a flexural element, find the equivalent mass terms at the nodes due to a concentrated mass M at point $x = L/3$.

9.2-4. Verify the terms in the consistent mass matrix $\mathbf{M}_t$ for beam translations given by Eq. (9.2-2).

9.2-5. Verify the terms in the consistent mass matrix $\mathbf{M}_r$ for beam rotations given by Eq. (9.2-5).

9.2-6. Derive the consistent mass matrix for a torsional element of uniform mass density, assuming that the polar moment of inertia of its cross section varies quadratically as $J = J_0[1 + (x/L)^2]$.

9.2-7. For a constant strain triangle [Fig. 9.3(a)], determine the equivalent nodal masses due to a concentrated mass M, located at the area coordinates $(\xi_1, \xi_2, \xi_3) = (\frac{1}{2}, \frac{1}{3}, \frac{1}{6})$.

9.2-8. Verify the terms in the first column of the consistent mass matrix $\mathbf{M}_1$ for the constant strain triangle [Fig. 9.3(a)] given by Eq. (9.2-12).

9.2-9. For a bilinear displacement rectangle [Fig. 9.3(b)], find the equivalent nodal masses due to a concentrated mass M at point $(\xi, \eta) = (\frac{1}{4}, \frac{1}{3})$.

9.2-10. Verify the terms in the first column of the consistent mass matrix $\mathbf{M}_1$ for the bilinear displacement rectangle [Fig. 9.3(b)] given by Eq. (9.2-14).

9.2-11. Derive the consistent mass matrix for a bilinear displacement rectangle [Fig. 9.3(b)] of uniform mass density, assuming that its thickness varies linearly as $t = t_0(1 + \xi/2)$.

9.2-12. For element Tet-4 [Fig. 9.4(a)], determine the equivalent nodal masses due to a concentrated mass M, located at the volume coordinates $(\xi_1, \xi_2, \xi_3, \xi_4) = (\frac{1}{12}, \frac{1}{6}, \frac{1}{4}, \frac{1}{2})$.

9.2-13. Verify the terms in the third column of the consistent mass matrix $\mathbf{M}_1$ for element Tet-4 [Fig. 9.4(a)] given by Eq. (9.2-16).

9.2-14. For element RS8 [Fig. 9.4(b)], determine the first column of the consistent mass matrix M_1 due to a concentrated mass M at point $(\xi, \eta, \zeta) = (-\frac{1}{2}, \frac{1}{3}, -\frac{3}{4})$.

9.2-15. Verify the terms in the fourth column of the consistent mass matrix $\mathbf{M}_1$ for element RS8 [Fig. 9.4(b)] given by Eq. (9.2-18).

9.2-16. Suppose that a concentrated mass M is located at point $(\xi, \eta) = (-\frac{1}{4}, \frac{1}{2})$ on the MZC rectangle [Fig. 9.5(a)]. Find the last two terms in column five of the consistent mass matrix due to this condition.

9.2-17. Derive the term $(M_t)_{2,3}$ in Table 9.1 for the MZC rectangle [Fig. 9.5(a)].

9.2-18. Derive the term $(M_{r1})_{2,2}$ for the MZC rectangle [Fig. 9.5(a)].

9.2-19. Derive the term $(M_{r2})_{3,3}$ for the MZC rectangle [Fig. 9.5(a)].

9.2-20. Let a concentrated mass M be located at point $(\xi, \eta) = (\frac{1}{4}, \frac{3}{4})$ on the BFS rectangle [Fig. 9.5(b)], and determine the first three terms in row ten of the consistent mass matrix due to this influence.

9.2-21. Derive the term $(M_t)_{9,9}$ for the BFS rectangle [Fig. 9.5(b)].

9.2-22. Derive the term $(M_{r1})_{9,11}$ for the BFS rectangle [Fig. 9.5(b)].

9.2-23. Derive the term $(M_{r2})_{9,12}$ for the BFS rectangle [Fig. 9.5(b)].

9.4-1. The figure shows a fixed-end prismatic beam composed of three flexural elements. Set up the stiffness matrix **S** and the consistent mass matrix **M** for the four displacement coordinates that are unrestrained. Then reduce these coefficient matrices by eliminating the rotations and retaining the translations. Solve the eigenvalue problem to obtain two angular frequencies and the corresponding mode shapes for the reduced system.

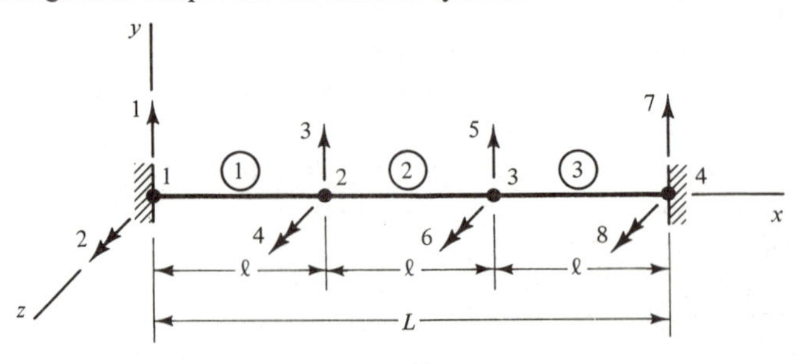

Problem 9.4-1

9.4-2. An unrestrained prismatic beam consisting of two flexural elements is shown in the figure. Construct the stiffness and consistent mass matrices **S** and **M** for the six unrestrained displacement coordinates. Reduce these matrices by eliminating the rotations and retaining the translations. Solve the eigenvalue problem to obtain angular frequencies and vibrational mode shapes for the reduced system.

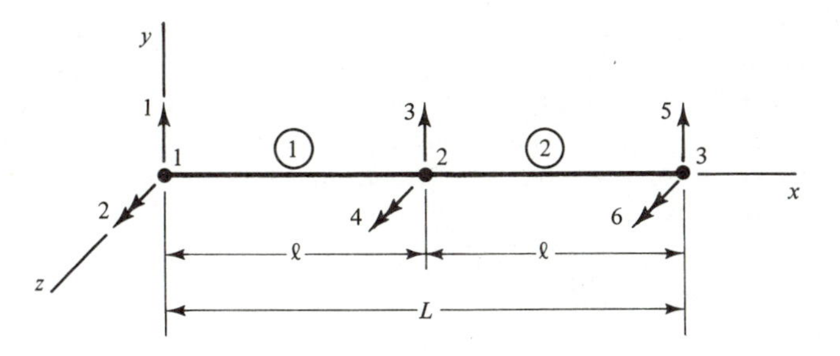

Problem 9.4-2

9.4-3. The two-element prismatic beam in the figure is fixed at point 1 and restrained against rotation (but not translation) at point 3. Assemble the stiffness and consistent mass matrices **S** and **M** for the unrestrained displacement coordinates. Reduce these matrices by eliminating the rotation and retaining the translations. Solve the eigenvalue problem to obtain angular frequencies and vibrational mode shapes for the reduced system.

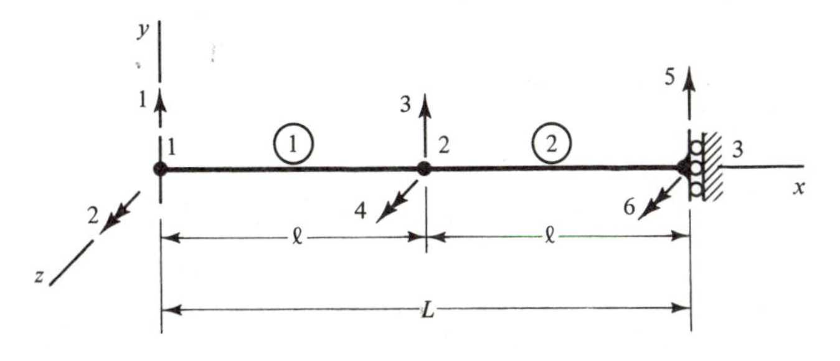

Problem 9.4-3

9.4-4. In the figure a two-element prismatic beam is fixed at the left end and restrained against translation (but not rotation) at the right end. Construct the stiffness and consistent mass matrices **S** and **M** for the unrestrained displacement coordinates. Reduce these matrices by eliminating the rotations and retaining the translation. Then find the angular frequency of vibration for the remaining system, which has only one degree of freedom.

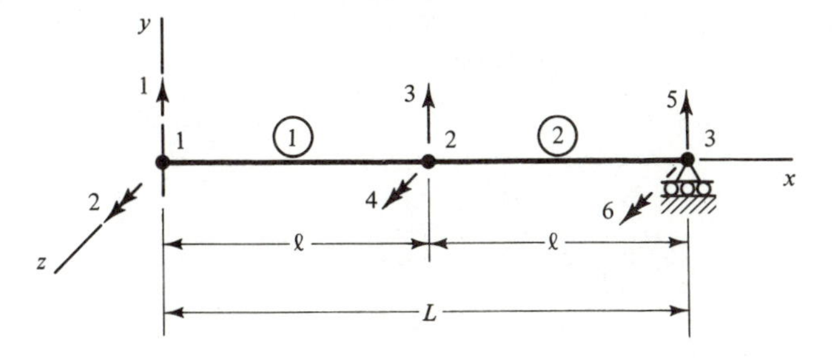

Problem 9.4-4

9.4-5. The figure shows a two-element prismatic beam that is simply supported at points 1 and 3. Assemble the stiffness and consistent mass matrices **S** and **M** for the unrestrained displacement coordinates. Reduce these matrices by eliminating the rotations and retaining the translation. Then find the angular frequency of vibration for the remaining system, which has only one degree of freedom.

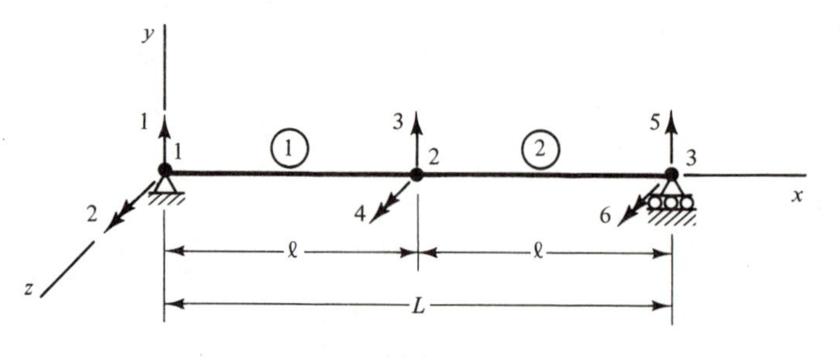

Problem 9.4-5

CHAPTER

10

Instability Analysis

10.1 INTRODUCTION

Because of nonlinear strain-displacement relationships, axial and membrane (or in-plane) stresses are found to have secondary influences that modify flexural stiffnesses. Such terms may be collected in an array that is referred to by the following names, depending upon the application: (a) geometric stiffness matrix, (b) initial stress stiffness matrix, or (c) stability matrix. In this chapter we will use the name *initial stress stiffness matrix*, $\mathbf{K}_0$, as a reminder of the cause of the influences. Matrix $\mathbf{K}_0$ is derived for various elements in Sec. 10.2.

If initial axial or membrane stresses are assumed not to be affected by flexural deformations, a *linear instability* analysis usually is feasible. In such a case *bifurcation buckling* will occur [see Fig. 10.1(a)], and a critical level of axial or membrane loading may be found in the context of an eigenvalue problem (see Sec. 10.3). In this case the load and the displacement in Fig. 10.1(a) are in orthogonal directions, so the collapse is unexpected. This situation arises with beams subjected to axial loads and with thin plates that are loaded in their own planes.

On the other hand, plane and space frames and shell structures ordinarily have interdependencies between axial (or membrane) and flexural stresses and strains. Elastic collapses of such structures are said to be *nonlinear instabilities*. Figure 10.1(b) shows a load-displacement curve for this type of problem. Here the critical buckling load occurs when the slope of the curve becomes zero. To analyze this type of instability, we need incremental or iterative techniques, as discussed in Sec. 10.4.

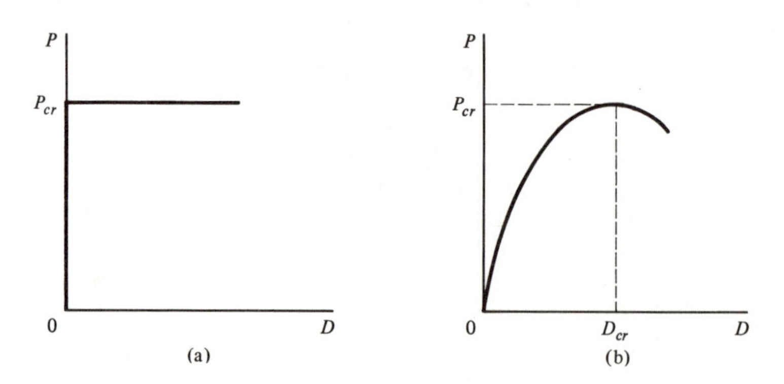

Figure 10.1 Instability Plots: (a) Linear (b) Nonlinear

When axial or membrane loads in beams or plates are slightly eccentric (or the structures have initial curvatures), we can compute secondary flexural stresses due to the loads. For such imperfect cases the buckling does not occur instantaneously, as implied by Fig. 10.1(a). Instead, the load-displacement diagram is curved, as indicated in Fig. 10.1(b). Although such problems are not as complicated as frames or shells, we can also solve them as described in Sec. 10.4.

10.2 INITIAL STRESS STIFFNESS MATRICES

We begin with a *flexural element* having an initial axial force P_0, which is shown acting in its positive sense in Fig. 10.2(a). For this element the strain-displacement relationship due to flexure can be written in two parts (1), as follows:

$$\epsilon_x = \epsilon_{x1} + \epsilon_{x2} \tag{10.2-1}$$

where the flexural strain is:

$$\epsilon_{x1} = -y \frac{d^2 v}{dx^2} \tag{10.2-2}$$

and the axial strain is:

$$\epsilon_{x2} = \frac{1}{2}\left(\frac{dv}{dx}\right)^2 \tag{10.2-3}$$

The expression for ϵ_{x2} in Eq. (10.2-3) is a first approximation of the axial strain due to flexure. That is, the incremental length ds in Fig. 10.2(a) may be written as:

$$ds = dx\sqrt{1 + \left(\frac{dv}{dx}\right)^2} \tag{a}$$

Expansion of the square root term with the binomial theorem gives:

$$ds = dx\left[1 + \frac{1}{2}\left(\frac{dv}{dx}\right)^2 + \cdots\right] \tag{b}$$

where the second term in the brackets is the approximate axial strain in Eq.

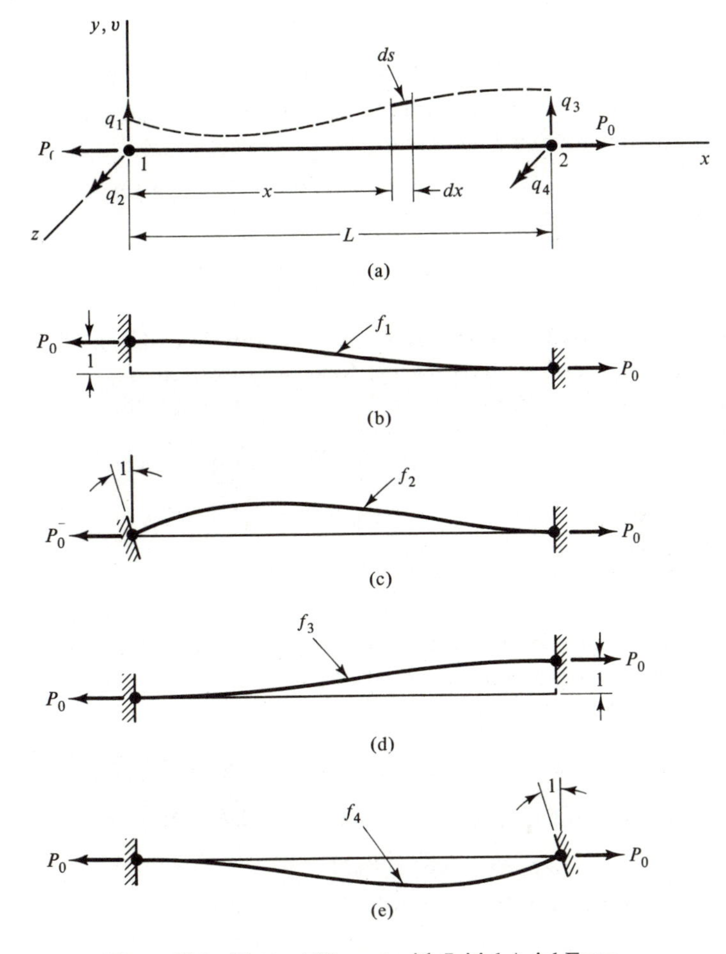

Figure 10.2 Flexural Element with Initial Axial Force

(10.2-3). Previously (in Sec. 1.5), we expressed the generic displacement v in terms of the nodal displacements $\mathbf{q}$ as:

$$v = \mathbf{f}\ \mathbf{q} \tag{c}$$

in which the shape functions are:

$$f_1 = 1 - 3\frac{x^2}{L^2} + 2\frac{x^3}{L^3} \qquad f_2 = x - 2\frac{x^2}{L} + \frac{x^3}{L^2}$$

$$f_3 = 3\frac{x^2}{L^2} - 2\frac{x^3}{L^3} \qquad f_4 = -\frac{x^2}{L} + \frac{x^3}{L^2} \tag{d}$$

Differentiation of Eq. (c) and substitution of the results into Eqs. (10.2-2) and (10.2-3) yields:

$$\epsilon_{x1} = -y\mathbf{f}_{,xx}\mathbf{q} \tag{10.2-4}$$

and

$$\epsilon_{x2} = \tfrac{1}{2}(\mathbf{f}_{,x}\mathbf{q})^2 = \tfrac{1}{2}\mathbf{q}^{\mathrm{T}}\mathbf{f}^{\mathrm{T}}_{,x}\mathbf{f}_{,x}\mathbf{q} \tag{10.2-5}$$

In a displaced configuration the sum of the potential energy of the flexural stress σ_{x1} and the initial axial stress σ_{x0} is:

$$U = U_1 + U_0 \tag{e}$$

The flexural part of this sum has the form:

$$U_1 = \tfrac{1}{2}\int_V \sigma_{x1}\epsilon_{x1}\, dV \tag{f}$$

Axial strain is omitted from this expression because integration of $\sigma_{x1}\epsilon_{x2}$ over the cross section produces zero energy. Next, we substitute $\sigma_{x1} = E\epsilon_{x1}$ and Eq. (10.2-4) into Eq. (f) and integrate over the cross section to find:

$$U_1 = \tfrac{1}{2}EI\mathbf{q}^{\mathrm{T}}\int_0^L \mathbf{f}^{\mathrm{T}}_{,xx}\mathbf{f}_{,xx}\, dx\, \mathbf{q} \tag{g}$$

The axial part of Eq. (e) is:

$$U_0 = \int_V \sigma_{x0}\epsilon_{x2}\, dV \tag{h}$$

Flexural strain is omitted from this formula because integration of $\sigma_{x0}\epsilon_{x1}$ over the cross section gives a zero result. Then substitution of Eq. (10.2-5) into Eq. (h) and integration over the cross-sectional area yields:

$$U_0 = \tfrac{1}{2}P_0\mathbf{q}^{\mathrm{T}}\int_0^L \mathbf{f}^{\mathrm{T}}_{,x}\mathbf{f}_{,x}\, dx\, \mathbf{q} \tag{i}$$

Application of the potential-energy theorem (see Sec. 1.9) produces:

$$(\mathbf{K} + \mathbf{K}_0)\mathbf{q} = \mathbf{p} \tag{10.2-6}$$

In this equation we have the previously derived stiffness matrix:

$$\mathbf{K} = EI\int_0^L \mathbf{f}^{\mathrm{T}}_{,xx}\mathbf{f}_{,xx}\, dx \tag{10.2-7}$$

And we also have the initial stress stiffness matrix:

$$\mathbf{K}_0 = P_0\int_0^L \mathbf{f}^{\mathrm{T}}_{,x}\mathbf{f}_{,x}\, dx \tag{10.2-8}$$

Evaluation of the latter expression gives:

$$\mathbf{K}_0 = \frac{P_0}{30L}\begin{bmatrix} 36 & 3L & -36 & 3L \\ 3L & 4L^2 & -3L & -L^2 \\ -36 & -3L & 36 & -3L \\ 3L & -L^2 & -3L & 4L^2 \end{bmatrix} \tag{10.2-9}$$

which is the initial stress stiffness matrix for a prismatic flexural element.

The matrix $\mathbf{K}_0$ in Eq. (10.2-9) evolves from the potential energy of initial axial stress and an approximation to the axial strain associated with flexure.

Terms in the matrix have the physical interpretation of actions required in addition to those in **K** to produce unit nodal displacements in the presence of the axial force P_0 [see Figs. 10.2(b) through (e)].

The strain-displacement relationship in Eq. (10.2-3) may be simplified by using the following approximation for the slope at any point (2):

$$\frac{dv}{dx} \approx \frac{1}{L}(q_3 - q_1) \tag{j}$$

which is the slope of the chord from node 1 to node 2 (see Fig. 10.3). Under this assumption, the matrix $\mathbf{K}_0$ becomes:

$$\mathbf{K}_0 = \frac{P_0}{L}\begin{bmatrix} 1 & 0 & -1 & 0 \\ 0 & 0 & 0 & 0 \\ -1 & 0 & 1 & 0 \\ 0 & 0 & 0 & 0 \end{bmatrix} \tag{10.2-10}$$

While this approximation appears to be rather crude, its advantage lies in the simplicity of Eq. (10.2-10) compared to Eq. (10.2-9).

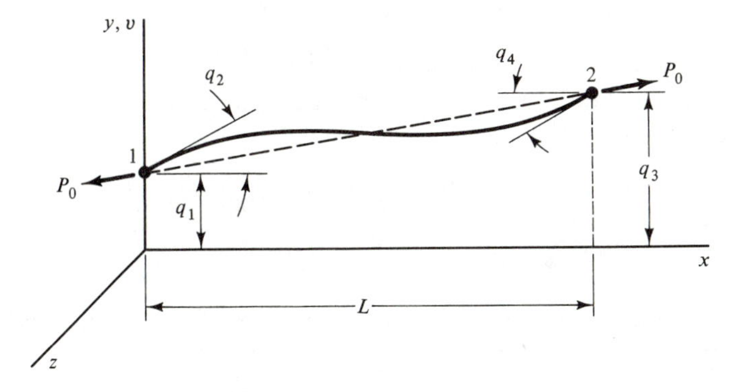

Figure 10.3 Simplified Geometry

Plate-bending elements provide very important applications of the concept of an initial stress stiffness matrix. For such elements the nonlinear strain-displacement relationships (in Cartesian coordinates) are usually taken to be:

$$\begin{aligned} \epsilon_x &= \epsilon_{x1} + \epsilon_{x2} \\ \epsilon_y &= \epsilon_{y1} + \epsilon_{y2} \\ \gamma_{xy} &= \gamma_{xy1} + \gamma_{xy2} \end{aligned} \tag{10.2-11}$$

In these expressions the first terms (see Sec. 6.1) are:

$$\begin{aligned} \epsilon_{x1} &= -zw_{,xx} \\ \epsilon_{y1} &= -zw_{,yy} \\ \gamma_{xy1} &= -2zw_{,xy} \end{aligned} \tag{10.2-12}$$

And the second terms become (3, 4):

$$\begin{aligned} \epsilon_{x2} &= \tfrac{1}{2}w_{,x}^2 \\ \epsilon_{y2} &= \tfrac{1}{2}w_{,y}^2 \\ \gamma_{xy2} &= \tfrac{1}{2}(w_{,x}w_{,y} + w_{,y}w_{,x}) \end{aligned} \tag{10.2-13}$$

The formulas in Eqs. (10.2-13) consist of first approximations to membrane strains due to flexure in a plate. Previously (in Chapter 6) we had:

$$w = \mathbf{f}\ \mathbf{q} \tag{k}$$

Differentiation of Eq. (k) and substitution of the results into Eqs. (10.2-12) and (10.2-13) yields*:

$$\boldsymbol{\epsilon}_1 = \begin{bmatrix} \epsilon_{x1} \\ \epsilon_{y1} \\ \gamma_{xy1} \end{bmatrix} = -z \begin{bmatrix} \mathbf{f}_{,xx} \\ \mathbf{f}_{,yy} \\ 2\mathbf{f}_{,xy} \end{bmatrix} \mathbf{q} = -z\bar{\boldsymbol{\phi}}\ \mathbf{q} \tag{10.2-14}$$

and

$$\boldsymbol{\epsilon}_2 = \begin{bmatrix} \epsilon_{x2} \\ \epsilon_{y2} \\ \gamma_{xy2} \end{bmatrix} = \tfrac{1}{2}\mathbf{q}^{\mathrm{T}} \begin{bmatrix} \mathbf{f}_{,x}^{\mathrm{T}}\mathbf{f}_{,x} \\ \mathbf{f}_{,y}^{\mathrm{T}}\mathbf{f}_{,y} \\ \mathbf{f}_{,x}^{\mathrm{T}}\mathbf{f}_{,y} + \mathbf{f}_{,y}^{\mathrm{T}}\mathbf{f}_{,x} \end{bmatrix} \mathbf{q} \tag{10.2-15}$$

To formulate stiffnesses, we assume that flexural deformations occur in the presence of membrane stresses. Then we let:

$$\boldsymbol{\sigma}_1 = \{\sigma_{x1}, \sigma_{y1}, \tau_{xy1}\} = \mathbf{E}\ \boldsymbol{\epsilon}_1 \tag{10.2-16}$$

Also,

$$\boldsymbol{\sigma}_0 = \{\sigma_{x0}, \sigma_{y0}, \tau_{xy0}\} \tag{10.2-17}$$

As for the flexural element, we write the potential energy of the flexural stresses and the initial membrane stresses as the sum of two parts. That is,

$$U = U_1 + U_0 \tag{e}$$

repeated

The flexural part of the potential energy is:

$$U_1 = \tfrac{1}{2}\int_V \boldsymbol{\sigma}_1^{\mathrm{T}}\boldsymbol{\epsilon}_1\, dV \tag{ℓ}$$

Substitution of Eqs. (10.2-14) and (10.2-16) into Eq. (ℓ) and integration through the thickness gives:

$$U_1 = \frac{t^3}{24}\mathbf{q}^{\mathrm{T}} \int_A \bar{\boldsymbol{\phi}}^{\mathrm{T}}\mathbf{E}\ \bar{\boldsymbol{\phi}}\, dA\, \mathbf{q} \tag{m}$$

When the potential energy theorem is applied, Eq. (m) leads to:

$$\mathbf{K} = \frac{t^3}{12}(\bar{\mathbf{K}}_1 + \bar{\mathbf{K}}_2 + \bar{\mathbf{K}}_3 + \bar{\mathbf{K}}_4) \tag{10.2-18}$$

*The symbol $\bar{\boldsymbol{\phi}}$ denotes $\boldsymbol{\phi}$ devoid of $\mathbf{q}$; that is, $\boldsymbol{\phi} = \bar{\boldsymbol{\phi}}\mathbf{q}$.

in which

$$\bar{\mathbf{K}}_1 = E_{11} \int_A \mathbf{f}^{\mathrm{T}}_{,xx}\mathbf{f}_{,xx}\, dA \tag{10.2-19a}$$

$$\bar{\mathbf{K}}_2 = E_{22} \int_A \mathbf{f}^{\mathrm{T}}_{,yy}\mathbf{f}_{,yy}\, dA \tag{10.2-19b}$$

$$\bar{\mathbf{K}}_3 = 2E_{12} \int_A \mathbf{f}^{\mathrm{T}}_{,xx}\mathbf{f}_{,yy}\, dA \tag{10.2-19c}$$

$$\bar{\mathbf{K}}_4 = 4E_{33} \int_A \mathbf{f}^{\mathrm{T}}_{,xy}\mathbf{f}_{,xy}\, dA \tag{10.2-19d}$$

These formulas were implied by the discussion in Chapter 6, but here the terms in matrix $\mathbf{E}$ connote either an isotropic or an orthotropic material.

The membrane part of the potential energy is:

$$U_0 = \int_V \boldsymbol{\sigma}_0^{\mathrm{T}}\boldsymbol{\epsilon}_2\, dV \tag{n}$$

Substitution of Eqs. (10.2-15) and (10.2-17) into Eq. (n) and integration through the thickness, followed by application of the potential-energy theorem, leads to:

$$\mathbf{K}_0 = t(\bar{\mathbf{K}}_{01} + \bar{\mathbf{K}}_{02} + \bar{\mathbf{K}}_{03}) \tag{10.2-20}$$

where the thickness t is constant. The three matrices in Eq. (10.2-20) are:

$$\bar{\mathbf{K}}_{01} = \sigma_{x0} \int_A \mathbf{f}^{\mathrm{T}}_{,x}\mathbf{f}_{,x}\, dA \tag{10.2-21a}$$

$$\bar{\mathbf{K}}_{02} = \sigma_{y0} \int_A \mathbf{f}^{\mathrm{T}}_{,y}\mathbf{f}_{,y}\, dA \tag{10.2-21b}$$

$$\bar{\mathbf{K}}_{03} = \tau_{xy0} \int_A (\mathbf{f}^{\mathrm{T}}_{,x}\mathbf{f}_{,y} + \mathbf{f}^{\mathrm{T}}_{,y}\mathbf{f}_{,x})\, dA \tag{10.2-21c}$$

For the expressions in Eqs. (10.2-21), it is assumed that the membrane stresses are constant (or averaged) throughout the element. However, when numerical integration is used, the membrane stresses can be calculated at the integration points and need not be averaged. Note that the matrices $\mathbf{f}^{\mathrm{T}}_{,x}\mathbf{f}_{,y}$ and $\mathbf{f}^{\mathrm{T}}_{,y}\mathbf{f}_{,x}$ in Eq. (10.2-21c) are not necessarily equal (see Example 2).

We can apply Eqs. (10.2-20) and (10.2-21) to the plate elements studied previously in Chapter 6. Table 10.1 contains the three parts of the initial stress stiffness matrix $\mathbf{K}_0$ for the MZC rectangle (5), as given by Eq. (10.2-20). Similarly, (6) lists $\mathbf{K}_0$ for the BFS rectangle.

On the other hand, the procedure for obtaining matrix $\mathbf{K}_{01}$ for the CKZ triangle (of variable thickness) may be expressed as:

$$\mathbf{K}_{01} = A \sum_{j=1}^{n} W_j \sigma_{x0j} t_j \mathbf{f}^{\mathrm{T}}_{,xj}\mathbf{f}_{,xj} \tag{10.2-22a}$$

where A is the area of the triangle. Equation (10.2-22a) is the numerical equivalent of Eq. (10.2-21a). The constants needed for this summation appear in Table

TABLE 10.1 Initial Stress Stiffness Matrix for MZC Rectangle

$$\mathbf{K}_0 = t(\bar{\mathbf{K}}_{01} + \bar{\mathbf{K}}_{02} + \bar{\mathbf{K}}_{03})$$

$$\bar{\mathbf{K}}_{01} = \frac{\sigma_{x0} b}{630a} \begin{bmatrix} 276 & & & & & & & & & & & \\ 66b & 24b^2 & & & & & & & & & & \\ -42a & 0 & 112a^2 & & & & & & & \text{Sym.} & & \\ -276 & -66b & 42a & 276 & & & & & & & & \\ -66b & -24b^2 & 0 & 66b & 24b^2 & & & & & & & \\ -42a & 0 & -28a^2 & 42a & 0 & 112a^2 & & & & & & \\ -102 & -39b & 21a & 102 & 39b & 21a & 276 & & & & & \\ 39b & 18b^2 & 0 & -39b & -18b^2 & 0 & -66b & 24b^2 & & & & \\ -21a & 0 & -14a^2 & 21a & 0 & 56a^2 & 42a & 0 & 112a^2 & & & \\ 102 & 39b & -21a & -102 & -39b & -21a & -276 & 66b & -42a & 276 & & \\ -39b & -18b^2 & 0 & 39b & 18b^2 & 0 & 66b & -24b^2 & 0 & -66b & 24b^2 & \\ -21a & 0 & 56a^2 & 21a & 0 & -14a^2 & 42a & 0 & -28a^2 & -42a & 0 & 112a^2 \end{bmatrix}$$

$$\bar{\mathbf{K}}_{02} = \frac{\sigma_{y0} a}{630b} \begin{bmatrix} 276 & & & & & & & & & & & \\ 42b & 112b^2 & & & & & & & & & & \\ -66a & 0 & 24a^2 & & & & & & & \text{Sym.} & & \\ 102 & 21b & -39a & 276 & & & & & & & & \\ 21b & 56b^2 & 0 & 42b & 112b^2 & & & & & & & \\ 39a & 0 & -18a^2 & 66a & 0 & 24a^2 & & & & & & \\ -102 & -21b & 39a & -276 & -42b & -66a & 276 & & & & & \\ 21b & -14b^2 & 0 & 42b & -28b^2 & 0 & -42b & 112b^2 & & & & \\ -39a & 0 & 18a^2 & -66a & 0 & -24a^2 & 66a & 0 & 24a^2 & & & \\ -276 & -42b & 66a & -102 & -21b & -39a & 102 & -21b & 39a & 276 & & \\ 42b & -28b^2 & 0 & 21b & -14b^2 & 0 & -21b & 56b^2 & 0 & -42b & 112b^2 & \\ 66a & 0 & -24a^2 & 39a & 0 & 18a^2 & -39a & 0 & -18a^2 & -66a & 0 & 24a^2 \end{bmatrix}$$

$$\bar{\mathbf{K}}_{03} = \frac{\tau_{xy0}}{90}\begin{bmatrix} 45 & & & & & & & & & & & \\ 0 & 0 & & & & & & & & & & \\ 0 & -5ab & 0 & & & & & & & & & \\ 0 & 18b & 0 & -45 & & & & & \text{Sym.} & & & \\ -18b & 0 & 5ab & 0 & 0 & & & & & & & \\ 0 & 5ab & 0 & 0 & -5ab & 0 & & & & & & \\ -45 & -18b & 18a & 0 & 0 & -18a & 45 & & & & & \\ 18b & 6b^2 & -5ab & 0 & 0 & 5ab & 0 & 0 & & & & \\ -18a & -5ab & 6a^2 & 18a & 5ab & 0 & 0 & -5ab & 0 & & & \\ 0 & 0 & -18a & 45 & 18b & 18a & 0 & -18b & 0 & -45 & & \\ 0 & 0 & 5ab & -18b & -6b^2 & -5ab & 18b & 0 & 5ab & 0 & 0 & \\ 18a & 5ab & 0 & -18a & -5ab & -6a^2 & 0 & 5ab & 0 & 0 & -5ab & 0 \end{bmatrix}$$

3.1. Similarly, from Eqs. (10.2-21b) and (10.2-21c) we obtain:

$$\mathbf{K}_{02} = A \sum_{j=1}^{n} W_j \sigma_{y0j} t_j \mathbf{f}^{\mathrm{T}}_{,yj} \mathbf{f}_{,yj} \tag{10.2-22b}$$

$$\mathbf{K}_{03} = A \sum_{j=1}^{n} W_j \tau_{xy0j} t_j (\mathbf{f}^{\mathrm{T}}_{,x} \mathbf{f}_{,y} + \mathbf{f}^{\mathrm{T}}_{,y} \mathbf{f}_{,x})_j \tag{10.2-22c}$$

Because the displacement shape functions for this element are expressed in area coordinates (see Table 6.3), the derivatives $\mathbf{f}_{,x}$ and $\mathbf{f}_{,y}$ require use of the chain rule, as stated in Eqs. (3.2-12).

Element PBQ8 is another type for which numerical integration is necessary. In this case the transformation of derivatives from local to global coordinates requires using the inverse of a 2×2 Jacobian matrix $\mathbf{J}$, as follows:

$$\begin{bmatrix} \mathbf{f}_{3,x} \\ \mathbf{f}_{3,y} \end{bmatrix} = \mathbf{J}^{-1} \begin{bmatrix} \mathbf{f}_{3,\xi} \\ \mathbf{f}_{3,\eta} \end{bmatrix} \tag{o}$$

The subscript 3 connotes the fact that only the third row of matrix $\mathbf{f}$ (corresponding to w) is to be used. The Jacobian matrix in Eq. (o) is:

$$\mathbf{J} = \begin{bmatrix} x_{,\xi} & y_{,\xi} \\ x_{,\eta} & y_{,\eta} \end{bmatrix} \tag{p}$$

Then we write the expression for numerical evaluation of $\mathbf{K}_{01}$ as:

$$\mathbf{K}_{01} = \sum_{k=1}^{n} \sum_{j=1}^{n} R_j R_k (\sigma_{x0})_{jk} t_{jk} (\mathbf{f}^{\mathrm{T}}_{3,x})_{jk} (\mathbf{f}_{3,x})_{jk} |\mathbf{J}_{jk}| \tag{10.2-23a}$$

Similarly,

$$\mathbf{K}_{02} = \sum_{k=1}^{n} \sum_{j=1}^{n} R_j R_k (\sigma_{y0})_{jk} t_{jk} (\mathbf{f}^{\mathrm{T}}_{3,y})_{jk} (\mathbf{f}_{3,y})_{jk} |\mathbf{J}_{jk}| \tag{10.2-23b}$$

and

$$\mathbf{K}_{03} = \sum_{k=1}^{n} \sum_{j=1}^{n} R_j R_k (\tau_{xy0})_{jk} t_{jk} (\mathbf{f}^{\mathrm{T}}_{3,x} \mathbf{f}_{3,y} + \mathbf{f}^{\mathrm{T}}_{3,y} \mathbf{f}_{3,x})_{jk} |\mathbf{J}_{jk}| \tag{10.2-23c}$$

Note that the plate thickness and the initial membrane stresses may be different at each integration point.

The initial stress stiffness matrix for the annular plate element (see Sec. 6.5) is not required, because instability analyses of circular and annular plates are adequately covered in the technical literature (3, 11).

Example 1

For the MZC rectangle, find the term $(K_0)_{2,4}$ in the initial stress stiffness matrix. From Eqs. (6.2-2b) the displacement shape functions f_2 and f_4 for this element are:

$$\begin{aligned} f_2 &= \frac{b}{8}(1 - \xi)(1 + \eta)(1 - \eta)^2 \\ f_4 &= \frac{1}{8}(1 + \xi)(1 - \eta)(2 + \xi - \eta - \xi^2 - \eta^2) \end{aligned} \tag{q}$$

Differentiation of these functions with respect to x and y produces:

$$f_{2,x} = -\frac{1}{8a}(1+\eta)(1-\eta)^2 b$$
$$f_{4,x} = \frac{1}{8a}[3(1-\xi^2) - (\eta+\eta^2)](1-\eta) \tag{r}$$

and

$$f_{2,y} = \frac{1}{8}(1-\xi)(-1-2\eta+3\eta^2)$$
$$f_{4,y} = \frac{1}{8b}(1+\xi)[(-\xi+\xi^2) - 3(1-\eta^2)] \tag{s}$$

Substitution of $f_{2,x}$ and $f_{4,x}$ into Eq. (10.2-21a), followed by integration, gives:

$$(\bar{K}_{01})_{2,4} = \frac{b^2}{64a}\sigma_{x0}\int_{-1}^{1}\int_{-1}^{1} -(1+\eta)(1-\eta)^2[3(1-\xi^2) - (\eta+\eta^2)](1-\eta)\,d\xi\,d\eta$$
$$= -\frac{11b^2}{105a}\sigma_{x0} \tag{t}$$

which is the same as that in Table 10.1. Similar substitutions into Eqs. (10.2-21b) and (10.2-21c) yield:

$$(\bar{K}_{02})_{2,4} = \frac{7a}{210}\sigma_{y0} \tag{u}$$

and

$$(\bar{K}_{03})_{2,4} = \left(\frac{1}{10} + \frac{1}{10}\right) b\tau_{xy0} = \frac{b}{5}\tau_{xy0} \tag{v}$$

From Eq. (v) we see that $f_{2,x}f_{4,y} = f_{2,y}f_{4,x}$; and their sum could be written as $2f_{2,x}f_{4,y}$.

Example 2

Determine $(K_0)_{9,11}$ for the BFS rectangle. From Table 6.2 we have:

$$f_9 = (3\xi^2 - 2\xi^3)(3\eta^2 - 2\eta^3) \qquad f_{11} = (\xi^2 - \xi^3)(3\eta^2 - 2\eta^3)a \tag{w}$$

The derivatives of these functions are:

$$f_{9,x} = \frac{6}{a}(\xi - \xi^2)(3\eta^2 - 2\eta^3) \qquad f_{11,x} = (2\xi - 3\xi^2)(3\eta^2 - 2\eta^3) \tag{x}$$

and

$$f_{9,y} = \frac{6}{b}(3\xi^2 - 2\xi^3)(\eta - \eta^2) \qquad f_{11,y} = \frac{6a}{b}(\xi^2 - \xi^3)(\eta - \eta^2) \tag{y}$$

Substituting $f_{9,x}$ and $f_{11,x}$ into Eq. (10.2-21a) and integrating, we find:

$$(\bar{K}_{01})_{9,11} = 6b\sigma_{x0}\int_{0}^{1}\int_{0}^{1}(\xi - \xi^2)(2\xi - 3\xi^2)(3\eta^2 - 2\eta^3)^2\,d\xi\,d\eta$$
$$= \frac{13b}{350}\sigma_{x0} \tag{z}$$

Similar substitutions into Eqs. (10.2-21b) and (10.2-21c) give:

$$(\bar{K}_{02})_{9,11} = \frac{11a^2}{175b}\sigma_{y0} \tag{a'}$$

and

$$(\bar{K}_{03})_{9,11} = \left(\frac{1}{20} - \frac{1}{20}\right) a\tau_{xy0} = 0 \tag{b'}$$

In this case we see that the two terms in Eq. (b′) cancel each other and produce zero.

10.3 LINEAR INSTABILITY AND VIBRATIONS

By summing contributions from elements to assemble terms for the whole structure, we can obtain the matrix $\mathbf{S}_{N0}$, which is the initial stress stiffness matrix for all of the nodes. Thus,

$$\mathbf{S}_{N0} = \sum_{i=1}^{n_e} \mathbf{K}_0 \tag{10.3-1}$$

As for vibrations in Sec. 9.3, we are primarily interested in the submatrix $\mathbf{S}_{FF0}$ pertaining to nodal degrees of freedom. For simplicity of notation, the subscripts FF will be dropped to obtain the symbol $\mathbf{S}_0$. Then the homogeneous equilibrium equations for the nodal degrees of freedom may be stated as:

$$(\mathbf{S} + c\mathbf{S}_0)\mathbf{D} = \mathbf{O} \tag{10.3-2}$$

As before, the vector $\mathbf{D}$ in this expression contains only the nodal displacements that are free to occur. The constant c in Eq. (10.3-2) is a scale factor that multiplies initial loads and stresses. That is, for flexural elements, P_0 is replaced by cP_0; and for plate-bending elements $\boldsymbol{\sigma}_0$ is replaced by $c\boldsymbol{\sigma}_0$.

Now let us consider increasing the scale factor c until bifurcation buckling occurs. Because the translations in vector $\mathbf{D}$ are orthogonal to the axial or membrane stresses, this type of collapse is abrupt and unexpected.

Comparing Eq. (10.3-2) with Eq. (9.3-5), we see that both are algebraic eigenvalue problems in nonstandard (but symmetric) forms. Therefore, Eq. (10.3-2) has nontrivial solutions only if the following condition is true:

$$|\mathbf{S} + c\mathbf{S}_0| = 0 \tag{10.3-3}$$

The smallest absolute value of c satisfying this equation represents the scale factor on the initial loads that produces buckling (7). This level of loading is called the *critical buckling load*, and the corresponding value of c may be referred to as the *critical eigenvalue*, c_{cr} (or *buckling load factor*). Back-substitution of c_{cr} into Eq. (10.3-2) yields the *buckling mode shape*. The resulting vector $\boldsymbol{\Phi}_{cr}$ is often called the *critical eigenvector*.

It should be emphasized that the scale factor c is multiplied with the results of an axial or plane-stress analysis that precedes the buckling analysis. To perform such a preliminary task, we need a pattern of loads that cause initial axial stresses in flexural members or initial membrane stresses in plate elements.

After the eigenvalue c_{cr} is found, it is used to scale the initial loads, yielding the critical buckling loads for the entire problem.

As explained in Sec. 9.3, Eq. (10.3-2) can be transformed to the standard, symmetric form of the eigenvalue problem [see Eq. (9.3-6)] by factoring matrix **S** and changing coordinates. Hence, we obtain:

$$(\mathbf{S}_{U0} + \lambda_i \mathbf{I})\mathbf{\Phi}_{Ui} = \mathbf{O} \tag{10.3-4}$$

where

$$\mathbf{S}_{U0} = \mathbf{U}^{-T}\mathbf{S}_0\mathbf{U}^{-1} \qquad \lambda_i = \frac{1}{c_i} \qquad \mathbf{\Phi}_{Ui} = \mathbf{U}\,\mathbf{D} \tag{10.3-5}$$

This manipulation is desirable because λ_{max} then corresponds to the smallest value c_{min} (or c_{cr}) of the scale factor. Thus, we have:

$$c_{cr} = \frac{1}{\lambda_{max}} \qquad \mathbf{\Phi}_{cr} = \mathbf{U}^{-1}\mathbf{\Phi}_{Ucr} \tag{10.3-6}$$

where $\mathbf{\Phi}_{cr}$ and $\mathbf{\Phi}_{Ucr}$ are the critical eigenvectors in original and standard coordinates, respectively.

Vibrations in the presence of initial axial or membrane stresses lead to the following eigenvalue problem (8):

$$(-\omega_i^2\mathbf{M} + \mathbf{S} + \mathbf{S}_0)\mathbf{\Phi}_i = \mathbf{O} \tag{10.3-7}$$

in which the variable time has been separated from those of space as before (see Sec. 9.3). Nontrivial solutions for Eq. (10.3-7) exist only if:

$$|-\omega_i^2\mathbf{M} + \mathbf{S} + \mathbf{S}_0| = 0 \tag{10.3-8}$$

This characteristic equation yields angular frequencies that are different from the values obtained when matrix $\mathbf{S}_0$ is not present. If initial stresses in $\mathbf{S}_0$ are positive, the frequencies are raised; and if the initial stresses are negative, the frequencies are lowered. In this formulation, a frequency of zero occurs when the initial stresses cause buckling.

The initial stress stiffness matrix $\mathbf{S}_0$ can be reduced in a manner that is analogous to the method described in Sec. 9.4 for the consistent mass matrix (9). Thus,

$$\mathbf{S}^*_{BB0} = \mathbf{T}_B^T\mathbf{S}_0\mathbf{T}_B \tag{10.3-9}$$

where the transformation matrix $\mathbf{T}_B$ is given by Eq. (9.4-9). As before, the advantage of such a reduction is to produce a smaller eigenvalue problem.

Example

The simple beam in Fig. 10.4 is composed of two prismatic flexural elements having the same value of *EI*. An axial force P_0 is applied to the beam, which results in an initial stress stiffness matrix for all the nodes, as follows:

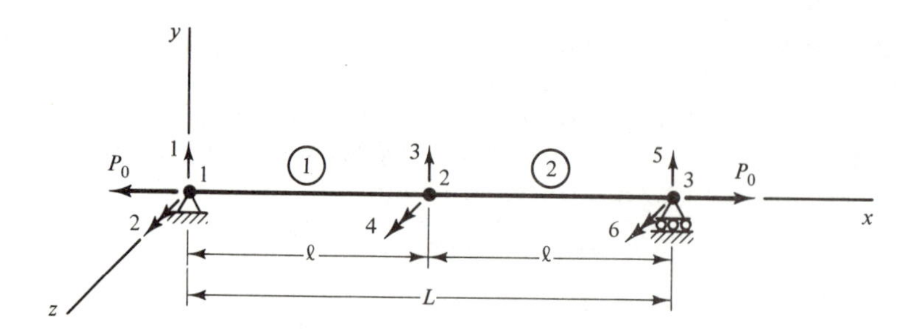

Figure 10.4 Two-Element Simple Beam

$$\mathbf{S}_{N0} = \frac{P_0}{30\ell}\begin{bmatrix} 36 & 3\ell & -36 & 3\ell & 0 & 0 \\ 3\ell & 4\ell^2 & -3\ell & -\ell^2 & 0 & 0 \\ -36 & -3\ell & 72 & 0 & -36 & 3\ell \\ 3\ell & -\ell^2 & 0 & 8\ell^2 & -3\ell & -\ell^2 \\ 0 & 0 & -36 & -3\ell & 36 & -3\ell \\ 0 & 0 & 3\ell & -\ell^2 & -3\ell & 4\ell^2 \end{bmatrix} \begin{matrix} 1 \\ 2 \\ 3 \\ 4 \\ 5 \\ 6 \end{matrix} \qquad \text{(a)}$$
$$\begin{matrix} 1 & 2 & 3 & 4 & 5 & 6 \end{matrix}$$

In Eq. (a) the contributions of elements 1 and 2 to matrix $\mathbf{S}_{N0}$ appear within dashed boxes. Also, the displacement coordinates for the problem (see Fig. 10.4) are numbered at the right and below the matrix. The nodal stiffness matrix $\mathbf{S}_N$ for the unrestrained structure was shown previously in the example of Sec. 9.4.

In the present example we will show how the rotational displacements can be eliminated while retaining the translational displacements for the purpose of calculating the critical buckling load P_{cr}. Toward that end, we remove the first and fifth rows and columns from matrices $\mathbf{S}_N$ and $\mathbf{S}_{N0}$, because displacements one and five in the figure are restrained by supports. Then the remaining 4×4 matrices are rearranged to put the rotational terms before the translational terms. Thus,

$$\mathbf{S} = \begin{bmatrix} \mathbf{S}_{AA} & \mathbf{S}_{AB} \\ \mathbf{S}_{BA} & \mathbf{S}_{BB} \end{bmatrix} = \frac{EI}{\ell^3}\left[\begin{array}{ccc|c} 4\ell^2 & 2\ell^2 & 0 & -6\ell \\ 2\ell^2 & 8\ell^2 & 2\ell^2 & 0 \\ 0 & 2\ell^2 & 4\ell^2 & 6\ell \\ \hline -6\ell & 0 & 6\ell & 24 \end{array}\right] \begin{matrix} 2 \\ 4 \\ 6 \\ 3 \end{matrix} \qquad \text{(b)}$$
$$\begin{matrix} 2 & 4 & 6 & 3 \end{matrix}$$

and

$$\mathbf{S}_0 = \begin{bmatrix} \mathbf{S}_{AA0} & \mathbf{S}_{AB0} \\ \mathbf{S}_{BA0} & \mathbf{S}_{BB0} \end{bmatrix} = \frac{P_0}{30\ell}\left[\begin{array}{ccc|c} 4\ell^2 & -\ell^2 & 0 & -3\ell \\ -\ell^2 & 8\ell^2 & -\ell^2 & 0 \\ 0 & -\ell^2 & 4\ell^2 & 3\ell \\ \hline -3\ell & 0 & 3\ell & 72 \end{array}\right] \begin{matrix} 2 \\ 4 \\ 6 \\ 3 \end{matrix} \qquad \text{(c)}$$
$$\begin{matrix} 2 & 4 & 6 & 3 \end{matrix}$$

The inverse of submatrix $\mathbf{S}_{AA}$ in Eq. (b) is:

$$\mathbf{S}_{AA}^{-1} = \frac{\ell}{24EI}\begin{bmatrix} 7 & -2 & 1 \\ -2 & 4 & -2 \\ 1 & -2 & 7 \end{bmatrix} \tag{d}$$

This array and the other submatrices of $\mathbf{S}$ from Eq. (b) may be substituted into Eq. (9.4-5) to find:

$$S_{BB}^{*} = \frac{24EI}{\ell^3} - \frac{18EI}{\ell^3} = \frac{6EI}{\ell^3} = \frac{48EI}{L^3} \tag{e}$$

which is only a single term.

To reduce the initial stress stiffness matrix, we construct the transformation matrix $\mathbf{T}_B$ by evaluating $-\mathbf{S}_{AA}^{-1}\mathbf{S}_{AB}$ and putting it into the upper part of the matrix in Eq. (9.4-9). Hence,

$$-\mathbf{S}_{AA}^{-1}\mathbf{S}_{AB} = \frac{3}{2\ell}\{1, 0, -1\} \tag{f}$$

and

$$\mathbf{T}_B = \frac{1}{2\ell}\{3, 0, -3, \mid 2\ell\} \tag{g}$$

Here the submatrix $\mathbf{I}_B$ in matrix $\mathbf{T}_B$ is simply 1, because there is only one translational displacement remaining (number 3). Substitution of matrix $\mathbf{S}_0$ from Eq. (c) and $\mathbf{T}_B$ from Eq. (g) into Eq. (10.3-9) produces:

$$S_{BB0}^{*} = \frac{12P_0}{5\ell} = \frac{24P_0}{5L} \tag{h}$$

which (like S_{BB}^{*}) is only a single term.

Using S_{BB}^{*} and S_{BB0}^{*} from Eqs. (e) and (h), we can set up the eigenvalue problem (Eq. 10.3-2) for the remaining system as:

$$\left(\frac{48EI}{L^3} + c\,\frac{24P_0}{5L}\right) D_3 = 0 \tag{i}$$

Thus, from Eq. (10.3-3) we have:

$$P_{cr} = c_{cr}P_0 = -\frac{10EI}{L^2} \tag{j}$$

The exact solution (3) is:

$$P_{cr} = -\frac{\pi^2 EI}{L^2} \tag{k}$$

Therefore, the percentage error in the approximate solution becomes:

$$e = \frac{100}{-\pi^2}[-10 - (-\pi^2)] = +1.3\% \tag{ℓ}$$

which is an upper bound.

TABLE 10.2 Instability Coefficients β for Prismatic Beams Modeled by Four Elements

Support Conditions		Exact (3)	TR	% Error	TO	% Error
Simple	P_0 ... L ... P_0	9.870	9.875	+0.051	9.876	+0.061
Fixed	P_0 ... P_0	39.48	39.78	+0.76	40.00	+1.3
Cantilever	P_0 ... P_0	2.467	2.467	+0.00	2.468	+0.041
Propped	P_0 ... P_0	20.19	20.23	+0.20	20.25	+0.30

$$P_{cr} = -\beta \frac{EI}{L^2}$$

TR: Translations and Rotations
TO: Translations Only

Instability coefficients β for prismatic beams with various end conditions are given in Table 10.2. In each case the beam is modeled by four flexural elements, and the results using the initial stress stiffness matrix (with and without elimination of rotations) are compared against exact values.

10.4 NONLINEAR INSTABILITY ANALYSIS

In this section we consider those problems where interdependencies exist between flexural and axial (or membrane) stresses and strains. As mentioned in Sec. 10.1, frames and shells fall into this category. In plane and space frames, the magnitudes of axial forces depend upon the level of loading and cannot be determined by a separate analysis (unless the structure happens to be statically determinate). Similarly, in general shells and axisymmetric shells, the values of membrane stresses usually cannot be determined beforehand from an independent analysis.

The procedure to be followed in analyzing such structures for collapse loads must be either *iterative* or *incremental* or both. The iterative approach (9) yields at best only the collapse load and the corresponding pattern of displacements. On the other hand, the incremental method (10) produces a continuous history of load against displacements. For this reason the latter technique is considered to be the better method and will be described further.

In the incremental approach to matrix analysis for nonlinear instability and large displacements, the loads (or in some cases the displacements) are increased in a series of increments; and in each step the structure is assumed to behave linearly. Nonlinear strain-displacement relationships are included in an approximate manner through the use of an initial stress stiffness matrix $\mathbf{S}_0$ augmenting the conventional stiffness matrix $\mathbf{S}$. In addition, the overall change in geometry is taken into account for each step by revising nodal coordinates in accordance with calculated displacements. Stresses computed in a given step are used in the initial stress stiffness matrix for the next step. The linearized *incremental equation* for the jth step is:

$$[\mathbf{S}(\mathbf{D}_{j-1}) + \mathbf{S}_0(\mathbf{D}_{j-1}, \boldsymbol{\sigma}_{j-1})]\,\Delta\mathbf{D}_j = \Delta\mathbf{A}_j \tag{10.4-1}$$

In this expression the symbol $\mathbf{S}(\mathbf{D}_{j-1})$ means that the stiffness matrix $\mathbf{S}$ is computed on the basis of the displacements $\mathbf{D}_{j-1}$ obtained in the preceding step. Similarly, $\mathbf{S}_0(\mathbf{D}_{j-1}, \boldsymbol{\sigma}_{j-1})$ implies that the initial stress stiffness matrix $\mathbf{S}_0$ is found using both displacements $\mathbf{D}_{j-1}$ and the stresses $\boldsymbol{\sigma}_{j-1}$ from the preceding step.

Table 10.3 shows the stepping procedure, which could be refined by iteration within each step. The point on a load-displacement curve where the slope is zero (or very close to zero) signifies the critical loading level for nonlinear instability [see Fig. 10.1(b)].

TABLE 10.3 Incremental Load Method

Step	Stiffnesses	Incremental Loads	Incremental Displacements	Total Displacements	Incremental Stresses	Total Stresses
1	$\mathbf{S}(0) + \mathbf{S}_0(0, \boldsymbol{\sigma}_0)$	$\Delta\mathbf{A}_1$	$\Delta\mathbf{D}_1$	$\mathbf{D}_1$	$\Delta\boldsymbol{\sigma}_1$	$\boldsymbol{\sigma}_1$
2	$\mathbf{S}(\mathbf{D}_1) + \mathbf{S}_0(\mathbf{D}_1, \boldsymbol{\sigma}_1)$	$\Delta\mathbf{A}_2$	$\Delta\mathbf{D}_2$	$\mathbf{D}_2$	$\Delta\boldsymbol{\sigma}_2$	$\boldsymbol{\sigma}_2$
3	$\mathbf{S}(\mathbf{D}_2) + \mathbf{S}_0(\mathbf{D}_2, \boldsymbol{\sigma}_2)$	$\Delta\mathbf{A}_3$	$\Delta\mathbf{D}_3$	$\mathbf{D}_3$	$\Delta\boldsymbol{\sigma}_3$	$\boldsymbol{\sigma}_3$
..	... ...	...	...	...	...	...
j	$\mathbf{S}(\mathbf{D}_{j-1}) + \mathbf{S}_0(\mathbf{D}_{j-1}, \boldsymbol{\sigma}_{j-1})$	$\Delta\mathbf{A}_j$	$\Delta\mathbf{D}_j$	$\mathbf{D}_j$	$\Delta\boldsymbol{\sigma}_j$	$\boldsymbol{\sigma}_j$

10.5 PROGRAM BUCPQ8 AND APPLICATIONS

In this last program we use again the plate-bending element PBQ8 to calculate the buckling load factors and mode shapes for isotropic plates in plane stress. Program BUCPQ8 first analyzes a plate with loads in its own plane using the plane-stress stiffness matrix for element Q8 from Sec. 3.4. Then it takes the flexural stiffness matrix from Eq. (6.4-7) and the initial stress stiffness matrix from Eqs. (10.2-23) to solve the buckling problem. The structural and load data for BUCPQ8 remain the same as in the static analysis program named PSQ8. However, each set of loads to be investigated must be accompanied by buckling data. These data have two parts. The first part consists of buckling parameters, as follows:

NRN = number of restrained nodes for flexure

NMOD = number of buckling modes desired

Normally, only the fundamental (or first) buckling mode is of interest. The second part of the buckling data gives the nodal restraints for flexure.

The logic in a program for instability analysis is different from that of the other programs in this book. First, it is necessary to perform an initial plane stress analysis for a particular set of loads in the plane of the plate. Then we must construct the flexural stiffness matrix as well as the initial stress stiffness matrix for all the flexural degrees of freedom in the structure. These arrays are formed as square matrices that are subsequently reduced by matrix condensation, as explained in Sec. 10.3. As for vibrational analysis, the rotational displacements at nodes are eliminated, whereas the translational displacements are retained. Then the desired eigenvalues and eigenvectors are calculated for the standard, symmetric form, as described in Sec. 10.3. The computer output includes the mode number, the buckling load factor, and the mode shape for each of the buckling modes desired. The following outline shows the logic to be added within the loading loop of Program PSQ8 to convert it to Program BUCPQ8:

7. Read and print buckling data
   (a) Buckling parameters
   (b) Nodal restraints for flexure
   (c) Calculate new displacement indexes
8. Generate stiffness and initial stress stiffness matrices for flexure
   (a) Element stiffness matrix
   (b) Element initial stress stiffness matrix
   (c) Transfer to structural stiffness and initial stress stiffness matrices
9. Reduce stiffness and initial stress stiffness matrices
   (a) Calculate reduced stiffness matrix

   (b) Determine transformation matrix
   (c) Calculate reduced initial stress stiffness matrix
10. Determine eigenvalues and eigenvectors
   (a) Factor stiffness matrix
   (b) Invert stiffness factor
   (c) Transform eigenvalue problem to standard, symmetric form
   (d) Solve for eigenvalues and eigenvectors
   (e) Back-transform eigenvectors
   (f) Normalize eigenvectors (with respect to largest terms)
11. Print results (for each mode)
   (a) Mode number
   (b) Buckling load factor
   (c) Buckling mode shape

Example 1

A square cantilever plate element PBQ8 appears in Fig. 10.5. It has a uniformly distributed line loading b_{x0} (force per unit length) acting in the x direction on edge 1-3. The nodal restraints in the figure are the same as those in Figs. 6.13 and 9.7. The following values of physical parameters are given:

$$a = 0.3 \text{ m} \qquad t = 4 \times 10^{-3} \text{ m} \qquad E = 7 \times 10^{7} \text{ kN/m}^2$$
$$\nu = 0.3 \qquad b_{x0} = 2 \text{ kN/m}$$

where the units are S.I. and the material is aluminum.

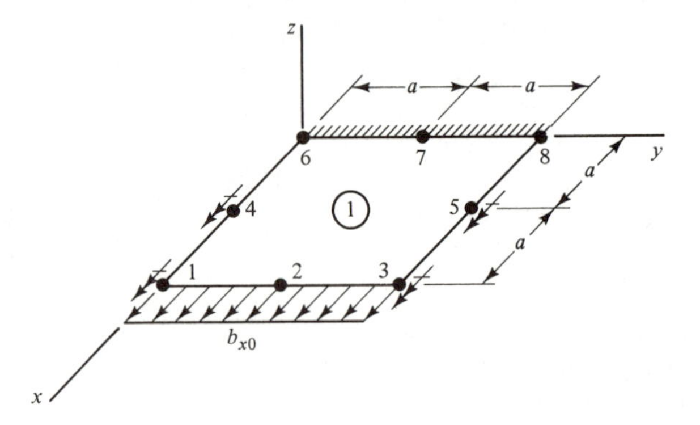

Figure 10.5 Example 1: Cantilever Element PBQ8 with Edge Load in x Direction

Table 10.4 contains the computer output from Program BUCPQ8 for the fundamental buckling mode. The buckling load factor in the output is $c_{cr} = -1.427$. Therefore, the critical buckling load intensity is found to be $(b_x)_{cr} = c_{cr} b_{x0} = -2.854$ kN/m. The theoretical value (3) for no shearing deformations is:

TABLE 10.4 Computer Output for Example 1

```
PROGRAM BUCPQ8

***  EXAMPLE 1:  CANTILEVER ELEMENT WITH EDGE LOAD IN X DIRECTION  ***

STRUCTURAL PARAMETERS
   NN   NE  NRN  NLS          E            PR            T
    8    1    7    1  7.0000D 07  3.0000D-01  4.0000D-03

COORDINATES OF NODES
 NODE          X           Y
    1  6.0000D-01  0.0000D-01
    2  6.0000D-01  3.0000D-01
    3  6.0000D-01  6.0000D-01
    4  3.0000D-01  0.0000D-01
    5  3.0000D-01  6.0000D-01
    6  0.0000D-01  0.0000D-01
    7  0.0000D-01  3.0000D-01
    8  0.0000D-01  6.0000D-01

ELEMENT INFORMATION
ELEM.   N1   N2   N3   N4   N5   N6   N7   N8
    1    1    3    8    6    2    5    7    4

NODAL RESTRAINTS FOR PLANE STRESS
 NODE   R1   R2
    1    0    1
    3    0    1
    4    0    1
    5    0    1
    6    1    1
    7    1    1
    8    1    1

NUMBER OF DEGREES OF FREEDOM =    6
NUMBER OF NODAL RESTRAINTS   =   10
NUMBER OF TERMS IN SN        =   39

**********  LOADING NUMBER    1  **********
  NLN  NEL  NEV  IPR
    3    0    0    0

ACTIONS AT NODES
 NODE        AN1         AN2
    1  2.0000D-01  0.0000D-01
    2  8.0000D-01  0.0000D-01
    3  2.0000D-01  0.0000D-01

NODAL DISPLACEMENTS
 NODE        DN1         DN2
    1  3.9000D-06 -0.0000D-01
    2  3.9000D-06 -2.6470D-23
    3  3.9000D-06 -0.0000D-01
    4  1.9500D-06 -0.0000D-01
    5  1.9500D-06  0.0000D-01
    6  0.0000D-01  0.0000D-01
    7  0.0000D-01  0.0000D-01
    8  0.0000D-01  0.0000D-01

ELEMENT STRESSES
ELEM. INT.    X STRESS    Y STRESS   XY STRESS
```

TABLE 10.4 (cont.)

```
      1     1  5.0000D 02   1.5000D 02   4.0592D-12
      1     2  5.0000D 02   1.5000D 02   1.5177D-11
      1     3  5.0000D 02   1.5000D 02  -4.0706D-12
      1     4  5.0000D 02   1.5000D 02  -1.5188D-11

SUPPORT REACTIONS
 NODE           AR1          AR2
    1  0.0000D-01 -6.0000D-02
    3  0.0000D-01  6.0000D-02
    4  0.0000D-01 -2.4000D-01
    5  0.0000D-01  2.4000D-01
    6 -2.0000D-01 -6.0000D-02
    7 -8.0000D-01  2.9490D-17
    8 -2.0000D-01  6.0000D-02

BUCKLING PARAMETERS
  NRN NMOD
    7    1

NODAL RESTRAINTS FOR FLEXURE
 NODE   R1   R2   R3
    1    0    1    0
    3    0    1    0
    4    0    1    0
    5    0    1    0
    6    1    1    1
    7    1    1    1
    8    1    1    1

NUMBER OF TRANSLATIONAL D.O.F            =    5
NUMBER OF ROTATIONAL D.O.F.              =    6
NUMBER OF TRANSLATIONAL RESTRAINTS       =    3
NUMBER OF ROTATIONAL RESTRAINTS          =   10

MODE     1
BUCKLING LOAD FACTOR -1.4270D 00
 NODE       DN1        NODE       DN1        NODE       DN1        NODE       DN1
    1  1.0000           2  1.0000           3  1.0000           4  0.3022
    5  0.3022           6  0.0000           7  0.0000           8  0.0000
```

$$(b_x)_{cr} = -\frac{\pi^2 E t^3}{(2a)^2 12(1 - \nu^2)} = -2.812 \text{ kN/m}$$

Hence, the finite-element solution is 1.49 percent higher than this value.

Example 2

Figure 10.6 shows a quarter of the same simply supported rectangular plate that was analyzed for vibrations in Example 2 of Sec. 9.5. However, the plate now has uniformly distributed edge loads of intensities b_{x0} and b_{y0} (force per unit length) applied in the x and y directions. Nodal restraints required to find the fundamental buckling mode are for the symmetric-symmetric case (see Fig. 10.6 and Sec. 9.5). We have values of physical parameters, as follows:

$$a = 1.5 \text{ in.} \qquad b = 1 \text{ in.} \qquad t = 0.15 \text{ in.}$$

$$E = 3 \times 10^4 \text{ k/in.}^2 \qquad \nu = 0.3 \qquad b_{x0} = b_{y0} = 3 \text{ k/in.}$$

in which the material is steel and the units are U.S.

The buckling load factor obtained from Program BUCPQ8 for the fundamental

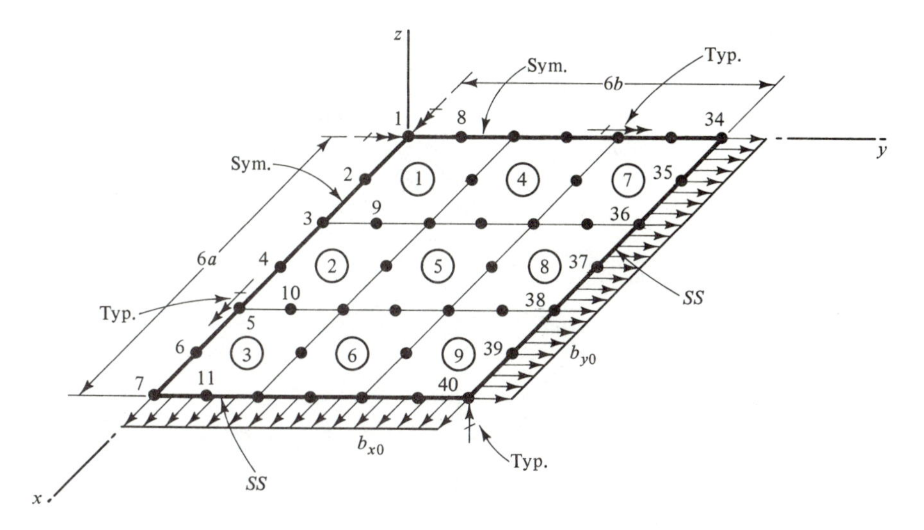

Figure 10.6 Example 2: Simply Supported Rectangular Plate with PBQ8 Network and Edge Loads in x and y Directions

mode is $c_{cr} = -0.30205$. Thus, the critical buckling load intensity is $(b_x)_{cr} = (b_y)_{cr} = c_{cr}b_{x0} = c_{cr}b_{y0} = -0.90615$ k/in. This value can be compared with the known formula (11):

$$(b_x)_{cr} = (b_y)_{cr} = -\frac{\left(c + \dfrac{1}{c}\right)^2}{1 + c^2}\pi^2 \frac{Et^3}{12(1 - \nu^2)L_y^2} = -0.91793 \text{ k/in.}$$

where

$$c = \frac{L_x}{L_y}$$

Thus, the relative error in the finite element solution is -1.28 percent.

Example 3

For a final example we consider the star-shaped plate with a circular opening shown in Fig. 10.7(a). This plate has four planes of symmetry and is subjected to forces equal to P_0 at each of its four points. For the purpose of an instability analysis, we take the shaded quarter of the plate, which is depicted in Fig. 10.7(b) with a PBQ8 network. A restraint against translation in the z direction exists at node 21, and points on the x and y axes possess symmetry restraints. Physical parameters in this problem are:

$$a = 0.05 \text{ m} \qquad t = 2 \times 10^{-3} \text{ m} \qquad E = 3.5 \times 10^8 \text{ kN/m}^2$$

$$\nu = 0.3 \qquad P_0 = 1 \text{ kN}$$

where the units are S.I. and the material is tungsten.

Upon running the data for this plate with Program BUCPQ8, we find that the buckling load factor for the first mode is $c_{cr} = -0.52126$. Therefore, the critical buckling load is $P_{cr} = c_{cr}P_0 = -0.52126$ kN. Of course, the buckling mode is symmetric with respect to all the planes of symmetry.

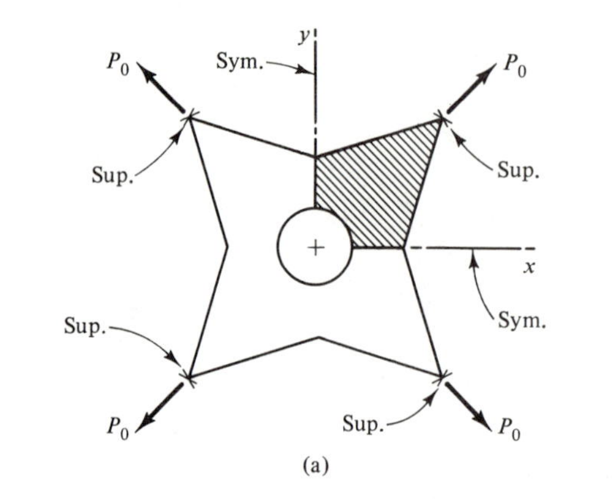

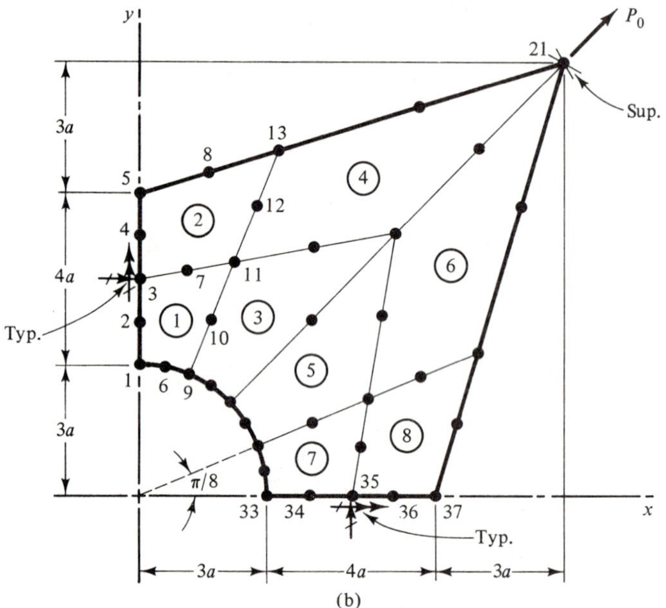

Figure 10.7 Example 3: (a) Star Plate with Circular Opening and Corner Forces (b) PBQ8 Network

REFERENCES

1. Rodden, W. P., Jones, J. P., and Bhuta, P. G., "A Matrix Formulation of the Transverse Structural Influence Coefficients of an Axially Loaded Timoshenko Beam," *AIAA Jour.*, Vol. 1, No. 1, 1963, pp. 225–227.
2. Turner, M. J., Dill, E. H., Martin, H. C., and Melosh, R. J., "Large Deflections of Structures Subjected to Heating and External Loads," *Jour. Aero. Sci.*, Vol. 27, 1960, pp. 97–106.

3. Timoshenko, S. P., and Gere, J. M., *Theory of Elastic Stability*, 2d ed., McGraw-Hill, New York, 1961.

4. Archer, J. S., "Consistent Matrix Formulations for Structural Analysis Using Finite-Element Techniques," *AIAA Jour.*, Vol. 3, No. 10, 1965, pp. 1910–1918.

5. Przemieniecki, J. S., "Discrete-Element Methods for Stability Analysis of Complex Structures," *Aero. Jour.*, Vol. 72, 1968, pp. 1077–1086.

6. Von Riesemann, W. A., "Large Deflections of Elastic Beams and Plates," *Ph.D. Thesis*, Department of Civil Engineering, Stanford University, 1968.

7. Rubinstein, M. F., *Structural Systems—Statics, Dynamics, and Stability*, Prentice-Hall, Englewood Cliffs, N.J., 1971.

8. Anderson, R. G., Irons, B. M., and Zienkiewicz, O. C., "Vibrations and Stability of Plates Using Finite Elements," *Int. Jour. Sol. Struc.*, Vol. 4, No. 10, 1968, pp. 1031–1035.

9. Gallagher, R. H., *Finite Element Analysis Fundamentals*, Prentice-Hall, Englewood Cliffs, N.J., 1975.

10. Przemieniecki, J. S., *Theory of Matrix Structural Analysis*, McGraw-Hill, New York, 1968.

11. Szilard, R., *Theory and Analysis of Plates*, Prentice-Hall, Englewood Cliffs, N.J., 1974.

PROBLEMS

10.2-1. Verify the terms in the initial stress stiffness matrix $\mathbf{K}_0$ for the flexural element given by Eq. (10.2-9).

10.2-2. Derive the three parts contributing to the term $(K_0)_{2,2}$ for the MZC rectangle in Table 10.1.

10.2-3. Derive the three parts contributing to the term $(K_0)_{2,3}$ for the MZC rectangle in Table 10.1.

10.2-4. Derive the three parts contributing to the term $(K_0)_{9,10}$ for the BFS rectangle.

10.2-5. Derive the three parts contributing to the term $(K_0)_{11,12}$ for the BFS rectangle.

10.3-1. The figure shows a fixed-end prismatic beam composed of three flexural elements. Construct the matrices $\mathbf{S}$ and $\mathbf{S}_0$ for the four displacement coordinates that are not restrained. Then reduce these matrices by eliminating the rotations

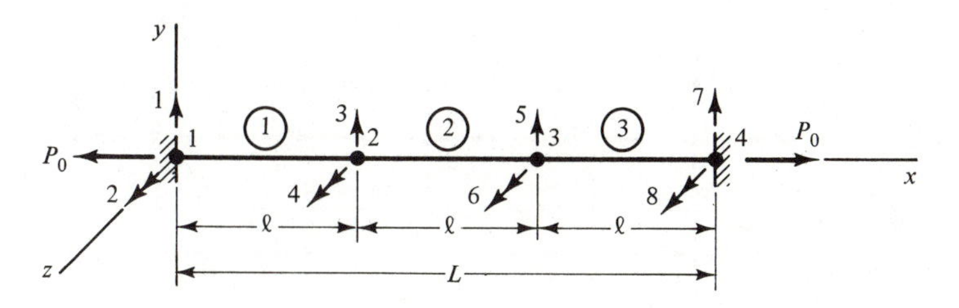

Problem 10.3-1

and retaining the translations. Solve the eigenvalue problem to obtain the critical buckling load P_{cr} and the corresponding mode shape for the reduced system.

10.3-2. A prismatic cantilever beam consisting of two flexural elements appears in the figure. Assemble the matrices $\mathbf{S}$ and $\mathbf{S}_0$, and reduce them by eliminating the rotations and keeping the translations. Find the critical buckling load and mode shape for the reduced system.

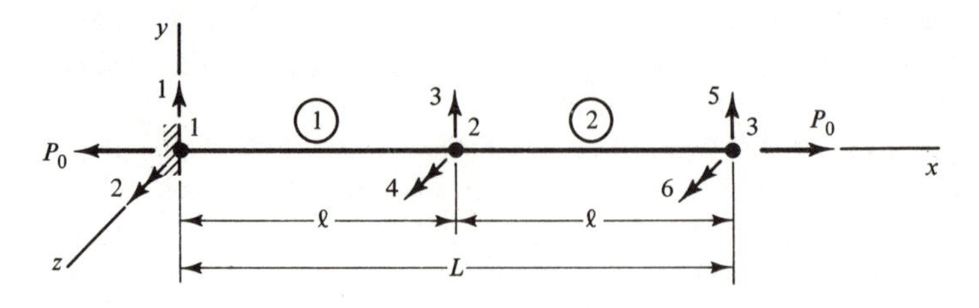

Problem 10.3-2

10.3-3. The two-element prismatic beam in the figure has three degrees of freedom (two translations and one rotation). Construct the matrices $\mathbf{S}$ and $\mathbf{S}_0$, and reduce them by eliminating the rotation. Determine the critical buckling load and mode shape for the reduced system.

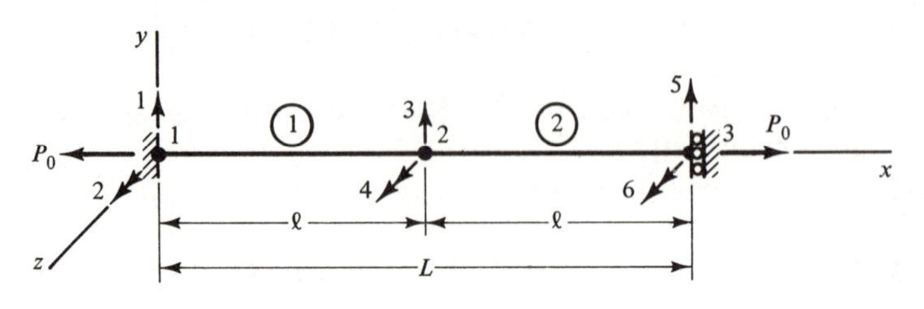

Problem 10.3-3

10.3-4. The figure shows a two-element prismatic beam with three degrees of freedom (one translation and two rotations). Assemble the matrices $\mathbf{S}$ and $\mathbf{S}_0$, and reduce them by eliminating the rotations. Find the critical buckling load for the remaining one-degree-of-freedom system.

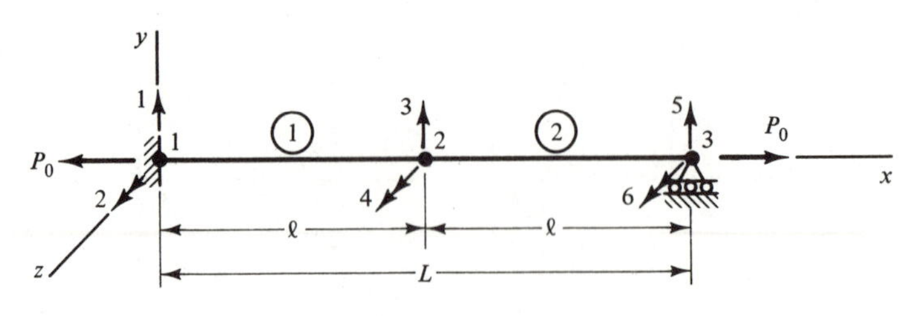

Problem 10.3-4

10.3-5. Set up the eigenvalue problem for vibrations of the beam in Prob. 10.3-1 (with reduction).

10.3-6. Set up the eigenvalue problem for vibrations of the beam in Prob. 10.3-2 (with reduction).

10.3-7. Set up the eigenvalue problem for vibrations of the beam in Prob. 10.3-3 (with reduction).

10.3-8. Set up the eigenvalue problem for vibrations of the beam in Prob. 10.3-4 (with reduction).

Notation

1. MATRICES AND VECTORS

Symbol	Definition
A	Action vector
B	Strain-displacement matrix
C	Strain-stress matrix
D	Displacement vector
E	Stress-strain matrix
F	Flexibility matrix
G	Constraint matrix
H	Operator relating $\mathbf{x}$ to $\boldsymbol{\xi}$
I	Identity matrix
J	Jacobian matrix
K	Element stiffness matrix
L	Lower triangular matrix
M	Mass matrix (also edge moments)
N	Edge forces
O	Null matrix
P	Concentrated force vector
Q	Shearing force vector
R	Rotation matrix
S	Stiffness matrix
T	Transformation matrix

$\mathbf{U}$	Upper triangular matrix
$\mathbf{V}$	Eigenvector matrix
$\mathbf{W}$	Submatrix of $\mathbf{g}^T\mathbf{g}$
$\mathbf{X}$	Vector of unknowns
$\mathbf{Y}$	Vector of unknowns
$\mathbf{Z}$	Vector of unknowns
$\mathbf{b}$	Body force vector for element
$\mathbf{c}$	Generalized displacement vector for element
$\mathbf{d}$	Linear differential operator for strain-displacement relationships
$\mathbf{f}$	Interpolation function matrix
$\mathbf{g}$	Generic-generalized displacement operator
$\mathbf{h}$	Nodal-generalized displacement operator
$\mathbf{p}$	Nodal load vector for element
$\mathbf{q}$	Nodal displacement vector for element
$\mathbf{u}$	Generic displacement vector for element

2. SUBSCRIPTS FOR MATRICES AND VECTORS

Symbol	Definition
A	Nodal displacements eliminated
B	Nodal displacements retained
F	Free
L	Lumped
N	Nodal
R	Restrained
T	Temperature
b	Body
e	Element
i	Index
j	Index
k	Index
ℓ	Index
m	Number
n	Number
$_0$	Initial
r	Radial direction
s	Structure
x	x Direction
y	y Direction
z	z Direction

3. SIMPLE VARIABLES

Symbol	Definition
A	Area
C	Constant
E	Young's modulus
G	Shearing modulus
I	Moment of inertia
J	Polar moment of inertia
L	Length
M	Moment
N	Normal force
P	Force
Q	Shearing force
R	Radius (also weighting factor)
T	Temperature
U	Strain energy
V	Potential energy
W	Work (also weighting factor)
X	Generalized action
a	Constant
b	Constant
c	Constant
d	Constant
e	Constant
f	Interpolation function
i	Index for . . .
j	Index for . . .
k	Index for . . .
ℓ	Index for . . .
m	Number of . . .
n	Number of . . .
p	Action at element node
q	Displacement of element node
r	Radius
s	Segment length
t	Thickness (also time)
u	Translation in x direction
v	Translation in y direction
w	Translation in z direction
x	Cartesian coordinate
y	Cartesian coordinate
z	Cartesian coordinate

4. GREEK LETTERS

Symbol	Definition
Δ	Increment
Σ	Summation
Φ	Function or mode
α	Thermal coefficient
β	Instability coefficient
γ	Shearing strain
δ	Increment
ϵ	Normal strain
ζ	Dimensionless coordinate
η	Dimensionless coordinate
θ	Rotation or angle
λ	Direction cosine (also eigenvalue)
μ	Frequency coefficient
ν	Poisson's ratio
ξ	Dimensionless coordinate
π	3.1416 . . . .
ρ	Mass density
σ	Normal stress
τ	Shearing stress
ϕ	Curvature
ψ	Twist ($d\theta_x/dx$)
ω	Angular frequency (also angular velocity)

APPENDIX

Integration Formulas

A.1 Line. Origin at centroid c (Fig. A.1)

$$x_1 + x_2 = 0 \qquad \int_L dL = L$$

$$\int_L x\, dL = 0 \qquad \int_L x^2\, dL = (x_1^2 + x_2^2)\frac{L}{6} = \frac{L^3}{12}$$

$$\int_L x^3\, dL = 0 \qquad \int_L x^4\, dL = (x_1^4 + x_2^4)\frac{L}{10} = \frac{L^5}{80}$$

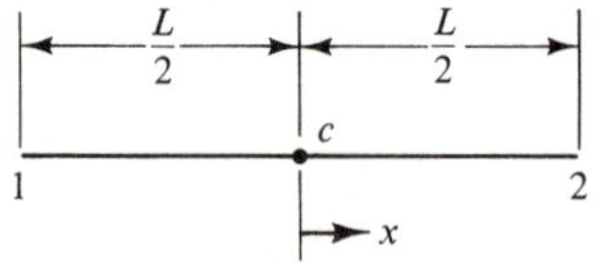

Figure A.1 Line

A.2 Triangle. Origin at centroid c (Fig. A.2)

$$x_1 + x_2 + x_3 = 0 \qquad y_1 + y_2 + y_3 = 0$$

$$\int_A dA = \frac{1}{2}\begin{vmatrix} 1 & x_1 & y_1 \\ 1 & x_2 & y_2 \\ 1 & x_3 & y_3 \end{vmatrix} = \frac{1}{2}(x_{ij}y_{ik} - x_{ik}y_{ij}) \begin{cases} i, j, k = 1, 2, 3 \\ i, j, k = 2, 3, 1 \\ i, j, k = 3, 1, 2 \end{cases}$$

$$\int_A x\, dA = \int_A y\, dA = 0 \qquad \int_A x^2\, dA = (x_1^2 + x_2^2 + x_3^2)\frac{A}{12}$$

$$\int_A xy\,dA = (x_1y_1 + x_2y_2 + x_3y_3)\frac{A}{12} \qquad \int_A x^3\,dA = (x_1^3 + x_2^3 + x_3^3)\frac{A}{30}$$

$$\int_A x^2y\,dA = (x_1^2y_1 + x_2^2y_2 + x_3^2y_3)\frac{A}{30} \qquad \int_A x^4\,dA = (x_1^4 + x_2^4 + x_3^4)\frac{A}{30}$$

$$\int_A x^3y\,dA = (x_1^3y_1 + x_2^3y_2 + x_3^3y_3)\frac{A}{30} \qquad \int_A x^2y^2\,dA = (x_1^2y_1^2 + x_2^2y_2^2 + x_3^2y_3^2)\frac{A}{30}$$

Interchange x and y for integrals not shown.

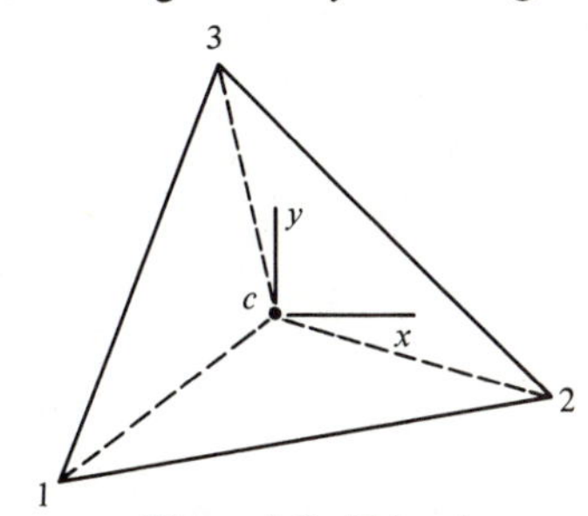

Figure A.2 Triangle

A.3 Tetrahedron. Origin at centroid c (Fig. A.3)

$$x_1 + x_2 + x_3 + x_4 = 0$$
$$y_1 + y_2 + y_3 + y_4 = 0$$
$$z_1 + z_2 + z_3 + z_4 = 0$$

$$\int_V dV = \frac{1}{6}\begin{vmatrix} 1 & x_1 & y_1 & z_1 \\ 1 & x_2 & y_2 & z_2 \\ 1 & x_3 & y_3 & z_3 \\ 1 & x_4 & y_4 & z_4 \end{vmatrix}$$

$$\int_V x\,dV = \int_V y\,dV = \int_V z\,dV = 0$$

$$\int_V x^2\,dV = (x_1^2 + x_2^2 + x_3^2 + x_4^2)\frac{V}{20}$$

Figure A.3 Tetrahedron

$$\int_V y^2 \, dV = (y_1^2 + y_2^2 + y_3^2 + y_4^2)\frac{V}{20}$$

$$\int_V z^2 \, dV = (z_1^2 + z_2^2 + z_3^2 + z_4^2)\frac{V}{20}$$

$$\int_V xy \, dV = (x_1 y_1 + x_2 y_2 + x_3 y_3 + x_4 y_4)\frac{V}{20}$$

$$\int_V yz \, dV = (y_1 z_1 + y_2 z_2 + y_3 z_3 + y_4 z_4)\frac{V}{20}$$

$$\int_V zx \, dV = (z_1 x_1 + z_2 x_2 + z_3 x_3 + z_4 x_4)\frac{V}{20}$$

APPENDIX

B

Gaussian Quadrature

B.1 Numerical Integration. The process of computing the value of a definite integral [see Fig. B.1(a)]

$$I_x = \int_{x_1}^{x_2} f(x)\,dx \tag{a}$$

from a set of numerical values of the integrand is called numerical integration. The problem is solved by representing the integrand by an interpolation formula and then integrating this formula between specified limits. When applied to the integration of a function of a single variable, the method is called *mechanical quadrature*.

If interpolation formulas for numerical integration are polynomials of sufficiently high order relative to those assumed for displacement (or other) functions in a finite element, the integrations will be exact. Otherwise, the process of numerical integration introduces an additional source of error into the analysis. The most accurate of the quadrature formulas in common usage is that of Gauss, which involves unequally spaced points that are symmetrically placed.

B.2 Gaussian Quadrature. The procedure for Gauss's method requires changing the variable from x to a dimensionless coordinate ξ with its origin at the center of the range of integration, as shown in Fig. B.1(b). The expression for x in terms of ξ is:

$$x = \tfrac{1}{2}(x_1 + x_2) + \tfrac{1}{2}(x_2 - x_1)\xi \tag{b}$$

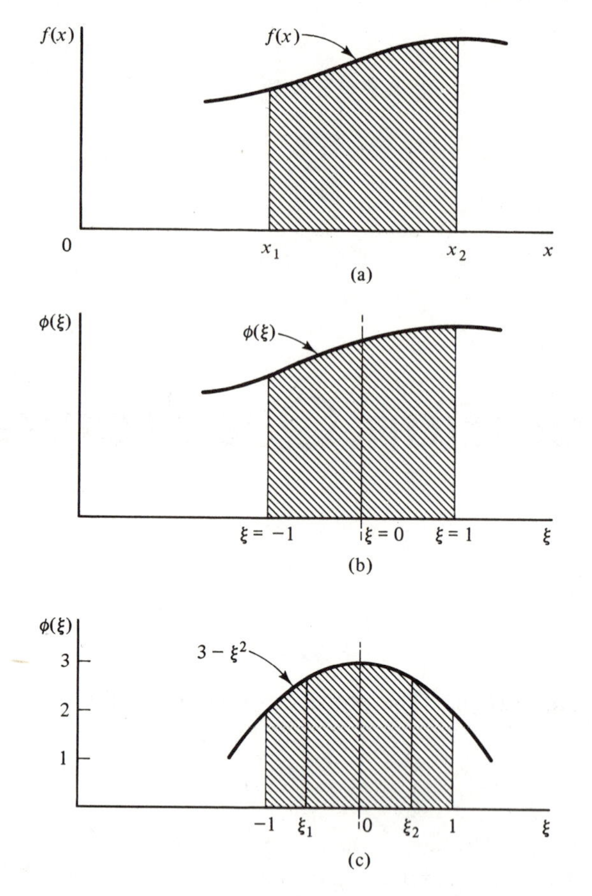

Figure B.1 Gaussian Quadrature

Substitution of Eq. (b) into the function in Eq. (a) gives:

$$f(x) = \phi(\xi) \tag{c}$$

Also,

$$dx = \tfrac{1}{2}(x_2 - x_1)\, d\xi \tag{d}$$

Then substituting Eqs. (c) and (d) into Eq. (a) and changing the limits of integration yields:

$$I_x = \tfrac{1}{2}(x_2 - x_1) \int_{-1}^{1} \phi(\xi)\, d\xi \tag{e}$$

Gauss's formula for determining the integral in Eq. (e) consists of summing the weighted values of $\phi(\xi)$ at n specified points, as follows:

$$\begin{aligned} I_\xi &= \int_{-1}^{1} \phi(\xi)\, d\xi = \sum_{j=1}^{n} R_j \phi(\xi_j) \\ &= R_1\phi(\xi_1) + R_2\phi(\xi_2) + \cdots + R_n\phi(\xi_n) \end{aligned} \tag{B-1}$$

In this expression R_j is a *weighting factor* for point j, ξ_j is the *location of point* j relative to the center, and n is the *number of points* at which $\phi(\xi)$ is to be evaluated.

To derive the quantities in Eq. (B-1), we assume that the integrand can be expanded in a convergent power series in the interval $\xi = -1$ to $\xi = 1$. Thus,

$$\phi(\xi) = a_0 + a_1\xi + a_2\xi^2 + \cdots + a_m\xi^m \tag{f}$$

Integration of Eq. (f) produces:

$$I_\xi = \int_{-1}^{1} \phi(\xi)\,d\xi = 2a_0 + \tfrac{2}{3}a_2 + \tfrac{2}{5}a_4 + \cdots \tag{g}$$

On the other hand, evaluating Eq. (f) at each point gives:

$$\begin{aligned} \phi(\xi_1) &= a_0 + a_1\xi_1 + a_2\xi_1^2 + \cdots + a_m\xi_1^m \\ \phi(\xi_2) &= a_0 + a_1\xi_2 + a_2\xi_2^2 + \cdots + a_m\xi_2^m \\ &\cdots\cdots\cdots\cdots\cdots\cdots\cdots\cdots \\ \phi(\xi_n) &= a_0 + a_1\xi_n + a_2\xi_n^2 + \cdots + a_m\xi_n^m \end{aligned} \tag{h}$$

Substitute Eqs. (h) into Eq. (B-1) and collect the coefficients of a_0, a_1, and so on.

$$\begin{aligned} I_\xi = a_0(&R_1 + R_2 + \cdots + R_n) \\ &+ a_1(R_1\xi_1 + R_2\xi_2 + \cdots + R_n\xi_n) \\ &+ a_2(R_1\xi_1^2 + R_2\xi_2^2 + \cdots + R_n\xi_n^2) \\ &+ \cdots\cdots\cdots\cdots\cdots\cdots \\ &+ a_m(R_1\xi_1^m + R_2\xi_2^m + \cdots + R_n\xi_n^m) \end{aligned} \tag{i}$$

The coefficients in Eqs. (g) and (i) must be identical. Therefore,

$$\begin{aligned} R_1 + R_2 + \cdots + R_n &= 2 \\ R_1\xi_1 + R_2\xi_2 + \cdots + R_n\xi_n &= 0 \\ R_1\xi_1^2 + R_2\xi_2^2 + \cdots + R_n\xi_n^2 &= \tfrac{2}{3} \\ \cdots\cdots\cdots\cdots\cdots\cdots \end{aligned} \tag{j}$$

If $m \geq 2n$, it is possible to write $2n$ equations of type (j) and solve them simultaneously for the $2n$ quantities ξ_1 to ξ_n and R_1 to R_n. Their values have been ascertained, and they are listed in Table B.1.

Example

Using Gaussian quadrature, integrate numerically the function $\phi(\xi) = 3 - \xi^2$ shown in Fig. B.1(c). First, let $n = 1$ and find from Table B.1 that:

$$\xi_1 = 0 \qquad R_1 = 2$$

Then from Eq. (B-1) we have:

$$I_\xi = R_1\phi(\xi_1) = (2)(3) = 6$$

which is approximate. Next, let $n = 2$ and obtain from Table B.1:

$$\xi_1 = -\xi_2 = -\frac{1}{\sqrt{3}} = -0.577\ldots \qquad R_1 = R_2 = 1$$

TABLE B.1 Coefficients for Gaussian Quadrature

n	$\pm\xi_i$	R_i
1	0.0	2.0
2	0.5773502692	1.0
3	0.7745966692 0.0	0.5555555556 0.8888888889
4	0.8611363116 0.3399810436	0.3478548451 0.6521451549
5	0.9061798459 0.5384693101 0.0	0.2369268851 0.4786286705 0.5688888889
6	0.9324695142 0.6612093865 0.2386191861	0.1713244924 0.3607615730 0.4679139346
7	0.9491079123 0.7415311856 0.4058451514 0.0	0.1294849662 0.2797053915 0.3818300505 0.4179591837
8	0.9602898565 0.7966664774 0.5255324099 0.1834346425	0.1012285363 0.2223810345 0.3137066459 0.3626837834

Now we find from Eq. (B-1):

$$I_\xi = \sum_{j=1}^{2} R_j \phi(\xi_j) = (1)(3 - \xi_1^2) + (1)(3 - \xi_2^2)$$
$$= (2)(2.666\ldots) = 5.333\ldots$$

which is exact.

Gaussian quadrature is admirably suited for numerical integration of polynomials for finite elements where dimensionless parameters such as ξ are used and the origin is located at the center. This arrangement is convenient (but not essential) for implementing the method. Solutions are exact for polynomials of degree $2n - 1$. That is, only one integration point is required for the exact integration of a linear function. Two integration points are required for a cubic polynomial, and so on.

APPENDIX

C

Computer Solution Routines

C.1 Factorization of Symmetric Matrices. In a previous book* the *Cholesky square-root method* was described for factoring a symmetric, positive-definite matrix **A** into the form:

$$\mathbf{A} = \mathbf{U}^{\mathrm{T}}\mathbf{U} \tag{C-1}$$

where **U** is an upper triangular matrix and $\mathbf{U}^{\mathrm{T}}$ is its transpose. In addition, it was shown that a symmetric matrix **A** can always be factored as follows:

$$\mathbf{A} = \bar{\mathbf{U}}^{\mathrm{T}}\mathbf{D}\,\bar{\mathbf{U}} \tag{C-2}$$

regardless of whether it is positive-definite. Factorization of the second type is called the *modified Cholesky method.* In Eq. (C-2) the matrix $\bar{\mathbf{U}}$ is an upper triangular matrix with a one in each diagonal position, and $\bar{\mathbf{U}}^{\mathrm{T}}$ is its transpose. The symbol **D** represents a diagonal matrix containing the squares of diagonal terms factored from the rows of $\bar{\mathbf{U}}$. Recurrence equations for $\bar{U}_{ij}$ and D_{jj}, generated columnwise, are (for $j = 2, 3, \ldots, n$):

$$\bar{U}_{ij} = \frac{1}{D_{ii}}\left(A_{ij} - \sum_{k=1}^{i-1} D_{kk}\bar{U}_{ki}\bar{U}_{kj}\right) \qquad (1 < i < j) \tag{a}$$

$$D_{jj} = A_{jj} - \sum_{k=1}^{j-1} D_{kk}\bar{U}^2_{kj} \qquad (1 < i = j) \tag{b}$$

Both of these equations contain the product $D_{kk}\bar{U}_{kj}$ within the summations. Let this product be:

*See Appendix D in *Matrix Analysis of Framed Structures*, 2d ed., by W. Weaver, Jr., and J. M. Gere, Van Nostrand-Reinhold, New York, 1980.

$$\bar{U}_{kj}^* = D_{kk}\bar{U}_{kj} \tag{c}$$

and obtain $\bar{U}_{ij}$ and D_{jj} as follows:

$$\bar{U}_{ij}^* = A_{ij} - \sum_{k=1}^{i-1} \bar{U}_{ki}\bar{U}_{kj}^* \qquad (1 < i < j) \tag{C-3}$$

$$D_{jj} = A_{jj} - \sum_{k=1}^{j-1} \bar{U}_{kj}\bar{U}_{kj}^* \qquad (1 < i = j) \tag{C-4}$$

(where $j = 2, 3, \ldots, n$) and

$$\bar{U}_{kj} = \frac{1}{D_{kk}} \bar{U}_{kj}^* \tag{C-5}$$

Thus, for column j the temporary result $\bar{U}_{ij}^*$ is obtained for each off-diagonal term after the first [see Eq. (C-3)]. Then the diagonal term D_{jj} is computed [Eq. (C-4)], during which calculation the final value of each off-diagonal term is also produced [Eq. (C-5)].

Assume that the following system of *linear algebraic equations* is to be solved:

$$\mathbf{A}\ \mathbf{X} = \mathbf{B} \tag{C-6}$$

in which $\mathbf{X}$ is a column vector of n unknowns and $\mathbf{B}$ is a column vector of constant terms. First, substitute Eq. (C-2) into Eq. (C-6) to obtain:

$$\bar{\mathbf{U}}^{\mathrm{T}}\mathbf{D}\ \bar{\mathbf{U}}\ \mathbf{X} = \mathbf{B} \tag{C-7}$$

Then define the vector $\mathbf{Y}$ to be:

$$\bar{\mathbf{U}}\ \mathbf{X} = \mathbf{Y} \tag{C-8}$$

In addition, define the vector $\mathbf{Z}$ to be:

$$\mathbf{D}\ \mathbf{Y} = \mathbf{Z} \tag{C-9}$$

Substitution of Eq. (C-8) into Eq. (C-9) and then the latter into Eq. (C-7) produces:

$$\bar{\mathbf{U}}^{\mathrm{T}}\mathbf{Z} = \mathbf{B} \tag{C-10}$$

Now we can obtain the original vector of unknowns $\mathbf{X}$ in three steps using Eqs. (C-10), (C-9), and (C-8). In the first step we solve for the vector $\mathbf{Z}$ in Eq. (C-10). Since $\bar{\mathbf{U}}^{\mathrm{T}}$ is a lower triangular matrix, the elements of $\mathbf{Z}$ can be calculated in a series of forward substitutions, as follows:

$$Z_i = B_i - \sum_{k=1}^{i-1} \bar{U}_{ki}Z_k \qquad (1 < i) \tag{C-11}$$

The second step consists of solving for the vector $\mathbf{Y}$ in Eq. (C-9). Because $\mathbf{D}$ is a diagonal matrix, the elements of $\mathbf{Y}$ are found by dividing terms in $\mathbf{Z}$ by corresponding diagonals of $\mathbf{D}$. Thus,

$$Y_i = \frac{Z_i}{D_{ii}} \qquad (i = 1, 2, \ldots, n) \tag{C-12}$$

In the third step we find the vector **X** from Eq. (C-8). Since $\bar{\mathbf{U}}$ is an upper triangular matrix, the elements of **X** are determined in a backward substitution procedure, as follows:

$$X_i = Y_i - \sum_{k=i+1}^{n} \bar{U}_{ik} X_k \qquad (i < n) \tag{C-13}$$

This step completes the solution of the original equations [Eq. (C-6)] for the unknown quantities. The factorization and solution procedures for the modified Cholesky method require the least possible number of arithmetic operations.*

If zeros appear in the upper part of a given column in matrix **A**, they are not affected by the factorization. To take advantage of this fact in a computer routine, it is necessary to store the matrix **A** columnwise as a singly subscripted variable. Subsequent sections in this appendix contain subprograms for factorization and solution with a matrix **A** that is either filled or partly filled with nonzero terms.

C.2 Subprogram VECFAC. Figure C.1(a) depicts the upper triangular part of a symmetric matrix **A**(,), and Fig. C.1(b) shows the same information stored columnwise as a singly subscripted vector **A**( ). In this section an algorithm for factoring the latter array into the form shown by Eq. (C-2) is given as a FORTRAN-oriented flow chart labeled VECFAC. The steps in this flow chart are the same as those in Subprogram FACTOR for the modified Cholesky method.† However, additional statements are required for the purpose of locating terms in **A** when stored as a vector instead of as a square matrix.

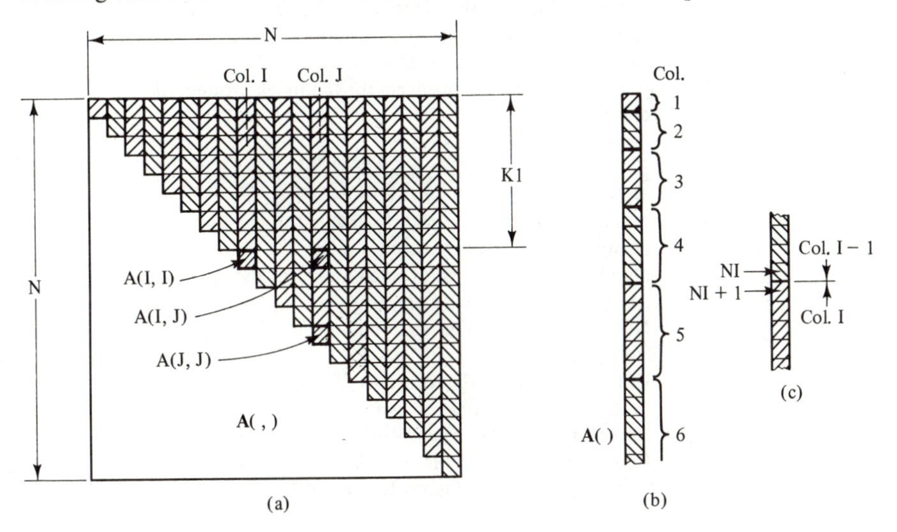

Figure C.1 Upper Triangle of Matrix **A** Stored as a Vector

**Ibid.*

†*Ibid.*

Neither Subprogram VECFAC nor Subprogram VECSOL (described in the next article) is utilized by the finite-element programs in this book. However, they are included in the series of subprograms because of their general usefulness. Furthermore, they serve as guides to understanding the more complicated subprograms called SKYFAC and SKYSOL, which are given in subsequent articles and used in the finite-element programs.

The formal name of the subprogram in Flow Chart C.1 is:

VECFAC(N,A,*)

The first term in the parentheses is the integer number N, which denotes the size of the matrix to be factored. The second identifier denotes a symmetric matrix **A** of real numbers, and the third symbol (an asterisk) signifies a nonstandard RETURN to an error message in the main program if **A** is found not to be positive-definite. In addition to this notation, the integer numbers I, J, and K correspond to i, j, and k in the recurrence equations. Also, the real variables SUM and TEMP are used for summation and temporary storage. Other integer identifiers used in the subprograms are defined in Table C.1.

TABLE C.1 Integer Identifiers Used in Subprograms

Identifier	Definition
II	= Index for locating the diagonal term A(I, I) when stored as a vector
IJ	= Index for locating the off-diagonal term A(I, J) when stored as a vector
JJ	= Index for locating the diagonal term A(J, J) when stored as a vector
JMI	= J – I; NCJI = NCJ – JMI
K1	= Number of nonzero products for the inner product of columns I and J
KI, KJ	= Similar to II, IJ, and so on
KMI	= K – I; NCKI = NCK – KMI
NCD	= NCJI – NCI = NCJ – JMI – NCI
NCI	= Number of nonzero terms in column I of the upper triangular skyline of matrix **A**
NCJ	= Number of nonzero terms in column J of the upper triangular skyline of matrix **A**
NCK	= Number of nonzero terms in column K of the upper triangular skyline of matrix **A**
ND( )	= List of indexes for the locations of diagonal elements in the compacted form of matrix **A**
NI	= Record pointer for column I [see Fig. C.1(c)]
NJ	= Record pointer for column J (similar to NI)
NK	= Record pointer for column K (similar to NI)

The first part of the flow chart implements Eq. (C-3), and the second part applies Eqs. (C-4) and (C-5) from the preceding section. Elements of the upper triangular matrix $\bar{\mathbf{U}}$ are generated columnwise in the storage locations originally

FLOW CHART C.1: SUBPROGRAM VECFAC(N,A,*)

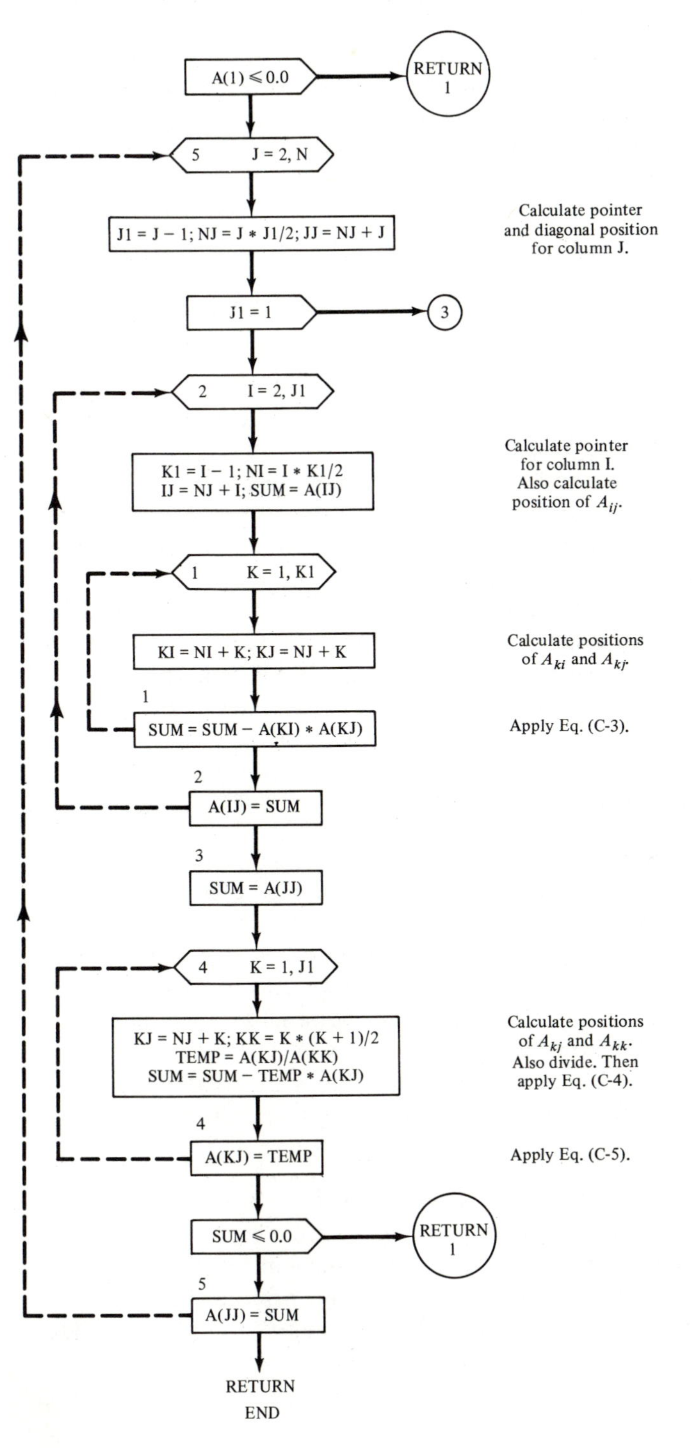

Calculate pointer and diagonal position for column J.

Calculate pointer for column I. Also calculate position of A_{ij}.

Calculate positions of A_{ki} and A_{kj}.

Apply Eq. (C-3).

Calculate positions of A_{kj} and A_{kk}. Also divide. Then apply Eq. (C-4).

Apply Eq. (C-5).

occupied by the upper triangular part of the matrix **A**. Thus, the identifier A remains in use throughout the flow chart. In addition, each diagonal element D_{ii} is stored in the diagonal position A_{ii} of matrix **A**. If a zero or negative value of D_{ii} is detected, control is transferred (by means of the nonstandard RETURN) to an error message in the main program.

C.3 Subprogram VECSOL. The second subprogram in this series accepts the factored matrix from Subprogram VECFAC and solves for the unknowns in the original set of equations [Eq. (C-6)]. The name of this subprogram is:

VECSOL(N,U,B,X)

The meaning of N is the same as before, and the symbol U denotes the matrix from Subprogram VECFAC. The identifiers B and X represent real vectors of constant terms and unknowns, respectively.

Flow Chart C.2 shows the logic for Subprogram VECSOL. In the first portion of the flow chart the intermediate vector **Z** is computed by forward substitutions, according to Eq. (C-11). Note that the vector **X** is used as temporary storage for **Z** in this part of the subprogram.

The second portion of the chart involves finding the vector **Y** by dividing each value of Z_i by the corresponding diagonal term D_{ii} [see Eq. (C-12)]. In this instance, the vector **X** is used as temporary storage for **Y**, and the term D_{ii} is known to be in the diagonal position U_{ii} (see Subprogram VECFAC).

In the last portion of the chart, the final values of the elements in the vector **X** are calculated by Eq. (C-13). This backward sweep completes the solution of the original equations.

C.4 Subprogram SKYFAC. A "skyline" pattern of nonzero terms appears in the upper triangle of the symmetric matrix **A**(,) in Fig. C.2(a). The same information is stored more compactly as a vector **A**( ), as shown in Fig. C.2(b). Provided in addition is an integer vector ND of length N [see Fig. C.2(c)], which contains indexes for the locations of diagonal elements in the compacted form of **A**. The vector ND must be generated in the program that calls Subprogram SKYFAC (see Program PSCST in Appendix D).

Flow Chart C.3 contains the steps for a subprogram that factors the matrix **A** in its compacted form. The name of this subprogram is:

SKYFAC(N,A,ND,*)

where the terms in parentheses have already been defined. Subprogram SKYFAC was devised by modifying Subprogram VECFAC to account for the presence of zeros in the upper triangle of matrix **A**. Note that the number of zeros at the top of either column I or column J [see Fig. C.2(a)] may determine the number of nonzero products for the inner product of these columns. Integer variables (such as JMI) labeled in Fig. C.2(a) are defined in Table C.1.

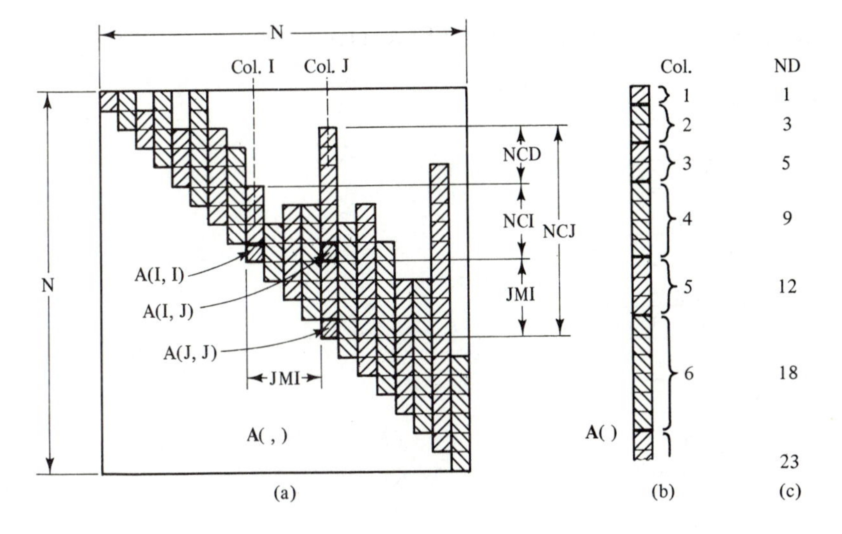

Figure C.2 "Skyline" of Matrix **A** Stored as a Vector

As in the earlier subprogram for factorization, the matrix $\bar{\mathbf{U}}$ is generated and placed in the storage locations originally occupied by the matrix **A**, but the identifier A remains in use. Furthermore, each diagonal element D_{ii} is stored in A_{ii}, and control is transferred to an error message in the main program if $D_{ii} \leq 0$.

Subprogram SKYSOL, given in the next section, is intended to be used in conjunction with Subprogram SKYFAC. They are both used in the finite-element programs in this book, which take advantage of the presence of zeros in the upper triangle of the nodal stiffness matrix.

C.5 Subprogram SKYSOL. The last subprogram in the series is analogous to Subprogram VECSOL (see Flow Chart C.2), except that it applies to a compacted coefficient matrix. This subprogram accepts the factor $\bar{\mathbf{U}}$ from Subprogram SKYFAC and solves for the unknowns in the original system of equations [Eq. (C-6)]. The name of the subprogram is:

SKYSOL(N,U,B,X,ND)

where all of the identifiers are familiar terms that have been defined previously.

As in the earlier solution subprogram, the intermediate vectors **Z** and **Y** are generated in the vector **X**, and final values of **X** are calculated in the backward sweep. The elements of $\bar{\mathbf{U}}$ and **B** are left unaltered by this solution routine, and it can be used repeatedly for the same matrix $\bar{\mathbf{U}}$ with different vectors of constant terms.

FLOW CHART C.2: SUBPROGRAM VECSOL(N,U,B,X)

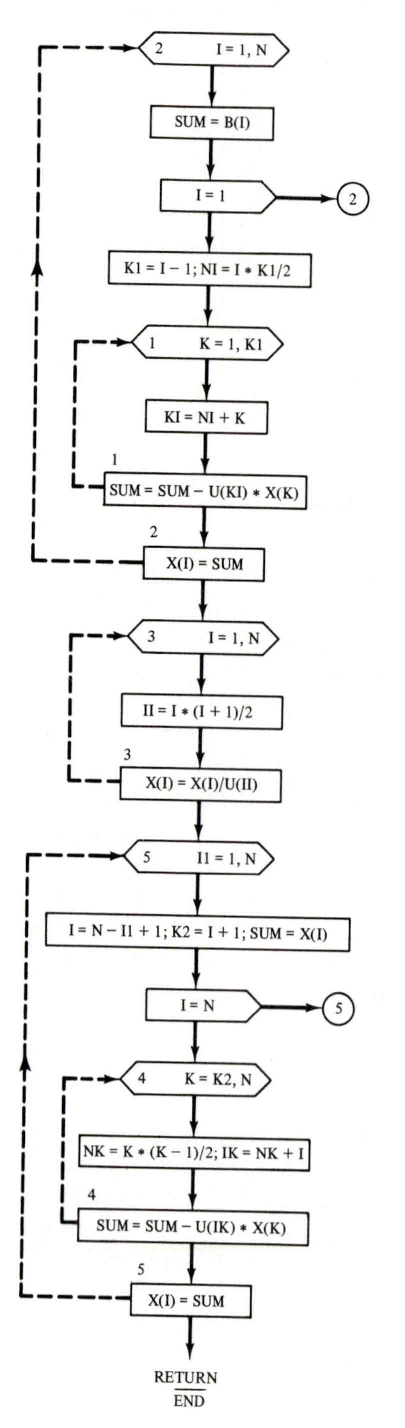

Calculate pointer for column I.

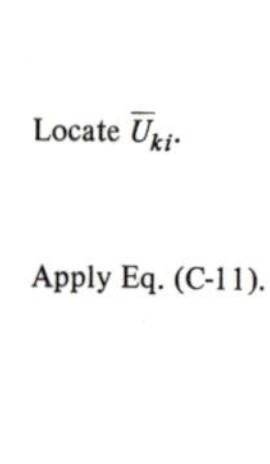

Locate $\overline{U}_{ii}$.

Apply Eq. (C-12).

Calculate pointer for column K, and locate $\overline{U}_{ik}$.

Apply Eq. (C-13).

FLOW CHART C.3: SUBPROGRAM SKYFAC(N,A,ND,*)

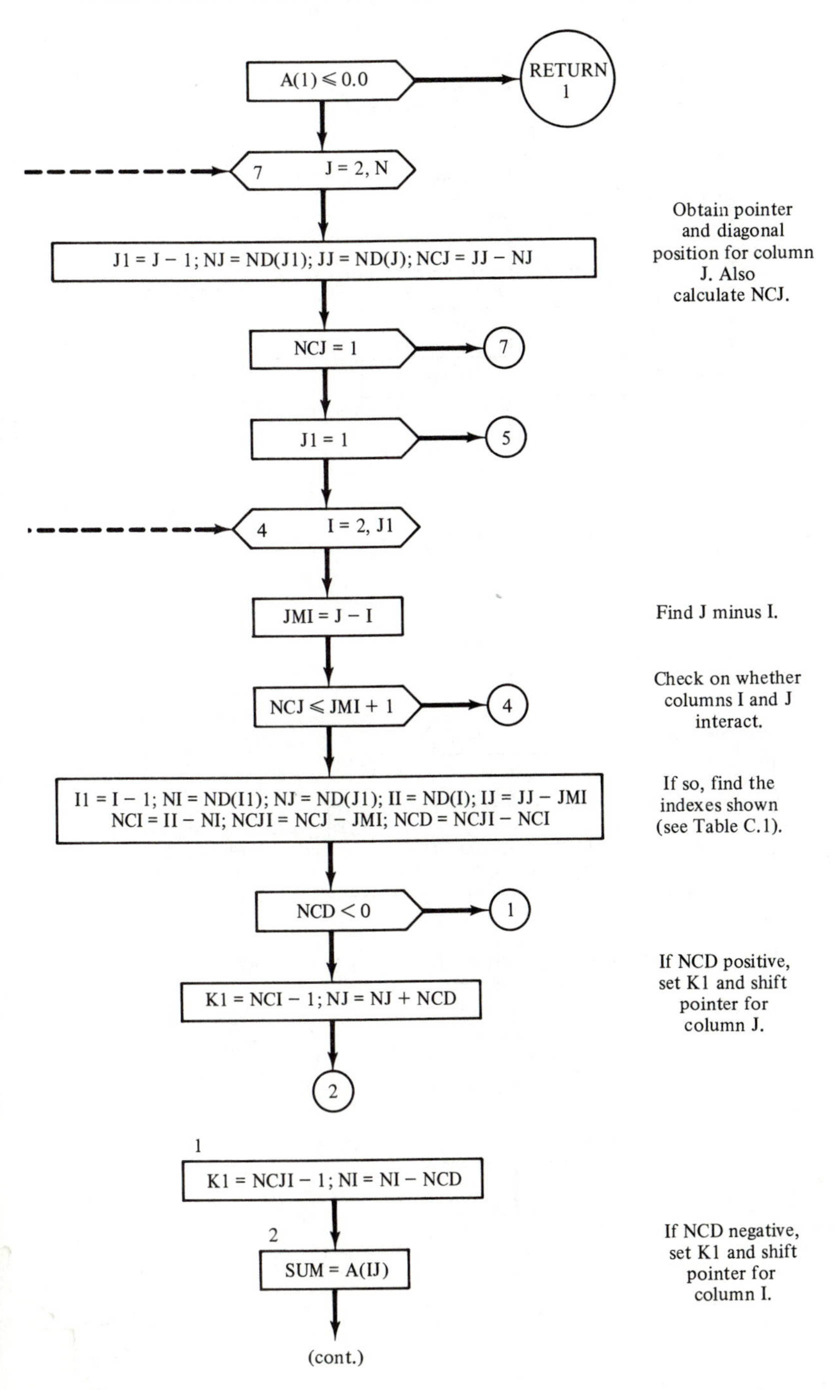

Obtain pointer and diagonal position for column J. Also calculate NCJ.

Find J minus I.

Check on whether columns I and J interact.

If so, find the indexes shown (see Table C.1).

If NCD positive, set K1 and shift pointer for column J.

If NCD negative, set K1 and shift pointer for column I.

FLOW CHART C.3: (cont.)

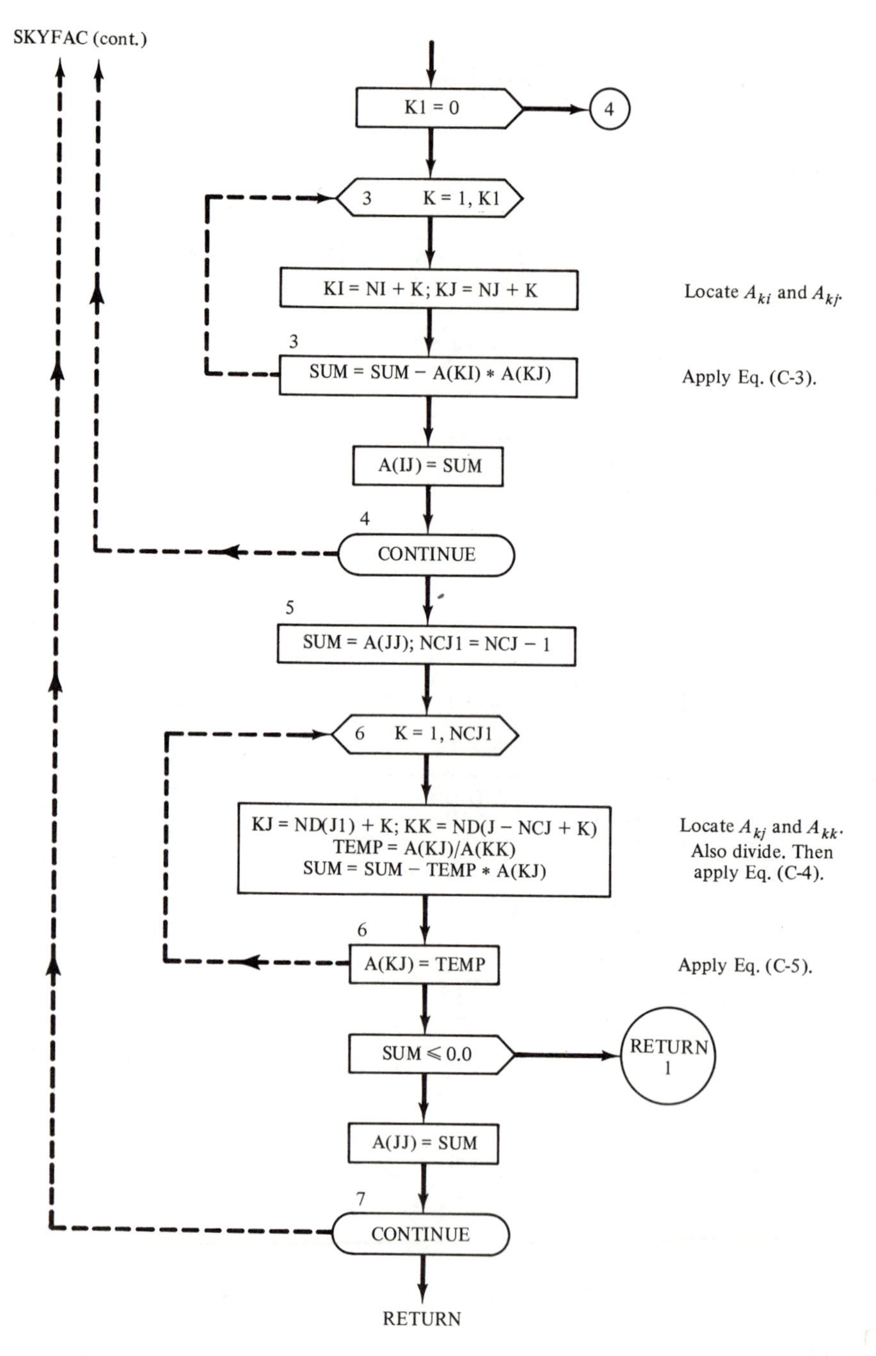

FLOW CHART C.4: SUBPROGRAM SKYSOL(N,U,B,X,ND)

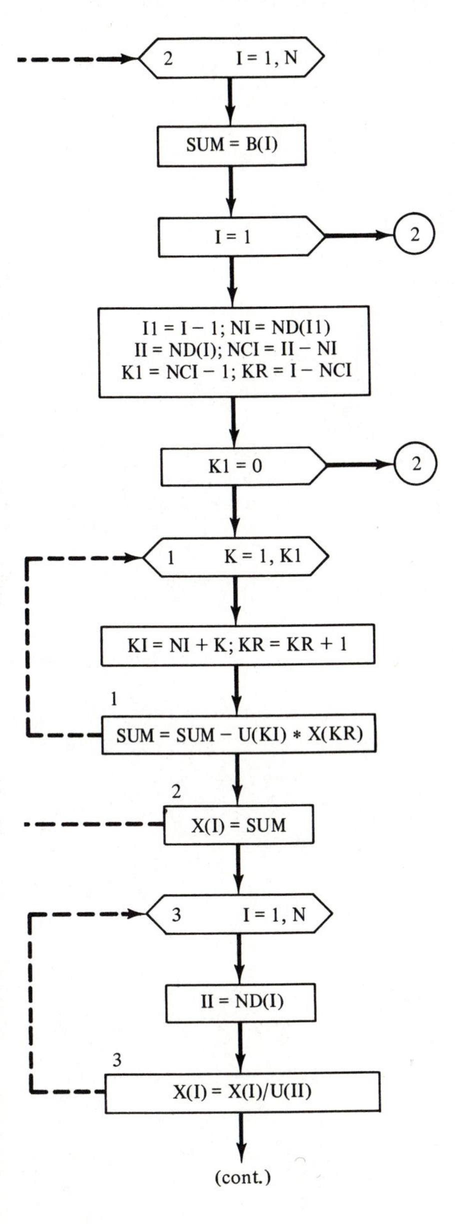

Obtain pointer and diagonal position for column I. Then find NCI, K1, and right-hand side pointer KR.

Locate $\overline{U}_{ki}$ and right-hand side-term.

Apply Eq. (C-11).

Locate $\overline{U}_{ii}$.

Apply Eq. (C-12).

FLOW CHART C.4: (cont.)

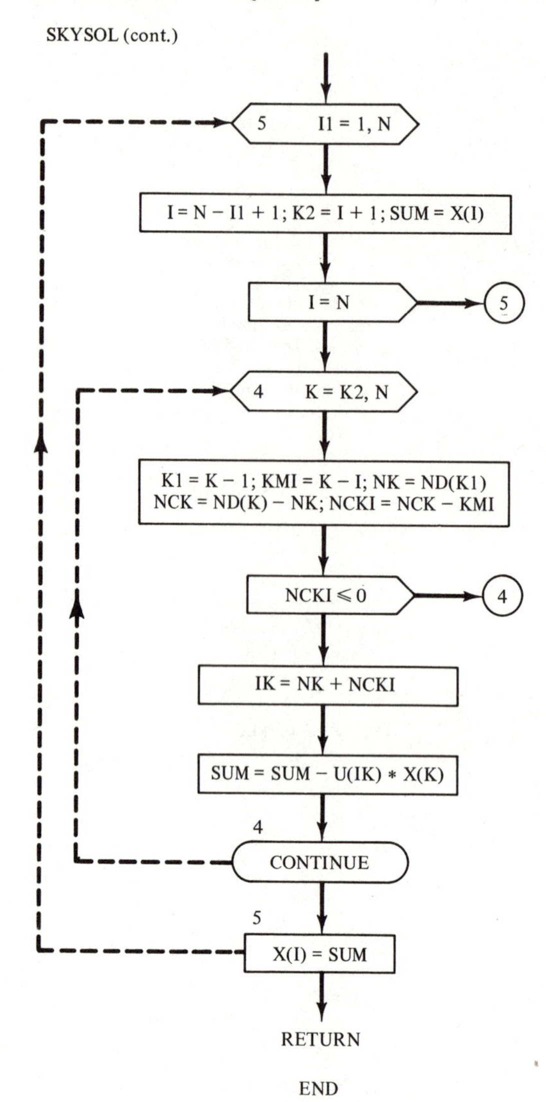

Calculate the indexes shown (see Table C.1).

Check on whether $\overline{U}_{ik}$ exists.

If so, locate $\overline{U}_{ik}$.

Apply Eq. (C-13).

APPENDIX D

Flow Chart for Program PSCST

(*See next page.*)

MAIN PROGRAM

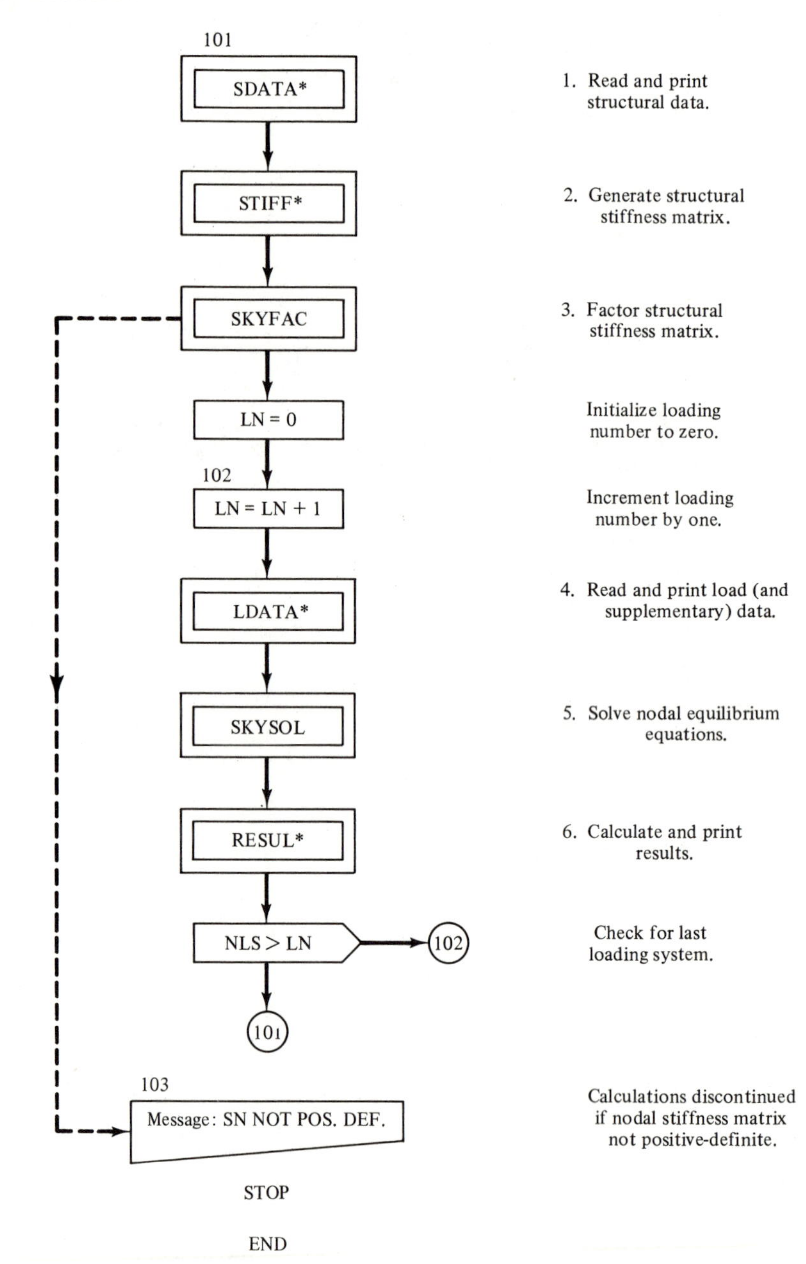

*Subprograms that differ for every program.

1. SUBPROGRAM SDATA FOR PROGRAM PSCST

a. Problem Identification

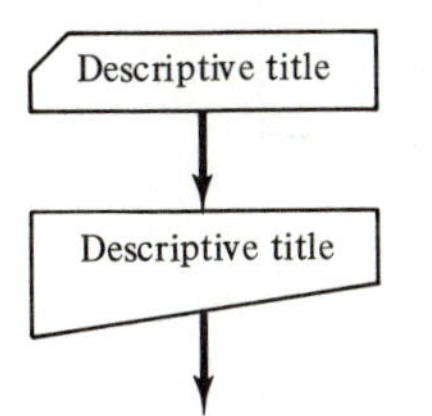

b. Structural Parameters

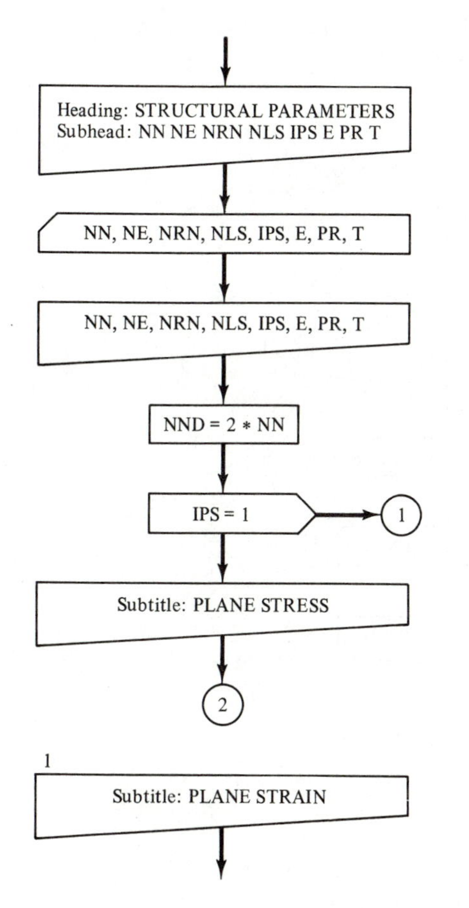

Read and print structural parameters.

Calculate number of nodal displacements possible.

Check for case of plane stress or plane strain.

c. Nodal Coordinates

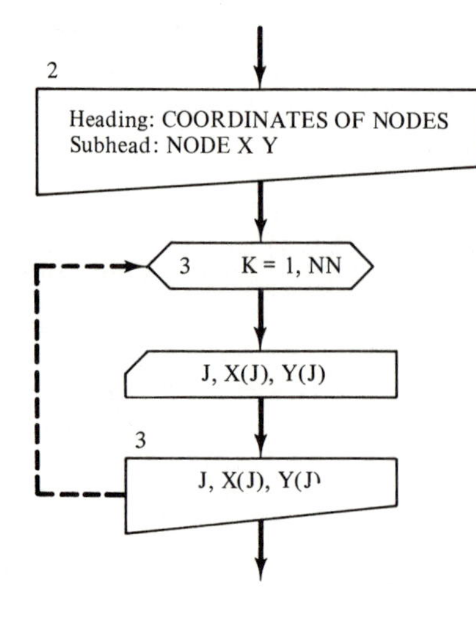

Read and print nodal coordinates.

d. Element Information

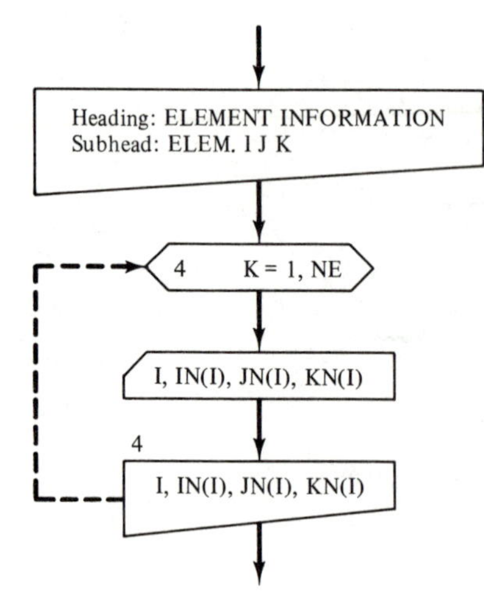

Read and print node numbers in counterclockwise sequence.

e. Nodal Restraints

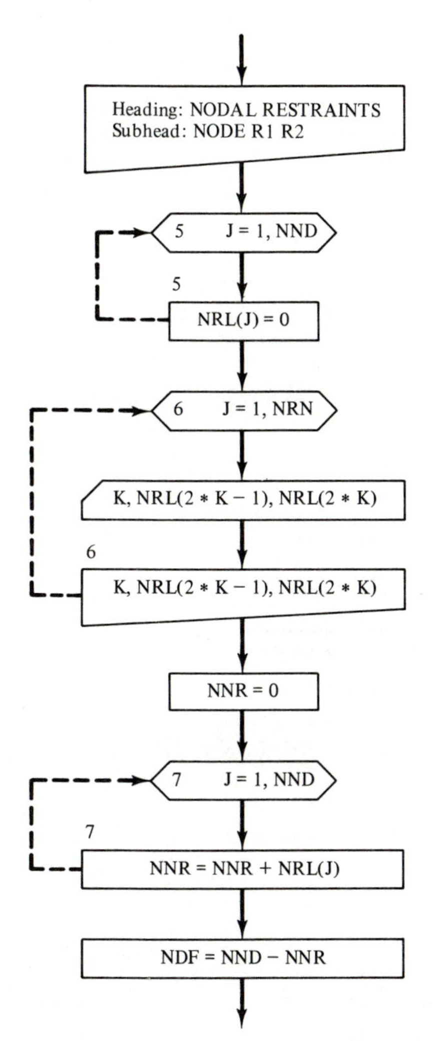

f. Calculate Displacement Indexes

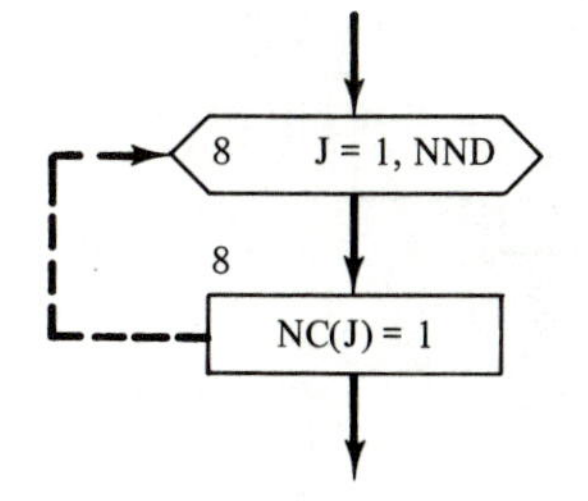

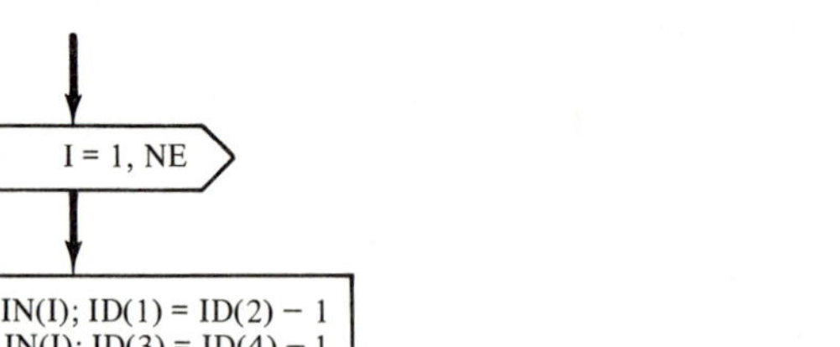
10 I = 1, NE
ID(2) = 2 * IN(I); ID(1) = ID(2) − 1
ID(4) = 2 * JN(I); ID(3) = ID(4) − 1
ID(6) = 2 * KN(I); ID(5) = ID(6) − 1
Calculate six displacement indexes.
N1 = NND
Initialize N1 to NND.
9 J = 1, 6
K = ID(J)
NRL(K) = 1
9
Check for restraint.
K < N1
N1 = K
Find the minimum displacement index.
9
CONTINUE
10 J = 1, 6
K = ID(J)
NRL(K) = 1
10
Check for restraint.
N2 = K − N1 + 1
Determine the number of nonzero terms.
NC(K) < N2
NC(K) = N2
Update vector NC.
10
CONTINUE

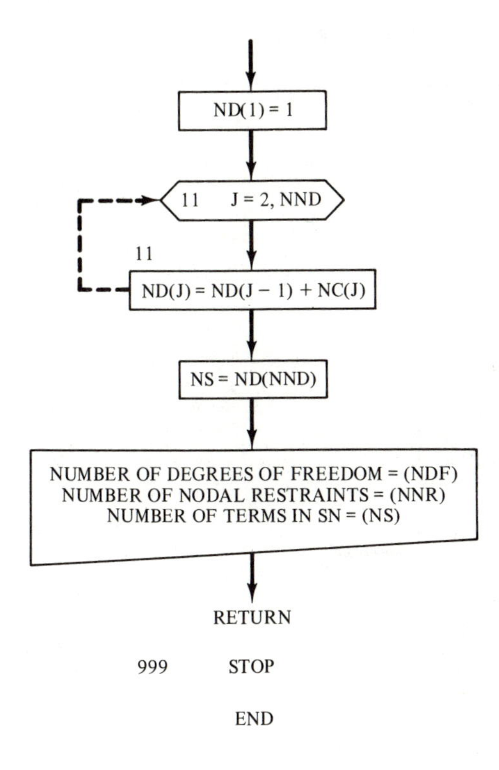
ND(1) = 1
11 J = 2, NND
11
ND(J) = ND(J − 1) + NC(J)
NS = ND(NND)
NUMBER OF DEGREES OF FREEDOM = (NDF)
NUMBER OF NODAL RESTRAINTS = (NNR)
NUMBER OF TERMS IN SN = (NS)
RETURN
999 STOP
END

2. SUBPROGRAM STIFF FOR PROGRAM PSCST

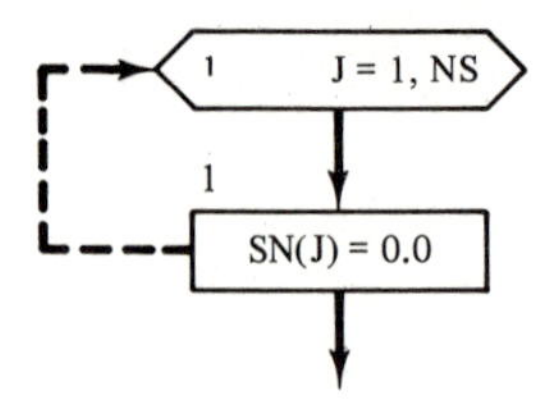

Clear nodal stiffness matrix.

a. Element Stiffness Matrix

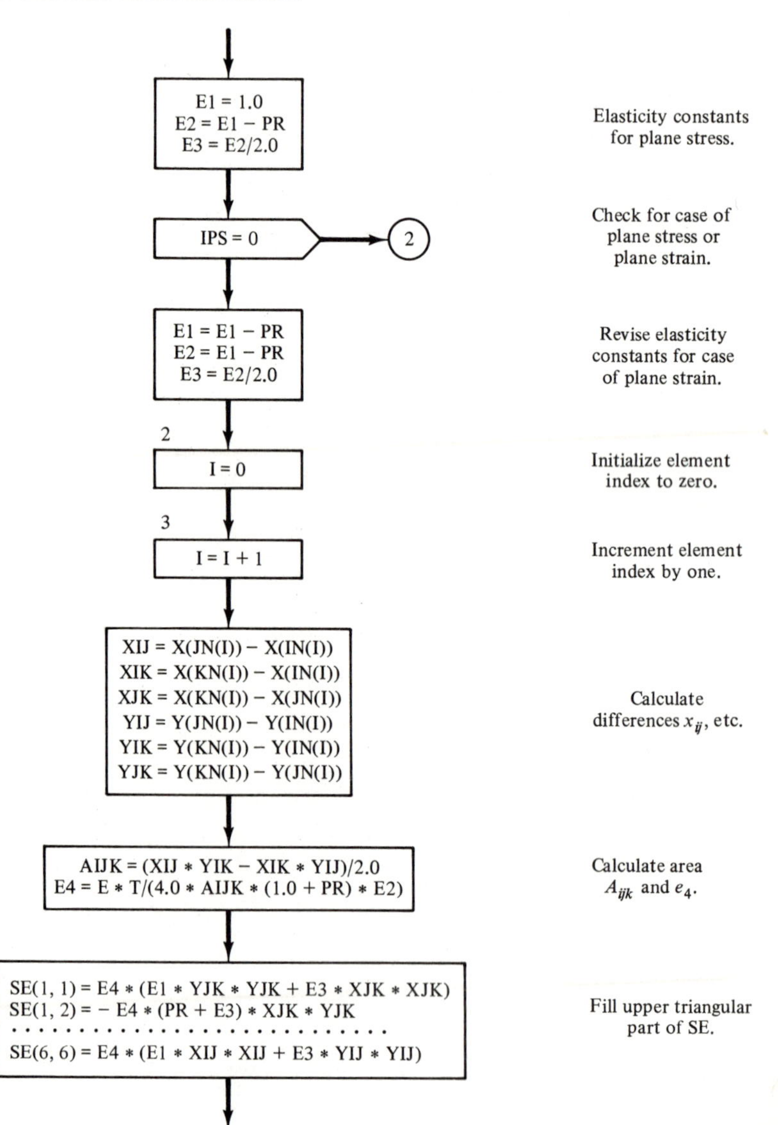

b. Transfer to Structural Stiffness Matrix

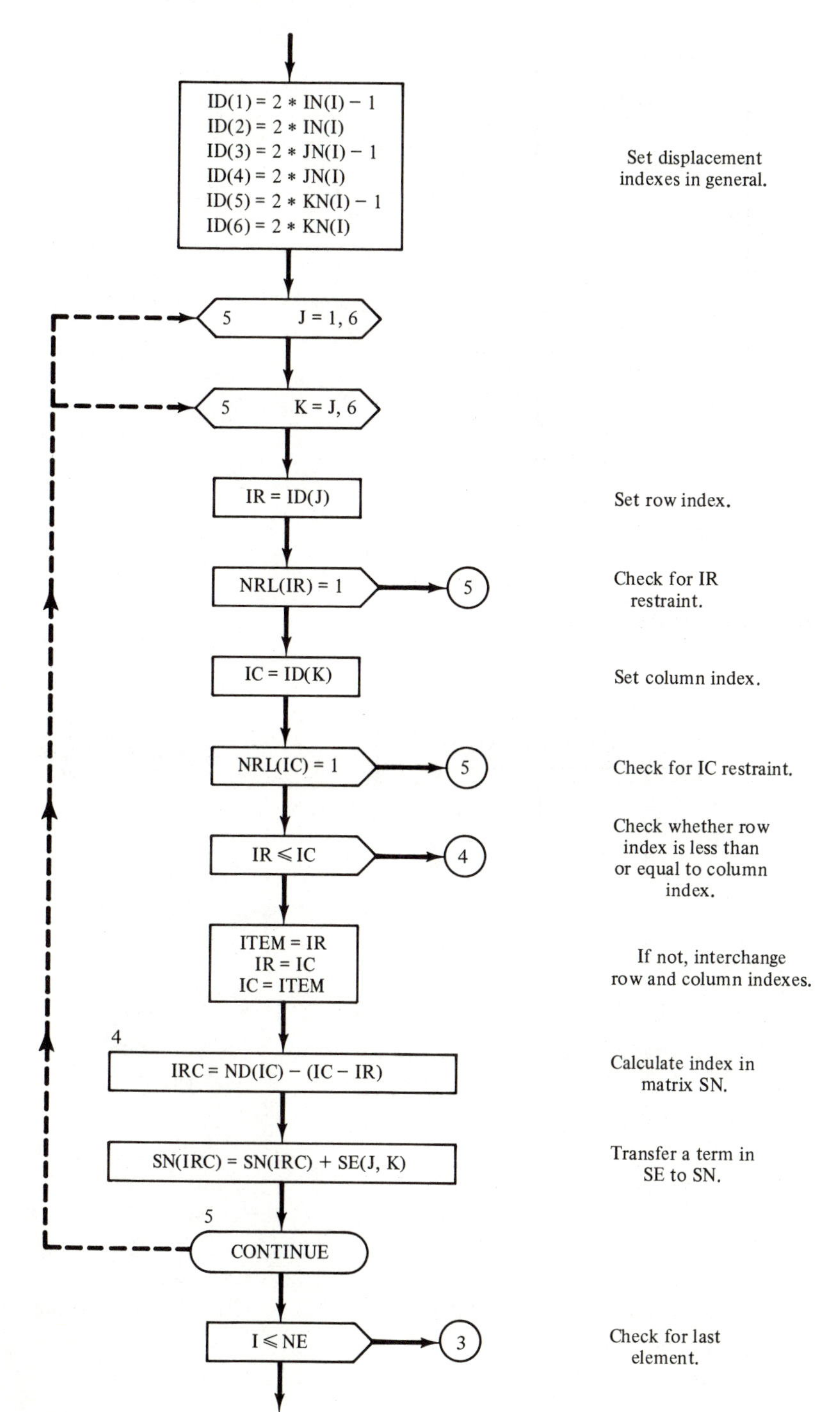

c. Modification of SN for Restraints

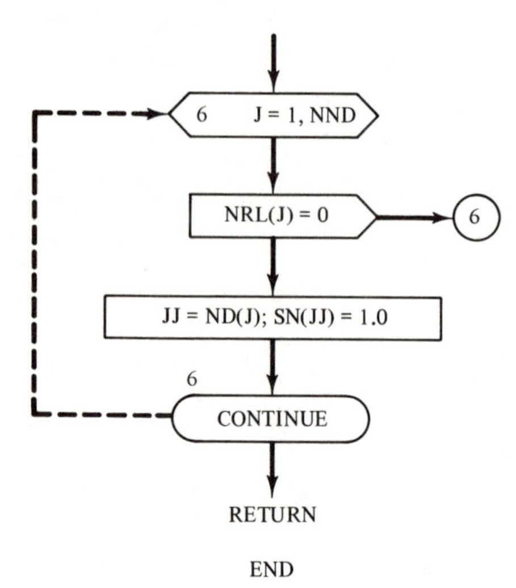

Check for restraint.

Set diagonal element of SN equal to unity.

3. SUBPROGRAM SKYFAC

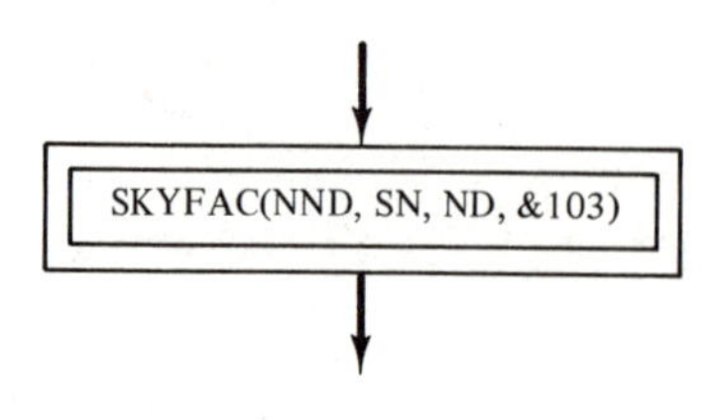

Factor the skyline of SN in place, without rearrangement.

4. SUBPROGRAM LDATA FOR PROGRAM PSCST

a. Load Parameters

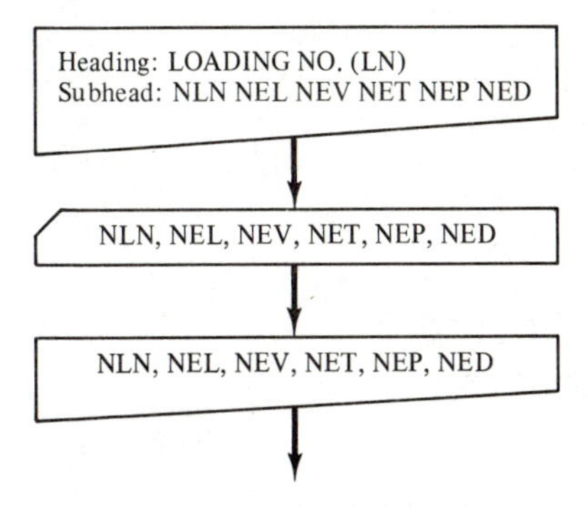

Read and print load parameters.

b. Nodal Loads

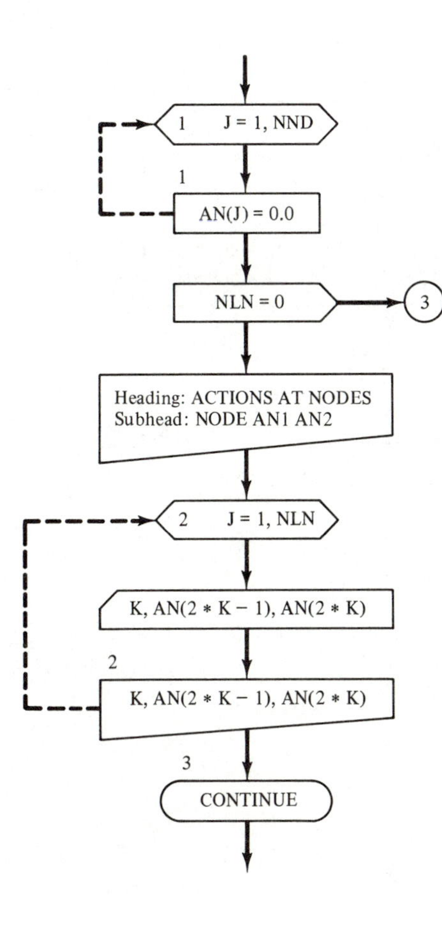

Clear nodal action vector.

Check to determine whether nodal loads exist.

Read and print nodal loads.

c. Line Loads*

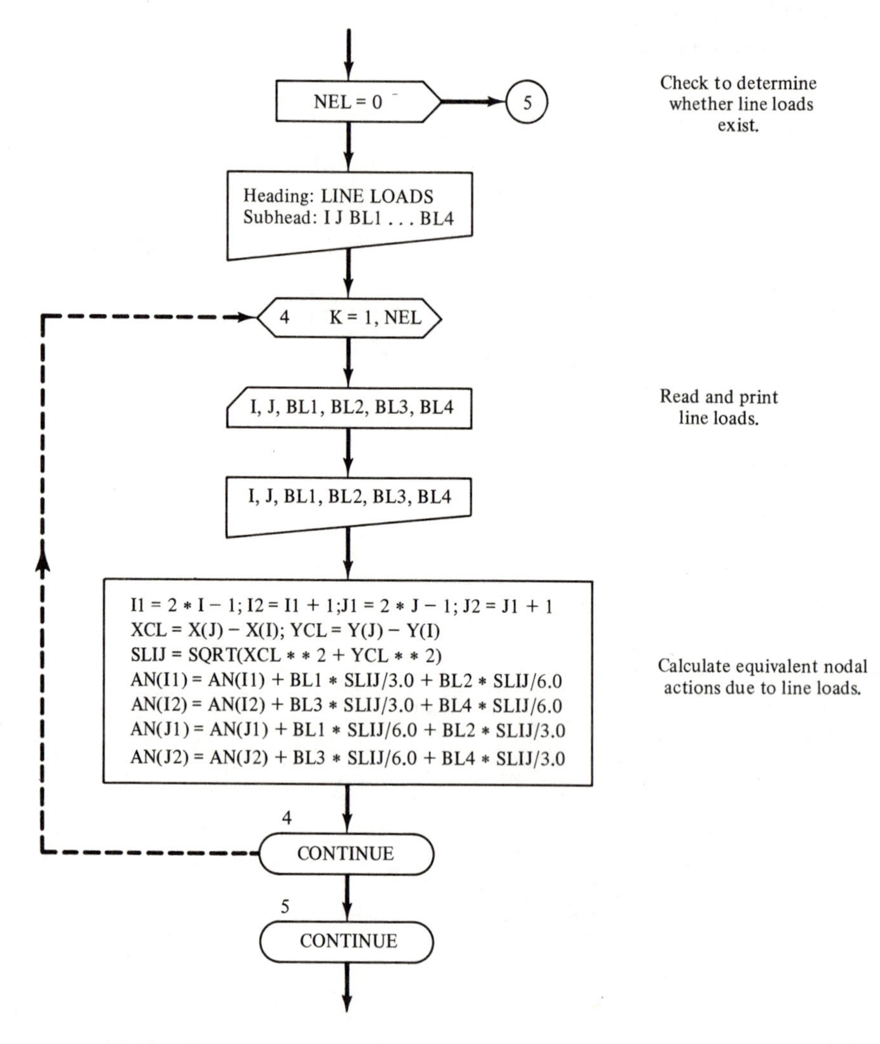

*Optional supplementary influence.

d. Volume Loads*

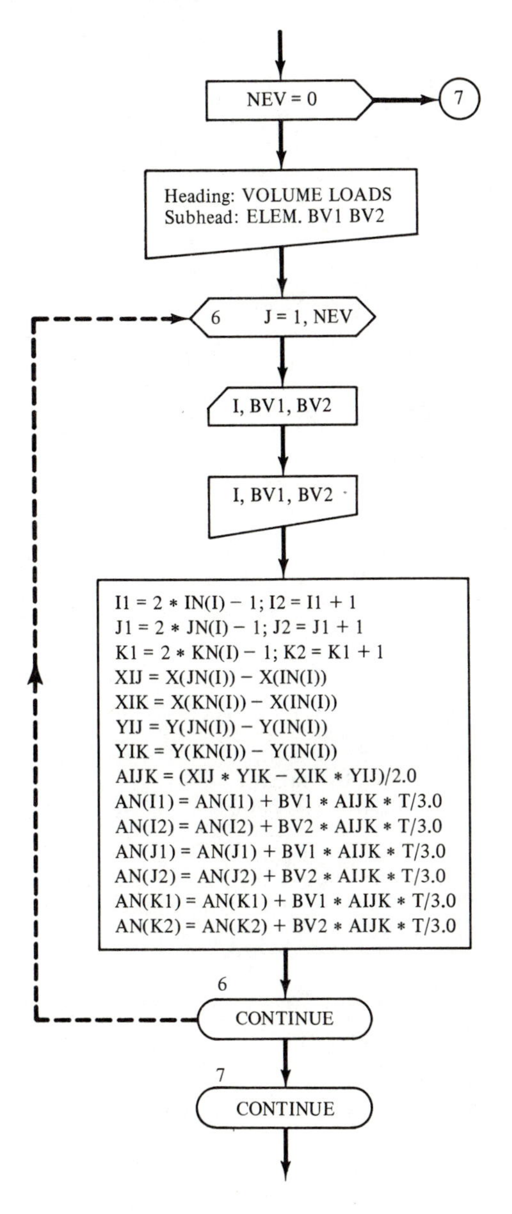

Check to determine whether volume loads exist.

Read and print volume loads.

Calculate equivalent nodal actions due to volume loads.

*Optional supplementary influence.

e. Temperature Strain*

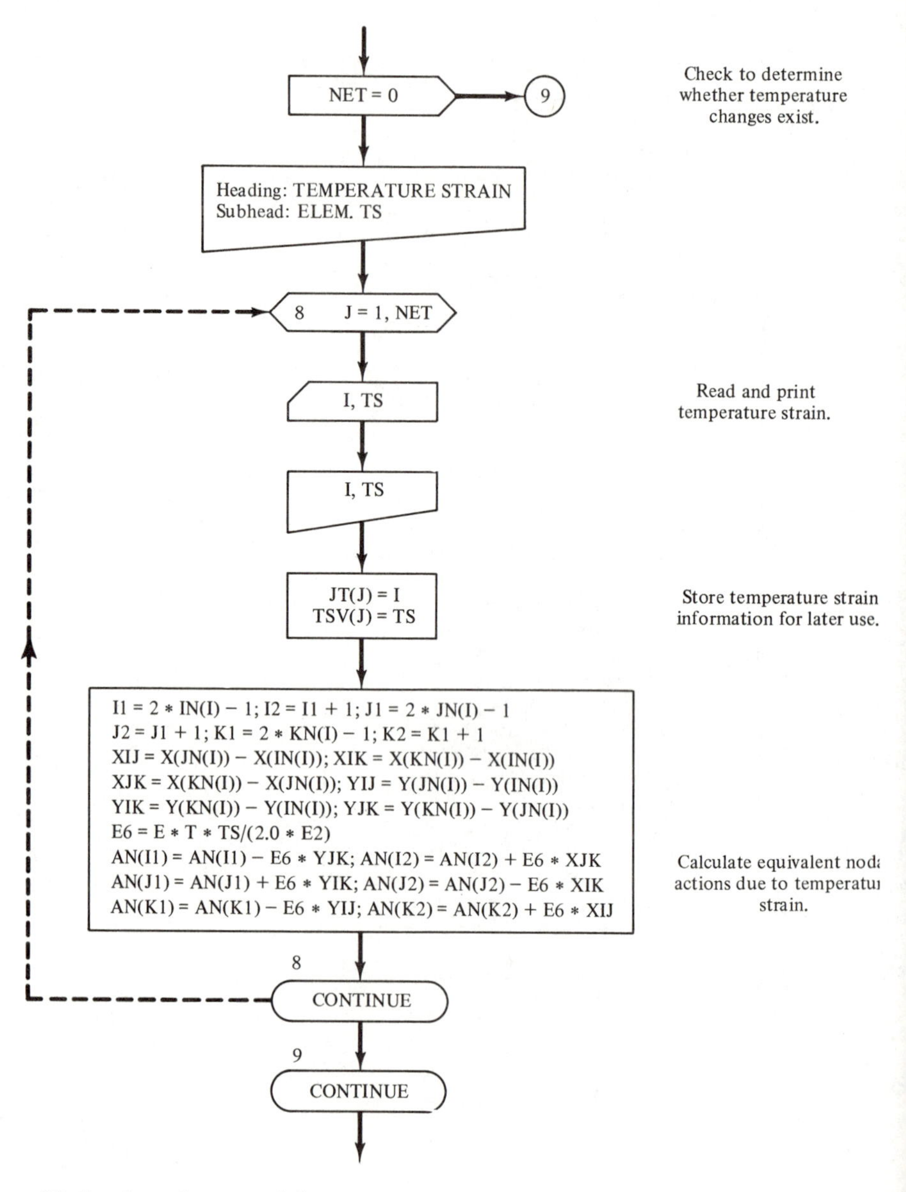

*Optional supplementary influence.

f. Prestrains*

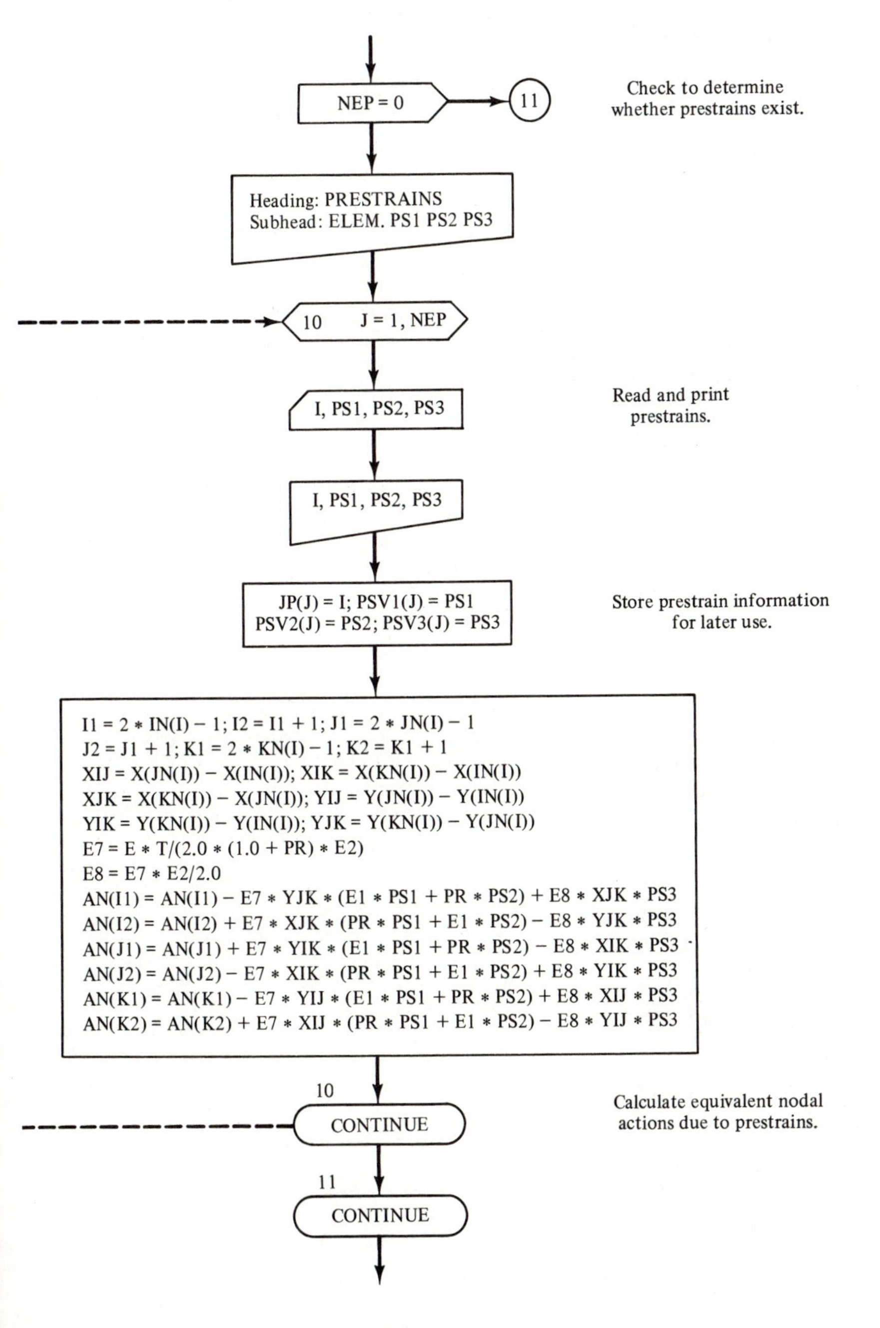

*Optional supplementary influence.

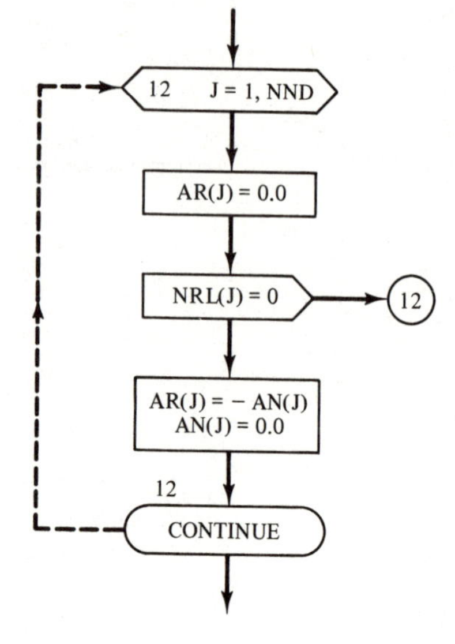

Clear reaction vector.

Check for restraint.

Transfer action directly to reaction, and replace former with zero.

g. Support Displacements*

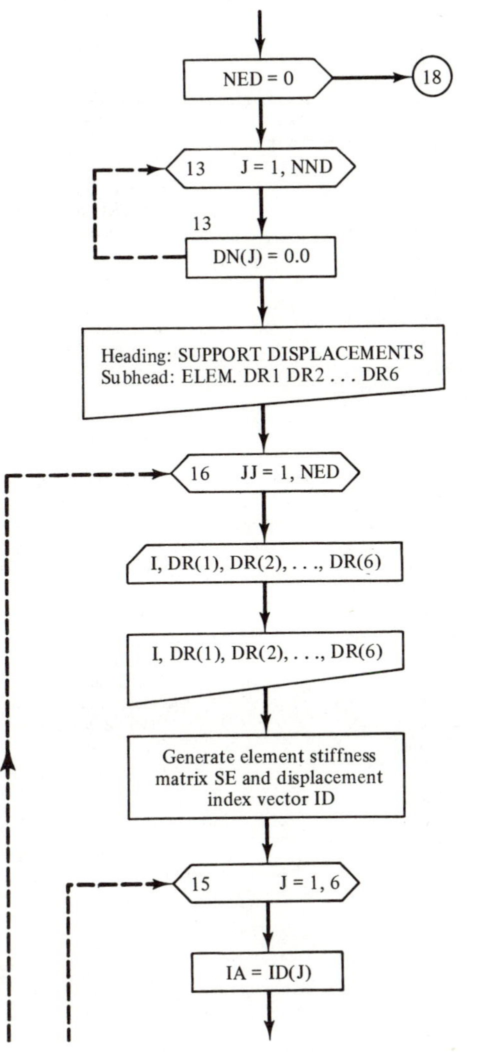

Check to determine whether support displacements exist.

Clear nodal displacement vector.

Read and print support displacements.

Fill both upper and lower triangular parts of SE (see part *b* of Subprogram RESUL).

Set action index.

*Optional supplementary influence.

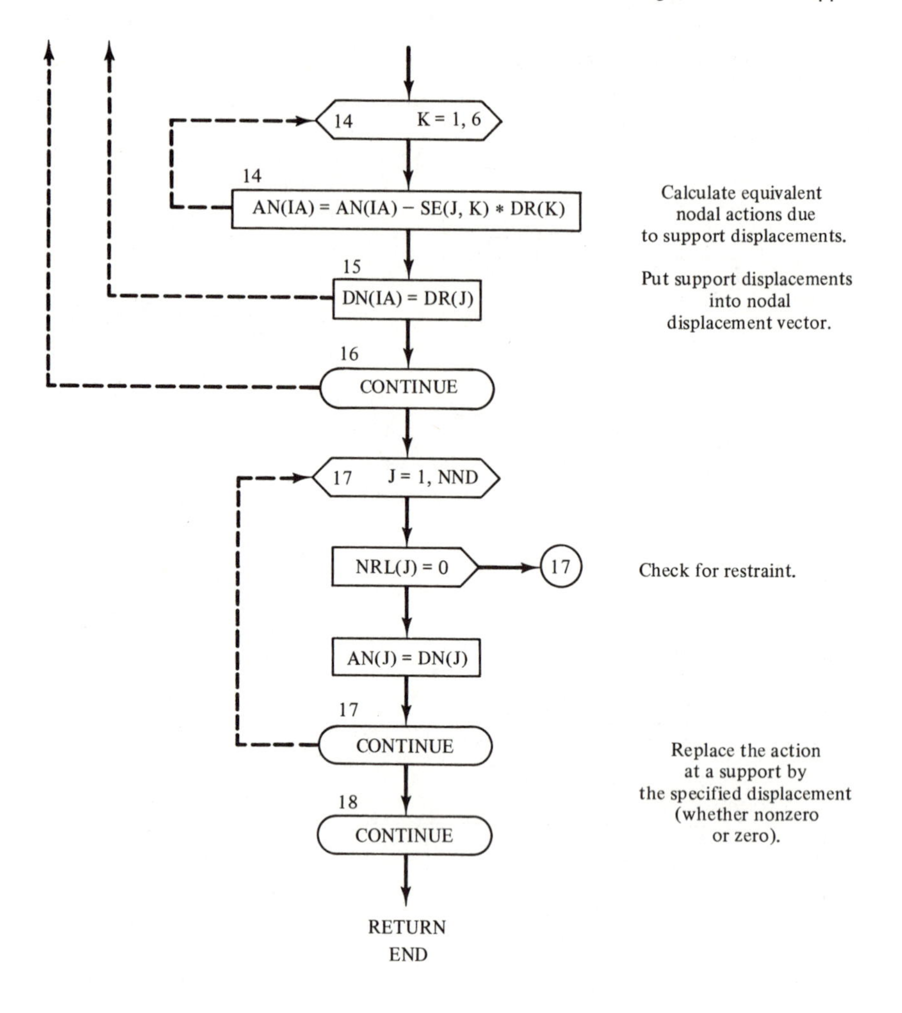

5. SUBPROGRAM SKYSOL

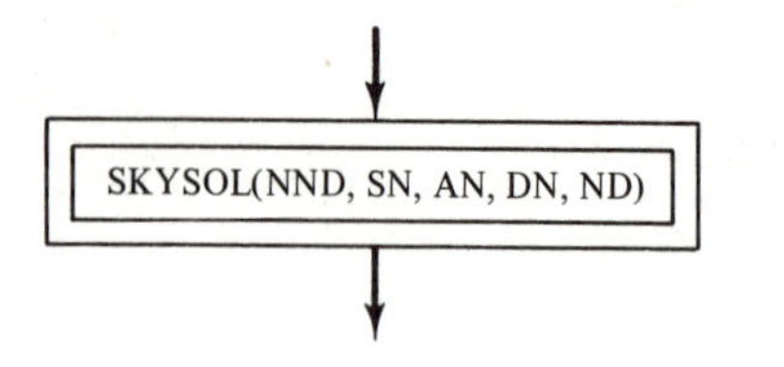

Solve the nodal equilibrium equations in place, without rearrangement.

6. SUBPROGRAM RESUL FOR PROGRAM PSCST

a. Nodal Displacements

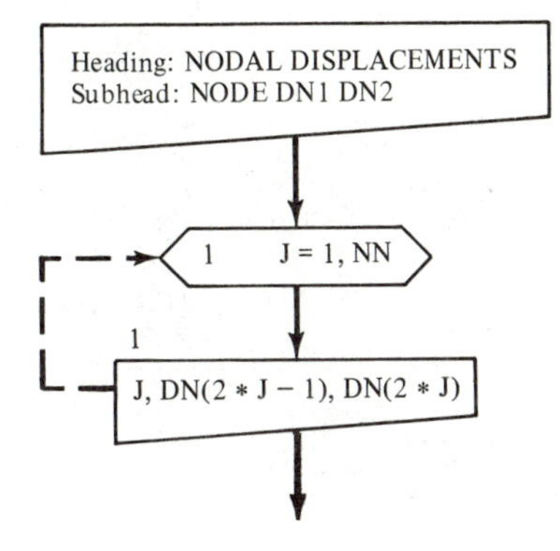

Print displacements.

b. Element Stresses

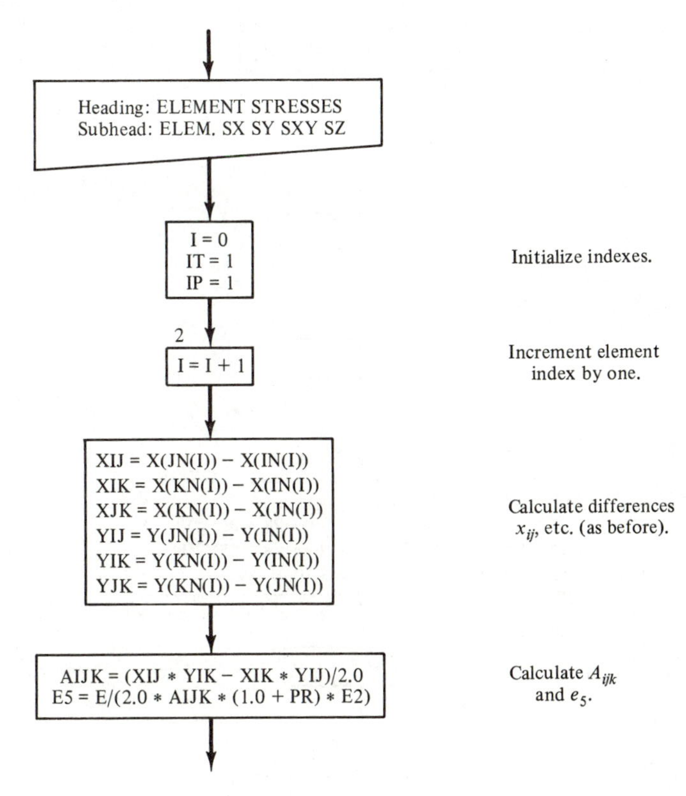

Initialize indexes.

Increment element index by one.

Calculate differences x_{ij}, etc. (as before).

Calculate A_{ijk} and e_5.

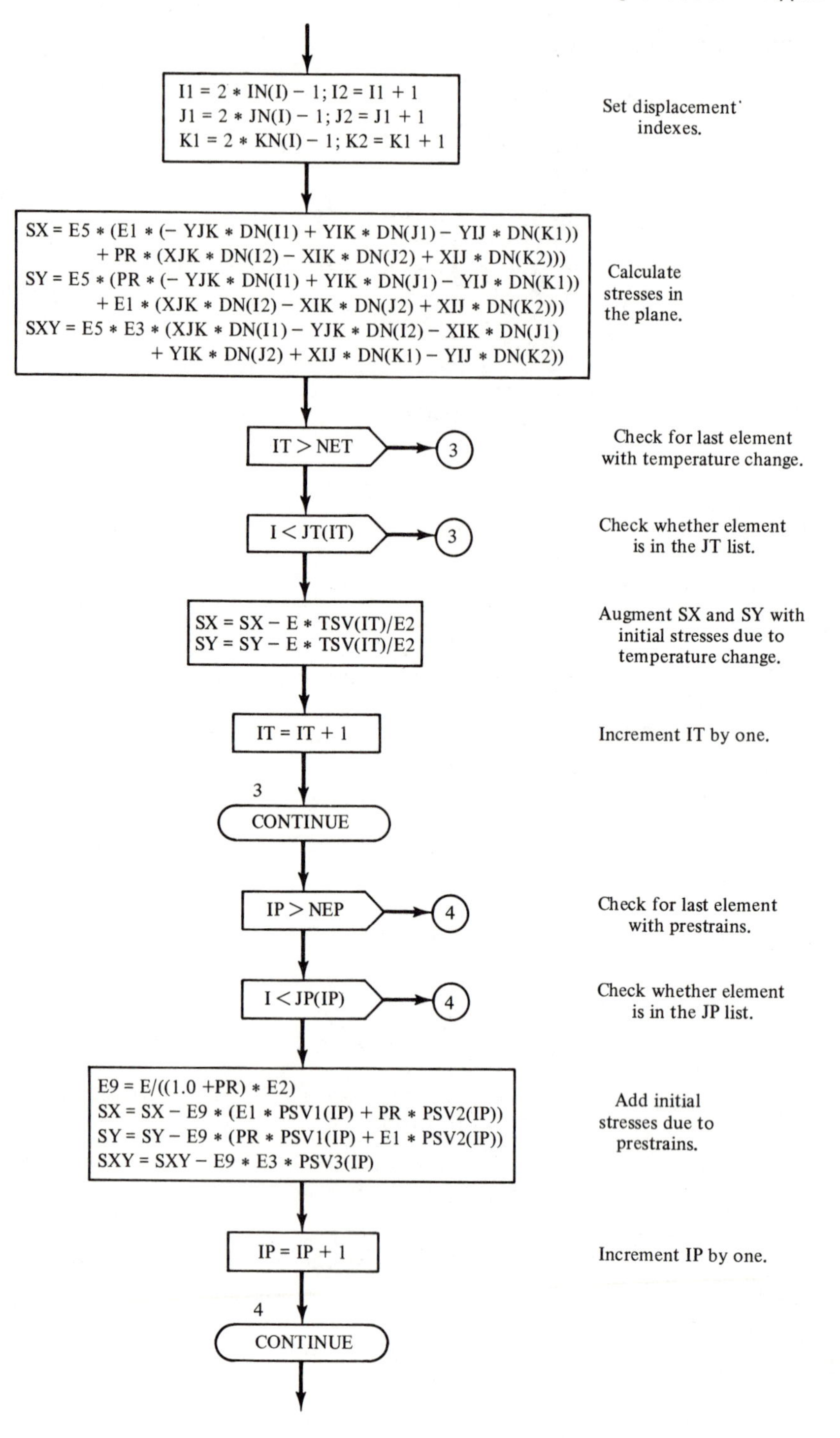
I1 = 2 * IN(I) − 1; I2 = I1 + 1
J1 = 2 * JN(I) − 1; J2 = J1 + 1
K1 = 2 * KN(I) − 1; K2 = K1 + 1
Set displacement indexes.
SX = E5 * (E1 * (− YJK * DN(I1) + YIK * DN(J1) − YIJ * DN(K1))
+ PR * (XJK * DN(I2) − XIK * DN(J2) + XIJ * DN(K2)))
SY = E5 * (PR * (− YJK * DN(I1) + YIK * DN(J1) − YIJ * DN(K1))
+ E1 * (XJK * DN(I2) − XIK * DN(J2) + XIJ * DN(K2)))
SXY = E5 * E3 * (XJK * DN(I1) − YJK * DN(I2) − XIK * DN(J1)
+ YIK * DN(J2) + XIJ * DN(K1) − YIJ * DN(K2))
Calculate stresses in the plane.
IT > NET
3
Check for last element with temperature change.
I < JT(IT)
3
Check whether element is in the JT list.
SX = SX − E * TSV(IT)/E2
SY = SY − E * TSV(IT)/E2
Augment SX and SY with initial stresses due to temperature change.
IT = IT + 1
Increment IT by one.
3
CONTINUE
IP > NEP
4
Check for last element with prestrains.
I < JP(IP)
4
Check whether element is in the JP list.
E9 = E/((1.0 +PR) * E2)
SX = SX − E9 * (E1 * PSV1(IP) + PR * PSV2(IP))
SY = SY − E9 * (PR * PSV1(IP) + E1 * PSV2(IP))
SXY = SXY − E9 * E3 * PSV3(IP)
Add initial stresses due to prestrains.
IP = IP + 1
Increment IP by one.
4
CONTINUE

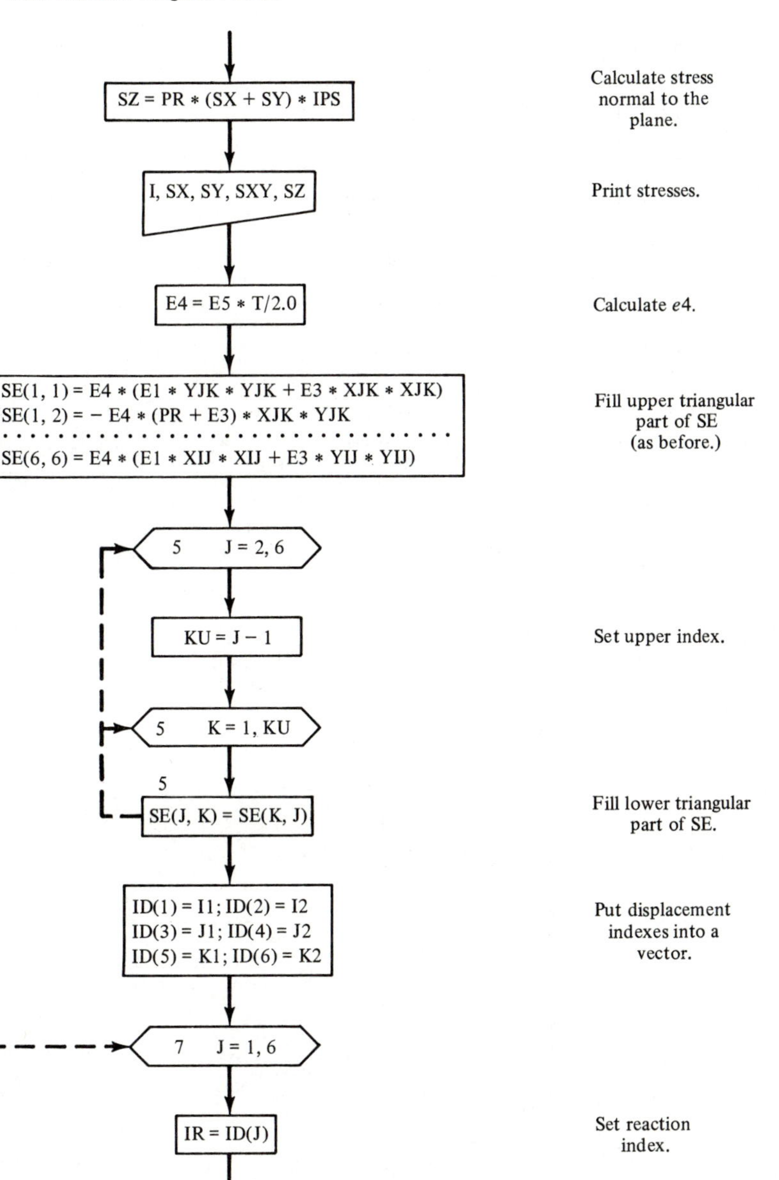
SZ = PR * (SX + SY) * IPS
Calculate stress normal to the plane.
I, SX, SY, SXY, SZ
Print stresses.
E4 = E5 * T/2.0
Calculate e4.
SE(1, 1) = E4 * (E1 * YJK * YJK + E3 * XJK * XJK)
SE(1, 2) = − E4 * (PR + E3) * XJK * YJK
SE(6, 6) = E4 * (E1 * XIJ * XIJ + E3 * YIJ * YIJ)
Fill upper triangular part of SE (as before.)
5 J = 2, 6
KU = J − 1
Set upper index.
5 K = 1, KU
5
SE(J, K) = SE(K, J)
Fill lower triangular part of SE.
ID(1) = I1; ID(2) = I2
ID(3) = J1; ID(4) = J2
ID(5) = K1; ID(6) = K2
Put displacement indexes into a vector.
7 J = 1, 6
IR = ID(J)
Set reaction index.
NRL(IR) = 0
7
Check for restraint.

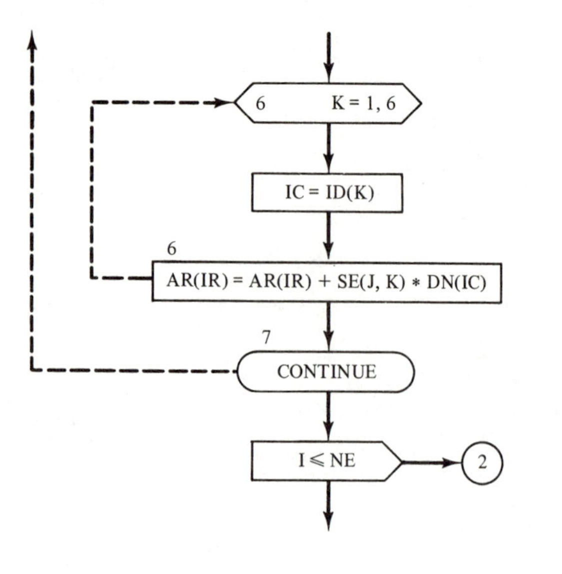

c. Support Reactions

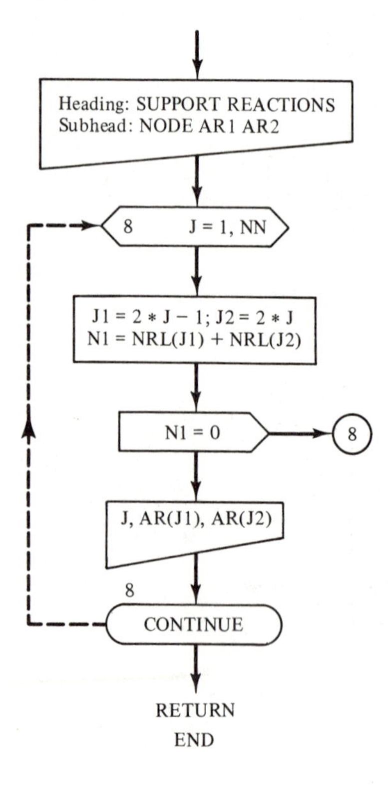

Answers to Problems

Chapter 1

1.5-1. $\mathbf{K} = \frac{EA}{3L}\begin{bmatrix} 7 & -8 & 1 \\ -8 & 16 & -8 \\ 1 & -8 & 7 \end{bmatrix}$ **1.5-2.** $\mathbf{p}_b = \frac{b_x L}{6}\begin{bmatrix} 1 \\ 4 \\ 1 \end{bmatrix}$

1.5-3. $\mathbf{p}_b = \{b_1, 2b_1 + 2b_3, b_3\}L/6$ **1.5-4.** $\mathbf{p}_T = \{0, -1, 1\}2EA\alpha\,\Delta T_3/3$

1.5-5. Given

1.5-6. $\mathbf{p}_b = \{6x^2 - 6xL, 3x^2L - 4xL^2 + L^3, -6x^2 + 6xL, 3x^2L - 2xL^2\}M_z/L^3$

1.5-7. $\mathbf{p}_T = \{-1, 0, 1, -L\}2EI\alpha\,\Delta T_2/Ld$ **1.5-8.** $\mathbf{p}_0 = \{6, L, -6, 5L\}EI\phi_2/6L$

1.5-9. $\mathbf{K}_r = \frac{kL}{420}\begin{bmatrix} 156 & 22L & 54 & -13L \\ 22L & 4L^2 & 13L & -3L^2 \\ 54 & 13L & 156 & -22L \\ -13L & -3L^2 & -22L & 4L^2 \end{bmatrix}$ **1.5-10.** Given

1.5-11. Given **1.6-1.** Given **1.6-2.** $\mathbf{p}_0 = \{9, 2L, -9, 7L\}EI\phi_2/10L$

1.6-3. Given **1.6-4.** $\mathbf{p}_0 = \{1, -1\}GJ\psi_0$

1.7-1. $\mathbf{p}_j = \{p_{k1}\cos\theta + p_{k2}\sin\theta, -p_{k1}\sin\theta + p_{k2}\cos\theta,$
$-p_{k1}R(1 - \cos\theta) + p_{k2}R\sin\theta + p_{k3}\}$

1.7-2. $\mathbf{q}_D = \{q_{A1} + z_{AD}q_{A5} - y_{AD}q_{A6},\ q_{A2} - z_{AD}q_{A4} + x_{AD}q_{A6},$
$q_{A3} + y_{AD}q_{A4} - x_{AD}q_{A5},\ q_{A4},\ q_{A5},\ q_{A6}\}$

1.7-3. $\mathbf{K}_{11} = \mathbf{K}_{22} = \frac{EA}{L}\begin{bmatrix} \cos^2\gamma & \sin\gamma\cos\gamma \\ \sin\gamma\cos\gamma & \sin^2\gamma \end{bmatrix} = -\mathbf{K}_{12} = -\mathbf{K}_{21}$

1.7-4. See Tables 4.26 and 4.27 of (2)

1.8-1. $S_{N11} = S_{N22} = S_{N77} = S_{N88} = 1$
$S_{N33} = 36EI/L^3$; $S_{N34} = -6EI/L^2$; $S_{N35} = -12EI/L^3$; etc.
$A_{N3} = P$; $A_{N6} = M$

1.8-2. $S_{N11} = S_{N22} = S_{N33} = S_{N44} = 1$
$S_{N55} = 0.506EA/L$; $S_{N59} = -0.128EA/L$; $S_{N5,10} = 0.096EA/L$; etc.
$A_{N8} = A_{N10} = -P$

1.8-3. $S_{N11} = S_{N22} = S_{N33} = S_{N10,10} = S_{N11,11} = S_{N12,12} = 1$
$S_{N44} = 112EI/L^3$; $S_{N46} = -6EI/L^2$; $S_{N47} = -12EI/L^3$; etc.
$A_{N5} = -P$; $A_{N9} = M$

1.8-4. $\mathbf{A}_N = \{0, 0, P, 0, -18\alpha/L, M - 6\alpha, 18\alpha/L, \theta_z\}$ where $\alpha = EI\theta_z/L$

1.8-5. $\mathbf{A}_N = \{0, 0.333\beta, \Delta_x, -0.096\beta, 0.128\beta, 0.096\beta, 0.25\beta, -P, 0, -P\}$
where $\beta = EA\,\Delta_x/L$

1.8-6. $\mathbf{A}_N = \{0, \Delta_y, -6\gamma, 0, -P + 12\gamma/L, 6\gamma, 0, 0, M, 0, 0, 0\}$ where $\gamma = EI\,\Delta_y/L^2$

Chapter 2

2.2-2. $\mathbf{p}_r = -[e_4\{e_1 y_{ki} y_{jk}, -\nu y_{ki} x_{jk}, e_1 y_{ki}^2, -\nu x_{ki} y_{ki}, e_1 y_{ij} y_{ki}, -\nu x_{ij} y_{ki}\}$
$+ e_5\{x_{jk} x_{ki}, -y_{jk} x_{ki}, x_{ki}^2, -x_{ki} y_{ki}, x_{ki} x_{ij}, -x_{ki} y_{ij}\}]u_{rj}$

2.2-4. $\mathbf{p}_b = \{7, 7, 5, 5, 11, 11\}P/23\sqrt{2}$

2.2-5. $\mathbf{p}_b = \{0, 0, 2b_{xj} + b_{xk}, 0, b_{xj} + 2b_{xk}, 0\}L_{jk}/6$ **2.3-1.** Given

2.3-3. $\mathbf{p}_r = -[\{-2s_1, -s_3, 2s_1, -s_3, s_1, s_3, -s_1, s_3\}$
$+ \{s_4, s_6, 2s_4, -s_6, -2s_4, -s_6, -s_4, s_6\}]u_{r2}$

2.3-5. $\mathbf{p}_b = \{0, 1, 0, 1, 0, 0, 0, 0\}2b_{y0}a/3$

2.3-6. $\mathbf{p}_T = \{-b, -2a/3, b, -4a/3, b, 4a/3, -b, 2a/3\}\dfrac{E\alpha\,\Delta T}{2(1-\nu)}t$

2.3-7. $\mathbf{p}_b = \{15, 0, 0, 9b, 5, 0, 0, 3b, 81, 0, 0, -27b, 27, 0, 0, -9b\}P/128$

2.3-8. $\mathbf{p}_T = \{0, -6a, -a^2, 0, 0, 6a, -a^2, 0, 0, 6a, a^2, 0, 0, -6a, a^2, 0\}\dfrac{E\alpha\,\Delta T}{15(1-\nu)}t$

Chapter 3

3.3-1. $\xi_1 = 0$; $R_1 = 2$ **3.3-2.** $\xi_1 = -\xi_2 = -\dfrac{1}{\sqrt{3}}$; $R_1 = R_2 = 1$

3.3-3. $\xi_1 = -\xi_2 = -\sqrt{\tfrac{3}{5}}$; $\xi_3 = 0$; $R_1 = R_2 = \tfrac{5}{9}$; $R_3 = \tfrac{8}{9}$

3.3-4. $\mathbf{p}_b = \{2b_1 + b_2, b_1 + 2b_2\}L/6$ **3.3-5.** $\mathbf{p}_b = \{1, 3\}m_{x2}L/12$

3.3-6. $K_{11} = \dfrac{tbE_{11}}{3a} + \dfrac{taE_{33}}{3b}$ **3.3-7.** Same as 2.3-6

3.3-8. $\mathbf{p}_T = \{0, -4a, -a^2, 0, 0, 4a, -a^2, 0, 0, 4a, a^2, 0, 0, -4a, a^2, 0\}\dfrac{E\alpha\,\Delta T}{9(1-\nu)}t$

3.3-9. through **3.3-15** Given **3.4-1.** $K_{11} = (0.1138E_{11} + 0.5723E_{33})t$

3.4-2. $K_{22} = (0.5723E_{22} + 0.1138E_{33})t$ **3.4-3.** $K_{56} = 0.2581(E_{12} + E_{33})t$

3.4-4. $K_{55} = (0.1929E_{11} + 0.5312E_{33})t$ **3.4-5.** $K_{66} = (0.5312E_{22} + 0.1929E_{33})t$

3.4-6. $\mathbf{p}_{b1} = \{0, -1\}6.583b_y t$

3.4-7. $\mathbf{p}_b = \{0, 0, 2b_{x2} + b_{x3}, 0, b_{x2} + 2b_{x3}, 0, 0, 0\}L_{23}/6$ **3.4-8.** Same as 3.4-7

3.4-9. $\mathbf{p}_{T2} = \{1, -1\}\dfrac{5E\alpha\,\Delta T}{2(1-\nu)}t$ **3.4-10.** Same as 3.4-9

3.4-11. $\mathbf{p}_{b1,2,5} = \{-1, 0, -1, 0, 4, 0\}abtb_x/3$ **3.4-12.** Same as 3.4-11

3.4-13. $x_g = -\frac{1}{4}\sum_{i=1}^{4} x_i + \frac{1}{2}\sum_{i=5}^{8} x_i;\ y_g = -\frac{1}{4}\sum_{i=1}^{4} y_i + \frac{1}{2}\sum_{i=5}^{8} y_i$

3.4-14. $\mathbf{p}_{b3,4,7} = \{-5P_x, -5P_y, -9P_x, -9P_y, 30P_x, 30P_y\}3/128$

3.4-15. $\mathbf{p}_{b2,3,6} = \{1, 0, 1, 0, 4, 0\}L_{23}b_x/6$

Chapter 4

4.1-1. $\sigma_{P1} = 22.10 \quad \sigma_{P2} = 15.00 \quad \sigma_{P3} = -20.10$
$$\mathbf{R}_P = \begin{bmatrix} 0.8722 & 0.4892 & 0 \\ 0 & 0 & 1 \\ 0.4892 & -0.8722 & 0 \end{bmatrix}$$

4.1-2. $\sigma_{P1} = 18.28 \quad \sigma_{P2} = 8.000 \quad \sigma_{P3} = -14.28$
$$\mathbf{R}_P = \begin{bmatrix} 0 & 0.9153 & -0.4027 \\ 1 & 0 & 0 \\ 0 & -0.4027 & -0.9153 \end{bmatrix}$$

4.1-3. $\sigma_{P1} = 9.000 \quad \sigma_{P2} = -11.00 \quad \sigma_{P3} = -25.00$
$$\mathbf{R}_P = \begin{bmatrix} 0.5145 & 0 & 0.8575 \\ 0 & 1 & 0 \\ -0.8575 & 0 & 0.5145 \end{bmatrix}$$

4.1-4. $\sigma_{P1} = 22.18 \quad \sigma_{P2} = 8.852 \quad \sigma_{P3} = -15.03$
$$\mathbf{R}_P = \begin{bmatrix} 0.3985 & 0.9155 & 0.05508 \\ 0.8948 & -0.4013 & 0.1958 \\ 0.2013 & -0.02874 & -0.9791 \end{bmatrix}$$

4.1-5. $\sigma_{P1} = 27.68 \quad \sigma_{P2} = -1.971 \quad \sigma_{P3} = -20.71$
$$\mathbf{R}_P = \begin{bmatrix} -0.2549 & -0.3274 & 0.9098 \\ 0.1736 & 0.9101 & 0.3762 \\ -0.9512 & 0.2539 & -0.1752 \end{bmatrix}$$

4.1-6. $\sigma_{P1} = 24.61 \quad \sigma_{P2} = -0.8222 \quad \sigma_{P3} = -9.786$
$$\mathbf{R}_P = \begin{bmatrix} 0.8733 & -0.4809 & -0.07855 \\ 0.4169 & 0.6539 & 0.6314 \\ -0.2523 & -0.5841 & 0.7715 \end{bmatrix}$$

4.4-1. $K_{12} = \dfrac{cE}{12(1+\nu)e_2}$ **4.4-2.** $K_{69} = \dfrac{abcE}{9(1+\nu)e_2}\left(\dfrac{e_1}{c^2} - \dfrac{2e_3}{b^2} + \dfrac{e_3}{a^2}\right)$

4.4-3. $p_{b4} = (1-\eta)(1-\zeta)P_x/4;\ p_{b7} = (1+\eta)(1-\zeta)P_x/4$
$p_{b16} = (1-\eta)(1+\zeta)P_x/4;\ p_{b19} = (1+\eta)(1+\zeta)P_x/4$
(others = 0)

4.4-4. $\boldsymbol{\sigma}_0 = \{1, 1, 1, 0, 0, 0\}\xi\eta\zeta E\alpha\,\Delta T_0/e_2$

4.4-5. $\mathbf{E}\ \mathbf{B}_{1,3} = -\{\nu(1-\xi)(1-\eta)/c,\ \nu(1-\xi)(1-\eta)/c,\ e_1(1-\xi)(1-\eta)/c,$
$$0,\ e_3(1-\xi)(1-\zeta)/b,\ e_3(1-\eta)(1-\zeta)/a\}\frac{E}{8(1+\nu)e_2}$$

4.4-6. $p_{b2} = p_{b5} = p_{b8} = p_{b11} = p_{b14} = p_{b17} = p_{b20} = p_{b23}$
$= -abcb_g$; (others = 0)

4.4-7. $\mathbf{p}_0 = \{-a, -b, 0, -a, b, 0, a, b, 0, a, -b, 0,$
$$-a, -b, 0, -a, b, 0, a, b, 0, a, -b, 0\}\frac{Ec\gamma_0}{2(1+\nu)}$$

4.4-8. $\mathbf{p}_T = \{b, a, 0, -b, a, 0, -b, -a, 0, b, -a, 0, -b, -a, 0, b, -a, 0, b, a, 0, -b, a, 0\} \dfrac{Ec\alpha\, \Delta T_0}{3e_2}$

4.5-1. $K_{11} = \dfrac{EV}{(1+\nu)e_2}\left(\dfrac{e_1}{a^2} + \dfrac{e_3}{b^2} + \dfrac{e_3}{c^2}\right)$ **4.5-2.** $K_{12} = \dfrac{EV}{2ab(1+\nu)e_2}$

4.5-3. $K_{13} = \dfrac{EV}{2ac(1+\nu)e_2}$ **4.5-4.** $K_{33} = \dfrac{EV}{(1+\nu)e_2}\left(\dfrac{e_1}{c^2} + \dfrac{e_3}{b^2} + \dfrac{e_3}{a^2}\right)$

4.5-5. $K_{23} = \dfrac{EV}{2bc(1+\nu)e_2}$ **4.5-6.** $\mathbf{p}_b = \{0, 0, 0, 4, 0, 0, 1, 0, 0, 1, 0, 0\}P_x/6$

4.5-7. $\mathbf{p}_b = \{0, 0, H^a_{11}, 0, 0, H^a_{21}, 0, 0, H^a_{31}, 0, 0, H^a_{41}\}b_z/6$

4.5-8. $\boldsymbol{\sigma}_T = -\{1, 1, 1, 0, 0, 0\}E\alpha\, \Delta T/e_2$

Chapter 5

5.2-1. $\mathbf{p}_b = 2\pi\{0, 1, 0, 0, 0, 1\}6.25P_z$

5.2-2. $\mathbf{p}_b = 2\pi\{0, 0, 48.81, 0, 22.78, 0\}b_{r2} + 2\pi\{0, 0, 22.78, 0, 42.31, 0\}b_{r3}$

5.2-3. $\mathbf{p}_b = -2\pi\{0, 4.583, 0, 5.417, 0, 5.000\}b_g A$

5.2-4. $\mathbf{p}_b = 2\pi\{(0, 0, 0, 0, 1, 1, 1, 1\}7.5P$

5.2-5. $\mathbf{p}_T = 2\pi\{1, 0, -2, 0, 2, 0, -1, 0\}40E\alpha\, \Delta T/3e_2$

5.2-6. $\mathbf{p}_b = 2\pi\{4, 0, 5, 0, 5, 0, 4, 0\}200b_r/3$

5.3-1. $\mathbf{p}_b = 2\pi\{1, 0, 0, 0, 0, 0, 1, 0\}5.25P_r$

5.3-2. $\mathbf{p}_b = 2\pi\{0, 0, 464, 116, 448, 112, 0, 0\}b_n/6$

5.3-3. $\mathbf{p}_{b1} = 2\pi\{2798, 0\}\rho\omega^2$

5.3-4. $p_{b2} = 2\pi[r_1(3b_1 + b_2) + r_2(b_1 + b_2)]L_{12}/12$
$p_{b4} = 2\pi[r_1(b_1 + b_2) + r_2(b_1 + 3b_2)]L_{12}/12$

Chapter 6

6.2-1. $\mathbf{K}_r = r_z \int_A \mathbf{f}^T\mathbf{f}\, dA$ **6.2-2.** $\mathbf{p}_{b1} = \{1, b/3, -a/3\}b_z ab$

6.2-3. $\mathbf{p}_{b2} = \{2, b, 0\}P/4$ **6.2-4.** $\mathbf{p}_{b4} = \{39, -12b, -11a\}b_z ab/48$

6.2-5. $K_{12} = \dfrac{Et^3}{12(1-\nu^2)}\left(\dfrac{a}{b^2} + \dfrac{\nu}{2a} + \dfrac{\lambda}{5a}\right)$

6.2-6. $K_{13} = \dfrac{Et^3}{12(1-\nu^2)}\left(-\dfrac{b}{a^2} - \dfrac{\nu}{2b} - \dfrac{\lambda}{5b}\right)$

6.2-7. $\mathbf{p}_{01} = \dfrac{Et^3\phi_{xx0}}{12(1-\nu^2)}\{0, -\nu a, b\}$ **6.2-8.** $\mathbf{p}_{T1} = \dfrac{Et^3\alpha\, \Delta T}{6(1-\nu)}\{0, -a, b\}$

6.2-9. $\mathbf{p}_{b3} = \{36, -6b, 6a, ab\}b_z ab/144$

6.2-10. $\mathbf{p}_{b1} = \{54, 9b, -12a, 2ab\}b_{z3}ab/720$ **6.2-11.** $\mathbf{p}_{b2} = \{0, 0, 4, -b\}M/8$

6.2-12. $\mathbf{p}_{04} = \dfrac{Et^3\phi_{xy0}}{12(1+\nu)}\{-1, 0, 0, 0\}$

6.2-13. $\mathbf{p}_{T3} = \dfrac{Et^2\alpha\, \Delta T}{72(1-\nu)}\{0, -6a, 6b, (a^2 + b^2)\}$

6.4-1. $K_{12} = \dfrac{aEt}{18(1+\nu)}$ **6.4-2.** $K_{13} = -\dfrac{bEt}{18(1+\nu)}$

6.4-3. $\{p_{b1}, p_{b4}, p_{b7}, p_{b10}\} = -\{6, 14, 11, 15\}P/72$

6.4-4. $\{p_{b13}, p_{b16}, p_{b19}, p_{b22}\} = \{18, 8, 9, 24\}P/36$

6.4-5. $\{p_{b7}, p_{b19}\} = \{1, 4\}b_{z4}ab/9$

6.4-6. $\mathbf{p} = \{0, 0, -1, 0, 0, -1, 0, 0, -1, 0, 0, -1, 0, 0, 2, 0, 0, 2, 0, 0, 2, 0, 0, 2\}M_y/4$

Chapter 9

9.2-1. $\mathbf{M} = \dfrac{\rho A_0 L}{12}\begin{bmatrix} 5 & 3 \\ 3 & 7 \end{bmatrix}$ **9.2-2.** $\mathbf{M} = \dfrac{M}{16}\begin{bmatrix} 9 & 3 \\ 3 & 1 \end{bmatrix}$

9.2-3. $\mathbf{M} = \dfrac{M}{729}\begin{bmatrix} 400 & 80L & 140 & -40L \\ & 16L^2 & 28L & -8L^2 \\ & \text{Sym.} & 49 & -14L \\ & & & 4L^2 \end{bmatrix}$ **9.2-4.** Given **9.2-5.** Given

9.2-6. $\mathbf{M} = \dfrac{\rho J_0 L}{90}\begin{bmatrix} 22 & 13 \\ 13 & 32 \end{bmatrix}$ **9.2-7.** $\mathbf{M}_1 = \dfrac{M}{36}\begin{bmatrix} 9 & 6 & 3 \\ 6 & 4 & 2 \\ 3 & 2 & 1 \end{bmatrix}$ **9.2-8.** Given

9.2-9. $\mathbf{M}_1 = \dfrac{M}{576}\begin{bmatrix} 9 & 15 & 30 & 18 \\ & 25 & 50 & 30 \\ & \text{Sym.} & 100 & 60 \\ & & & 36 \end{bmatrix}$ **9.2-10.** Given

9.2-11. $\mathbf{M}_1 = \dfrac{\rho V}{72}\begin{bmatrix} 6 & 4 & 2 & 3 \\ & 10 & 5 & 2 \\ & \text{Sym.} & 10 & 4 \\ & & & 6 \end{bmatrix}$ **9.2-12.** $\mathbf{M}_1 = \dfrac{M}{144}\begin{bmatrix} 1 & 2 & 3 & 6 \\ & 4 & 6 & 12 \\ & \text{Sym.} & 9 & 18 \\ & & & 36 \end{bmatrix}$

9.2-13. Given **9.2-14.** $\mathbf{M}_{1,1} = \dfrac{7M}{3072}\{21, 7, 14, 42, 3, 1, 2, 6\}$

9.2-15. Given **9.2-16.** $\{M_{11,5}, M_{12,5}\} = -\{4b, 5a\}405Mb/262144$

9.2-17. $(M_t)_{2,3} = -\rho ta^2b^2/25$ **9.2-18.** $(M_{r1})_{2,2} = \rho t^3b^3/315a$

9.2-19. $(M_{r2})_{3,3} = \rho t^3a^3/315b$

9.2-20. $\{M_{10,1}, M_{10,2}, M_{10,3}\} = \{-30, -9b, 5a\}405Mb/4194304$

9.2-21. $(M_t)_{9,9} = 169\rho tab/1225$ **9.2-22.** $(M_{r1})_{9,11} = 13\rho t^3b/4200$

9.2-23. $(M_{r2})_{9,11} = -11\rho t^3a^2/25200$

9.4-1. $\omega_{1,2} = 22.51, 63.26\dfrac{1}{L^2}\sqrt{\dfrac{EI}{\rho A}}$; $\mathbf{\Phi} = \begin{bmatrix} 1 & 1 \\ 1 & -1 \end{bmatrix}$

9.4-2. $\omega_{1,2,3} = 0, 0, 22.47\dfrac{1}{L^2}\sqrt{\dfrac{EI}{\rho A}}$; $\mathbf{\Phi} = \begin{bmatrix} 1 & 1 & 1 \\ 1 & 0 & -0.6 \\ 1 & -1 & 1 \end{bmatrix}$

9.4-3. $\omega_{1,2} = 5.603, 31.19\dfrac{1}{L^2}\sqrt{\dfrac{EI}{\rho A}}$; $\mathbf{\Phi} = \begin{bmatrix} 0.5435 & -0.9364 \\ 1.0000 & 1.0000 \end{bmatrix}$

9.4-4. $\omega = \dfrac{15.69}{L^2}\sqrt{\dfrac{EI}{\rho A}}$ **9.4-5.** $\omega = \dfrac{9.941}{L^2}\sqrt{\dfrac{EI}{\rho A}}$

Chapter 10

10.2-1. Given **10.2-2.** Given **10.2-3.** Given

10.2-4. $(\bar{K}_{01})_{9,10} = -\dfrac{11b^2}{175a}\sigma_{x0}$; $(\bar{K}_{02})_{9,10} = -\dfrac{13a}{350}\sigma_{y0}$; $(\bar{K}_{03})_{9,10} = 0$

10.2-5. $(\bar{K}_{01})_{11,12} = -\dfrac{11ab^2}{1575}\sigma_{x0}$; $(\bar{K}_{02})_{11,12} = -\dfrac{a^3}{1050}\sigma_{y0}$; $(\bar{K}_{03})_{11,12} = 0$

10.3-1. $P_{cr} = -41.54\dfrac{EI}{L^2}$; $\boldsymbol{\Phi}_{cr} = \{1, 1\}$ **10.3-2.** $P_{cr} = -2.469\dfrac{EI}{L^2}$; $\boldsymbol{\Phi}_{cr} = \{0.2929, 1\}$

10.3-3. $P_{cr} = -10\dfrac{EI}{L^2}$; $\boldsymbol{\Phi}_{cr} = \{0.5, 1\}$ **10.3-4.** $P_{cr} = -21.54\dfrac{EI}{L^2}$

General References

TEXTBOOKS ON FINITE ELEMENTS*

1. Przemieniecki, J. S., *Theory of Matrix Structural Analysis*, McGraw-Hill, New York, 1968.
2. Desai, C. S., and Abel, J. F., *Introduction to the Finite Element Method*, Van Nostrand-Reinhold, New York, 1972.
3. Oden, J. T., *Finite Elements of Nonlinear Continua*, McGraw-Hill, New York, 1972.
4. Martin, H. C., and Carey, G. F., *Introduction to Finite Element Analysis*, McGraw-Hill, New York, 1973.
5. Norrie, D. H., and deVries, G., *The Finite Element Method*, Academic Press, New York, 1973.
6. Strang, G., and Fix, G. J., *An Analysis of the Finite Element Method*, Prentice-Hall, Englewood Cliffs, N.J., 1973.
7. Gallagher, R. H., *Finite Element Analysis Fundamentals*, Prentice-Hall, Englewood Cliffs, N.J., 1975.
8. Huebner, K. H., *The Finite Element Method for Engineers*, Wiley, New York, 1975.
9. Bathe, K. J., and Wilson, E. L., *Numerical Methods in Finite Element Analysis*, Prentice-Hall, Englewood Cliffs, N.J., 1976.
10. Segerlind, L. J., *Applied Finite Element Analysis*, Wiley, New York, 1976.
11. Hinton, E., and Owen, D. R. J., *Finite Element Programming*, Academic Press, London, 1977.
12. Zienkiewicz, O. C., *The Finite Element Method*, 3d ed., McGraw-Hill, Ltd., London, 1977.

*Listed in chronological order.

13. Desai, C. S., *Elementary Finite Element Method*, Prentice-Hall, Englewood Cliffs, N.J., 1979.
14. Cheung, Y. K., and Yeo, M. F., *A Practical Introduction to Finite Element Analysis*, Pitman, London, 1979.
15. Hinton, E., and Owen, D. R. J., *An Introduction to Finite Element Computations*, Pineridge Press, Swansea (UK), 1979.
16. Owen, D. R. J., and Hinton, E., *Finite Elements in Plasticity*, Pineridge Press, Swansea (UK), 1980.
17. Cook, R. D., *Concepts and Applications of Finite Element Analysis*, 2d ed., Wiley, New York, 1981.
18. Becker, E. B., Carey, G. F., and Oden, J. T., *Finite Elements: An Introduction*, Prentice-Hall, Englewood Cliffs, N.J., 1981.
19. Bathe, K. J., *Finite Element Procedures in Engineering Analysis*, Prentice-Hall, Englewood Cliffs, N.J., 1982.

Index